Studies in Mechanobiology, Tissue Engineering and Biomaterials

Volume 7

Series Editor
Amit Gefen, Ramat Aviv, Israel

Further volumes of this series can be found on our homepage:
http://www.springer.com/series/8415

Tim McGloughlin
Editor

Biomechanics and Mechanobiology of Aneurysms

Springer

Tim McGloughlin
Department of Mechanical, Aeronautical
and Biomedical Engineering
Centre for Applied Biomedical
Engineering Research and MSSI
University of Limerick
Limerick
Ireland
e-mail: tim.mcgloughlin@ul.ie

ISSN 1868-2006
ISBN 978-3-642-18094-1
DOI 10.1007/978-3-642-18095-8
Springer Heidelberg Dordrecht London New York

e-ISSN 1868-2014
e-ISBN 978-3-642-18095-8

Library of Congress Control Number: 2011936143

Cover design: WMXDesign GmbH, Heidelberg

Printed on acid-free paper

Springer is part of Springer Science+Business Media (www.springer.com)

Foreword

Mechanobiology, the linkage of biology with mechanics, is a field of investigation that is rapidly increasing our fundamental understanding of how structure and function at various scales from the molecular to the organ level play a synergistic role not only in maintaining a state of health, but also in the genesis and progression of disease. No area is more closely connected to mechanobiology than that of cardiovascular diseases, in which the interplay between mechanical forces and biological function is so intimately coupled. "Biomechanics and Mechanobiology of Aneurysms" is a compilation of thirteen chapters that examine the latest knowledge of this interplay within the very important clinical context of aneurysms. These chapters provide a comprehensive treatment of various manifestations of aneurysmal disease—abdominal aortic, thoracic aortic and intracranial aneurysms—covering topics ranging from basic pathophysiology of arteries to clinical treatment. While the areas presented are interdisciplinary and are presented as such, investigators from several fields will find this book both interesting and informative, including biologists, imaging scientists, bioengineers and clinicians. The power of imaging and of computational modeling is demonstrated, as is their relevance to disease detection and treatment. The topic of risk factors is by no means neglected, nor are discussions of the advantages, challenges and complications of current therapeutic vascular interventions. Several chapters delve deeply into tissue biomechanics, and the difficult mathematical coupling of interactions between fluid and solid mechanics is a topic that is treated. The book concludes with a chapter on biomechanical considerations of animal models for aneurysms, thus completing a quite comprehensive volume of material that represents the latest and best in our understanding of this important disease.

Professor McGloughlin and his colleagues have assembled a book that is an extremely valuable component in the series of volumes in "Studies in Mechanobiology, Tissue Engineering and Biomaterials." They are to be congratulated for producing a volume that is a must for the desk of every investigator exploring the causes, detection and treatment of aneurysms.

Atlanta, Georgia

Don P. Giddens
Lawrence L. Gellerstedt
Jr. Chair in Bioengineering and Georgia
Research Alliance Eminent Scholar
Georgia Institute of Technology

Preface

Aneurysms have claimed the lives of many famous people including Albert Einstein and former French President Charles De Gaulle and yet a complete understanding of the causes of aneurysms remains remarkable elusive. The clinical investigation of aneurysm behaviour has been a highly topical subject for almost 250 years with reports in Philosophical Transactions (1683–1775) entitled "Concerning aneurysms of the Thigh" by Benjamin Gooch Surgeon at Norwich in 1775 and "An account of an aneurysm of the aorta dissected" in St. Bartholomew's Hospital London by Surgeon Pierce Dod MD and in 1785 John Hunter was the first to successfully ligate a popliteal artery aneurysm, these being perhaps some of the earliest clinical case reports of modern times.

On observation of middle aged males one frequently concludes that they are healthy and fit, yet amongst those over 65 years of age it is estimated that almost 10% of them may have an aneurysm of one form or another. With the remarkable success of statin therapy in dealing with coronary artery disease and the steady decline in incidence of coronary artery disease, the impact of other cardiovascular diseases on mortality and morbidity has grown steadily over the last 20 years.

Accompanied by this increased incidence has come more regular and systematic screening of subjects and a remarkable suite of greatly improved diagnostic tools. In this monograph, the current thinking on aneurysm diagnostics, aneurysm mechanobiology and the range of therapeutic options are explored from a biomechanics perspective. All of the authors are currently active in cutting edge research related to aneurysm behaviour or treatment. I hope that this book will provide fresh insights into aneurysm diagnostics and therapeutics for the clinical, bioengineering and scientific communities and inspire them to work even more collaboratively than here-to-fore and thereby lead to improved outcomes for the many aneurysm sufferers.

The book tries to capture key topics in aneurysm biomechanics and mechanobiology in a fashion that is accessible to the readership and yet provides a good overview of the current state-of-the-art from some of the world's leading experts in aneurysm behaviour modelling and treatment.

Abdominal aortic aneurysm (AAA) disease is a chronic degenerative disorder and is an important cause of preventable deaths in older patients. Prevalence rates are estimated between 1.3 and 8.9% in men and between 1.0 and 2.2% in women. Rupture of an AAA is responsible for 1.3% of deaths in men aged 65–85 years, and accounts for 10,000 deaths annually in the United Kingdom. Survival after ruptured AAA is in the region of 10–20%. In the case of small aneurysms (diameter <5.5 cm), the usual treatment is careful and frequent observation. However, if the aneurysm diameter exceeds 5.5 cm or is rapidly growing, aneurysm repair is usually recommended. However, with the aging of the population and the increasing number of smokers, the incidence of the AAA is rising. The classical risk factors for atherosclerosis, such as tobacco smoking, male sex, age, hypertension, and hyperlipidaemia have all been found to be also risk factors for AAA. The degradation of the aortic wall is also influenced by tobacco smoking, sex, age, hypertension, chronic obstructive pulmonary disease, hyperlipidaemia, and family history of the disorder. With recent advances in non-invasive diagnostic imaging more AAAs are being detected. Approximately 150,000 new cases are diagnosed each year in the USA, and as many as 15,000 deaths per year are attributed to AAA rupture.

The prevalence and incidence of thoracic aortic aneurysms (TAA) is more difficult to assess than is the abdominal portion of the aorta due to poorer access to screening. The overall incidence rate of TAAs is estimated at 10.4 per 100.000 person-years. Intracranial aneurysms (IA) are found in 2–6% percent of the population and can present occasionally with a devastating and often fatal intracranial bleed. Ruptured IA is estimated to account for 8% of all strokes. These high levels of aneurysm occurrence and the associated clinical morbidity and mortality rates highlight the need for a greater understanding of the biomechanical behaviour of all forms of aneurysms.

Since the seminal work of Paul and others at the Univeristy of Strathclyde on the biomechanics of hip replacement in the late 1960s, biomechanics has grown to be a major scientific discipline encompassing many diverse aspects of medical technology and associated therapies. With the development of improved imaging techniques such as computer tomography (CT) and magnetic resonance imaging (MRI), biomechanists have been able to create very high quality realistic models of skeletal and soft tissue structures. These high quality models have enabled engineers and scientists to develop physiologically realistic simulations both in silico and in vitro which allow the behaviour of the biological systems to be extensively studied. The development of computational analysis of these structures is becoming increasingly valuable in a clinical setting and there is now frequent interaction between engineers and surgeons in the planning phase of major interventions. Furthermore the models allow advanced simulations of the in vivo situation to be analysed and the influence of pressure, temperature, flow and force in healthy, diseased and treated situations can be thoroughly investigated. In the context of this book, this is particularly the case and many of the studies of aneurysmal behaviour presented utilise advanced modelling methods to provide new insights into their response to the phyiological action of arterial blood.

In Chap. 1, the pathophysiology of the aorta is considered and the contribution of genetic risk factors to the development, growth and rupture of aneurysms in different segments of the arterial tree is considered. The role of inflammation and matrix metalloproteinases (MMPs) in the pathogenesis of AAA by causing proteolytic degradation of structural proteins is also presented as are aneurysm size and growth rate. Remaining with the diagnostic theme, Chap. 2 presents a wide range of approaches for imaging of aneurysms. These are divided into techniques, which principally provide information on structure (X-ray projection imaging including angiography, CT, MRI, ultrasound) and those which provide information on biological function. The principal example of the latter is positron emission tomography (PET), though it suffers from low spatial resolution. Simple measurement of the diameter made using 2D real-time ultrasound has to date been sufficient for the purposes of selection of patients for repair of abdominal aortic aneurysm (AAA) with CT, MRI and X-ray angiography being utilised when more detailed evaluation of anatomy is required. Information on blood velocity and related quantities such as flow rate can also be obtained using MRI and ultrasound. Measurements of wall distension occurring during the cardiac cycle can be obtained and may then be used to estimate the stiffness of AAA. Typically 3D imaging data may be obtained in AAA using CT or MRI, and in cerebral aneurysms using CT, MRI or spiral angiography. Data on inlet velocities may be obtained using Doppler ultrasound or MRI. The extraordinary advances in medical imaging technology both in terms of spatial and temporal resolution allows very high quality anatomical data to be gathered and plays an important role in many of the computational modelling approaches described in subsequent chapters.

Chapter 3 further examines many of the strength issues in aneurysm tissues considering the dissection issues more common in the thoracic aorta and the rupture events associated with the abdominal aorta. Particular attention is focussed on the importance of the role of local wall strength and local wall stress and the mechanobiology of the failure mechanism. The pathobiologic mechanisms which may pre-dispose patients to both aortic and thoracic aneurysms are considered and their differing pathophysiologies are discussed in detail. This chapter articulates the state-of-the-art of aortic biomechanics noting a relative paucity of data relating to TAA. The importance of mechanical properties and stress patterns in the tissue and the clinical value of such biomechanical parameters in predicting rupture behaviour are also considered. In Chap. 4 the growing role of computer-aided diagnosis (CAD) in the clinical arena in recent years and how such tools can be utilised in the assessment of aneurysm behaviour is discussed. Dividing CAD into detection and quantification techniques enables the engineering analyst to develop usable quantities aimed at helping identify aneurysms that may be at risk of rupture, with the primary focus being on the more common AAA. The role of peak wall stress, and vascular asymmetry and rupture indices based on finite element analysis and their potential use as clinical adjuncts to aneurysm diameter are discussed. The development of advanced computational modelling tools with the capability to identify rupture locations and incidence has given rise to an increasing demand for knowledge of the mechanical properties of the healthy and diseased arterial tissue. In Chap. 5 this important topic is considered with a view to providing a better understanding of the mechanical behaviour of aortic tissue. An improved understanding of the mechanical properties of the vascular tissue will enhance our ability to predict aneurismal behaviour and to design implants that can stay in place and/or protect the aneurysm wall from blood pressure. The authors review the current state of literature on the mechanical properties of AAA tissue, and the challenges faced in determining these properties.

In Chap. 6 the highly successful and modern technique of endovascular repair of abdominal aortic aneurysms (AAAs) is examined in detail. Once again the benefits of computational modelling are highlighted and the combination of the power of computational fluid dynamics with the investigations of structural issues in models of post-operative cases is presented. This approach, known as fluid structure interaction (FSI), is particularly useful as both the fluid forces acting on the graft and stresses on the aneurysm wall following the placement of the device can be investigated. The stresses and hemodynamics in healthy, diseased and treated aneurysms are presented and the magnitude, direction and possible clinical role of the forces, moments and pressures acting on the implanted devices are discussed. In addition the authors consider some patient specific influences, which may need to be further investigated. Continuing with the FSI approach Chap. 7 examines a range of numerical simulations of normal aortas, ruptured and non-ruptured AAAs again utilising patient specific geometries reconstructed from CT scans. The models incorporated material properties of the arterial wall, intraluminal thrombus (ILT), and associated calcifications and also considered the role of iliac bifurcation and neck angulations and hypertension in rupture behaviour. The analysis focuses on the importance of material properties in the numerical investigations and considers the importance of directional strength of the tissue as well as the influence of intraluminal thrombus (ILT), suggesting that the ILT may alleviate stresses in the aneurysm wall. The position of peak wall stresses and the usefulness rupture potential index (RPI) in AAA simulations are shown to offer clinical potential as are geometrical features in the aneurysms.

Expanding further on the biomechanical theme Chap. 8 further explores the clinical complications associated with the endovascular approach to treatment of AAAs and TAAs. The major bioengineering design challenges associated with endovascular stent-grafts including endograft migration (i.e., loss of positional stability), stent fractures and endoleaks (i.e., persistence of blood flow into the aneurysm sac after device placement) are all examined. Since these complications are frequently life-threatening and costly events, and can give rise to the need for secondary procedures and life-long follow-up with imaging studies a more comprehensive understanding of the biomechanical environment experienced by endografts in vivo is identified as a critical factor in improving their performance. The influence of tortuosity and size of the endograft on the mechanical loads experienced by the devices is examined as is the hemodynamic state of the patient. Fixation mechanics (radial pressure vs. hooks and barbs), the size of aortic area to which the stent-graft can attach, and the quality of the vessel wall in the attachment zone are all investigated and a succinct summary of the state-of-the-art in computational patient-specific modelling is presented. In Chap. 9 experimental investigations of parameters, which are likely to reduce the ability of an implanted stent-graft for the treatment of Abdominal Aortic Aneurysm (AAA) to resist migration are presented. Stent-graft rigidity, geometry and attachment forces in models with differing compliance and with pulsatile wall motions induced by physiological flow are all presented and magnitudes of migration forces in endovascular stent-grafts are quantified. The magnitudes of these forces are of considerable importance in device design for EVAR applications. Iliac bifurcation angle and placement of the devices in tortuous configurations were found to also influence fixation behaviour. The test methods described could be of value in the future preclinical evaluation of stent-grafts and in the design phase of next generation EVAR devices.

Returning to the clinical theme, Chap. 10 examines thoracic aortic aneurysms in detail and discusses how these aneurysms can both rupture and dissect. The value of diagnostic CT is presented and a variety of treatment approaches are considered. These include open repair for proximal thoracic aneurysms (ascending and arch) which are predominantly treated using an open approach, and open or an endovascular approach in the descending aorta. The value of biomechanics and computational flow dynamics in gaining an insight into the magnitude and direction of the forces acting on these thoracic endografts and the subsequent patient benefits are also described.

The theme of Chap. 11 moves up the body to the head and examines the challenging issues associated with intracranial aneurysms and while less frequent in occurrence, rupture can have similar devastating consequences to rupture in the other aneurysm locations. Advances in imaging techniques in recent years have led to the development of microneurosurgical techniques which have made permanent exclusion of the aneurysm from the circulation a valuable treatment option with a remarkable safety profile. In keeping with the rupture risk issues at other aneurismal sites, biomechanics research on device performance and flow mechanics and structural behaviour of IAs remain important topics. Chapter 12 examines the structural issues in cerebral aneurysms in considerable detail and discusses the development of biomechanical models of arterial wall behaviour that accounts for the structural arrangement of collagen fibres in load bearing configurations. Patient-specific geometry of the internal carotid artery is examined and growth and remodelling (G&R) of the aneurismal section is presented in what is believed to be the first patient-specific model of cerebral aneurysm evolution that incorporates a realistic constitutive model of the arterial wall and explicitly links G&R to the pulsatile mechanical environment.

In Chap. 13, the role of animal models in determining underlying causes of aneurysms and in the development of new treatment options for aneurysm sufferers is presented. The animal models may be used for both refining surgical techniques and stent-graft device characterization and are typically implemented in large animals (dog, pig, and sheep). Recent efforts have also aimed at determining both the biomechanical alterations that occur with aneurysm formation and the potential for rupture. This chapter considers the appropriateness of these aortic aneurysms models for biomechanical investigation and points to future options.

I would like to thank the authors and co-authors who have enabled me to put together a truly up-to-the minute book on biomechanics and mechanobiology of aneurysms which I hope will provide new insights to a wide readership. I would also like to pay tribute to my many colleagues who reviewed the various chapters and who provided many additional valuable inputs to the already excellent contributions. I would also like to pay tribute to the Series Editor, Professor Amit Gefen for his ongoing support and encouragement and to Christoph Baumann at Springer who has provided invaluable and continuous publishing support.

Ireland, May 2011 Tim McGloughlin

Contents

Aneurysm: Epidemiology Aetiology and Pathophysiology 1
Natzi Sakalihasan, Helena Kuivaniemi, Betty Nusgens,
Rodolphe Durieux and Jean-Olivier Defraigne

Imaging of Aneurysms. 35
Peter Hoskins, Scott Semple, Phil White and Jennifer Richards

Biomechanics and Pathobiology of Aortic Aneurysms. 67
Julie A. Phillippi, Salvatore Pasta and David A. Vorp

Computer-Aided Diagnosis of Abdominal Aortic Aneurysms. 119
Barry J. Doyle and Timothy M. McGloughlin

Mechanical Properties of AAA Tissue 139
Madhavan L. Raghavan and Erasmo Simão da Silva

Fluid–Structure Interaction in Healthy, Diseased and Endovascularly Treated Abdominal Aortic Aneurysms 163
David S. Molony, Stephen Broderick, Anthony Callanan,
Timothy M. McGloughlin and Michael T. Walsh

Biomechanical Aspects of Abdominal Aortic Aneurysm (AAA) and its Risk of Rupture: Fluid Structure Interaction (FSI) Studies . . . 181
M. Xenos and D. Bluestein

Computational Analysis of Displacement Forces Acting on Endografts Used to Treat Aortic Aneurysms 221
C. Alberto Figueroa and Christopher K. Zarins

Experimental Analysis of Endovascular Treatment of AAA and Predictors of Long Term Outcomes 247
Timothy Corbett, David S. Molony, Eamon G. Kavanagh, Pierce A. Grace, Michael T. Walsh and Timothy McGloughlin

Thoracic Aortic Aneurysms—Clinical Assessment and Treatment 285
Moqueet A. Qureshi, Brian D. Conway and Roy K. Greenberg

Intracranial Aneurysms: Clinical Assessment and Treatment Options 331
Manik Mehra, Gabriela Spilberg, Matthew J. Gounis and Ajay K. Wakhloo

Modelling Cerebral Aneurysm Evolution 373
Paul N. Watton, Yiannis Ventikos and Gerhard A. Holzapfel

Biomechanical Considerations of Animal Models of Aortic Aneurysm 401
Darren Haskett, Mohamad Azhar and Jonathan P. Vande Geest

Author Index 423

Aneurysm: Epidemiology Aetiology and Pathophysiology

Natzi Sakalihasan, Helena Kuivaniemi, Betty Nusgens, Rodolphe Durieux and Jean-Olivier Defraigne

Abstract Abdominal aortic aneurysm (AAA) disease is a chronic degenerative disorder and is an important cause of preventable deaths in older patients. Prevalence rates are estimated between 1.3 and 8.9% in men and between 1.0 and 2.2% in women. However, with the aging of the population and the increasing number of smokers, the incidence of the AAA is rising. The prevalence and incidence of thoracic aortic aneurysms (TAA) is more difficult to assess than for the abdominal portion of the aorta due to poorer access to screening. The overall incidence rate of TAAs is estimated at 10.4 per 100,000 person-years. The classical risk factors for atherosclerosis, such as tobacco smoking, male sex, age, hypertension, and hyperlipidemia have all been found to be also risk factors for AAA. The pathophysiology of the aorta above and below the diaphragm has shown significant differences in biomechanical properties, atherosclerotic distribution, proteolytic pattern, and cell signaling pathways that have implications in the development of an aortic aneurysm. During the last decades an overwhelming amount of evidence has been accumulated in support of genetic risk factors contributing to the

N. Sakalihasan (✉) · R. Durieux · J.-O. Defraigne
Department of Cardiovascular Surgery, University Hospital of Liège, Liège, Belgium
e-mail: nsaka@chu.ulg.ac.be

R. Durieux
e-mail: rodolphedurieux@hotmail.com

J.-O. Defraigne
e-mail: jo.defraigne@ulg.ac.be

H. Kuivaniemi
Sigfried and Janet Weis Center for Research, Geisinger Clinic, Danville, PA, USA
e-mail: shkuivaniemi@geisinger.edu

B. Nusgens
Laboratory of Connective Tissues Biology, University of Liège, Liège, Belgium
e-mail: betty.nusgens@ulg.ac.be

Stud Mechanobiol Tissue Eng Biomater (2011) 7: 1–33
DOI: 10.1007/8415_2010_47

Published Online: 8 February 2011

development, growth and rupture of aneurysms in different segments of the arterial tree. Inflammation and matrix metalloproteinases (MMPs) also play a key role in the pathogenesis of AAA by causing proteolytic degradation of structural proteins. The size of an aneurysm is a universally recognized factor in predicting the probability of rupture; the risk of rupture increases as the diameter of the aneurysm increases. Rupture occasionally occurs in small aneurysms. The risk of rupture and dissection of TAAs also increase with increasing diameter. In addition, not only the size but also the growth rate of the aneurysm has been consistently shown to be critical in predicting rupture.

1 Introduction

The word "aneurysm" derives from the Greek *ανευρυσμα* (aneurusma), meaning widening, and can be defined as a permanent and irreversible localized dilatation of a vessel. This "abnormal" dilatation involves the three layers of the vascular wall: the intima, the media, and the adventitia. This definition differentiates an aneurysm from a false aneurysm, which is a perivascular pulsatile hematoma secondary to a vessel injury. In a pseudoaneurysm the capsule is devoid of any vascular structure: the external limit of the pulsatile dilatation is made of amorphous fibrous material. Similarly, the infiltration of blood within the vascular wall as in aortic dissection is not an aneurysm in the strict meaning of the term.

In terms of morphology, reviewed in detail by Slaney [152], two types of dilatation involving all the layers of the vessel wall can be recognized. Most aneurysms are fusiform since the whole circumference of the artery is affected. In contrast, an aneurysm is designed "saccular" if it involves only a part of the circumference.

The normal diameter of the abdominal aorta varies with age, sex, and bodyweight, [11] and decreases progressively from its entry into the abdominal cavity to the iliac bifurcation. In elderly men, the infrarenal abdominal aortic diameter measures between 15 and 24 mm [91]. McGregor and colleagues [105] defined an abdominal aortic aneurysm (AAA) as an aorta with an infrarenal diameter greater than 30 mm.

2 Epidemiology of Aortic Aneurysms

2.1 Abdominal Aortic Aneurysm

Aneurysmal dilation of the abdominal aorta (AAA) represents a chronic degenerative disease and is an important cause of preventable deaths in elderly patients. In fact the mortality rate due to AAA is about 1.3% of all deaths among men aged between 65 and 85 years in the western world and constitutes the 14th leading cause of death in the United States and the tenth leading cause of death in older men, who are the principal victims [86]. The incidence of aortic aneurysm seems to have increased over the past two decades. This increased frequency can be explained either by improved detection and/or by an actual increase in the number of aneurysms as suggested by some recent data [45, 122]. Moreover, determining the true incidence of aortic aneurysms is challenging, mainly at the thoracic level. Many cases remain undetected because aneurysmal disease is usually silent until it becomes complicated either by dissection, rupture or embolisation. The second reason for underestimating the incidence of the disease is that the lethal aortic dissections or ruptures are often misdiagnosed as myocardial infarction [44].

The incidence of AAA has increased due in part to the ageing of the population, the rise in the number of smokers, the introduction of screening programmes and improved diagnostic tools. Abdominal aortic aneurysms are diagnosed each year in 200,000 patients in the United States. The total number of people with AAA is estimated at around 2 million in the United States. Rupture of these aneurysms causes roughly between 6,000 and 10,000 deaths per year in the USA [56, 143, 155]. A higher incidence of AAA has been reported in northern Scandinavia and the incidence of rupture may be increasing in the European Union population [1]. The disorder is more common in men than in women and frequency of the disease is higher with increasing age. Prevalence rates are estimated at between 1.3 and 8.9% in men and between 1.0 and 2.2% in women [87, 88, 94, 95, 99, 151, 170]. However, since smoking is one of the most important risk factors for abdominal aortic aneurysm [99, 170] and the number of female smokers is rising [17] the sex ratios for the prevalence of the disorder will probably change in the future [151]. The prevalence of abdominal aortic aneurysms in tobacco smokers is more than four times that in life-long non-smokers [170].

Most aneurysms discovered by screening are of small size and do not need immediate intervention [9, 167]. However, they can enlarge with time with a mean rate that is initially slow and then increases exponentially [93]. Epidemiological screening studies suggest a strong association between AAA and atherosclerosis, coronary artery disease, and peripheral arterial disease, although the etiologic relation between atherosclerosis and aneurysmal disease remains controversial. So it is not surprising that the frequency of AAA in subjects >60 years of age with coronary artery disease was reported as high as 14% [100].

2.2 Thoracic Aortic Aneurysms

Most thoracic aortic aneurysms (TAAs) occur in the ascending aorta followed by the descending thoracic aorta, the aortic arch and the thoraco-abdominal aorta. The average age at the time of diagnosis is around 70 years, with women being significantly older than men [27]. The leading cause of mortality from this type of aneurysm is aortic rupture, accounting for 60% of deaths [27, 129]. Thoracic aortic aneurysm rupture is associated with an exceedingly high mortality of

94–100% [12, 71]. The prevalence and incidence of aneurysms of the thoracic aorta is more difficult to assess than for the abdominal portion due to poorer access to screening.

Similar to abdominal aortic aneurysms, the incidence of thoracic aneurysms increased significantly during the past two decades in both sexes partly because of the increasing use of CT-scanning.

In a cohort study Clouse and colleagues reported an overall incidence of TAAs of 10.4 per 100,000 per year with the incidence increasing with age [27]. Olsson and co-workers studied the incidence of thoracic aortic aneurysms and dissections in the Swedish population and reported an annual incidence of 16.3 per 100,000 in men and 9.1 per 100,000 in women [122]. The incidence is higher than in the previous study but dissections were excluded in the study led by Clouse. The incidence has increased by 52% in men and 28% in women over a 15-year-period while the median age at diagnosis has decreased from 73 to 71 years.

3 Aetiology and Risk Factors

3.1 Bicuspid Aortic Valve

Bicuspid aortic valve (BAV) is the most common congenital heart malformation, affecting 1–2% of the population, and constitutes an important risk factor for the development of aortic valve disease (regurgitation, stenosis or both). Surgery for fibrocalcific stenosis will be eventually required in more than 75% of patients with BAV. Several studies have established familial clustering of bicuspid aortic valve, presumably indicating genetic inheritance [29, 68]. Dilatation of the aortic root and ascending thoracic aorta is common in patients with bicuspid aortic valves and the risk of acute dissection is increased. The prevalence of aortic dilatation in BAV has been estimated at between 33 and 80% and the tubular portion of the ascending aorta is involved twice as often as the aortic root [37, 116]. The precise mechanisms underlying the development and progression of thoracic aortic disease in BAV patients have not been clearly established. Two hypotheses have been proposed to explain the development of aortic dilatation [28]. The first theory is a poststenotic dilatation associated with increased wall shear stress due to modified flow profile, as can be seen in tricuspid valve stenosis. The other hypothesis for the high incidence of aortic dilatation in BAV patients is that the disease is associated with a congenital aortic fragility responsible for cystic medial degeneration in the aortic wall of the ascending aorta. There is considerable evidence to support the second hypothesis. Aortic aneurysm can occur in BAV patients in absence of valve dysfunction [123]. Yasuda and co-workers showed that aortic valve replacement alone could not prevent progressive aortic dilation in BAV, suggesting an intrinsic abnormality of the aortic wall [186]. Moreover cystic medial necrosis was present in the wall of the ascending aorta of 75% of patients with BAV compared to 14% of patients with tricuspid aortic valve undergoing the same surgery. A congenital deficiency in fibrillin-1 in the aortic wall of BAV patients could be the underlying cause of this aortic wall fragility [50].

3.2 Thoracic Aortic Aneurysms

The aetiology of TAAs is thought to be multifactorial, with genetic, environmental, and physiologic influences identified. A combination of factors may play a role in weakening the aortic wall which results in aortic dilatation and aneurysm formation. Aneurysmal disease of the aorta is characterized by regional heterogeneity between the thoracic and the abdominal aorta and even within the thoracic region. The thoracic aorta is itself divided in four segments (aortic root, ascending aorta, arch and descending aorta). 60% of thoracic aortic aneurysms involve the aortic root and/or ascending aorta, 40% involve the descending aorta, 10% involve the arch, and 10% involve the thoracoabdominal aorta (with some involving more than 1 segment) [69]. The pathophysiology of the aorta above and below the diaphragm has shown significant differences in biomechanical properties, atherosclerotic distribution, proteolytic pattern, and cell signalling pathways that have implications in the development of an aortic aneurysm [137]. As described above, AAA is strongly associated with atherosclerosis; although some TAAs are also associated with atherosclerotic disease, many occur in the complete absence of plaque deposition [70]. Achneck and colleagues have even reported that aortic root pathology (annulo-aortic ectasia or type A dissection) is associated with decreased systemic atherosclerosis compared to control subjects [2].

Nevertheless some risk factors are the same for both aortic regions: male sex, advanced age, cigarette smoking, hypertension, chronic obstructive pulmonary disease, and coronary artery disease. Moreover an abdominal aortic aneurysm is diagnosed in approximately 20% of patients with large TAA [127].

The majority of aneurysms of the descending thoracic aorta are associated with atherosclerosis and could be related to abdominal aneurysms in terms of physiopathology and risk factors [69].

Aneurysms of the aortic root and ascending aorta are most commonly related to gradual degenerative changes (loss of elastic fibers) in the medial layer of the aortic wall. This process is known as cystic media necrosis and is characterized by smooth muscle cell ‘drop out’, elastic fibre degeneration and the accumulation of mucoid material within cystic spaces in the media. Cystic medial degeneration occurs normally to some extent with aging, and the process is accelerated by hypertension. These changes within the muscular layers of the aorta may be associated with some genetic syndromes (Marfan’s syndrome, Ehlers–Danlos syndrome, bicuspid aortic valve, Loeys–Dietz syndrome and Familial Thoracic Aortic Aneurysm syndrome), acquired defects in the aortic media or idiopathic causes.

3.3 Abdominal Aortic Aneurysms

There are many causes of aneurysmal dilatation, but few abdominal aortic aneurysms are the direct consequence of specific causes such as trauma, acute infection (brucellosis, salmonellosis), chronic infection (tuberculosis), inflammatory diseases and connective tissue disorders (Marfan Syndrome, Ehlers–Danlos type IV) [160]. Some causes are quite exceptional (Behçet and Takayasu disease) [48, 103]; others have a historical context (syphilis).

Most abdominal aortic aneurysms are non-specific in origin. Moreover, because this disorder is invariably associated with severe atherosclerotic damage of the aortic wall, it has been traditionally regarded that AAAs develop as a consequence of atherosclerosis [131]. This conventional view has been increasingly challenged in recent years. Clinical and basic research studies indicate that aneurysms arise through pathogenic mechanisms that differ, at least in part, from those responsible for athero-occlusive disease [158, 182]. Much published work lends support to this concept [23, 55]. Defawe and colleagues [33] showed that two physiological inhibitors of proteases (TIMP-2 and PAI-1) were expressed less in abdominal aortic aneurysms than in athero-occlusive disease, suggesting a significant role for protease inhibitors during the divergent evolution of the initial atherosclerotic plaque towards either abdominal aortic aneurysm or athero-occlusive disease. Moreover, since not all patients with atherosclerosis develop an abdominal aortic aneurysm, even if atherosclerosis does have a role in the pathogenesis of the disorder, additional factors are probably involved in aneurysm development. The classic risk factors for atherosclerosis, such as tobacco smoking, male sex, age, hypertension, and hyperlipidemia have all been found to also be risk factors for AAA [14, 18, 88, 151, 156, 170]. Moreover several studies report a familial clustering of the disorder supporting a genetic contribution to the development of AAA [16, 25, 72, 81, 101, 125, 135, 148, 171]. There is a strong clinical association between tobacco smoking and aneurysm development [14, 17, 96, 170]. A report that compared relative risks for different diseases in chronic cigarette smokers showed that the risk of developing abdominal aortic aneurysms is threefold greater than the risk of developing coronary artery disease and nearly fivefold greater than the risk of developing cerebrovascular disease [89]. Based on these clinical observations, chronic tobacco smoking is probably the most important environmental risk factor for development and progression of aortic aneurysms. The rate of growth of AAAs has been reported to be higher in current smokers than in non-smokers (2·83 mm per year vs. 2·53 mm per year, respectively) [18, 99]. The mechanisms by which smoking could promote aneurysm formation remain unknown and are being investigated [19].

The recently published data of the 7-year prospective TromsØ study reported male sex, increasing age, smoking, hypertension, hypercholesterolemia and low high-density lipoprotein cholesterol as risk factors for developing an AAA [53]. Surprisingly, the use of statins was associated in this study with increased risk of AAA although experimental data show down regulation of major mediators of

AAA progression by statins. [39, 149] Perhaps only individuals with very high cholesterol levels were treated, and treatment most likely was suboptimal.

Although diabetes is a major risk factor for developing atherosclerosis, large epidemiology screening studies showed either a negative association or no association between diabetes and AAA [39, 179].

4 Genetics of Aneurysms

During the past 10 years an overwhelming amount of evidence has been accumulated in support of genetic risk factors contributing to the development, growth and rupture of aneurysms in different arteries of the body [83, 84, 109, 114, 115, 162, 178].. Aneurysms in different locations such as those in the thoracic aorta, the abdominal aorta and the intracranial arteries do not often occur in the same individual or even in different members of the same family. It has, therefore, been concluded that genetic risk factors differ based on the location of the aneurysm (Fig. 1) [114, 115, 162, 164]. There are some exceptions to this observation, one of the most important ones being the discovery of a common polymorphism located on the 9p21 chromosomal region and associated with both intracranial (IA) and abdominal aortic aneurysms (AAA) [63]. The genetics of thoracic aortic aneurysms and dissections (TAAD] appears to differ from that of the other types of aneurysms in the sense that they are more often single gene disorders where a mutation in one single gene explains the disease and is highly penetrant [109].

Many different approaches have been used to identify genetic risk factors for aneurysms. These methods include (1) DNA sequencing of candidate genes based on known pathobiology of the diseases; (2) genetic association studies with polymorphisms in candidate genes; (3) family-based DNA linkage studies to identify chromosomal regions likely to harbor a gene contributing to the disease; (4) DNA linkage studies using affected-relative pairs (ARPs) to reduce diagnostic uncertainty; and (5) genome-wide genetic association studies (GWAS) with cases and controls to analyze the entire genome with single nucleotide polymorphisms (SNPs). For a discussion on advantages and disadvantages of these approaches in studying aneurysms, the reader is referred to previously published reviews [80, 82, 92, 114, 115, 161, 162]. The choice of methods used depends largely on the number and type of available samples as well as the expected results and research questions asked. The cost of experiments also varies widely between the different approaches. For example, if the investigator has collected blood samples from 50 to 100 members of a large family in which individuals with the disease have been diagnosed in several generations, the obvious choice would be to carry out a genome-wide DNA linkage analysis using genomic DNA isolated from the blood samples. On the other hand, if samples are available from only a very limited number of patients, the only feasible approach would be DNA sequencing of

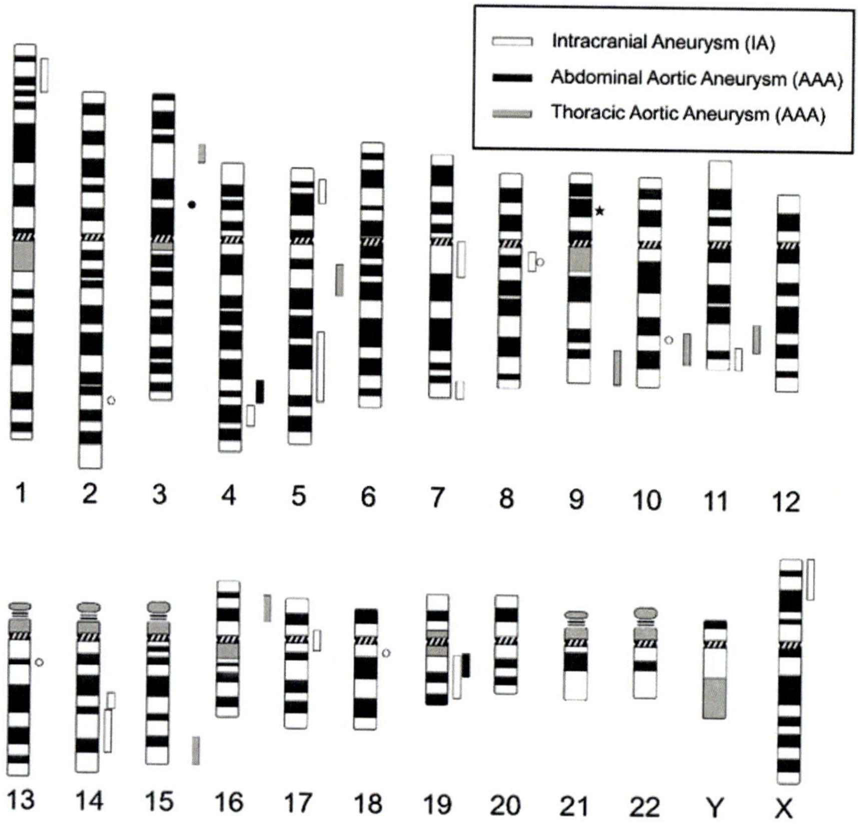

Fig. 1 Chromosomal localizations of the genetic susceptibility loci for AAA, IA and TAAD. Altogether four, sixteen and seven loci for AAA (*black symbols*), IA (*open symbols*) and TAAD (*grey symbols*), have been reported in the literature. Vertical lines adjacent to the chromosome ideograms indicate regions identified by DNA linkage studies and round symbols indicate locations of SNPs found in genetic association studies. The chromosome 9p21 region which contains the genetic variant rs10757278 associated with coronary artery disease, AAA and IA [63] is indicated with a star. Modified from [114, 115]. For details see text. The ideograms can be obtained from "Idiogram Album: Human" (copyright© 1994 David Adler, University of Washington, Department of Pathology) at http://www.pathology.washington.edu/research/cytopages/idiograms/human/

candidate genes to find out if the gene harbors mutations in the biologically plausible candidate gene. Two or more of the approaches can also be combined. For example, the investigator might consider DNA sequencing of candidate genes after first identifying a genomic region by DNA linkage study and then selecting genes located in that genomic region and having a potentially plausible role in disease pathogenesis.

4.1 Genetic Risk Factors Associated with Intracranial Aneurysms

Several DNA linkage analyses [54, 178] and two GWASs [13, 187] have been carried out for intracranial aneurysms (IA). These studies have identified 16 IA susceptibility regions on chromosomes 1, 2, 4, 5, 7, 8, 9, 10, 11, 12, 13, 14, 17, 18, 19, and X (Fig. 1), suggesting vast genetic heterogeneity [13, 54, 178, 187]. Similar to AAA, the association to the SNP located on the chromosome 9p21 has been replicated in many studies and in many different populations and it can now be considered a true genetic risk factor for IA [36, 62, 63, 187]. In a recent study smoking was shown to interact with the 9p21 locus providing evidence for gene-environment interaction [36].

4.2 Mutations Leading to Thoracic Aortic Aneurysms and Dissections

Altogether seven susceptibility loci for TAAD are currently known (Fig. 1), but they do not explain the familial aggregation of TAAD in all the families that have been studied, suggesting that additional loci will be found [109]. There is considerable overlap between the syndromic and non-syndromic forms of TAAD at the molecular level. Two TAAD loci on 3p24-25 and 9q33-q34 are the same as the genetic loci for the Loeys–Dietz syndrome, a rare autosomal dominant syndrome with a variety of clinical features, including generalized arterial tortuosity with ascending aortic aneurysm and dissection [109]. Also, the 16p13.13-p13.12 TAAD locus was identified using a large 178-member French family with TAAD and patent *ductus arteriosus* [75].

Four genes have been found to harbor mutations in patients with TAAD. These are the genes for: (1) transforming growth factor β receptor 1 (*TGFBR1*); (2) transforming growth factor β receptor 2 (*TGFBR2*); (3) smooth muscle myosin heavy chain 11 (*MYH11*); and (4) smooth muscle alpha actin 2 (*ACTA2*) [60, 109, 189].

These genetic studies have helped to dissect the pathobiology of TAAD in that dysregulation of TGFβ signaling appears to be one of the potential mechanisms leading to TAAD. Another molecular mechanism involved in TAAD and revealed by the genetic studies is smooth muscle contractility, since two genes, MYH11 in oblique please and ACTA2 in oblique please from this functional class, harbor mutations in patients with TAAD [109].

4.3 Genetic Risk Factors Associated with Abdominal Aortic Aneurysms

Population-based studies have demonstrated that family history of AAA is an important risk factor for AAA with an odds ratio (OR) of 1.96 (95% CI: 1.68–2.28) [114, 115]. The lifetime prevalence of AAA among siblings of AAA patients is estimated to be about eightfold higher than in the general population [120]. A recent twin study also strongly supported the involvement of genetic factors in AAA [176].

Two formal segregation analyses revealed statistically significant evidence for a genetic model with a major gene effect in AAA [101, 171]. Several studies have also reported on collections of AAA families [81, 119, 159], the largest one with 233 multiplex families [81]. In this study 72% of the AAA families were consistent with an autosomal recessive inheritance pattern, while 25% showed an autosomal dominant pattern of inheritance. Finding different modes of inheritance in separate collections of families is consistent with AAA being a multifactorial disease where different loci can have distinct modes of inheritance.

Several approaches have been used to study the genetics of AAA. DNA linkage study on 235 ARPs from 119 AAA families found significant linkage on chromosomes 4q31 (LOD score = 3.73, $p = 0.0012$) and 19q13 (LOD score = 4.75, $p = 0.00014$) [148]. In this study sex and number of affected relatives, and their interactions were used as covariates, allowing for genetic heterogeneity. These genomic regions were designated as AAA2 and AAA1 susceptibility loci, respectively, in Online Mendelian Inheritance in Man (OMIM) (Fig. 1). The 19q13 locus was found also in another DNA linkage study for AAA using Dutch AAA families [169]. The chromosome 4 and 19 candidate regions contain a large number of plausible and physiologically relevant positional candidate genes such as low density lipoprotein receptor-related protein 3 (*LRP3*), peptidase D (*PEPD*), hepsin (*HPN*), interleukin 15 (*IL15*), GRB2-associated binding protein 1 (*GAB1*) and endothelin receptor type A (*EDNRA*), but detailed analyses on them have not yet been reported.

A genome-wide genetic association study for AAAs using pooled DNA samples and a case–control design found an AAA-associated haplotype on chromosome 3p12.3 (Fig. 1) [46]. One SNP in this region (rs7635818) was genotyped in a total of 502 cases and 736 controls from the original study population ($P = 0.017$) and 448 cases and 410 controls from an independent replication sample ($P = 0.013$; combined $P = 0.0028$; combined $OR = 1.33$). An even stronger association with AAA was observed in a subset of smokers (391 cases, 241 controls, $P = 0.00041$, $OR = 1.80$), which represent the highest risk group for AAA. The AAA-associated haplotype is located 200 kbp upstream of the transcription start site for the contactin 3 gene (*CNTN3)*, a member of a family of cell adhesion molecules. It is a plausible candidate gene, but further studies are needed to establish what role CNTN3 plays in AAA pathophysiology.

To date, the largest genetic study on AAA used 2,836 AAA cases and 16,732 controls to test a SNP (rs10757278) discovered in a separate GWAS on coronary artery disease, in five different vascular phenotypes: coronary artery disease, peripheral artery disease, atherosclerotic stroke, IAs and AAA [63]. The SNP (rs10757278), located on chromosome 9p21, was associated with AAA ($P = 1.2 \times 10^{-12}$; $OR = 1.31$, 95% CI: 1.22–1.42). It was also associated with all other phenotypes except atherosclerotic stroke. This is the first genetic variant common to AAA, IAs and other cardiovascular diseases and it suggests shared pathophysiology in these vascular diseases. The SNP is located only 10 kbp away from rs10811661, a SNP known to be associated with type 2 diabetes. The two SNPs are not associated with one another, and AAA or IA are not associated with rs10811661. This finding is in agreement with epidemiological data demonstrating that diabetes is not a risk factor for AAA or IA [146].

4.4 How Does the Genetic Information on Aneurysms Help us in Explaining the Disease Process?

Discovering an SNP associated with a disease provides the start of a research project to functional studies to explain the variant's contribution to the disease pathogenesis. Identification of two key mechanisms, defects in TGFβ signaling and smooth muscle contractility, leading to TAAD are examples of success stories on this front. DNA linkage studies were used to identify the genomic regions likely to harbor mutations in patients with TAAD in these families. Candidate genes in the regions were analyzed in detail and sequence changes were found. The role of these mutations was then tested in cell-culture systems. Future work will include studies on mice genetically engineered to carry the same mutations identified in humans to close the loop into causality.

One of the more difficult situations was to explain the contribution of the SNP associated with AAA and IA on chromosome 9p21, since the closest gene to this SNP is a non-coding RNA gene, called *ANRIL*, and very little is known about the function of non-coding RNAs. Interestingly, *ANRIL* has at least three alternatively spliced transcripts, the expression of which varies in tissues [66]. Recent studies on mice with a deletion of a 70 kbp region encompassing the SNP and portions of *ANRIL*, but not the two closest protein coding genes *CDKN2A* and *CDKN2B*, showed that cardiac and vascular expression of *CDKN2A* and *CDKN2B* was affected [172]. Cells cultured from the aortae of the mutant mice proliferated faster than those from control mice, and did not show signs of senescence. This phenotype fits with risk for atherosclerosis and possibly occlusive disease, but it is more difficult to explain how it contributes to AAA and IA, which are characterized by smooth muscle cell apoptosis and loss.

Another result from genetic studies that requires further work is the correlation of genetic and clinical findings. How should we interpret the findings when the same gene harbors mutations in patients with two seemingly different diseases? Should molecular diagnosis using "genomic nosology" be adopted [162], where all individuals with a mutation in the same gene are classified as having the same disease or should the clinical manifestations be the basis for diagnosis? The case for molecular diagnosis is that the spectrum of clinical manifestations of rare genetic diseases can be wide, and often overlaps those seen in more common diseases; consequently, presence of a mutation in a given gene would classify the disease more accurately. For example, one could argue that all patients with mutations in

TGFBR2 gene should be classified as having Loeys–Dietz syndrome. The case for diagnosis based on clinical manifestations is that although genetic analyses are useful for explaining the underlying pathogenesis, they do not accurately predict the clinical phenotype. Thus mutations in the same gene could lead to different but overlapping phenotypes, e.g., non-syndromic TAAD versus Loeys–Dietz syndrome for mutations in *TGFBR2* [162, 164].

In many cases genetic studies alone are not enough, and multidisciplinary approaches are needed to explain disease process. To this end, genome-wide microarray-based RNA expression studies provide an unbiased way to assess the differences in diseased and undiseased tissue samples [162]. This approach has been used for AAA with aortic tissue [24, 90] and blood [57]. In an approach called "transcriptional genomics" the microarray results are then used to study the coordinated expression changes by analyzing the promoter regions of the genes with altered expression levels [114, 115]. The results indicated an important role for transcription factors belonging to the Ets family, and provided important leads for future studies and even suggestions for ways to interfere with the disease process pharmacologically [15, 114, 115].

5 Pathophysiology of Abdominal Aortic Aneurysms

The development of abdominal aortic aneurysms is clearly associated with alterations of the connective tissue in the aortic wall. The role of inflammation in the pathogenesis of AAA and the specific biological remodelling involved in the enlargement of arteries has been investigated only in the last twenty years. Over the past two decades, it became clear that several distinct but interrelated processes contribute to the pathologic alterations observed in human AAA tissue. Some of the most important of these processes include chronic inflammation, increased production of matrix degrading proteinases and their inhibitors, and localized degradation of structurally important connective tissue proteins, especially elastin and collagen [21, 107, 134, 138]. More recently, several groups have focused on the role of immunity in the development of AAA. [30, 121].

The new pathophysiological challenge is to understand the cascade of biological events which initiate, aggravate and are associated with aneurysmal remodeling [108].

Aortic aneurysms are characterized by thinning of the aortic media resulting from proteolytic injury to the extracellular matrix and smooth muscle cell disappearance, allowing further proteolytic injury leading to dilation and rupture. However, AAA and TAA present some striking differences: secondary (AAA) versus primary (TAA) pathology of vascular smooth muscle cells; linkage (AAA) or not (TAA) to atheroma; monogenic (TAA) versus polygenic (AAA) determinants; age and gender issues. Defining the similarities and differences are of great importance for the understanding of these vascular remodeling processes involved in the pathogenesis of aneurysms.

5.1 The Extracellular Matrix Proteins in Normal and Aneurysmal Aorta

The extracellular matrix (ECM) is a complex assembly of fibrillar macromolecules and associated glycoproteins embedded in a hydrated ground substance made of glycosaminoglycans and proteoglycans. In the vascular system, as in any other connective tissue, besides providing the architectural framework and mechanical properties required for vessel function, the ECM plays an informational role towards resident cells. The ECM macromolecules are recognized by specific cell surface receptors, such as integrins and associated signaling molecules. These specialized transmembrane structures physically connecting the ECM to the cell cytoskeleton act as sensors and transducers of mechanical signals into biochemical messages that regulate many cell functions such as proliferation, migration, gene expression, tissue architectural organization, survival and programmed cell death. The ECM components also operate by their capacity to sequestrate and modulate the availability of soluble effector molecules such as growth factors and cytokines. Furthermore, proteolytic fragments of ECM macromolecules or cryptic domains may have distinct functions in various processes.

The large blood vessels are made of three compartments: the *intima*, made of a single layer of endothelial cells lining the lumen; the *media* consisting of 50–80 concentric layers of lamellar units rich in elastic and collagen fibers, and the *adventitia* made of a loose collagen-rich tissue containing fibroblasts and vascularized by *vasa vasorum*. The main structural components of the large vessels such as aorta are collagens and the elastic fibers that result from the deposition of elastin on a scaffold of microfibrils, mainly fibrillins. These fibrillar structures deposited by smooth muscle cells (SMC) in the media are the major components of the arterial wall that contribute to its mechanical properties and that provide the elastic recoil required in a closed circulatory system [175]. The medial lamellar units are formed of two parallel thick lamellae of elastic fibers enveloping SMC with numerous resistant interlamellar protrusions rich in elastin. Most of the fibrillar collagens, I, III and V are concentrated in compact fibers concentrically oriented, closely associated with the elastic lamellae, and decorated with small patches of biglycan, a small leucin-rich proteoglycan (SLRP). The interlamellar matrix contains many microfibrils rich in fibrillin 1 and collagen VI. Versican, a large chondroitin sulfate proteoglycan that can associate with hyaluronic acid, is found in the interstitial space and is presumably aimed at sustaining the compression generated by pulsatile forces [40]. Although collagen and elastic fibers largely contribute to the mechanical performances of the vessels, the associated glycoproteins and proteoglycans are also significant actors in the maintenance of the vessels as indicated by the vascular pathologies related to their structural alteration or signaling dysfunction as observed for fibrillin 1 in Marfan's syndrome. Production of a matrix, mechanically competent and endowed with adequate signaling properties, involves a large number of biosynthetic steps and post-transcriptional modifications that are under the control of regulatory mechanisms issued from the environmental matrix as well as from soluble factors [85].

Three processes play a key role in the progression of AAA: inflammation and oxidative stress, degradation of the ECM of the aortic wall and impairment of its reconstruction associated to SMC depletion and apoptosis. A highly significant loss of elastin seems to be an early event in aneurysm formation resulting in elastic lamellae disruption and medial rarefaction without any substantial loss of collagen. The adventitial tissue that predominantly contains collagen may provide some mechanical resistance to the aorta in the absence of competent medial ECM. Collagen degradation seems to be the ultimate cause of rupture. The aneurysmal tissue is characterized by a large production of ECM-degrading enzymes (MMPs), proinflammatory cytokines and chemokines by the lympho-monocytic infiltrate and resident vascular wall cells, SMC and fibroblasts. Inhibitors of MMPs such as TIMPs and PAI-1 are also increased although with a balance in favor of proteolysis [33]. A strong increase of MMP-9, and of its activated form, in human and experimental aneurysms has been consistently reported. Knowledge of the intricate signaling pathways converging to AAA formation and progression as well as the events leading to its rupture should help at the identification of pharmacological agents aimed at preventing AAA progression.

5.2 Role of Metalloproteinases During the Development of AAA

In the aortic wall, elastin and collagen can be degraded by specific proteases displaying elastase and collagenase activity. These proteases are produced by resident cells of the vascular wall (medial SMCs and adventitial fibroblasts) and by the cells of the lympho-monocytic infiltrate. These inflammatory cells in the media and adventitia arise from the aortic blood but also from the medial neovascularisation which is characteristic of AAAs [65, 67, 145]. Leukocyte recruitment into the aortic wall is promoted by elastin degradation fragments as well as pro-inflammatory cytokines, chemokines, and prostaglandin derivates released by both resident mesenchymal cells and the inflammatory cells themselves [61, 78, 112, 113, 177] The proteolytic enzymes degrading elastin and collagen are defined as Matrix Metallo-Proteinases (MMPs) locally activated by either other MMPs or by plasmin generated by plasminogen activators [22, 23, 31, 32, 33, 49, 68, 98, 102, 117, 130, 133, 139, 140, 147, 150, 157, 163]. The role of MMPs and plasmin in AAA development has been confirmed in animal models [3, 8, 22, 38, 41, 98, 128, 150]. In aneurysmal and other sclerotic aortic diseases, an inflammatory infiltrate is present. In AAA, the inflammatory cells are present in the media and the adventitia, while this infiltration occurs mainly in the intima during aortic occlusive disease, as mentioned above by Defawe et al. Decreased gene expression of TIMP-2 and PAI-1 is observed in aneurismal tissues compared to occlusive aortic tissue. However, Thompson et al. observed an increased amount of Tissue Inhibitors of Matrix metalloProteinases (TIMPs) in the aneurysmal wall [156]; and Knox et al. as well as Tamarina et al. observed that the balance proteases/antiproteases was in favour of proteolysis [77, 153]. The significance of this imbalance during AAA development is reinforced by experimental studies in which the antiproteases are overexpressed or genetically inactivated [5, 6, 136].

5.3 *Role of Thrombus During the Development of the Aneurysm*

The development of AAA is associated with a mural thrombus in the majority of cases. In contrast to arterial occlusive diseases (AOD), blood flow is maintained in AAA resulting in continuous remodeling of the thrombus components. Aneurysm diameter has been reported to correlate with the plasma concentrations of fibrin formation and degradation products [185] as well as with the concentration of the circulating complex, plasmin-α2-anti-plasmin [96], which is potentially related to thrombus turnover. The role of an adherent thrombus in aneurysmal degeneration has also been investigated. While the thrombus may significantly reduce aneurysmal wall stress, its increasing thickness leads to local hypoxia at the inner layer of the media; this may induce increased medial neovascularisation and inflammation [174]. The idea that thrombus may be a source of proteases contributing to aneurysmal evolution has also been suggested following the initial report of high MMP-9 activity in thrombus [139, 140]. Recent results suggest that platelet activation and fibrin formation play a critical role in AAA development and rupture [73, 74]. Intra Luminal Thrombus (ILT) activities are polarized: initiated at the luminal interface with circulating blood, these activities are progressively conveyed towards the aneurysmal wall. Luminal platelet activation and fibrin formation convey zymogens from the blood towards the aneurysmal wall, thus participating in the extracellular matrix degradation [52]. On the abluminal pole, ILT was completely proteolysed and resembled the necrotic core of vulnerable plaque. Therefore, AAAs represent an accessible spatio-temporal pathophysiological model of human atherothrombosis, linking biological activities to clinical expression. In addition, Fontaine [51] provided evidence of polymorphonuclear neutrophil (PMN) trapping and MMP-9 storing within the aneurysmal thrombus. They also demonstrated the presence of plasminogen in the thrombus and its activator (u-PA) in the aneurysmal wall. This might result in local generation of plasmin, an activator of MMPs, in the aneurysmal wall.

6 Oxidative Stress and Aneurysm Development

Abdominal aortic aneurysm may have a specific cause, but in most cases no specific cause can be identified. As mentioned previously, AAA results from a chronic inflammatory disease characterized by extensive extracellular matrix degradation and vascular smooth muscle cell (VSCM) senescence or apoptosis, leading to weakening and dilatation of the aortic wall [142]. Aside from increased local production of cytokines, increased matrix metalloproteinases activity is a key element in the development or progression of AAA [4]. In this process, a causative or a contributive role for oxidative stress has been suggested [118].

6.1 Definition of Oxidative Stress

Reactive oxygen species (superoxide anion ($O_2^{-\cdot}$), hydrogen peroxide (H_2O_2), hydroxyl radical ($OH^{\cdot}$)) are continuously produced in the organism. This production depends on several sources such as membrane NADPH oxidase and myeloperoxidase in phagocytic cells (leukocytes and macrophages), mitochondria, prostaglandin synthesis by the cyclo-oxygenase pathway, hemic protein, and xanthine oxidase. Copper (Cu) and iron (Fe) catalyse radical chain reactions leading for example the production of hydroxyl radical from superoxide anion and hydrogen peroxide [79].

Normally, Reactive Oxygen Species (ROS) are produced at a very low level and intervene in normal physiology, controlling for example apoptosis, fertilization and other important processes. ROS are released by phagocytic cells to destroy bacteria. The cellular redox state is precisely balanced to preserve cell homeostasis. Thus, ROS are counteracted by an array of cellular and extracellular antioxidants, some of them being enzymatic (Cu- Zn- and Mn-superoxide dismutases, catalase, glutathione peroxydase and associated glutathione transferase and reductase,...) and others non-enzymatic (vitamins E and C, carotene, uric acid,...) [142]. Consequently, the balance between production and neutralization of ROS not only depends on the activity of ROS-generating systems, but also on levels of endogenous antioxidant systems. So oxidative stress and tissue damage may occur secondary to increased production and/or decreased neutralization of ROS. In these circumstances, ROS interact with cellular and extracellular molecular targets, causing damage to essential and structural proteins (oxidation and nitrosylation), to lipid and to DNA. Inflammation is one example of increased ROS production overwhelming local or systemic antioxidant defences, resulting in tissue damage.

The deleterious role exerted by Reactive Nitrogen Species (RNS) must also be mentioned [188]. Whereas constitutive endothelial NO synthase releases physiological amounts of nitric oxide (NO) contributing to normal vascular tone because of its relaxing properties on VSMCS, inducible NO synthase expressed in inflammatory cells and in VSMCs releases large amounts of NO. Reaction of NO with the superoxide anion gives rise to highly toxic peroxynitrites.

6.2 Role of Oxidative Stress in AAA

Oxidative stress plays an important role in the pathogenesis of numerous diseases and in several cardiovascular diseases, including atherosclerosis, hypertensive vascular disease, and coronary artery disease [181]. As an example, oxidative stress promotes lipoprotein oxidation, a key element in the development of atherosclerosis [104].

It is clear that inflammation is also a key element in the pathogenesis of AAA. During the development of AAA, infiltrating inflammatory cells release matrix metalloproteinases (MMPs), causing, as mentioned previously, the proteolytic degradation of structural proteins (such as elastin, collagen and laminin). MMPs which comprise a large family of several isoforms are secreted by various cell types and especially by inflammatory cell infiltrating the vascular wall and by resident VSCMs. MMPs are involved in many cellular processes, including migration, extracellular matrix deposition, and apoptosis, and are clearly implicated in AAA formation [64]. Aside from release of MMPs, inflammation contributes to the tissue damage and to the development of AAA by production of cytokines (interleukin-6, MCP-1, osteopontin) and activation of the cyclo-oxygenase-2 leading to further recruitment of immune cells and induction of apoptotic cell death pathways in VSMcs, such as Fas and perforin.

The link between oxidative stress and matrix degradation is well documented. ROS produced by inflammatory leukocytes contribute to tissue destruction observed in a variety of immunologic and infectious disorders. Higher superoxide anion ($O_2^{-\cdot}$) levels were demonstrated in the AAA segments of aneurysms undergoing surgical repair compared to the adjacent non-dilated aorta (although also atherosclerotic). Lipid peroxides and nitrosylated proteins were also increased and the major changes were closely localized to inflammatory cells and to SMCs. Thus a greater oxidative stress is observed in AAA compared to atherosclerotic but non-aneurysmal aorta [104].

ROS and RNS have been shown to regulate transcription of MMPs and to modulate their activity. $O_2^{-\cdot}$ and H_2O_2 activate secretion of MMPs by VSCMs [110]. In vitro, H_2O_2 promote secretion of MMP-9 by polymorphonuclear cells, and this process is completely prevented by the addition of catalase, an antioxidant enzyme transforming H_2O_2 to oxygen and water. ROS are required for the induction of apoptosis by various proinflammatory mediators, including cytokines. Thus ROS induce the apoptosis of VSMCs, which is a recently described feature of human AAA.

Several potential sources for ROS production have been identified in the vascular wall. In addition to inflammatory cells, $O_2^{-\cdot}$ is produced by 3 major cell types present in the vascular wall: SMCs, fibroblasts and endothelial cells. Mechanical stretch increases ROS production by NADPH oxidase in SMCs, which in turn activate NF-KB and increase matrix metalloproteinases. Infiltrating inflammatory cells produce ROS and generate proatherogenic factors that prime SMCs for ROS release which in turn recruit inflammatory cells. A positive feedback and vicious cycle are thus instituted (Fig. 2).

The major sources of ROS production are the NAD(P)H oxidase and the Ras-related C3 botulinum toxin substrate 1 (Rac1) [43, 154]. The phagocytic form of the enzyme uses predominantly NADPH whereas the vascular form uses NADPH or NADH as substrates. Expression of NADPH oxidase in SMCs is increased by several factors, including mechanical stress, cytokines, growth factors (angiotensin II and platelet derived growth factor), and oxidized LDL. Evidence for a role of NAD(P)H oxidase came also from experiments performed in apolipoprotein E-deficient mice where prolonged infusion of angiotensin II resulted in AAA formation, independently of the blood pressure [104]. Angiotensin II exerts various cardiovascular effects through activation of macrophages, induction of cytokines, and stimulation of expression and activation of NAD(P)H oxidase in SMCs, resulting in increased ROS production [181]. In this model, angiotensin II receptor blockers prevented upregulation of the $p22^{phox}$ NAD(P)H oxidase subunit and in knock-out mice for the $p47^{phox}$ subunit the development of AAA was prevented. Expression of subunits of NAD(P)H oxidase ($p47^{phox}$ and $p22^{phox}$) and NAD(P)H activity are increased in AAA and also in aneurysmal thoracic aorta especially in the regions where monocytes and angiotensin converting enzyme – positive macrophages accumulated [104, 154]. This localization closely corresponds with activity of matrix metalloproteinases. Taken collectively, these data

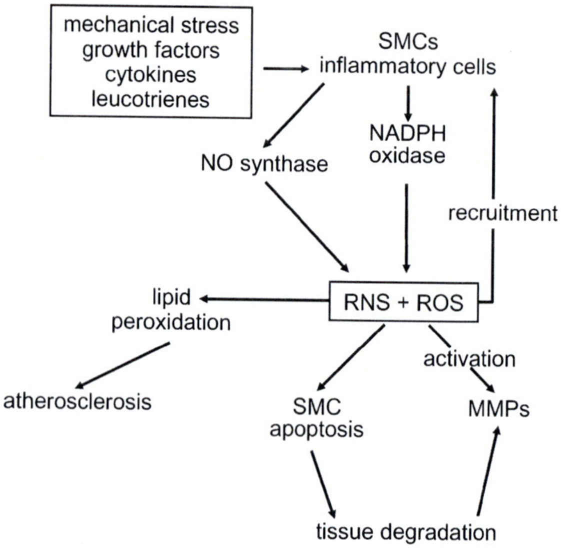

Fig. 2 The NADPH present in inflammatory cells and in vascular cells is activated by several mechanism. ROS are release leading to several consequences and ultimately to tissue destructive

illustrate the major role played by NAD(P)H oxidase in AAA and demonstrate the complex interactions between inflammation, oxidative stress and mechanical forces.

Changes in expression of genes implicated in oxidative stress may also be involved in AAA pathogenesis [184]. By means of cDNA microarray technique in an experimental model of elastase-induced AAA in mice, an upregulation of genes coding for enzymes (inducible NO synthase, e.g.) involved in oxidative stress and of genes coding for MMPS was demonstrated. In contrast, downregulation of genes coding for antioxidant enzymes (such as superoxide dismutase and glutathione-S-transferase) was observed. This distinct expression profile of antioxidant and pro-oxidant genes could result in an imbalance between ROS production and free radical scavengers, with an impaired redox state accelerating AAA formation. In the same studies, genes encoding for VSMCs contractile proteins and for extracellular matrix protein were downregulated, reflecting apoptosis of VSMCs, with progressive weakening of the vascular wall.

In humans, levels of vitamin C, and activities of Zn superoxide dismutase (SOD), glutathione peroxidase, and glutathione reductase were reduced in tissue samples from patients with AAA and atherosclerotic occlusive disease compared with non-diseased aorta from a different group of patients [42]. Similarly, plasma levels of vitamin E are reduced in patients with AAA compared to patients with coronary artery disease in the absence of AAA [139, 140]. It is worth noting that ascending aortic aneurysms of bicuspid aortic valve patients displayed a significantly lower expression of the metallothioneins family genes as compared to tricuspid valve aneurismal aorta [124].

Aside from ROS, RNS are increased in AAA tissue. Expression of the inducible NO synthase by macrophages and cytokines could generate large quantities of NO, which might also contribute to tissue damage. NO synthase is also a potential source of O_2^-. Inducible iNOS play a significant role in cardiovascular homeostasis, atherosclerosis, and inflammatory responses. NO mediates tissue damage by an amplification of the effects of proinflammatory cytokines on various cell types, which in turn may increase the expression and enzymatic activity of MMPs. In addition NO promotes nitration of proteins and can induce VSMC apoptosis.

The above data help to explain the deleterious role of smoking and the potential protective effects of statins in vascular disease [104]. By inhibiting the synthesis of cholesterol, statins also inhibit the synthesis of isoprenoid intermediates implicated in the post-translational modifications of intracellular signaling proteins (Ras and Rho proteins including Rac1) crucial in a variety of cellular events [168]. These pleiotropic effects of statins partially explain their positive influence on the endothelial dysfunction that is an early manifestation of atherosclerosis. In contrary, smoking leads to endothelial dysfunction, with a subsequent inflammatory process and plaque formation. Smoking also influences hemodynamic stress and oxidant injury and several large studies identified tobacco smoking as major contributive factor to an increased risk AAA development.

7 Haemodynamic Factors Involved in the Formation of Abdominal Aortic Aneurysm

In addition to atherosclerotic factors which have been discussed above, certain haemodynamic factors must be considered as acting in a synergistic fashion with atherosclerotic degeneration in the development of AAAs.

Once the process of aneurysmal dilatation has begun, the true haemodynamic process will start: Laplace's Law, which states that the intraparietal tension increases as a function of the growth of the radius is the substrate for a positive feedback process which leads inevitably to progressing dilatation. In terms of haemodynamics, the importance of the work of Vollmar [173] must be underlined here. In a particular population made of survivors of the Russian front who had undergone, in their youth, amputation of a limb, it was possible to see an inverse relationship between the convexity of AAA and the side of the limb amputated [173] The haemodynamic aspect of AAA will be discussed in the following chapters.

8 Risk of Rupture

The size of the aneurysm is a universally recognised factor to forecast rupture, and the risk of rupture increases as the diameter of the aneurysm enlarges [58, 111]. The overall mortality rate for patients with ruptured AAA is between 65 and 85% [76, 155], and about half of deaths attributed to rupture occur before the patient reaches the surgical room [144, 180].

Similar to AAA, the risk of rupture and dissection of TAAs increases as the diameter of the aneurysm increases [44]. Not only the size but also the rate of growth of TAA has been consistently shown to be critical in predicting rupture [20, 59, 97]. Moreover, rapid rate of expansion was found to be an independent risk factor for TAA rupture [20]. The incidence of ruptured TAAs is reported between three and five per 100,000 persons and is similar to the incidence of aortic dissection [26, 71].

The general consensus is that patients with a large aneurysm should undergo surgery. Therefore few data exist on rupture risk of large AAA in healthy patients with aneurysm diameter greater than 5.5 cm. However the rupture rate for patients with large aneurysm for whom elective repair was not planned because of medical contraindications or patient refusal was assessed by Lederle in a cohort prospective study [89]. The 1-year incidence of probable rupture by initial AAA diameter was 9.4% for AAA of 5.5–5.9 cm and more than 25% at 6 months for AAA 8.0 cm or greater. The real controversy surrounds the management of small aneurysms. The rupture rate for AAAs of 4.0–5.5 cm in diameter is 0.7–1.0% per year [86], UK [166, 7], higher than that of AAAs less than 4.0 cm. The rupture rate for small AAA is higher in women than in men [17]. Men and women have equivalent rupture rates for AAAs greater than 5.5 cm [126]. A study was undertaken in which patients with small aneurysms (diameter between 4·0 and 5·5 cm) were randomly assigned to two groups that underwent either early elective surgery or delayed repair after the diameter of the aneurysm had reached or exceeded 5·5 cm [166]. The results show closely similar survival curves for the two groups of patients. A US Veterans Administration study led to similar findings despite a lower operative mortality with early than with delayed repair (2·7 vs. 5·8%). The conclusions of these two studies were similar: rigorous surveillance of infrarenal aortic aneurysms smaller than 5·5 cm in diameter is safe, whereas early surgery is not associated with improved long-term survival.

Rapid expansion of the aortic diameters preceding fissuration and rupture has been observed in abdominal aortic aneurysms independently of their initial size [58, 93] which suggests that the size of the aneurysm, whatever its practical significance, is probably not the sole useful determinant for risk of rupture. Active investigations have been and still are being done to identify markers other than size that would predict a risk of rupture. A possible candidate is the level of serum MMP-9, which has been directly implicated in the proteolytic degradation of the extracellular matrix of the aortic wall [139, 140]. The amount of circulating MMP-9 has not only been reported to be significantly higher in patients with abdominal aortic aneurysm [106], but the presence of its activated form has also been significantly associated with the size and expansion rate of these aneurysms [94, 95]. However, heterogeneity of the activity of MMPs and the degree of infiltration inside the ruptured aortic wall have been demonstrated by Defawe et al. [34], as well as increased cell infiltration and metalloproteinases expression at the site of rupture.

Another factor that has been investigated as a potential serum marker is the reduced level of α1-antitrypsin (α1-AT) since it is one of the most abundant serum inhibitors of proteases. However, the importance of this marker for the prognosis of abdominal aortic aneurysms has not been defined because of contradictory findings [47]. Family history represents a risk factor for aneurysm rupture. A study of 313 pedigrees showed a four-fold higher rate of rupture in *familial* cases than in sporadic cases [171]. Additionally, a significantly earlier age at rupture (65 years vs. 75 years) was also reported in these familial cases. Another potential risk factor for rupture could be related to the sex of the patient. A report from the UK Small Aneurysm Trial has shown that the risk of rupture for small AAA in women was four-times higher than in men [167].

In a preliminary study, we observed an association between 18-FDG uptake by the aneurysm wall in some cases and rapid expansion of the AAA [141]. This 18-FDG uptake in the aneurysm wall probably reflects the presence of a large density of inflammatory cells (macrophages, lymphocytes) in the adventitia as previously described. These inflammatory cells could correspond to the increased metabolic activity seen on PET imaging [35] (Fig. 3). These preliminary observations were recently confirmed by a study performed by Reeps et al. in [132]. In their study, they observed a correlation between the increased FDG uptake and the patients with a very high macrophage activity with symptomatic AAA.

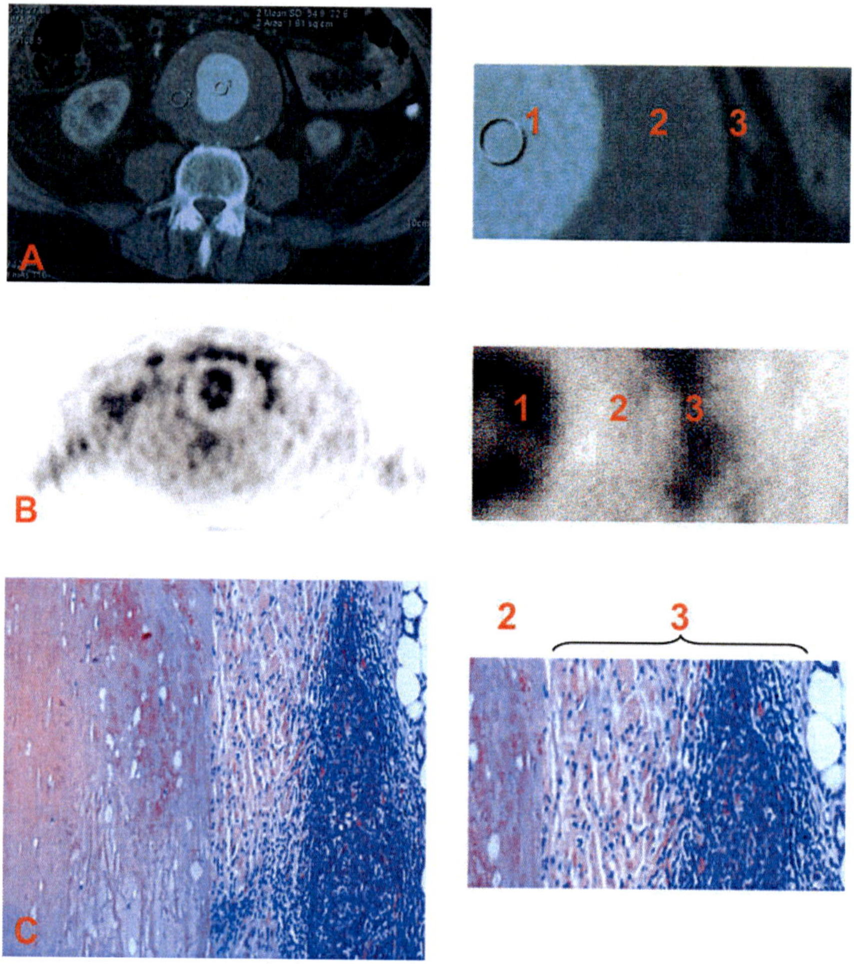

Fig. 3 **a** In this patient, a CT-scan demonstrating the presence of the large AAA (A). (A1) Aortic lumen, (A2) Parietal thrombus, (A3) Aortic wall. **b** A transaxial image shows a thin area of ^{18}F-FDG uptake (B3) corresponding to aneurysmal wall. Inside, a rim without significant uptake corresponds to parietal thrombus (B2) while the luminal area (B1) shows mild uptake of ^{18}F-FDG possibly associated with activated macrophages primed for several days by circulating mediators released at the time of the surgery (38). **c** Microscopic features of sample collected in an abdominal aortic aneurysm wall. Hematoxylin-eosin (200×) staining shows a gradient of inflammation from the adventitia to the parietal thrombus with some focally clusters of inflammatory cells

However, in accordance with Sakalihasan and Truijers's reports they did not find any correlation between maximum standard uptake value (SUV) and maximum cross-sectional infrarenal AAA diameter [141, 165]. Very recently [183], observed a positive association between high wall stress and increased metabolic activities in aneurysmal wall evaluated by PET-CT [132, 165].

Therefore these studies suggest a possible correlation between FDG uptake by the aneurysm wall and the triggering of proteolytic activities leading to rupture. The uptake of 18F-FDG is regarded as a functional image of the inflammatory infiltrate and thus as a potential non-invasive technique to identify unstable aneurysms that are prone to rupture.

9 Conclusion

We have summarized above the current knowledge on aortic aneurysms epidemiology, aetiology and pathophysiology. The aetiology of TAAs and AAAs is multifactoriel with genetic, environmental and physiologic determinants and the disease is characterized by regional heterogeneity between the thoracic and the abdominal aorta and even within the thoracic region in terms of biomechanical properties, atherosclerotic distribution, proteolytic pattern and cell signalling pathways.

Tobacco use, male sex, hypertension, chronic obstructive pulmonary disease and coronary artery disease seem to be the main risk factors for the development of the aneurysmal disease for both aortic regions.

A familial clustering of the disorder is currently well recognized supporting a genetic predisposition to the disease. The genetics of TAAs and dissections appears to be more often monogenic while many genetic risk factors are involved for the abdominal level, each with a small effect size.

Physiopathology of the aneurysmal disease is characterized by thinning of the aortic media resulting from proteolytic injury to the extracellular matrix and smooth muscle cell disappearance, allowing further proteolytic injury leading to dilation and rupture. Aneurysms of the aortic root and ascending aorta are most commonly related to cystic medial degeneration. Physiopathology of the aneurysms of the thoracic descending aorta and the abdominal aorta seems to be different and inflammation and oxidative stress play a key role in the development and the progression of the disease at this level.

The size of the aneurysm is a recognised factor to forecast rupture. The rate of growth of the aneurysm has also been shown to be critical in predicting rupture supporting the fact that the size is not the sole useful determinant for risk of rupture. Active investigations are being done to identify markers other than size that would predict a risk of rupture.

Acknowledgments The original work (Genetics) carried out in the Kuivaniemi laboratory was funded in part by the National Heart, Lung, and Blood Institute of the NIH (grants HL045996 and HL064310 to H.K.).The Department of Cardiovascular surgery University hospital of Liège is supported by the European Union integrated project ''Fighting Aneurysmal Disease'' (FAD, http://www.fighting-aneurysm.org/).

References

1. Acosta, S., Ogren, M., Bengtsson, H., et al.: Increasing incidence of ruptured abdominal aortic aneurysm: a population-based study. J. Vasc. Surg. **44**(2), 237–243 (2006)
2. Achneck, H., Modi, B., Shaw, C., et al.: Ascending thoracic aneurysms are associated with decreased systemic atherosclerosis. Chest **128**(3), 1580–1586 (2005)
3. Ailawadi, G., Eliason, J.L., Roelofs, K.J., et al.: Gender differences in experimental aortic aneurysm formation. Arterioscler. Thromb. Vasc. Biol. **24**(11), 2116–2122 (2004)
4. Alexander, J.J.: The pathobiology of aortic aneurysms. J. Surg. Res. **117**(1), 163–175 (2004)
5. Allaire, E., Forough, R., Clows, M., et al.: Local overexpression of TIMP-1 prevent aortic aneurysm degeneration and rupture in a rat model. J. Clin. Invest. **102**, 1413–1420 (1998)
6. Allaire, E., Hasenstab, D., Kenagy, R.D., et al.: Prevention of aneurysm development and rupture by local overexpression of plasminogen activator inhibitor-1. Circulation **98**(3), 249–255 (1998)
7. Aneurysm Detection and Management Veterans Affairs Cooperative Study Group: Immediate repair compared with surveillance of small abdominal aortic aneurysms. N. Engl. J. Med. **346**, 1437–1444 (2002)
8. Anidjar, S., Salzman, L., Gentric, D., et al.: Elastase-induced experimental aneurysms in rats. Circulation **82**, 973–981 (1990)
9. Ashton, H.A., Buxton, M.J., Day, N.E., et al.: The Multicentre Aneurysm Screening Study (MASS) into the effect of abdominal aortic aneurysm screening on mortality in men: a randomised controlled trial. Lancet **360**(9345), 1531–1539 (2002)
10. Barker, D.J.: Fetal origins of coronary heart disease. BMJ **311**(6998), 171–174 (1995)
11. Bengtsson, H., Sonesson, B., Bergqvist, D.: Incidence and prevalence of abdominal aortic aneurysms, estimated by necropsy studies and population screening by ultrasound. Ann. NY Acad. Sci. **18**, 1–24 (1996)
12. Bickerstaff, L.K., Pairolero, P.C., Hollier, L.H., et al.: Thoracic aortic aneurysms: a population-based study. Surgery **92**, 1103–1108 (1982)
13. Bilguvar, K., Yasuno, K., Niemela, M., et al.: Susceptibility loci for intracranial aneurysm in European and Japanese populations. Nat. Genet. **40**, 1472–1477 (2008)
14. Blanchard, J.F., Armenian, H.K., Friesen, P.P.: Risk factors for abdominal aortic aneurysm: results of a case-control study. Am. J. Epidemiol. **151**(6), 575–583 (2000)
15. Boddy, A.M., Lenk, G.M., Lillvis, J.H., et al.: Basic research studies to understand aneurysm disease. Drug News Perspect. **21**, 142–148 (2008)
16. Borkett-Jones, H.J., Stewart, G., Chilvers, A.S.: Abdominal aortic aneurysms in identical twins. J. Royal Soc. Med. **81**, 471–473 (1988)
17. Brown, L.C., Powell, J.T.: Risk factors for aneurysm rupture in patients kept under ultrasound surveillance. UK small aneurysm trial participants. Ann. Surg. **230**(3), 289–296 (1999). discussion 296–7
18. Brady, A.R., Thompson, S.G., Fowkes, F.G., et al.: Abdominal aortic aneurysm expansion: risk factors and time intervals for surveillance. Circulation **110**(1), 16–21 (2004)
19. Buckley, C., Wyble, C.W., Borhani, M., et al.: Accelerated enlargement of experimental abdominal aortic aneurysms in a mouse model of chronic cigarette smoke exposure. J. Am. Coll. Surg. **199**(6), 896–903 (2004)
20. Cambria, R.A., Glovicki, P., Stanson, A.W., et al.: Outcome and expansion rate of 57 thoraco-abdominal aortic aneurysms managed nonoperatively. Am. J. Surg. **170**(2), 213–217 (1995)
21. Campa, J.S., Greenhalgh, R.M., Powel, J.T.: Elastin degradation in abdominal aortic aneurysms. Atherosclerosis **65**, 13–21 (1987)
22. Carmeliet, P., Moons, L., Lijnen, R., et al.: Urokinase-generated plasmin activates matrix metalloproteinases during aneurysm formation. Nat. Genet. **17**(4), 439–444 (1997)

23. Carrell, T.W., Burnand, K.G., Wells, G.M., et al.: Stromelysin-1 (matrix metalloproteinase-3) and tissue inhibitor of metalloproteinase-3 are overexpressed in the wall of abdominal aortic aneurysms. Circulation **105**(4), 477–482 (2002)
24. Choke, E., Cockerill, G.W., Laing, K., et al.: Whole genome-expression profiling reveals a role for immune and inflammatory response in abdominal aortic aneurysm rupture. Eur. J. Vasc. Endovasc. Surg. **37**, 305–310 (2009)
25. Clifton, M.A.: Familial abdominal aortic aneurysms. Br. J. Surg. **64**(11), 765–766 (1977)
26. Clouse, W.D., Hallett Jr, J.W., Schaff, H.V., et al.: Acute aortic dissection: population-based incidence compared with degenerative aortic aneurysm rupture. Mayo Clin. Proc. **79**(2), 176–180 (2004)
27. Clouse, W.D., Hallett Jr, J.W., Schaff, H.V., et al.: Improved prognosis of thoracic aortic aneurysms: a population-based study. JAMA **280**(22), 1926–1929 (1998)
28. Coady, M.A., Stockwell, P.H., Robich, M.P., et al.: Should aortas in patients with bicuspid aortic valve really be resected at an earlier stage than tricuspid? CON. Cardiol. Clin. **28**(2), 299–314 (2010)
29. Cripe, L., Andelfinger, G., Martin, L.J., et al.: Bicuspid aortic valve is heritable. J. Am. Coll. Cardiol. **44**(1), 138–143 (2004)
30. Curci, J.A., Thompson, R.W.: Adaptive cellular immunity in aortic aneurysms: cause, consequence, or context? J. Clin. Invest. **114**(2), 168–171 (2004)
31. Curci, J.A., Liao, S., Huffman, M.D., et al.: Expression and localization of macrophage elastase (matrix metalloproteinase-12) in abdominal aortic aneurysms. J. Clin. Invest. **102**(11), 1900–1910 (1998)
32. Davis, V., Persidskaia, R., Baca-Regen, L., et al.: Matrix metalloproteinase-2 production and its binding to the matrix are increased in abdominal aortic aneurysms. Arterioscler. Thromb. Vasc. Biol. **18**(10), 1625–1633 (1998)
33. Defawe, O.D., Colige, A., Lambert, C.A., et al.: TIMP-2 and PAI-1 mRNA levels are lower in aneurysmal as compared to athero-occlusive abdominal aortas. Cardiovasc. Res. **60**(1), 205–213 (2003)
34. Defawe, O.D., Colige, A., Deroanne, C., et al.: Mechanical strain modulates monocyte chemotic protein-1 and IL-1beta mRNA expression by cultured human aortic smooth muscle cells. Presented at the XIIIth International Vascular Biology Meeting, June 1–5, Toronto (2004)
35. Defawe, O.D., Hustinx, R., Defraigne, J.O., et al.: Distribution of F-18 fluorodeoxyglucose (F-18 FDG) in abdominal aortic aneurysm: high accumulation in macrophages seen on PET imaging and immunohistology. Clin. Nucl. Med. **30**(5), 340–341 (2005)
36. Deka, R., Koller, D.L., Lai, D., et al.: The relationship between smoking and replicated sequence variants on chromosome 8 and 9 with familial intracranial aneurysm. Stroke **41**(6), 1132–1137 (2010)
37. Della Corte, A., Bancone, C., Quarto, C., et al.: Predictors of ascending aortic dilatation with bicuspid aortic valve: a wide spectrum of disease expression. Eur. J. Cardiothorac. Surg. **37**, 397–405 (2007)
38. Deng, G.G., Martin-McNulty, B., Sukovich, D.A., et al.: Urokinase-type plasminogen activator plays a critical role in angiotensin II-induced abdominal aortic aneurysm. Circ. Res. **92**(5), 510–517 (2003)
39. Diehm, N., Baumgartner, I.: Determinants of aneurysmal aortic disease. Circulation **119**(16), 2134–2135 (2009)
40. Dingemans, K.P., Teeling, P., Lagendijk, J.H., Becker, A.E.: Extracellular matrix of the human aortic media: an ultrastructural, histochemical and immunohistochemical study of the adult aortic media. Anat. Rec. **258**, 1–14 (2000)
41. Dobrin, P.B.: Animal models of aneurysms. Ann. Vasc. Surg. **13**(6), 641–648 (1999)
42. Dubick, M.A., Keen, C.L., DiSilvestro, R.A., et al.: Antioxidant enzyme activity in human abdominal aortic aneurysmal and occlusive disease. Proc. Soc. Exp. Biol. Med. **220**(1), 39–45 (1999)

43. El-Benna, J., Dang, P.M., Périanin, A.: Peptide-based inhibitors of the phagocyte NADPH oxidase. Biochem. Pharmacol. **80**(6), 778–785 (2010)
44. Elefteriades, J.A., Farkas, E.A.: Thoracic aortic aneurysm clinically pertinent controversies and uncertainties. J. Am. Coll. Cardiol. **55**, 841–857 (2010)
45. Elefteriades, J.A., Rizzo, J.A.: Epidemiology, prevalence, incidence, trends. In: Elefteriades, J.A. (ed.) Acute Aortic Disease, pp. 89–98. Informa Healthcare, New York, NY (2008)
46. Elmore, J.R., Obmann, M.A., Kuivaniemi, H., et al.: Identification of a genetic variant associated with abdominal aortic aneurysms on chromosome 3p12.3 by genome wide association. J. Vasc. Surg. **49**, 1525–1531 (2009)
47. Elzouki, A.N., Ryden Ahlgren, A., Lanne, T., et al.: Is there a relationship between abdominal aortic aneurysms and alpha1-antitrypsin deficiency (PiZ)? Eur. J. Vasc. Endovasc. Surg. **17**(2), 149–154 (1999)
48. Erentug, V., Bozbuga, N., Omeroglu, S.N., et al.: Rupture of abdominal aortic aneurysms in Behcet's disease. Ann. Vasc. Surg. **17**(6), 682–685 (2003)
49. Eriksson, P., Jones, K.G., Brown, L.C., et al.: Genetic approach to the role of cysteine proteases in the expansion of abdominal aortic aneurysms. Br. J. Surg. **91**(1), 86–89 (2004)
50. Fedak, P.W., de Sa, M.P., Verma, S., et al.: Vascular matrix remodeling in patients with bicuspid aortic valve malformations: implications for aortic dilatation. J. Thorac. Cardiovasc. Surg. **126**(3), 797–806 (2003)
51. Fontaine, V., Jacob, M.P., Houard, X., et al.: Involvement of the mural thrombus as a site of protease release and activation in human aortic aneurysms. Am. J. Pathol. **161**(5), 1701–1710 (2002)
52. Fontaine, V., Touat, Z., Mtairag el, M., et al.: Role of leukocyte elastase in preventing cellular re-colonization of the mural thrombus. Am. J. Pathol. **164**(6), 2077–2087 (2004)
53. Forsdahl, S.H., Singh, K., Solberg, S., Jacobsen, B.K.: Risk factors for abdominal aortic aneurysms: a 7-year prospective study: the Tromsø study, 1994–2001. Circulation **119**(16), 2202–2208 (2009)
54. Foroud, T., Sauerbeck, L., Brown, R., et al.: FIA Study Investigators. Genome screen to detect linkage to intracranial aneurysm susceptibility genes: the Familial Intracranial Aneurysm (FIA) study. Stroke **39**, 1434–1440 (2008)
55. Ghorpade, A., Baxter, B.T.: Biochemistry and molecular regulation of matrix macromolecules in abdominal aortic aneurysms. Ann. N. Y. Acad. Sci. **800**, 138–150 (1996)
56. Gillum, R.F.: Epidemiology of aortic aneurysm in the United States. J. Clin. Epidemiol. **48**(11), 1289–1298 (1995)
57. Giusti, B., Rossi, L., Lapini, I., et al.: Gene expression profiling of peripheral blood in patients with abdominal aortic aneurysm. Eur. J. Vasc. Endovasc. Surg. **38**, 104–112 (2009)
58. Glimaker, H., Holmberg, L., Elvin, A., et al.: Natural history of patients with abdominal aortic aneurysm. Eur. J. Vasc. Surg. **5**(2), 125–130 (1991)
59. Griepp, R.B., Ergin, M.A., Galla, J.D., et al.: Natural history of descending thoracic and thoracoabdominal aneurysms. Ann. Thorac. Surg. **67**(6), 1927–1930 (1999)
60. Guo, D.C., Pannu, H., Tran-Fadulu, V., et al.: Mutations in smooth muscle α-actin (ACTA2) lead to thoracic aortic aneurysms and dissections. Nat. Genet. **39**, 1488–1493 (2007)
61. Hance, K.A., Tataria, M., Ziporin, S.J., et al.: Monocyte chemotactic activity in human abdominal aortic aneurysms: role of elastin degradation peptides and 67-kD cell surface elastin receptor. J. Vasc. Surg. **35**(2), 254–261 (2002)
62. Hashikata, H., Liu, W., Inoue, K., et al.: Confirmation of an association of single-nucleotide polymorphism rs1333040 on 9p21 with familial and sporadic intracranial aneurysms in Japanese patients. Stroke **41**(6), 1138–1144 (2010)
63. Helgadottir, A., Thorleifsson, G., Magnusson, K.P., et al.: The same sequence variant on 9p21 associates with myocardial infarction, abdominal aortic aneurysm and intracranial aneurysm. Nat. Genet. **40**, 217–224 (2008)
64. Henderson, E.L., Geng, Y.J., Sukhova, G.K., et al.: Death of smooth muscle cells and expression of mediators of apoptosis by T lymphocytes in human abdominal aortic aneurysms. Circulation **99**(1), 96–104 (1999)

65. Herron, G.S., Unemori, E., Wong, M., et al.: Connective tissue proteinases and inhibitors in abdominal aortic aneurysms. Atheroscl. Thromb. **11**, 1667–1677 (1991)
66. Holdt, L.M., Beutner, F., Scholz, M., et al.: ANRIL expression is associated with atherosclerosis risk at chromosome 9p21. Arterioscler. Thromb. Vasc. Biol. **30**, 620–627 (2010)
67. Holmes, D.R., Liao, S., Parks, W.C., Thompson, R.W.: Medial neovascularization in abdominal aortic aneurysms: a histopathologic marker of aneurysmal degeneration with pathophysiologic implications. J. Vasc. Surg. **21**(5), 761–771 (1995). discussion 771-772
68. Huntington, K., Hunter, A.G., Chan, K.L.: A prospective study to assess the frequency of familial clustering of congenital bicuspid aortic valve. J. Am. Coll. Cardiol. **30**(7), 1809–1812 (1997)
69. Isselbacher, E.M.: Thoracic and abdominal aortic aneurysms. Circulation **111**(6), 816–828 (2005)
70. Ito, S., Akutsu, K., Tamori, Y., et al.: Differences in atherosclerotic profiles between patients with thoracic and abdominal aortic aneurysms. Am. J. Cardiol. **101**(5), 696–699 (2008)
71. Johansson, G., Markström, U., Swedenborg, J.: Ruptured thoracic aortic aneurysms: a study of incidence and mortality rates. J. Vasc. Surg. **21**, 985–988 (1995)
72. Jones, K., Powell, J., Brown, L., et al.: The influence of 4G/5G polymorphism in the plasminogen activator inhibitor-1 gene promoter on the incidence, growth and operative risk of abdominal aortic aneurysm. Eur. J. Vasc. Endovasc. Surg. **23**(5), 421–425 (2002)
73. Kazi, M., Thyberg, J., Religa, P.: Influence of intraluminal thrombus on structural and cellular composition of abdominal aortic aneurysm wall. J. Vasc. Surg. **38**(6), 1283–1292 (2003)
74. Kazi, M., Zhu, C., Roy, J., et al.: Difference in matrix-degrading protease expression and activity between thrombus-free and thrombus-covered wall of abdominal aortic aneurysm. Arterioscler. Thromb. Vasc. Biol. **25**(7), 1341–1346 (2005)
75. Khau Van Kien, P., Mathieu, F., Zhu, L., et al.: Mapping of familial thoracic aortic aneurysm/dissection with patent ductus arteriosus to 16p12.2–p13.13. Circulation **112**, 200–206 (2005)
76. Kniemeyer, H.W., Kessler, T., Reber, P.U., et al.: Treatment of ruptured abdominal aortic aneurysm, a permanent challenge or a waste of resources? Prediction of outcome using a multi-organ-dysfunction score. Eur. J. Vasc. Endovasc. Surg. **19**(2), 190–196 (2000)
77. Knox, J.B., Sukhova, G.K., Whittemore, A.D., Libby, P.: Evidence for altered balance between matrix metalloproteinases and their inhibitors in human aortic diseases. Circulation **95**, 205–212 (1997)
78. Koch, A.E., Kunkel, S.L., Pearce, W.H., et al.: Enhanced production of the chemotactic cytokines interleukin-8 and monocyte chemoattractant protein-1 in human abdominal aortic aneurysms. Am. J. Pathol. **142**(5), 1423–1431 (1993)
79. Koksal, C., Ercan, M., Boskurt, A.K., et al.: Abdominal aortic aneurysm or aortic occlusive disease: role of trace element imbalance. Angiology **58**, 191–195 (2007)
80. Kuivaniemi, H., Tromp, G.: Search for the aneurysm susceptibility gene(s). In: Keen, R.R., Dobrin, P.B. (eds.) Development of Aneurysms, pp. 219–233. Landes Bioscience, Georgetown (2000)
81. Kuivaniemi, H., Shibamura, H., Arthur, C., et al.: Familial abdominal aortic aneurysms: collection of 233 multiplex families. J. Vasc. Surg. **37**, 340–345 (2003)
82. Kuivaniemi, H., Kyo, Y., Lenk, G., Tromp, G.: Genome-wide approach to finding abdominal aortic aneurysm susceptibility genes in humans. Ann. N. Y. Acad. Sci. **1085**, 270–281 (2006)
83. Kuivaniemi, H., Platsoucas, C.D., Tilson, M.D., 3rd: Aortic aneurysms: an immune disease with a strong genetic component. Circulation **117**, 242–252 (2008)
84. Kuivaniemi, H., Boddy, A.M., Lillvis, J.H., et al.: Abdominal aortic aneurysms are deep, deadly and genetic. In: Sakalihasan, N., Kuivaniemi, H., Michel Liège, J.B. (eds.) Aortic

aneurysms, new insights into an old problem, pp. 299–323. Liège University Press, Belgium (2008)

85. Lapière, Ch.M., Courtois, A., Nusgens, B.: Extracellular matrix proteins in normal and aneurysmal aorta. In: Sakalihasan, N., Kuivaniemi, H., Michel, J.B. (eds.) Aortic Aneurysms, New Insights into an Old Problem, pp. 67–83. Les Editions de l'Université de Liège, Liège (2008)
86. Lederle, F.A.: In the clinic. Abdominal aortic aneurysm. Ann Intern Med 150:ITC5-1-15 (2009)
87. Lederle, F.A., Johnson, G.R., Wilson, S.E., et al.: The aneurysm detection and management study screening program: validation cohort and final results. Aneurysm detection and management veterans affairs cooperative study investigators. Arch. Intern. Med. **160**(10), 1425–1430 (2000)
88. Lederle, F.A., Johnson, G.R., Wilson, S.E.: Abdominal aortic aneurysm in women. J. Vasc. Surg. **34**(1), 122–126 (2001)
89. Lederle, F.A., Gary, R.J., Samuel, E.W., et al.: Rupture rate of large abdominal aortic aneurysms in patients refusing or unfit for elective repair. JAMA **287**(22), 2968–2972 (2002)
90. Lenk, G.M., Tromp, G., Weinsheimer, S., et al.: Whole genome expression profiling reveals a significant role for immune function in human abdominal aortic aneurysms. BMC Genomics **8**, 237 (2007)
91. Liddington, M.I., Heather, B.P.: The relationship between aortic diameter and body habitus. Eur. J. Vasc. Surg. **6**(1), 89–92 (1992)
92. Lillvis, J.H., Lenk, G.M., Kuivaniemi, H.: Genetics of Abdominal Aortic Aneurysms. In: Upchurch, G., Criado, E. (eds.) Aortic Aneurysms: pathogenesis and treatment, pp. 1–26. Humana Press Inc., Totowa, NJ (2009)
93. Limet, R., Sakalihassan, N., Albert, A.: Determination of the expansion rate and incidence of rupture of abdominal aortic aneurysms. J. Vasc. Surg. **14**(4), 540–548 (1991)
94. Lindholt, J.S., Vammen, S., Fasting, H., Henneberg, E.W., Heickendorff, L.: The plasma level of matrix metalloproteinase 9 may predict the natural history of small abdominal aortic aneurysms. A preliminary study. Eur. J. Vasc. Endovasc. Surg. **20**(3), 281–285 (2000)
95. Lindholt, J.S., Vammen, S., Juul, S., et al.: Optimal interval screening and surveillance of abdominal aortic aneurysms. Eur. J. Vasc. Endovasc. Surg. **20**(4), 369–373 (2000)
96. Lindholt, J.S., Heegaard, N.H., Vammen, S., et al.: Smoking, but not lipids, lipoprotein(a) and antibodies against oxidised LDL, is correlated to the expansion of abdominal aortic aneurysms. Eur. J. Vasc. Endovasc. Surg. **21**(1), 51–56 (2001)
97. Lobato, A.C., Puech-LeLeão, P.: Predictive factors for rupture of thoraco-abdominal aortic aneurysm. J. Vasc. Surg. **27**(3), 446–453 (1998)
98. Longo, G.M., Xiong, W., Greiner, T.C., et al.: Matrix metalloproteinases 2 and 9 work in concert to product aortic aneurysms. J. Clin. Invest. **110**(5), 625–632 (2002)
99. MacSweeney, S.T., Ellis, M., Worrell, P.C., et al.: Smoking and growth rate of small abdominal aortic aneurysms. Lancet **344**(8923), 651–652 (1994)
100. Madaric, J., Vulev, I., Bartunek, J., et al.: Frequency of abdominal aortic aneurysm in patients >60 years of age with coronary artery disease. Am. J. Cardiol. **96**(9), 1214–1216 (2005)
101. Majumder, P.P., St Jean, P.L., Ferrell, R.E., et al.: On the inheritance of abdominal aortic aneurysm. Am. J. Hum. Genet. **48**, 164–170 (1991)
102. Mao, D., Lee, J.K., Van Vickle, S.J., Thompson, R.W.: Expression of collagenase-3 (MMP-13) in human abdominal aortic aneurysms and vascular smooth muscle cells in culture. Biochem. Biophys. Res. Commun. **261**(3), 904–910 (1999)
103. Matsumura, K., Hirano, T., Takeda, K., et al.: Incidence of aneurysms in Takayasu's arteritis. Angiology **42**(4), 308–315 (1991)
104. McCormick, M.L., Gavrila, D., Weintraub, N.L.: Role of oxidative stress in the pathogenesis of abdominal aortic aneurysms. Arterioscler. Thromb. Vasc. Biol. **27**(3), 461–469 (2007)

105. McGregor, J.C., Pollock, J.G., Anton, H.C.: The value of ultrasonography in the diagnosis of abdominal aortic aneurysms. Scott. Med. J. **20**, 133–137 (1975)
106. McMillan, W.D., Pearce, W.H.: Increased plasma levels of metalloproteinase-9 are associated with abdominal aortic aneurysms. J. Vasc. Surg. **29**(1), 122–127 (1999). discussion 127-9
107. Menashi, S., Campa, J.S., Greenhalgh, R.M., Powel, J.T.: Collagen in abdominal aortic aneurysm: typing, content and degradation. J. Vasc. Surg. **6**, 578–582 (1987)
108. Michel, J.B., Thaunat, O., Houard, X., et al.: Topological determinants and consequences of adventitial responses to arterial wall injury. Arterioscler. Thromb. Vasc. Biol. **27**(6), 1259–1268 (2007)
109. Milewicz, D.M., Guo, D.C., Tran-Fadulu, V., et al.: Genetic basis of thoracic aortic aneurysms and dissections: focus on smooth muscle cell contractile dysfunction. Annu. Rev. Genomics Hum. Genet. **9**, 283–302 (2008)
110. Miller Jr, F.J., Sharp, W.J., Fang, X., et al.: Oxidative stress in human abdominal aortic aneurysms: a potential mediator of aneurysmal remodeling. Arterioscler. Thromb. Vasc. Biol. **22**(4), 560–565 (2002)
111. Nevitt, M.P., Ballard, D.J., Hallett Jr, J.W.: Prognosis of abdominal aortic aneurysms. A population-based study. N. Engl. J. Med. **321**(15), 1009–1014 (1989)
112. Newman, K.M., Jean-Claude, J., Li, H., et al.: Cellular localization of matrix metalloproteinases in the abdominal aortic aneurysm wall. J. Vasc. Surg. **20**, 814–820 (1994)
113. Newman, M.P., Jean-Claude, J., Li, H., et al.: Cytokines that activate proteolysis are increased in abdominal aortic aneurysms. Circulation **90**(5Pt 2), 1224–1227 (1994)
114. Nischan, J., Gatalica, Z., Curtis, M., et al.: Binding sites for ETS family of transcription factors dominate the promoter regions of differentially expressed genes in abdominal aortic aneurysms. Circ. Cardiovasc. Genet. **2**, 565–572 (2009)
115. Nischan, J., Lenk, G.M., Boddy, A.M., et al.: Abdominal Aortic Aneurysms – a Complex Genetic Disease. In: Laurent, A., Morel, E. (eds.) Aneurysms: Types, Risks, Formation and Treatment (ISBN: 978-1-60741-557-2), pp. 35–93. Nova Science Publishers, Inc, Hauppauge, NY (2009)
116. Nistri, S., Sorbo, M.D., Marin, M., et al.: Aortic root dilatation in young men with normally functioning bicuspid aortic valves. Heart **82**, 19–22 (1999)
117. Nollendorfs, A., Greiner, T.C., Nagase, H., Baxter, B.T.: The expression and localization of membrane type-1 matrix metalloproteinase in human abdominal aortic aneurysms. J. Vasc. Surg. **34**(2), 316–322 (2001)
118. Nordon, I.M., Hinchliffe, R.J., Holt, P.J., et al.: Review of current theories for abdominal aortic aneurysm pathogenesis. Vascular **17**(5), 253–263 (2009)
119. Norrgård, Ö., Rais, O., Ängquist, K.A.: Familial occurrence of abdominal aortic aneurysms. Surgery **95**, 650–656 (1984)
120. Ogata, T., MacKean, G.L., Cole, C.W., et al.: The lifetime prevalence of abdominal aortic aneurysms among siblings of aneurysm patients is eightfold higher than among siblings of spouses: an analysis of 187 aneurysm families in Nova Scotia; Canada. J. Vasc. Surg. **42**, 891–897 (2005)
121. Ogata, T., Gregoire, L., Goddard, K.A., et al.: Evidence for association between the HLA-DQA locus and abdominal aortic aneurysms in the Belgian population: a case control study. BMC Med. Genet. **7**, 67 (2006)
122. Olsson, C., Thelin, S., Ståhle, E., et al.: Thoracic aortic aneurysm and dissection: increasing prevalence and improved outcomes reported in a nationwide population-based study of more than 14 000 cases from 1987 to 2002. Circulation **114**, 2611–2618 (2006)
123. Pachulski, R.T., Weinberg, A.L., Chan, K.L.: Aortic aneurysm in patients with functionally normal or minimally stenotic bicuspid aortic valve. Am. J. Cardiol. **67**(8), 781–782 (1991)
124. Phillippi, J.A., Klyachko, E.A., Kenny, J.P., 4th, Eskay, M.A., Gorman, R.C., Gleason, T.G.: Basal and oxidative stress–induced expression of metallothionein is decreased in

ascending aortic aneurysms of bicuspid aortic valve patients circulation **119**, 2498–2506 (2009)
125. Powell, J.T.: Familial clustering of abdominal aortic aneurysm–smoke signals, but no culprit genes. Br. J. Surg. **90**(10), 1173–1174 (2003)
126. Powell, J.T., Brown, L.C., Greenhalgh, R.M., et al.: The rupture rate of large abdominal aortic aneurysms: is this modified by anatomical suitability for endovascular repair? Ann. Surg. **247**, 173–179 (2008)
127. Pressler, V., McNamara, J.J.: Aneurysm of the thoracic aorta. Review of 260 cases. J. Thorac. Cardiovasc. Surg. **89**(1), 50–54 (1985)
128. Pyo, R., Lee, J.K., Shipley JM, J., et al.: Targeted gene disruption of matrix metalloproteinase-9 (gelatinase B) suppresses development of experimental abdominal aortic aneurysms. J. Clin. Invest. **105**(11), 1641–1649 (2000)
129. Ramanath, V.S., Oh, J.K., Sundt 3rd, T.M., Eagle, K.A.: Acute aortic syndromes and thoracic aortic aneurysm. Mayo Clin. Proc. **84**(5), 465–481 (2009)
130. Rao, S.K., Reddy, K.V., Cohen, J.R.: Role of serine proteases in aneurysm development. Ann. N. Y. Acad. Sci. **800**, 131–137 (1996)
131. Reed, D., Reed, C., Stemmermann, G., Hayashi, T.: Are aortic aneurysms caused by atherosclerosis? Circulation **85**(1), 205–211 (1992)
132. Reeps, C., Essler, M., Pelisek, J., et al.: Increased 18F-fluorodeoxyglucose uptake in abdominal aortic aneurysms in positron emission/computed tomography is associated with inflammation, aortic wall instability, and acute symptoms. J. Vasc. Surg. **48**(2), 417–423 (2008)
133. Reilly, J.M.: Plasminogen activors in abdominal aortic aneurysmal disease. Ann. N. Y. Acad. Sci. **800**, 151–156 (1996)
134. Rizzo, R.J., McCarthy, W.J., Dixit, S.N., et al.: Collagen types and matrix protein content in human abdominal aortic aneurysms. J. Vasc. Surg. **10**, 365–373 (1989)
135. Rossaak, J.I., Van Rij, A.M., Jones, G.T., Harris, E.L.: Association of the 4G/5G polymorphism in the promoter region of plasminogen activator inhibitor-1 with abdominal aortic aneurysms. J. Vasc. Surg. **31**(5), 1026–1032 (2000)
136. Rouis, M., Adamy, C., Duverger, N., et al.: Adenovirus-mediated overexpression of tissue inhibitor of metalloproteinase-1 reduces atherosclerotic lesions in apolipoprotein E-deficient mice. Circulation **100**(5), 533–540 (1999)
137. Ruddy, J.M., Jones, J.A., Spinale, F.G., Ikonomidis, J.S.: Regional heterogeneity within the aorta: relevance to aneurysm disease. J. Thorac. Cardiovasc. Surg. **136**(5), 1123–1130 (2008)
138. Sakalihasan, N., Heyeres, A., Nusgens, B.V., et al.: Modifications of the extracellular matrix of aneurysmal abdominal aortas as a function of their size. Eur. J. Vasc. Surg. **7**, 633–637 (1993)
139. Sakalihasan, N., Delvenne, P., et al.: Activated forms of MMP2 and MMP9 in abdominal aortic aneurysms. J. Vasc. Surg. **24**(1), 127–133 (1996)
140. Sakalihasan, N., Pincemail, J., Defraigne, J.O., et al.: Decrease of plasma vitamin E (alpha-tocopherol) levels in patients with abdominal aortic aneurysm. Ann. N. Y. Acad. Sci. **800**, 278–282 (1996)
141. Sakalihasan, N., Van Damme, H., Gomez, P., et al.: Positron emission tomography (PET) evaluation of abdominal aortic aneurysm (AAA). Eur. J. Vasc. Endovasc. Surg. **23**(5), 431–436 (2002)
142. Sakalihasan, N., Limet, R., Defawe, O.D.: Abdominal aortic aneurysm. Lancet **365**(9470), 1577–1589 (2005)
143. Schlösser, F.J., Vaartjes, I., van der Heijden, G.J., et al.: Mortality after elective abdominal aortic aneurysm repair. Ann. Surg. **251**(1), 158–164 (2010)
144. Scott, R.A., Ashton, H.A., Kay, D.N.: Abdominal aortic aneurysm in 4237 screened patients: prevalence, development and management over 6 years. Br. J. Surg. **78**(9), 1122–1125 (1991)

145. Shah, P.K.: Inflammation, metalloproteinases, and increased proteolysis: an emerging pathophysiological paradigm in aortic aneurysm. Circulation **96**(7), 2115–2117 (1997)
146. Shantikumar, S., Ajjan, R., Porter, K.E., Scott, D.J.: Diabetes and the abdominal aortic aneurysm. Eur. J. Vasc. Endovasc. Surg. **39**, 200–207 (2010)
147. Shi, G.P., Sukhova, G.K., Grubb, A., et al.: Cystatin C deficiency in human atherosclerosis and aortic aneurysm. J. Clin. Invest. **104**(9), 1191–1197 (1999)
148. Shibamura, H., Olson, J.M., van Vlijmen-Van Keulen, C., et al.: Genome scan for familial abdominal aortic aneurysm using sex and family history as covariates suggests genetic heterogeneity and identifies linkage to chromosome 19q13. Circulation **109**, 2103–2108 (2004)
149. Shiraya, S., Miyake, T., Aoki, M., et al.: Inhibition of development of experimental aortic abdominal aneurysm in rat model by atorvastatin through inhibition of macrophage migration. Atherosclerosis **202**(1), 34–40 (2009)
150. Silence, J., Lupu, F., Collen, D., Lijnen, H.R.: Persistence of atherosclerotic plaque but reduced aneurysm formation in mice with Stromelysin-1 (MMP-3) gene inactivation. Arterioscler. Thromb. Vasc. Biol. **21**(9), 1440–1445 (2001)
151. Singh, K., Bonaa, K.H., Jacobsen, B.K., Bjork, L., Solberg, S.: Prevalence of and risk factors for abdominal aortic aneurysms in a population-based study: the Tromso Study. Am. J. Epidemiol. **154**(3), 236–244 (2001)
152. Slaney, S.G.: A history of aneurysm surgery. In: Greenhalgh, R.M., Mannick, J.A. (eds.) The Cause and Management of Aneurysm, pp. 1–19. WB Saunders Co., London (1990)
153. Tamarina, N.A., McMillan, W.D., Shively, V.P., Pearce, W.H.: Expression of matrix metalloproteinases and their inhibitors in aneurysms and normal aorta. Surgery **122**, 264–271 (1997)
154. Thomas, M., Gavrila, D., McCormick, M.L., et al.: Deletion of p47phox attenuates angiotensin II-induced abdominal aortic aneurysm formation in apolipoprotein E-deficient mice. Circulation **114**, 404–413 (2006)
155. Thompson, M.M.: Controlling the expansion of abdominal aortic aneurysms. Br. J. Surg. **90**(8), 897–898 (2003)
156. Thompson, R.W., Geraghty, P.J., Lee, J.K.,: Abdominal aortic aneurysms: basic mechanisms and clinical implications. Curr. Probl. Surg. **39**, 110–230 (2002).
157. Thompson, R.W., Parks, W.C.: Role of matrix metalloproteinases in abdominal aortic aneurysms. Ann. N. Y. Acad. Sci. **800**, 157–174 (1996). Review
158. Tilson, M.D.: Aortic aneurysms and atherosclerosis. Circulation **85**(1), 378–379 (1992)
159. Tilson, M.D., Seashore, M.R.: Fifty families with abdominal aortic aneurysms in two or more first-order relatives. Am. J. Surg. **147**, 551–553 (1984)
160. Towbin, J.A., Casey, B., Belmont, J.: The molecular basis of vascular disorders. Am. J. Hum. Genet. **64**(3), 678–684 (1999)
161. Tromp, G., Kuivaniemi, H.: How does one study genetic risk factors in a complex disease such as aneurysm. In: Sakalihasan, N., Kuivaniemi, H., Michel, J.B. (eds.) Aortic Aneurysms, New insights into an old problem, pp. 115–144. Liege University Press, Liege, Belgium (2008)
162. Tromp, G., Kuivaniemi, H.: Developments in genomics to improve understanding, diagnosis and management of aneurysms and peripheral artery disease. Eur. J. Vasc. Endovasc. Surg. **38**, 676–682 (2009)
163. Tromp, G., Gatalica, Z., Skunca, M., et al.: Elevated expression of matrix metalloproteinase-13 in abdominal aortic aneurysms. Ann. Vasc. Surg. **18**(4), 414–420 (2004)
164. Tromp, G., Kuivaniemi, H., Hinterseher, I., Carey, D.J.: Novel genetic mechanisms for aortic aneurysms. Curr. Atheroscler. Rep. **12**, 259–266 (2010)
165. Truijers, M., Kurvers, H.A., Bredie, S.J., et al.: In vivo imaging of abdominal aortic aneurysms; increased FDG uptake suggests inflammation in the aneurysmal wall. J. Endovasc. Ther. **15**(4), 462–467 (2008)

166. UK Small Aneurysm Trial Participants: Mortality results for randomised controlled trial of early elective surgery or ultrasonographic surveillance for small abdominal aortic aneurysms. Lancet **352**, 1649–1655 (1998)
167. UK Small Aneurysm Trial Participants: Long-term outcomes of immediate repair compared with surveillance of small abdominal aortic aneurysms. N. Engl. J. Med. **346**(19), 1445–1452 (2002)
168. Van Kuijk, Jp.P., Flu, W.J., Witteveen, O.P., et al.: The influence of statins on the expansion rate and rupture risk of abdominal aortic aneurysms. J. Cardiovasc. Surg. (Torino) **50**(5), 599–609 (2009)
169. Van Vlijmen-Van Keulen, C.J., Rauwerda, J.A., Pals, G.: Genome-wide linkage in three Dutch families maps a locus for abdominal aortic aneurysms to chromosome 19q13.3. Eur. J. Vasc. Endovasc. Surg. **30**, 29–35 (2005)
170. Vardulaki, K.A., Walker, N.M., Day, N.E., et al.: Quantifying the risks of hypertension, age, sex and smoking in patients with abdominal aortic aneurysm. Br. J. Surg. **87**(2), 195–200 (2000)
171. Verloes, A., Sakalihasan, N., Koulischer, L., et al.: Aneurysms of the abdominal aorta: familial and genetic aspects in three hundred thirteen pedigrees. J. Vasc. Surg. **21**, 646–655 (1995)
172. Visel, A., Zhu, Y., May, D., et al.: Targeted deletion of the 9p21 non-coding coronary artery disease risk interval in mice. Nature **464**, 409–412 (2010)
173. Vollmar, J.F., Pauschinger, P., Paes, E., et al.: Aortic aneurysms as a late sequellae of above-knee amputation. Lancet **ii**, 834–835 (1989)
174. Vorp, D.A., Lee, P.C., Wang, D.H., et al.: Association of intraluminal thrombus in abdominal aortic aneurysm with local hypoxia and wall weakening. J. Vasc. Surg. **34**(2), 291–299 (2001)
175. Wagenseil, J.E., Mecham, R.P.: Vascular extracellular matrix and arterial mechanics. Physiol. Rev. **89**, 957–989 (2009)
176. Wahlgren, C.M., Larsson, E., Magnusson, P.K., et al.: Genetic and environmental contributions to abdominal aortic aneurysm development in a twin population. J. Vasc. Surg. **51**, 3–7 (2010)
177. Walton, L.J., Franklin, I.J., Bayston, T., et al.: Inhibition of prostaglandin E2 synthesis in abdominal aortic aneurysms: implications for smooth muscle cell viability, inflammatory processes, and the expansion of abdominal aortic aneurysms. Circulation **100**(1), 48–54 (1999)
178. Weinsheimer, S., Lenk, G.M., van der Voet, M., et al.: Integration of expression profiles and genetic mapping data to identify candidate genes in intracranial aneurysm. Phys. Genomics **32**, 45–57 (2007)
179. Weiss, J.S., Sumpio, B.E.: Review of prevalence and outcome of vascular disease in patients with diabetes mellitus. Eur. J. Vasc. Endovasc. Surg. **31**(2), 143–150 (2006)
180. Wilmink, T.B., Quick, C.R., Hubbard, C.S., Day, N.E.: The influence of screening on the incidence of ruptured abdominal aortic aneurysms. J. Vasc. Surg. **30**(2), 203–208 (1999)
181. Xiong, W., Mactaggart, J., Knispel, R., et al.: Inhibition of reactive oxygen species attenuates aneurysm formation in a murine model. Atherosclerosis **202**(1), 128–134 (2009)
182. Xu, C., Zarins, C.K., Glagov, S.: Aneurysmal and occlusive atherosclerosis of the human abdominal aorta. J. Vasc. Surg. **33**(1), 91–96 (2001)
183. Xu, X.Y., Borghi, A., Nchimi, A., et al.: High levels of 18F-FDG uptake in aortic aneurysm wall are associated with high wall stress. Eur. J. Vasc. Endovasc. Surg. **39**(3), 295–301 (2010)
184. Yajima, N., Masuda, M., Miyazaki, M., et al.: Oxidative stress is involved in the development of experimental abdominal aortic aneurysm: a study of the transcription profile with complementary DNA microarray. J. Vasc. Surg. **36**(2), 379–385 (2002)
185. Yamasumi, K., Ojiro, M., Okumura, H., Aikou, T.: An activated state of blood coagulation and fibrinolysis in patients with abdominal aortic aneurysm. Am. J. Surg. **175**(4), 297–301 (1998)

186. Yasuda, H., Nakatani, S., Stugaard, M., et al.: Failure to prevent progressive dilation of ascending aorta by aortic valve replacement in patients with bicuspid aortic valve: comparison with tricuspid aortic valve. Circulation **108**(Suppl 1), II291–II294 (2003)
187. Yasuno, K., Bilguvar, K., Bijlenga, P., et al.: Genome-wide association study of intracranial aneurysm identifies three new risk loci. Nat. Genet. **42**, 420–425 (2010)
188. Zhang, J., Schmidt, J., Ryschich, E., et al.: Inducible nitric oxide synthase is present in human abdominal aortic aneurysm and promotes oxidative vascular injury. J. Vasc. Surg. **38**(2), 360–367 (2003)
189. Zhu, L., Vranckx, R., Khau Van Kien, P., et al.: Mutations in myosin heavy chain 11 cause a syndrome associating thoracic aortic aneurysm/aortic dissection and patent ductus arteriosus. Nat. Genet. **38**, 343–349 (2006)

Imaging of Aneurysms

Peter Hoskins, Scott Semple, Phil White and Jennifer Richards

Abstract There are several methods for imaging aneurysms. These can be divided into techniques which principally provide information on structure (X-ray projection imaging including angiography, CT, MRI, ultrasound) and those which provide information on biological function. The principal example of the latter is positron emission tomography (PET), though it suffers from low spatial resolution. The development of targeted contrast agents for use with MRI and ultrasound leads to the possibility of providing high-resolution information on biological function. There is no one imaging technique which is perfect, and in clinical practice different techniques are used to address different questions. Simple measurement of the diameter made using 2D real-time ultrasound is sufficient for the purposes of selection of patients for repair of abdominal aortic aneurysm (AAA). CT, MRI and X-ray angiography are used when more detailed evaluation of anatomy is required for the purposes of planning of interventional procedures in AAA and cerebral aneurysms. Information on blood velocity and related quantities such as flow rate can be obtained using MRI and ultrasound. The wall distension occurring during the cardiac cycle may be measured using ultrasound, MRI and multi-slice CT. Wall distension measurements may then be used to estimate the stiffness of AAA. For use with image guided modelling, data on the geometry and inlet velocity can be obtained from imaging. Typically 3D imaging data may be

P. Hoskins (✉)
Medical Physics Unit, University of Edinburgh, Edinburgh, UK
e-mail: P.Hoskins@ed.ac.uk

S. Semple
Clinical Research Imaging Centre, University of Edinburgh, Edinburgh, UK

P. White
Clinical Neurosciences, University of Edinburgh, Edinburgh, UK

J. Richards
Centre of Cardiovascular Science, University of Edinburgh, Edinburgh, UK

Stud Mechanobiol Tissue Eng Biomater (2011) 7: 35–65
DOI: 10.1007/8415_2010_64

Published Online: 29 January 2011

obtained in AAA using CT or MRI, and in cerebral aneurysms using CT, MRI or spiral angiography. Data on inlet velocities may be obtained using Doppler ultrasound or MRI. Imaging techniques are pivotal to the diagnosis, evaluation treatment and post procedure care of patients with aneurysms, and an essential component of methods in development such as image guided modelling. In this chapter imaging techniques will be considered briefly with respect to current clinical use, but mainly with respect to the theme of this book, the mechanobiology and biomechanics of abdominal aneurysms. A glossary of abbreviations used in this chapter is provided in an appendix.

1 Principles of Medical Imaging

Imaging techniques provide 2D or 3D information on the position of tissues and organs within the body. This section will briefly describe the basic principles of operation of the major imaging systems. Further details may be found in specialised texts on medical imaging [3, 14]. There are 2 main types of imaging system.

- *Structural imaging* techniques provide information on the geometry of tissues and organs, and in some cases on the change in geometry with time. In other words, these systems mainly provide information on tissue morphology and tissue motion such as blood velocity and changes in aneurysm size with time. The main structural imaging techniques are X-ray imaging including CT, MRI and ultrasound imaging.
- *Molecular imaging* techniques provide information related to the biological function of tissues and organs. These involve delivery into the patient of tracers, whose uptake is dependent on biological function, which are detectable by the imaging system. The dominant example of molecular imaging technique is positron emission tomography or PET. This provides images of lower spatial resolution (compared with structural imaging). For MRI and ultrasound there has been considerable work performed on the development of targeted-contrast agents in order to provide biological information at high resolution.

1.1 X-ray Imaging

This is the oldest clinical imaging technique. Within 2 years of the discovery of X-rays by Rontgen, X-ray imaging was widely used for diagnosis in hospitals. The basic technique involves production of X-rays from a tube, passage of the rays in straight lines through the body, and detection of the X-rays. The X-ray intensity at the detector is dependent on the attenuation by the different tissues along the path of the X-ray. This is most marked in skeletal imaging, where the bones which have high attenuation, are clearly differentiated from the soft tissue which has low attenuation. There are several types of modern X-ray system:

- *Projection radiography.* A static X-ray tube irradiates the patient producing a 2D projection image, such as a chest X-ray. Modern radiographic systems use flat plate technology to detect the X-rays so that the image is stored digitally.
- *Fluoroscopy.* Low intensity X-rays are produced continuously and detected by an image intensifier or flat plate solid state detector producing an image with higher noise than projection radiography, but allowing real-time use. Fluoroscopic imaging is usually used in conjunction with contrast agents such as barium for highlighting the gastrointestinal system, or iodine for highlighting the vascular system (called 'angiography').
- *Computed tomography (CT).* This consists of collection of data from different angles around the body, by rotation of the tube and the detector. The 1000 or so projection datasets are combined within the computer using a method called filtered back projection to produce a 2D cross sectional image in which the displayed image is related to the tissue attenuation coefficient. Collection of many 2D slices enables production of a 3D image.
- *Rotational angiography.* Rotational angiography is a variant of the CT imaging technique where the fluoroscopic system is rotated through 180°. The projection images can be displayed as a video to produce 3D perception, or CT reconstruction techniques can be used to produce a 3D image of the vascular system. Modern angiographic systems use flat panel solid state detectors rather than the image intensifier based systems (with or without charged couple device detectors) used pre 2005.

Projection radiography is dominated by low image contrast making it difficult to distinguish soft tissue lesions without the aid of contrast materials such as air, barium or iodine. With the advent of CT in the 1970s the improved image contrast revolutionised the detection of soft tissue lesions. Though the modern radiology department has other imaging techniques available, the vast majority of patient studies are performed using X-ray imaging, a fact which is often neglected when discussions take place on the 'best' imaging technique.

1.2 Ultrasound Imaging

High frequency sound waves (typically 2–18 MHz) are transmitted into the tissue along a beam. These are scattered by the tissue in all directions. A portion of the ultrasound is scattered or reflected by the tissues and returns back to the transducer. The depth (d), from which the echoes arose, is calculated from the time (t) between transmission of the ultrasound pulse and reception of the echo, and the assumed speed of sound (c) in tissue of 1540 m s^{-1}, using the formula $d = c.t/2$. This is called the 'pulse-echo' technique. Scanning of the beam through the tissues is used to build up a 2D image, and this can be performed sufficiently quickly to allow real-time imaging. The ultrasound beam is attenuated by passage through the tissue, requiring a depth-dependent gain to be used on acquisition. The image brightness is related to the echo amplitude. The range of echo amplitudes produced is extremely large with high amplitudes produced from the boundaries between organs and within organs, and lower level echoes produced by scattering from within the tissue parenchyma. Echo compression is used in order to enable display of the high and low level echoes at the same time.

Clinical ultrasound is based on real-time 2D scanning. Three-D ultrasound images may be obtained by sequential collection of 2D ultrasound images by movement of the transducer by the operator [26, 68]. In recent years 2D array transducers have become available, mainly for cardiac imaging, whereby the beam is steered within the 3D volume while the transducer is held against the skin [49].

The detection of motion using ultrasound may be performed using the Doppler effect. This has been applied to the detection of motion of fluids, principally blood flow, and the motion of tissues, principally that of the heart [23]. Typical display formats include the time-velocity waveform from an individual location, and a 2D real time colour display of the moving blood or tissues in which the colour is related to the local velocity.

Ultrasound is an inexpensive widely available technique whose strength lies with its non-invasive real-time nature accepted by the majority of patients. Though the resolution is comparable with other techniques such as MRI and CT, the image quality is not isotropic and there is an inherent noise ('speckle') which can make the image more difficult to interpret than for other techniques. Diagnostic ultrasound does have the potential to cause tissue damage via tissue heating and cavitation, and there are guidelines on the use of ultrasound to minimise the risk to the patient from these effects [12].

1.3 MRI

Magnetic Resonance Imaging (MRI) is a three-dimensional high resolution technique which gives excellent contrast between soft tissue types. Unlike X-ray, CT or PET, MRI does not involve ionising radiation, and so is particularly useful where ionising radiation dose reduction is desired, such as in longitudinal monitoring of AAA progression.

A powerful magnetic field is used to align the nuclear magnetisation of protons in hydrogen atoms in water. Common clinical magnetic field strengths are 1.5 tesla and 3 tesla although some lower or higher field strength clinical systems are available. A radiofrequency field is periodically applied to perturb the protons' spins from their alignment, producing a rotating magnetic field. When the radiofrequency field is switched off, the protons 'relax' back to their equilibrium position aligned with the main magnetic field. In doing so, they give off energy at a specific rotational frequency (dependent on the magnetic field strength). Through application of additional field gradients in three orthogonal directions, this rotational frequency is systematically varied across the direction of each gradient to encode the rotational frequencies of the protons with spatial information in three dimensions. Thus, a three dimensional image may be acquired in any plane through selective application of these factors.

The rate of proton relaxation back to equilibrium is governed by various factors, such as the chemical environment and binding of the local water molecules, density of water molecules etc., thus giving different signal strengths for different tissues. By altering the application of field gradients, radiofrequency fields and acquisition timings, image contrast can be manipulated to optimise contrast between different tissue types, or between normal and pathological tissues. Image contrast may be further altered with the application of MRI contrast agents, commonly paramagnetic or superparamagnetic agents which alter the magnetic properties of blood and tissue according to their distribution characteristics (largely defined by the contrast particle size).

MRI is unsuitable for some patients as a result of patient claustrophobia, or contraindications to MRI scanning, such as presence of pacemakers, or ferrous implants (pacemakers, aneurysm clips etc.).

MRI is a versatile imaging technique which may be acquired either with focus on high resolution anatomical detail, or in a dynamic fashion to determine functional information in normal and abnormal tissues.

1.4 Comparison of Structural Imaging Techniques

There is no single perfect imaging technique, and in practice each has its own limitations and strengths. X-ray projection images provide high resolution 2D data. MRI and CT provide high quality 3D data, and are useful in image guided modelling where detailed 3D geometry is needed. CT images are low-noise, relatively free from artefact, and an abdominal study can be obtained in 10–20 s, that is within a single breath-hold. However there is an associated radiation dose with CT which in general mitigates against its repeated use due to the risk of radiation induced tissue damage, especially increased risk of cancer. MRI acquisition may require repeated breath holds which makes the technique more challenging for the patient, with typical overall scan times of 20–40 min. Movement during acquisition, such as induced by cardiac/arterial pulsation, respiratory motion or motion of the GI tract, may substantially degrade the MR image. Ultrasound imaging is a real-time technique which allows for rapid measurement of, for example, diameter for use in stratifying risk of aneurysm rupture. Table 1 provides a comparison of the techniques for imaging the abdomen.

Table 1 Summary of the features of the main clinical imaging modalities relevant to aneurysms

	X-ray	MRI	Ultrasound
Main physical property relevant to image	X-ray attenuation coefficient	Magnetic relaxation times	Acoustic impedance
Real-time acquisition	Yes (Fluoroscopy)	No	Yes (2D/3D modes)
3D acquisition time	10–20 s (CT)	20–30 min	0.2–0.5 s
Full 3D cross-section of the body?	Yes (CT)	Yes	No
Spatial resolution (typical)	1 by 1 mm (*x*, *y*) (CT) 1–3 mm (*z*)	1 by 1 mm (*x*, *y*) 3–10 mm (*z*)	2 by 2 mm (*x*, *y*) 4 mm (*z*)
Ionising radiation dose to patient?	Yes	No	No
Patient acceptance	>99%	>90%	>99%

1.5 Positron Emission Tomography (PET)

This is based on the injection of radioactive materials, which on decay produce an electron (e^-) and an anti-particle called a positron (e^+). The positron is unstable and travels for a few mm before encountering an electron, which results in annihilation of both electron and positron, with the production of 2 high energy gamma photons which travel in opposite directions. The gamma photons are detected by sensors arranged in a doughnut fashion. Data is collected continuously while the subject is passed slowly through the scanner. The isotopes which are used in PET, such as O-15, N-13, C-11 and F-18, have half-lives of 2–100 min. Longer half-life isotopes may be manufactured using an off-site cyclotron and transported to the PET scanner, however short half-life isotopes require the use of an on-site cyclotron. These isotopes are incorporated into physiological molecules, such as water (^{13}O), acetate (^{11}C), FDG (^{13}F) and ammonia (^{13}N). The image parameter is the concentration of the isotope in the tissue. Studies of perfusion, metabolism and neurotransmitters are possible [74].

The image quality in PET is limited by the distance which the positron travels before annihilation, consequently it is not possible to obtain PET spatial resolution better than about 1 mm, and typically the spatial resolution in the abdomen is 4–5 mm. In practice a CT scan is usually performed immediately prior or following the PET scan in order to provide details of anatomy corresponding to regions of PET uptake.

1.6 Contrast Agents for MRI and Ultrasound

The original intent for contrast agents was to act to enhance the signal from MRI or ultrasound, and to improve image contrast, usually within the vascular system, hence the term ‘contrast agent’. This involves injection of a chelated heavy metal such as gadolinium for MRI [58], and of micro-bubbles for ultrasound [31]. A more sophisticated approach which has been developed in recent years is to provide the contrast agent with a means of targeting specific biological sites. This is achieved by the attachment of molecular probes such as an antibody, peptide or polysaccharide [6, 63, 76]. After injection in the vascular system the contrast agent will then adhere to the vascular wall where there are receptors for the molecular probe.

Recent developments have been made in the development of more targeted MR contrast agents, primarily utilising super-paramagnetic iron oxide particles. These are generally classified according to the diameter of the iron oxide particles in suspension, either small (SPIO with particles size commonly in the range 80–150 nm) or ultra-small (USPIO with particle size 10–30 nm). These contrast agents were initially developed for liver and spleen imaging where they are taken up by Kupffer cells distinguishing normal from abnormal parenchyma. It has been demonstrated in preliminary studies using animal models that these iron oxide particles accumulate in atherosclerotic plaques with high macrophage content [79]. Recently, preliminary reports have indicated the potential for SPIO and USPIO contrast agents to image inflammation, histologically demonstrated to be coincident with areas of increased levels of matrix metalloproteinases in aneurysms [40, 64, 91]. These biological processes have been linked to aneurysm rupture sites [88, 115]. This class of super-paramagnetic contrast agent therefore has the potential to be developed as a biomarker to predict further expansion and rupture in aortic aneurysms.

2 Imaging of Abdominal Aortic Aneurysms

In this section the imaging of abdominal aneurysms will be described, broadly separated into imaging-only measurements, and measurements made for the purpose of image guided modelling. The section mainly refers to abdominal aortic aneurysm (AAA) as these constitute the vast bulk of clinical workload. In addition thoracic aneurysms, which are much less prevalent, will also be considered.

2.1 Imaging Techniques in Clinical Use

Ultrasound and computed tomography (CT) scanning are the most commonly used clinical imaging modalities for the assessment of patients with aneurysmal disease of the aorta. Magnetic resonance imaging (MRI) does not currently have a role in routine clinical practice, although with the rapidly increasing availability of MRI, and faster imaging sequences on standard clinical MR systems, MR may well become regularly used in the diagnosis of AAA in the near future. The choice of modality is dependent on the indication for scanning and the extent of the aneurysm. The first consideration is whether the aneurysm is accessible by ultrasound, which in practice means that it must be confined to the infrarenal aorta. And secondly whether a diameter measurement is all that is required, or whether additional information is needed, for example to exclude rupture or to facilitate planning of operative intervention.

2.1.1 Aneurysm Surveillance

Surveillance data have shown no advantage for early surgical intervention over continued surveillance for aneurysms of less than 55 mm in diameter [67]. Such patients are therefore entered into a surveillance programme involving serial ultrasound measurement of the maximum antero-posterior (AP) diameter of the AAA and clinical examination, typically at six-monthly intervals. Whilst inter-operator variability can be problematic in ultrasound scanning, a high degree of consistency is achieved when scans are performed by accredited vascular technologists [51]. Figure 1 shows a typical image. Ultrasound images offer excellent discrimination of the wall, thrombus and lumen, but these details are not of any particular clinical or prognostic value. However ultrasound imaging can be compromised in patients with a high body mass index, where there is excessive bowel gas and where the aneurysm extends into the thorax. Patients with thoracic aneurysms are therefore monitored using serial CT scanning.

2.1.2 Aneurysm Screening

Aneurysm screening has been shown to reduce aneurysm-related mortality by 54% and a national screening programme has been introduced in the United Kingdom and in other countries [7, 92]. In the UK an ultrasound scan is offered to men at the

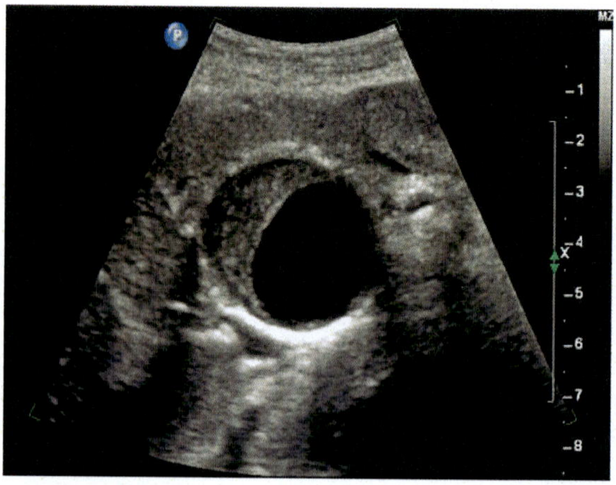

Fig. 1 Ultrasound image of a cross section of an AAA. The vessel lumen is seen as a dark central region with a crescent shaped area of thrombus. There is a coarse noise across the image which is speckle, a natural feature of all ultrasound images (This image provided courtesy of Siemens Healthcare)

age of 65 years, following which patients in whom an aneurysm has been diagnosed continue in a surveillance programme or are referred to a vascular surgeon for further management, whilst the remaining patients who have no AAA are discharged.

2.1.3 Diagnosis of Rupture

AAA are usually diagnosed incidentally following a clinical examination or a scan performed for another reason. It is common for AAA to remain undiagnosed until the point of rupture which is therefore associated with considerable mortality. For many patients presenting with ruptured AAA, the diagnosis is clearly evident on clinical examination and further imaging would only serve to delay transfer to theatre for definitive intervention. However, in patients presenting with symptoms consistent with a possible rupture, a contrast CT is the investigation of choice to secure the diagnosis. Extravasation of contrast into a surrounding haematoma is diagnostic of frank rupture, whilst more subtle signs such as discontinuity of mural calcification and a high-attenuation crescent sign in the thrombus may explain pain, suggest a contained leak or impending rupture.

2.1.4 Assessment of Patients for Surgery

Patients with an asymptomatic AAA of more than 55 mm in maximum diameter are considered for elective intervention following a thorough assessment of their general health. As part of this assessment, a CT scan with contrast is performed to help plan the appropriate intervention by characterising fully the extent of the aneurysm, assessing the relationship to key aortic branches, identifying anomalous anatomy (such as accessory renal vessels) and evaluating the aortic wall for adverse factors such as excessive calcification which may make surgery hazardous. Figure 2 shows a typical CT image of an AAA.

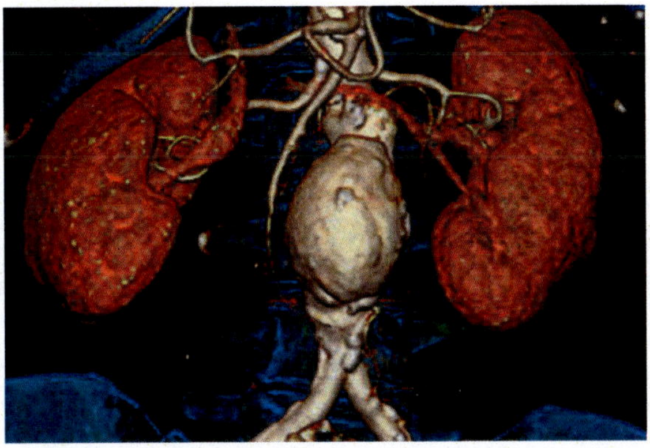

Fig. 2 CT image of an AAA. This is a surface shaded image showing the AAA below the level of the renal arteries. (This image provided courtesy of Siemens Healthcare)

2.1.5 Endovascular Aneurysm Repair

Endovascular aneurysm repair (EVAR) is increasingly replacing open surgical repair of AAA [48]. EVAR is a minimally invasive technique in which access is gained through the femoral arteries to place an expanding stent-graft in the aorta under fluoroscopic guidance to exclude the AAA from the circulation. These procedures involve considerable use of X-ray in theatre, often involving mobile C-arm fluoroscopy but increasingly utilizing interventional angiography suites, capable of high frame rate dynamic imaging, rotational angiography and cone beam CT imaging. A CT scan identifying the length and shape of the neck of the aneurysm is vital since the neck constitutes the proximal landing zone for the stent. Additionally the iliac arteries require evaluation since they must be of sufficient calibre to transmit the endoprosthesis. Following the procedure patients are followed-up after three months and then annually using CT scanning to detect any movement or loss of integrity of the stent. There is concern regarding the cumulative radiation dose from the procedure itself and the subsequent stent surveillance protocol [103], and both MRI and ultrasound have been proposed as alternative imaging modalities in these settings [19, 43, 75].

2.2 *Aneurysm Dynamics and Elastic Modulus*

The distension of arteries which accompanies the change in blood pressure throughout the cardiac cycle is typically up to 10% of the diameter. The distension occurs as a result of the elastic nature of the arterial wall. The measured change in diameter from imaging may be used to infer the elastic modulus. The most widely measured modulus which has been used in arterial imaging is the pressure-strain elastic modulus (E_p) as shown in Eq. 1 [66].

$$E_p = \frac{P_s - P_d}{(d_s - d_d)/d_d} \tag{1}$$

where d is diameter, P is pressure, and subscripts s and d are systole and diastole.

The pressure-strain elastic modulus does not account for wall thickness, and is an example of an index of 'structural stiffness'. Hayashi [37] categorised elastic moduli in arteries as being of 2 kinds. Indices of 'structural stiffness' describe the elasticity of the artery as a whole, such as E_p. Indices of material stiffness are a more fundamental property of the material of the vessel wall, mainly the Youngs elastic modulus E. If the imaging system is able to measure the wall thickness then E and E_p are related by Eq. 2.

$$E = \frac{d_d}{2h} E_p \tag{2}$$

Ultrasound imaging systems have been used to measure the distension from diastole to systole of the aorta and in AAA. This is achieved using a variety of techniques such as tracking of the RF data or from measured velocity obtained from Doppler ultrasound [38]. Extremely small motions can be detected, typically less than 10 μm, making the measurement of arterial distension relatively straightforward. A review of published data on the use of ultrasound and other techniques in AAA is provided by Van keulen [96]. Figure 3 shows typical distension waveforms from an AAA. Distension as a fraction of diameter (strain) decreases with age in normal aorta from a value of 8–10% in 20–30 year olds to 5% in subjects in their seventh decade [50]. In AAA at the maximum diameter strain is decreased compared to healthy arteries with values from 0 to 5% [96], and distension is in the range 0.5–3 mm (median 1 mm).

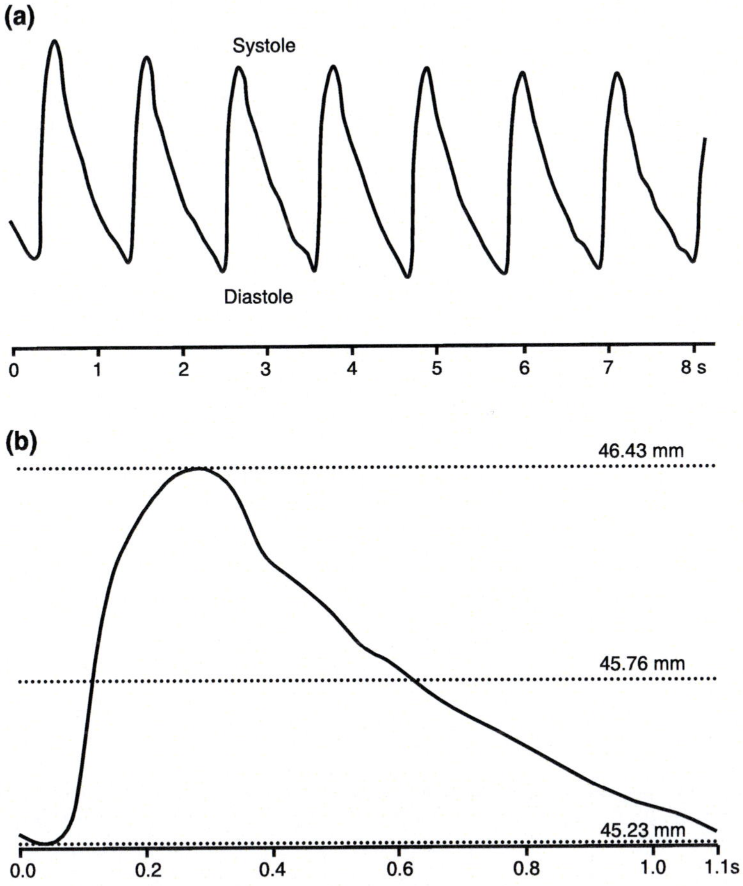

Fig. 3 **a** Distension waveforms from an AAA taken using a wall-track ultrasound system. **b** Average distension waveform. The median distension for AAA is about 1 mm, with a variation from about 0.5 to 3 mm in different patients

When distension and diameter measurements are combined with the blood pressure measured at the brachial artery using an arm cuff, E_p may be estimated. In normal aorta E_p increases with age, from a value of 0.7 Pa in 20–30 year olds [50] to values of 3.7 Pa in subjects in their seventh decade. In AAA E_p has been measured by several groups [50, 54, 59, 110], with mean values of 3.1–3.4 Pa (range 0.5–9.5 Pa). There was no dependence on maximum AAA diameter found in some studies [51, 58], however Wilson [55] found a weak positive correlation. The technique has a reproducibility of 18% [111]. Stiffer aneurysms were associated with decreased medical elastin as measured from biopsy [59] and increased collagen turnover [112], whereas more elastic aneurysms had increased elastolysis [112]. There was no difference in E_p between patients who went onto rupture compared with those who did not [50], however a later study did show that a decrease in E_p in the months immediately prior to rupture was associated with a greater risk of rupture. The overall finding from these studies is that it was not possible to use this measure of elastic modulus in a patient-specific sense as a predictor of rupture [113].

Distension and elastic modulus may also be assessed using ultrasound in the context of endovascular repair [53, 60, 78, 56]. After repair distension at the maximum diameter reduced from 1.0 mm to 0.24–0.3 mm (see [96].

These attempts at measuring elastic modulus from the distension waveform have the advantage of speed and simplicity which are positive features of a test in clinical practice. However they are based on an assumed model of the aneurysm as an isolated elastic ring. The method does not account for variations in elastic modulus and wall thickness at different locations, or account for asymmetry. Later work, described in detail elsewhere in this book, concentrates on the technique of image guided modelling which is able to use an assumed physical model which is more realistic, accounting for both 3D geometry and the presence of thrombus.

The earliest attempts which accounted for asymmetry of expansion were made using CT by Drangova et al. [21] in isolated specimens of AAA. Measurement of the Youngs modulus were made from the internal and external circumferences at different pressures, following a procedure developed by Bergel [11]. At physiologic pressures E was 275 times greater in the aneurysm than that of the normal wall proximal to it. There appears to be no attempt to use gated CT or MRI to measure elastic modulus in vivo using this approach.

2.3 Dynamic CT and MRI

Measurement of distension in vivo may also be performed using time-resolved MRI or CT. This is performed using techniques originally developed for imaging of the movement of the heart throughout the cardiac cycle. The spatial resolution of MRI and CT is 0.5–1 mm. Some modern multi-detector CT scanners are now capable of cover up to 16 cm in a single rotation with a resolution of approximately 0.5 mm [52]. The median AAA distension is 1 mm, so that in most cases the distension will be comparable with the resolution. The measurement of distension using MRI and CT is therefore much more challenging than for ultrasound which as noted above can easily detect motions of less than 10 μm. Despite these limitations, single-plane 2D data in AAA has been acquired using MRI [25, 94, 95, 99, 100] and CT [32, 86, 87]. Distension may be measured direct, or via the change in circumference. Figure 4 shows the change in cross sectional area with time for a patient with an AAA of diameter less than 5 cm. Collection of several 2D gated slices allows the 3D change in geometry of the AAA to be studied [98], from

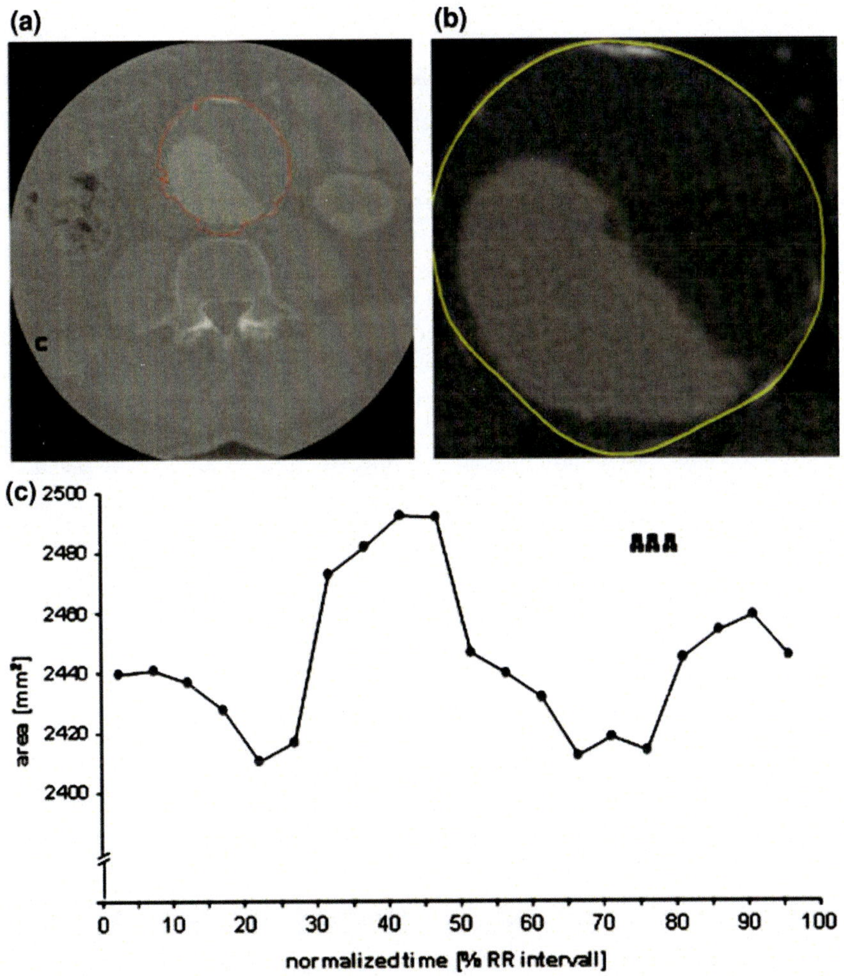

Fig. 4 Estimation of the area-time waveform in a patient with an AAA from gated CT. **a** the CT images are segmented using an active contour method, **b** an example of the segmented AAA contour, **c** area-time waveform. (Reprinted from [32], with permission from Springer)

which the change in volume with time may be measured. If this is combined with pressure then AAA compliance may be estimated (Fig. 5).

These techniques have potential in the pre-operative assessment and post-operative follow up of endovascular repair. A systematic review in this area [96] demonstrated significant distension of key points such as the AAA neck and the thoracic aorta which were maintained after endovascular repair. However the paper concluded that the clinical utility of dynamic measurements in endovascular repair was yet to be established.

2.4 Flow Imaging

It is possible to visualise 2D blood flow motion using commercially available colour flow ultrasound in real-time, however clinically there is little call for this in diagnosis of aneurysm disease. However it has been used to investigate leaking after endoluminal stent graft repair [84, 109].

Magnetic resonance angiography (MRA) is increasingly being used in the work-up for endovascular repair [8, 41], and for monitoring post-repair [77]. MRA commonly involves the use of paramagnetic contrast agents (usually gadolinium based) to view the luminal aortic tree [69]. For MRA focussing on AAA a 3D angiography sequence is acquired with suitable field of view and slab thickness to view the abdominal aortic lumen, commonly within an inspired breath-hold (to prevent breath motion artefact on resultant images). Rapid serial imaging of the aorta is then undertaken during injection of the MR contrast agent using a thinner slab (which allows more rapid data acquisition). When the contrast agent is observed in the aortic lumen, the 3D inspired MRA sequence is immediately

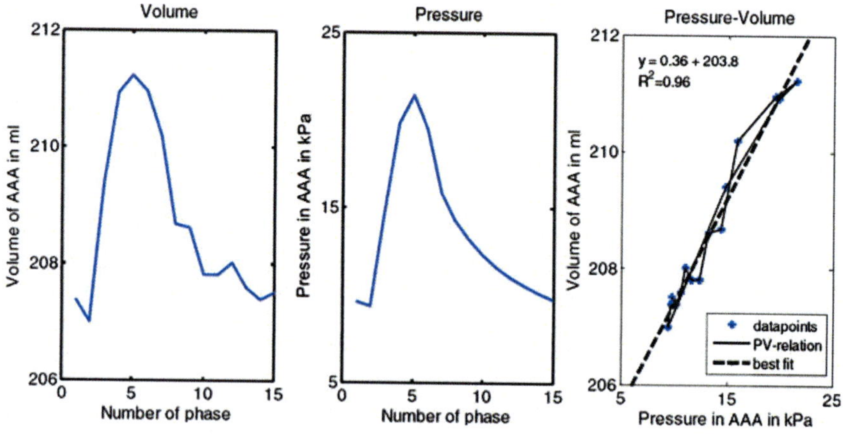

Fig. 5 Estimation of the volume-time waveform in a patient with an AAA from gated MRI. **a** volume-time waveform, **b** pressure waveform, **c** volume-pressure curve from which compliance may be estimated from the slope. Reprinted from [98], with permission from Elsevier)

repeated. Through subtraction of the pre-contrast MRA from the post-contrast MRA, visualisation of the lumen is achieved. The limitations of MRA for AAA imaging are a sensitivity to cardiac and respiratory motion, metallic objects or tissue/air interfaces, all of which may cause some degree of image artefacts. Areas of calcification are also not well shown. As such, MRA acquisitions of AAA (either for monitoring or prior to surgical repair) are commonly acquired with additional T1, and T2-weighted MRI acquisitions.

2.5 *Imaging of Inflammation*

Positron emission tomography (PET) offers opportunities to study the biological behaviour of AAA, especially inflammation. The first studies which identified uptake of the commonly used tracer fluorodeoxyglucose (FDG) in inflammation were in atherosclerotic plaque [73, 85]. Increased accumulation of FDG is seen in acute thoracic aneurysms [46], symptomatic AAA [72] and in inflammatory aneurysms [44] compared with asymptomatic aneurysms. There is a correlation between FDG uptake and histological evidence of inflammation and MMP-9 expression [72]. A review of PET-FDG imaging in vascular disease is provided by van der Vaart et al. [93].

The relationship between PET-FDG and wall stress was studied by Xu [118], in which five patients were studied. It was shown that the site of maximum stress co-localised with elevated FDG uptake in each case, suggesting that inflammation may have a role in the triggering of aneurysm rupture.

2.6 *Imaging of Geometry for Image Guided Modelling*

Image guided modelling consists of the integration of 3D imaging with computational modelling. Integration of 3D geometry with computational fluid dynamics (CFD) is used for estimation of the 3D time-varying blood flow-field and related quantities such as wall shear rate. Integration of geometry data with solid modelling is used for estimation of vessel wall distension, strain and stress patterns.

For the purposes of image guided modelling data is required on the 3D geometry of the aneurysm. The 3 structural imaging techniques of CT, MRI and 3D-ultrasound may, in principle, all be used to provide this data. In practice the success rate with 3D ultrasound is low for several reasons; the presence of bowel gas, registration artefacts caused by movement of the probe and patient, and lack of image contrast at the lateral edge of the vessel wall [36].

Most image guided modelling studies have used CT, where there is good image contrast between the flowing blood and the thrombus, and between the thrombus/wall and the surrounding tissues [27, 28, 35, 70, 90]. The low noise and good image contrast allows the inner and outer surfaces to be easily segmented (Fig. 6)

Most studies imposed a fixed wall thickness (e.g. [20]), on the basis that the contrast between the vessel wall and the thrombus was insufficient to allow visualisation of the wall. Martufi et al. [61] used an automated segmentation method to show that the true wall location could be identified.

MRI has also been used in image guided modelling to measure the aneurysm geometry [13]. The higher field strength of 3T imaging inherently improves signal to noise levels and thereby allows either higher resolution images or more rapid imaging to be acquired. However common artefacts are enhanced, such as occur during movement of the patient, the presence of foreign bodies, or susceptibility-related artefacts. This requires greater care in the acquisition of 3T data compared with 1.5T data. However the benefits include shorter breath-hold durations, acquisition of multiple MRI anatomical slices within single breath-holds and improved visualisation of the internal components of the aneurysm. This may be very useful in the advancement of image guided modelling of AAA by allowing thickness and elastic properties of the aneurysm geometry to be more accurately estimated. The higher resolution of 3T imaging also allows better visualisation of the aortic wall than 1.5T, particularly in areas of 'normal' no-thickened aortic wall.

2.7 *Imaging for Input Flow Boundary Conditions for Image Guided Modelling*

For computational fluid dynamics, data is required on the flow waveform at the inlet to the AAA. Published studies have used 3 methods for this; an assumed flow

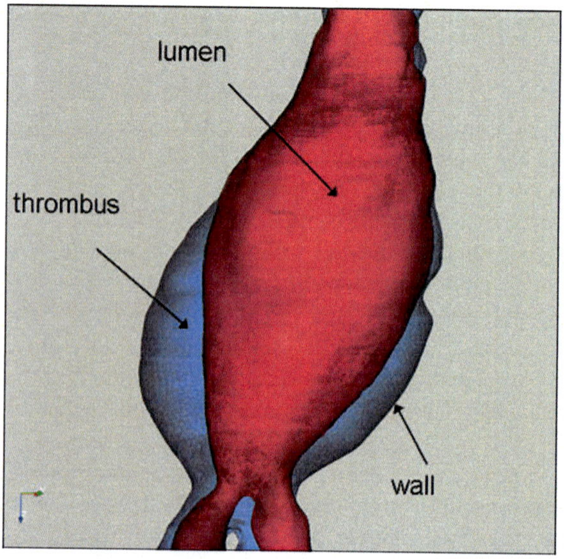

Fig. 6 Segmented inner and outer surfaces of an AAA, suitable for computational modelling. (Reprinted from [29]; with permission from Edinburgh University Library)

waveform, a waveform derived from Doppler ultrasound, and a waveform derived from MRI.

Published studies used an aorta waveform obtained in healthy volunteers from a paper published in 1970. More recently Fraser et al. [30] constructed an average velocity waveform from Doppler ultrasound data collected from patients with AAA. In the absence of any velocity data available from imaging, this average waveform could be used, recognising that the resultant CFD datasets may have some error associated with differences between the typical and the true flow waveforms.

Doppler ultrasound waveforms may be obtained using a conventional colour-flow ultrasound system. The operator places the Doppler sample volume centrally within the aorta just above the inlet to the AAA, and velocity versus time waveforms are acquired (Fig. 7). The velocity which is obtained is that from the centre of the lumen. By using a modification of the Womersley equations [117], which allow entry of the diameter and the centre-line velocity, the variation in velocity profile as a function of time through the cardiac cycle can be obtained. This time-varying 2D velocity profile may then be used as the inflow input in CFD.

Using MRI there is the opportunity to collect the time-varying 2D velocity profile directly. Most studies acquire only a single component of velocity, either along the z-axis, or along the vessel axis. With the recognition that flow in the aorta is spiral in nature it would be possible to collect 3-component 2D velocity profiles.

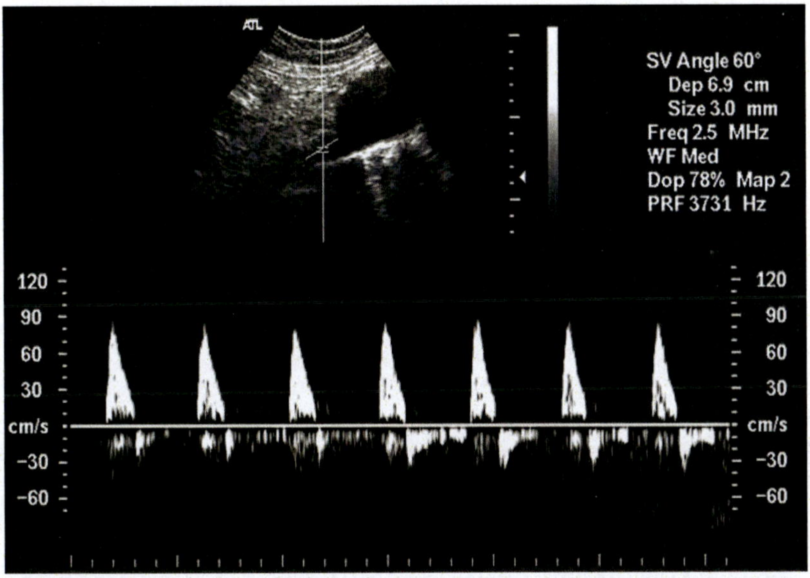

Fig. 7 Doppler ultrasound waveforms acquired proximal to an AAA. (Reprinted from: [29]; with permission from Edinburgh University Library)

3 Imaging of Cerebral Aneurysms

In this section the imaging of cerebral aneurysms will be described, broadly separated into imaging-only measurements, and measurements made for the purpose of image guided modelling.

3.1 Static Imaging

Traditional visualisation of cerebral aneurysms was performed using single or bi-plane angiography, involving direct intra-arterial injection of an iodinated contrast agent to highlight the aneurysm. A number of non-invasive techniques have been used in recent years for primary diagnosis of the presence of cerebral aneurysms; these are CT angiography (CTA), rotational angiography, MRA and Doppler ultrasound.

CTA and MRA techniques enable acquisition of 3D data showing the aneurysm structure and location. A systematic review [104] showed that these two techniques were similar in their diagnostic accuracy, with accuracies of 89% for MRA and 90% for CT. Technologies have improved substantially in the last decade particularly with regard to speed of acquisition and bone subtraction algorithms for CT and widespread use of higher field strengths (3T) in MR. The main MR sequences used in clinical practice for MR Angiography are 3D time-of-flight and ultrafast gradient echo based (contrast enhanced) MRA with images obtained pre/post contrast for digital subtraction.

Ultrasound images may be obtained with the transducer placed over a region of the skull which is relatively thin and flat, such as the temporal bone, where the distortion of the ultrasound beam produced by the skull is minimal. For ultrasound the aneurysm is visualised by detecting the Doppler ultrasound signal produced from the blood flow within the aneurysm. It was however found that for ultrasound accuracy varied from 71 to 82%, dependent on location, and was very poor for aneurysms less than 5 mm in diameter [105]. Furthermore up to 20% of patients will not have an adequate bone window [102].

A method for 3D data acquisition which is especially suitable for planning interventional procedures on aneurysms is rotational angiography. Rotational angiography can produce a series of individual projection images which can be displayed in a volume rendered display creating 3D perception, or alternatively, the images can be used in 3D—cone beam reconstruction and then produce a 3D Computed Tomography data set (see below).

Interventional procedures such as placement of coils in an aneurysm, are performed using fluoroscopy, which allows for real-time imaging typically accomplished using roadmapping technology- both 2D and 3D. Roadmapping is a well established technique whereby a digital subtraction unit subtracts out the background during screening and then contrast is injected—typically into an artery of vein in region of interest. This creates a "map" of the contrast containing structures against the background. On subsequent fluoroscopic screening (until the "roadmap" facility is cancelled) the operator can see a catheter/wire/coil or other device against the vessel map (hence the term roadmapping). The angiography system is typically referred to as a C-arm, with the X-Ray tube at one end of the arm and at the other end of the arm a Caesium Iodide flat panel X-Ray detector. These machines can obtain projectional 2D digital subtraction angiographic (DSA) images at variable frame speeds (on latest machines up to 30 frames/s is standard) Rotation of the C-arm through 180° or more allows collection of data which may be reconstructed in the same way as CT data using tomographic techniques [4, 24]. Typically the C-arm rotates at a rate of 40° per second over a 5 s period. During rotational angiography iodinated contrast agent is injected intra arterially into the patient whilst the C arm rotates and the acquisition is timed to occur as the contrast agent opacifies the selected intracranial vessel and aneurysm. In practice typically both pre and intra contrast rotations are performed to enable a digital subtraction 3D angiogram to be produced. Comparison against conventional angiography demonstrated that rotational angiography detects more small (<3 mm) additional aneurysms [97] (Fig. 8). Rotational angiographic CT-like imaging can also be a valuable adjunct during interventional procedures to assess stent placement and opening, assess ventricular size, exclude new haemorrhage etc. [10, 106]. This technique has inherent contrast but is subject to movement artefact (not such a problem in interventional neuroradiology procedures, which are often performed under general anaesthesia). Whole brain perfusion imaging is also potentially

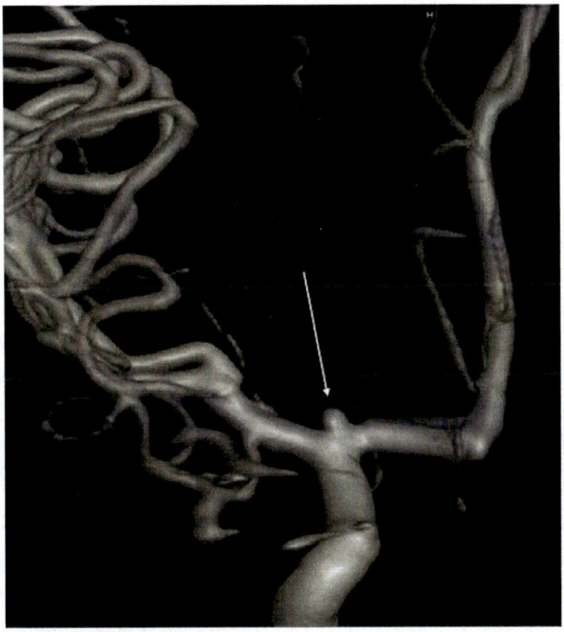

Fig. 8 Subtracted 3D rotational angiogram (volume rendered) demonstrating ruptured small right internal carotid artery bifurcation aneurysm

achievable using angiographic CT techniques and development of this tool in humans is in progress [82].

3.2 Dynamic Imaging

Early studies demonstrating the pulsation of cerebral aneurysms during the cardiac cycle used ultrasound imaging [101]. This was validated using a pulsating aneurysm phantom [39], and further studies showed that the degree of pulsation increased with intracranial pressure [102]. Dynamic CTA, sometimes called 4D CTA, is also now possible and has been used to study changes in aneurysm size and morphology during the cardiac cycle in an attempt to elicit which aneurysms may be inherently more unstable and therefore prone to rupture/rerupture. In Dynamic CTA a series of 3D volumes through the cardiac cycle is collected as described above [34, 42, 45], validated in a pulsating phantom [119]. Ishida [42] noted pulsation in 9 of 28 saccular aneurysms and in 3 of 5 non-saccular aneurysms. Hayakawa et al. [34] investigated 23 patients with ruptured aneurysms, of which 4 showed pulsation whose location corresponded to the site of rupture in all 4 cases. It was hypothesised in these studies that the pulsation occur where the wall is thin, and that this is a site that is at risk of rupture.

Dynamic imaging has also been used to investigate the passage of contrast agent through the vascular system, in order to observe the direction of filling of the aneurysm, and to look for abnormal filling through collateral vessels which could indicate vascular occlusion or spasm [65].

3.3 Imaging Techniques in Clinical Use

Non invasive imaging techniques have developed in clinical practice over the last 10–15 years to the extent that they are now the main initial diagnostic tool for intracranial aneurysms. CTA is particularly used in ruptured aneurysms due to its availability & speed in sick often restless patients [33, 89]. Either MRA or CTA are used in aneurysm screening with MRA often preferred as it avoids irradiation [108]. In aneurysms incidentally detected on anatomical brain imaging either CTA or MRA may be used next to confirm diagnosis, location and more detailed anatomy of the aneurysm to enable treatment planning [1, 107]; see Fig. 9. DSA remains the definitive diagnostic tool for intracranial aneurysms but it is invasive, expensive (compared to CTA/MRA) and more time consuming. The addition of rotational subtraction technique (sometimes called 3D DSA) further increases its accuracy [97]. However, DSA is key in most instances to planning definitive aneurysm treatment either neurosurgical or endovascular. Although in some units neurosurgical clipping may be performed

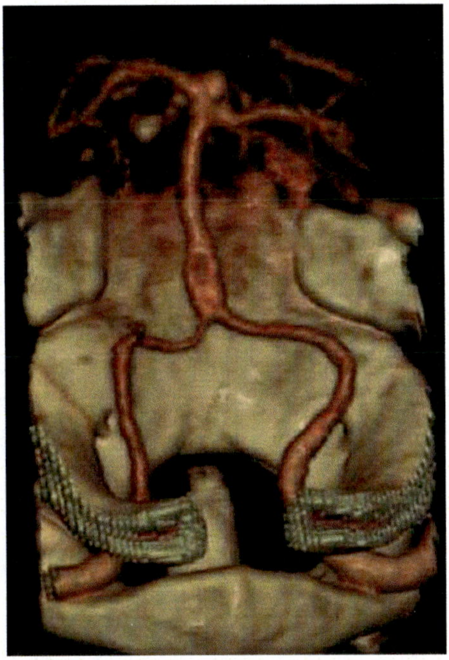

Fig. 9 Volume rendered CTA reconstruction demonstrating ruptured small basilar tip aneurysm (*white arrow*) with incidental fenestrated basilar artery (*black arrow*)

on the basis of CTA alone for simple aneurysms [33] this is by no means universal practice and in most circumstances DSA will be required prior to definitive aneurysm treatment.

In endovascular treatment diagnostic 2D/3D DSA is essential and is performed first to establish working projection(s) for the endovascular treatment (most commonly endosaccular coil placement within aneurysm). Then, as outlined above, fluoroscopic techniques including subtracted roadmap and sometimes a 3D roadmap technique as well as unsubtracted fluoroscopy are used to control the placement of devices. Typically a biplane angiographic unit (comprising 2 C arms) is utilised in neurointerventional procedures to enable a good visualisation of the aneurysm (especially the neck), parent artery and distal run off vasculature throughout. During or at the end of a coiling or stenting procedure angiographic CT may be obtained to help assist procedure or subsequent patient management (Fig. 10).

In ruptured aneurysm patients a number of complications can occur as a result of the rupture causing subarachnoid haemorrhage (SAH) and imaging techniques are widely used to detect/monitor these. These include CT for hydrocephalus and Transcranial Doppler and CT Perfusion to detect or monitor post SAH vasospasm [57, 81, 116]. TCD has the advantage of being a bedside test and can be repeated with no radiation burden though it gives a surrogate measure of brain perfusion rather than the absolute measurement of cerebral blood flow provided by most

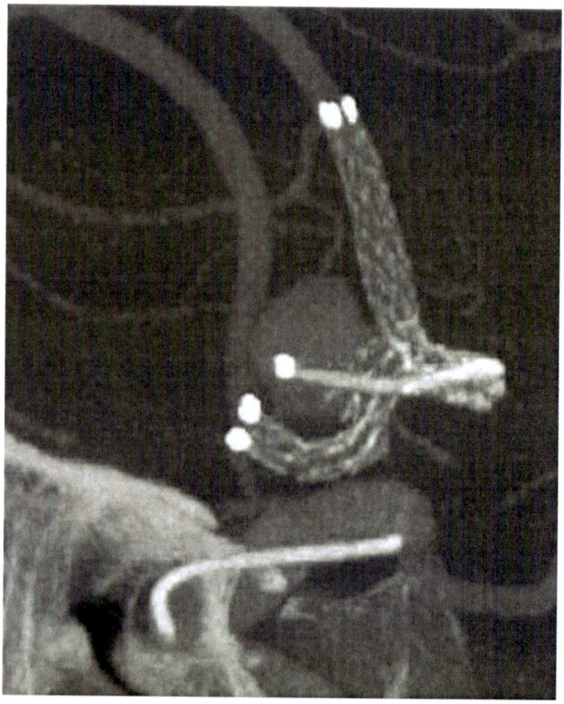

Fig. 10 AngioCT MIP reconstruction of AngioCT obtained during dilute intra-arterial contrast injection demonstrating stents in situ across neck of an anterior communicating aneurysm with microcatheter placed inside aneurysm preparatory to coiling

CTP techniques. If vasospasm is severe and poorly responsive to medical therapy then DSA controlled balloon angioplasty of the affected major vessel segments may be performed [22].

Following endovascular aneurysm treatment DSA/3D DSA is often used in the initial early follow up of coiled aneurysms and sometimes for clipped ones too. Use of CTA is possible in clipped aneurysms but still impractical in coiled aneurysms due to the extensive metal artefact obscuring detail at the critical region of the aneurysm neck. More delayed long term imaging follow-up after endovascular aneurysm treatment is increasingly with MRA techniques though the optimal strategy is probably to use both time of flight and fast contrast enhanced techniques [2, 47]. There is some evidence that 3T may be more accurate here than 1–1.5T [5]—see Fig. 11.

3.4 Techniques Used for Image Guided Modelling

Acquisition of 3D geometry for image guided modelling has been performed using rotational angiography [16–18, 83], CT angiography [9, 80] or MRI [15, 71]. Published studies have mostly involved a standard inlet flow waveform obtained

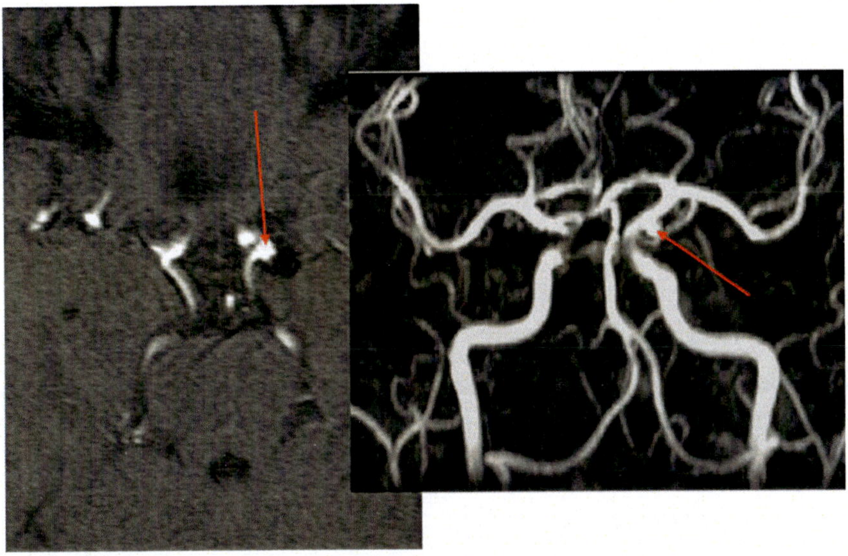

Fig. 11 Time-of-flight MRA base image and MIP reconstructions at 18 months post coiling of left posterior communicating artery aneurysm demonstrating neck recurrence (*red arrows*)

from healthy volunteers using MRA or transcranial Doppler [9, 16, 18, 80, 83]. Image guided modelling based on patient-specific inlet flow obtained from MRI has also been reported [15, 71].

Modelling of aneurysmal flow is being utilised in the development of new therapeutic devices [71] such as stents (to divert flow away from the aneurysm) as well as to advance the understanding and prediction of aneurysmal development, growth and rupture [62]. Ultimately in the future flow image guided modelling may enable aneurysm treatment to be appropriately optimised to the individual patient and aneurysm. In practical terms these techniques will be much easier to achieve in the elective treatment of unruptured aneurysms than in the emergency treatment of ruptured aneurysms.

4 Conclusions

Imaging techniques are pivotal to the diagnosis, evaluation treatment and post procedure care of patients with aneurysms. At each stage of this patient pathway different modalities have key roles to play but in most patients CT/CTA, DSA, TCD and MRA will all be utilised at some stage because of their different capabilities and underlying physical properties.

Glossary of Abbreviations

AAA Abdominal aortic aneurysm
AP Antero-posterior
BMUS British Medical Ultrasound Society
CFD Computational fluid dynamics
CT Computed tomography
CTA CT angiography
CTP CT perfusion
DSA Digital subtraction angiography
EVAR Endovascular aneurysm repair
FDG Fluorodeoxyglucose
MHz Mega-hertz
MRA Magnetic resonance angiography
MRI Magnetic resonance imaging
PET Positron emission tomography
SAH Subarachnoid haemorrhage
SPIO Small paramagnetic iron oxide (particles)
TCD Transcranial Doppler ultrasound
USPIO Ultrasmall paramagnetic iron oxide (particles)

References

1. Agid, R., Lee, S.K., Willinisky, R.A., et al.: Acute SAH: using 64-slice multidetector CT angiography to triage patients' treatment. Neuroradiology **48**, 787–794 (2006)
2. Agid, R., Willinsky, R.A., Lee, S.K., TerBrugge, K.G., Farb, R.I.: Characterization of aneurysm remnants after endovascular treatment: contrast-enhanced MR angiography versus catheter digital subtraction angiography. Am. J. Neuroradiol. **29**, 1570–1574 (2008)
3. Allisy-Roberts, P., Williams, J.R.: Farr's Physics for Medical Imaging. Elsevier (2008)
4. Anxionnat, R., Bracard, S., Macho, J., Da Costa, E., Vaillant, R., Launay, L., Trousset, Y., Romeas, R., Picard, L.: 3D angiography—Clinical interest. First applications in interventional neuroradiology. J. Neuroradiol. **25**, 251–262 (1998)
5. Anzalone, N., Scomazzoni, F., Cirillo, M., Righi, C., Simionato, F., Cadioli, M., Iadanza, A., Kirchin, M.A., Scotti, G.: Follow-up of coiled cerebral aneurysms at 3T: comparison of 3D time-of-flight MR angiography and contrast-enhanced MR angiography. Am. J. Neuroradiol. **29**, 1530–1536 (2008)

6. Artemov, D., Mori, N., Okollie, B., Bhujwalla, Z.M.: MR molecular imaging of the Her-2/*neu* receptor in breast cancer cells using targeted iron oxide nanoparticles. Magn. Reson. Med. **49**, 403–408 (2003)
7. Ashton, H.A., Buxton, M.J., Day, N.E., Kim, L.G., Marteau, T.M., Scott, R.A.P., Thompson, S.G., Walker, N.M.: The Multicentre Aneurysm Screening Study (MASS) into the effect of abdominal aortic aneurysm screening on mortality in men: a randomised controlled trial. Lancet **360**, 1531–1539 (2002)
8. Atar, E., Belenky, A., Hadad, M., Ranany, E., Baytner, S., Bachar, G.N.: MR angiography for abdominal and thoracic aortic aneurysms: assessment before endovascular repair in patients with impaired renal function. Am. J. Roentgenol. **186**, 386–393 (2006)
9. Baek, H., Jayaraman, M.V., Karniadakis, G.E.: Wall shear stress and pressure distribution on aneurysms and infundibulae in the posterior communicating artery bifurcation. Ann. Biomed. Eng. **37**, 2469–2487 (2009)
10. Benndorf, G., Strother, C.M., Claus, B., Naeini, R., Morsi, H., Klucznik, R., Mawad, M.E.: Angiographic CT in Cerebrovascular Stenting. Am. J. Neuroradiol. **26**, 1813–1818 (2005)
11. Bergel, D.H.: The static elastic properties of the arterial wall. J. Physiol. **156**, 44–57 (1961)
12. BMUS. The British Medical Ultrasound Society: Guidelines for the safe use of diagnostic ultrasound equipment. Ultrasound **18**, 52–59 (2010)
13. Breeuwer, M., de Putter, S., Kose, U., Speelman, L., Visser, K., Gerritsen, F., et al.: Towards patient-specific risk assessment of abdominal aortic aneurysm. Med. Biol. Eng. Comput. **46**, 1085–1095 (2008)
14. Brown, B.H., Smallwood, R.H., Barber, D.C., Lawford, P.V., Hose, D.R.: Medical Physics and Biomedical Engineering. Institute of Physics Publishing, Philadelphia (1999)
15. Boussel, L., Rayz, V., Martin, A., Acevedo-Bolton, G., Lawton, M.T., Higashida, R., Smith, W.S., Young, W.L., Saloner, D.: Phase-contrast magnetic resonance imaging measurements in intracranial aneurysms in vivo of flow patterns, velocity fields, and wall shear stress: comparison with computational fluid dynamics. Magn. Reson. Med. **61**, 409–417 (2009)
16. Cebral, J.R., Castro, M.A., Appanaboyina, S., Putman, C.M., Millan, D., Frangi, A.F.: Efficient pipeline for image-based patient-specific analysis of cerebral aneurysm hemodynamics: technique and sensitivity. IEEE Trans. Med. Imaging **24**, 457–467 (2005)
17. Chen, J.L., Wang, S.Z., Ding, G.H., Yang, X.J., Li, H.Y.: Patient-specific blood dynamic simulations in assessing endovascular occlusion of intracranial aneurysms. J. Hydrodynamics **21**, 271–276 (2009)
18. Chien, A., Tateshima, S., Castro, M., Sayre, J., Cebral, J., Vinuela, F.: Patient-specific flow analysis of brain aneurysms at a single location: comparison of hemodynamic characteristics in small aneurysms. Med. Biol. Eng. Comput. **46**, 1113–1120 (2008)
19. Cohen, E.I., Weinreb, D.B., Siegelbaum, R.H., Honig, S., Marin, M., Weintraub, J.L., Lookstein, R.A.: Time-resolved MR angiography for the classification of endoleaks after endovascular aneurysm repair. J. Magn. Reson. Imaging **27**, 500–503 (2008)
20. Di Martino, E.S., Guadagni, G., Fumero, A., Ballerini, G., Spirito, R., Biglioli, P., et al.: Fluid-structure interaction within realistic three-dimensional models of the aneurysmatic aorta as a guidance to assess the risk of rupture of the aneurysm. Med. Eng. Phys. **23**, 647–655 (2001)
21. Drangova, M., Holdsworth, D.W., Boyd, C.J., Dunmore, P.J., Roach, M.R., Fenster, A.: Elasticity and geometry measurements of vascular specimens using a high-resolution laboratory CTt scanner. Physiol. Meas. **14**, 277–290 (1993)
22. Eskridge, J.M., Song, J.K.: A practical approach to the treatment of vasospasm. Am. J. Neuroradiol. **18**, 1653–1660 (1997)
23. Evans, D.H., McDicken, W.N.: Doppler Ultrasound, 2nd ed edn. Wiley, Chichester (2000)
24. Fahrig, R., Fox, A.J., Lownie, S., Holdsworth, D.W.: Use of a C-arm system to generate true three-dimensional computed rotational angiograms: preliminary in vitro and in vivo results. Am. J. Neuroradiol. **18**, 1507–1514 (1997)

25. Faries, P.L., Agarwal, G., Lookstein, R., Bernheim, J.W., Cayne, N.S., Cadot, H., et al.: Use of cine magnetic resonance angiography in quantifying aneurysm pulsatility associated with endoleak. J. Vasc. Surg. **38**, 652–656 (2003)
26. Fenster, A., Downey, D.B., Cardinal, H.N.: Three-dimensional ultrasound imaging. Phys. Med. Biol. **46**, R67–R99 (2001)
27. Fillinger, M.F., Raghavan, M.L., Marra, S.P., Cronenwett, J.L., Kennedy, F.E.: In vivo analysis of mechanical wall stress and abdominal aortic aneurysm rupture risk. J. Vasc. Surg. **36**, 589–597 (2002)
28. Fillinger, M.F., Marra, S.P., Raghavan, M.L., Kennedy, F.E.: Prediction of rupture risk in abdominal aortic aneurysm during observation: wall stress versus diameter. J. Vasc. Surg. **37**, 724–732 (2002)
29. Fraser, K.H.: Computational estimation of haemodynamics and tissue stresses in abdominal aortic aneurysms. PhD thesis (2007)
30. Fraser, K.H., Meagher, S., Blake, J.R., Easson, W.J., Hoskins, P.R.: Characterisation of an abdominal aortic velocity waveform in patients with abdominal aortic aneurysm. Ultrasound Med. Biol. **34**, 73–80 (2008)
31. Frinking, P.J.A., Bouakaz, A., Kirkhorn, J., Ten Cate, F.J., de Jong, N.: Ultrasound contrast imaging: current and new potential methods. Ultrasound Med. Biol. **26**, 965–975 (2000)
32. Ganten, M.K., Krautter, U., von Tengg-Kobligk, H., Bockler, D., Schumacher, H., Stiller, W., et al.: Quantification of aortic distensibility in abdominal aortic aneurysm using ECG-gated multidetector computed tomography. Eur. Radiol. **18**, 966–973 (2008)
33. Goddard, A.J.P., Tan, G., Becker, J.: Computed tomography angiography for the detection and characterization of intra-cranial aneurysms: current status. Clin. Radiol. **60**, 1221–1236 (2005)
34. Hayakawa, M., Katada, K., Anno, H., Imizu, S., Hayashi, J., Irie, K., Negoro, M., Kato, Y., Kanno, T., Sano, H.: CT angiography with electrocardiographically gated reconstruction for visualizing pulsation of intracranial aneurysms: identification of aneurysmal protuberance presumably associated with wall thinning. Am. J. Neuroradiol. **26**, 1366–1369 (2005)
35. Heng, M.S., Fagan, M.J., Collier, J.W., Desai, G., McCollum, P.T., Chetter, I.C.: Peak wall stress measurement in elective and acute abdominal aortic aneurysms. J. Vasc. Surg. **47**, 17–22 (2008)
36. Hammer, S., Jeays, A., MacGillivray, T.J., Allan, P.L., Hose, R., Barber, D., Easson, W.J., Hoskins, P.R.: Acquisition of 3D arterial geometries and integration with computational fluid dynamics. Ultrasound Med. Biol. **35**, 2069–2083 (2009)
37. Hayashi, K.: Experimental approaches on measuring the mechanical-properties and constitutive laws of arterial-walls. J. Biomech. Eng. **115**, 481–488 (1993)
38. Hoeks, A.P.G., Brands, P.J., Willigers, J.M., Reneman, R.S.: Non-invasive measurement of mechanical properties of arteries in health and disease. J. Eng. Med. **213**, 195–202 (1999)
39. Hoskins, P.R., Prattis, J., Wardlaw, J.: A flow model of cerebral aneurysms for use with power Doppler studies. Br. J. Radiol. **71**, 76–80 (1998)
40. Howarth, S.P.S., Tang, T.Y., Graves, M.J., U-King-Im, J.M., Li, Z.Y., Walsh, S.R., Gaunt, M.E., Gillard, J.H.: Non-invasive MR imaging of inflammation in a patient with both asymptomatic carotid atheroma and an abdominal aortic aneurysm: a case report. Ann. Surg. Innov. Res. **1**, 4 (2007)
41. Iozzelli, A., D'Orta, G., Aliprandi, A., Secchi, F., Di Leo, G., Sardanelli, F.: The value of true-FISP sequence added to conventional gadolinium-enhanced MRA of abdominal aorta and its major branches. Eur. J. Radiol. **72**, 489–493 (2008)
42. Ishida, F., Ogawa, H., Simizu, T., Kojima, T., Taki, W.: Visualizing the dynamics of cerebral aneurysms with four-dimensional computed tomographic angiography. Neurosurgery **57**, 460–470 (2005)
43. Kopp, R., Zurn, W., Weidenhagen, R., Meimarakis, G., Clevert, D.A.: First experience using intraoperative contrast-enhanced ultrasound during endovascular aneurysm repair for infrarenal aortic aneurysms. J. Vasc. Surg. **51**, 1103–1110 (2010)

44. Kotze, C.W., Menezes, L.J., Endozo, R., Groves, A.M., Ell, P.J., Yusuf, S.W.: Increased metabolic activity in abdominal aortic aneurysm detected by 18F-fluorodeoxyglucose (18F-FDG) positron emission tomography/computed tomography (PET/CT). Eur. J. Vasc. Endovasc. Surg. **38**, 93–99 (2009)
45. Krings, T., Willems, P., Barfett, J., Ellis, M., Hinojosa, N., Blobel, J., Geibprasert, S.: Pulsatility of an intracavernous aneurysm demonstrated by dynamic 320-detector row CTA at high temporal resolution. Cen. Eur. Neurosurg. **70**, 214–218 (2009)
46. Kuehl, H., Eggebrecht, H., Boes, T., Antoch, G., Rosenbaum, S., Ladd, S., Bockisch, A., Barkhausen, J., Erbel, R.: Detection of inflammation in patients with acute aortic syndrome: comparison of FDG-PET/CT imaging and serological markers of inflammation. Heart **94**, 1472–1477 (2008)
47. Kwee, T.C., Kwee, R.M.: MR angiography in the follow-up of intracranial aneurysms treated with Guglielmi detachable coils: systematic review and meta-analysis. Neuroradiology **49**, 703–713 (2007)
48. Lall, P., Gloviczki, P., Agarwal, G., Duncan, A.A., Kalra, M., Hoskin, T., Oderich, G.S., Bower, T.C.: Comparison of EVAR and open repair in patients with small abdominal aortic aneurysms: can we predict results of the PIVOTAL trial? J. Vasc. Surg. **49**, 52–59 (2009)
49. Lang, R.M., Mor-Avi, V., Sugeng, L., Nieman, P.S., Sahn, D.J.: Three-dimensional echocardiography: the benefits of the additional dimension. J. Am. Coll. Cardiol. **48**, 2053–2069 (2006)
50. Lanne, T., Sonesson, B., Bergqvist, D., Bengtsson, H., Gustafsson, D.: Diameter and compliance in the male human abdominal aorta: influence of age and aortic aneurysm. Eur. J. Vasc. Surg. **6**, 178–184 (1992)
51. Lanne, T., Sandgren, T., Mangell, P., Sonesson, B., Hansen, F.: Improved reliability of ultrasonic surveillance of abdominal aortic aneurysms. Eur. J. Vasc. Endovasc. Surg. **13**, 149–153 (1997)
52. Lembcke, A., Hein, P.A., Borges, A.C., Rogalla, P.: One-stop-shop cardiac diagnosis in a single heart beat using 320-slice computed tomography: ascending aortic aneurysm, hypertrophic cardiomyopathy and mixed valvular heart disease. Eur. J. Cardiothorac. Surg. **35**, 726 (2009)
53. Lindblad, B., Dias, N., Malina, M., Ivancev, K., Resch, T., Hansen, F., et al.: Pulsatile wall motion (PWM) measurements after endovascular abdominal aortic aneurysm exclusion are not useful in the classification of endoleak. Eur. J. Vasc. Endovasc. Surg. **28**, 623–628 (2004)
54. Long, A., Rouet, L., Bissery, A., Rossignol, P., Mouradian, D., Sapoval, M.: Compliance of abdominal aortic aneurysms: evaluation of tissue Doppler imaging. Ultrasound Med. Biol. **30**, 1099–1108 (2004)
55. Long, A., Rouet, L., Bissery, A., Rossignol, P., Mouradian, D., Sapoval, M.: Compliance of abdominal aorticaneurysms evaluated by tissue Doppler imaging: correlation with aneurysm size. J. Vasc. Surg. **42**, 18-26 (2005)
56. Long, A., Rouet, L., Vitry, F., Albertini, J.N., Marcus, C., Clement, C.: Compliance of Abdominal Aortic Aneurysmsbefore and after Stenting with Tissue Doppler Imaging: Evolution during Follow-Up and Correlation withAneurysm Diameter. Ann. Vasc. Surg**23**, 49-59 (2009)
57. McGirt, M.J., Blessing, R.P., Goldstein, L.B.: Transcranial doppler monitoring and clinical decision-making after subarachnoid hemorrhage. J. Stroke Cerebrovasc. Dis. **12**, 88–92 (2003)
58. McRobbie, D.W., Moore, E.A., Graves, M.J., Prince, M.R.: From Picture to Proton. Cambridge University Press, Cambridge (2003)
59. MacSweeney, S.T., Young, G., Greenhalgh, R.M., Powell, J.T.: Mechanical properties of the aneurysmal aorta. Br. J. Surg. **79**, 1281–1284 (1992)
60. Malina, M., Lanne, T., Ivancev, K., Lindblad, B., Brunkwall, J.: Reduced pulsatile wall motion of abdominal aortic aneurysms after endovascular repair. J. Vasc. Surg. **27**, 624–631 (1998)

61. Martufi, G., Di Martino, E.S., Amon, C.H., Muluk, S.C., Finol, E.A.: Three-dimensional geometrical characterization of abdominal aortic aneurysms: image-based wall thickness distribution. J. Biomech. Eng. 131: article Number: 061015 (2009)
62. Mitsos, A.P., Kakalis, N.M.P., Ventikos, Y.P., Byrne, J.V.: Haemodynamic simulation of aneurysm coiling in an anatomically accurate computational fluid dynamics model: technical note. Neuroradiology **50**, 341–347 (2008)
63. Moran, C.M., Ross, J.A., Cunningham, C., Butler, M., Anderson, T., Newby, D., Fox, K.A.A., McDicken, W.N.: Manufacture and acoustical characterisation of a high-frequency contrast agent for targeting applications. Ultrasound Med. Biol. **32**, 421–428 (2006)
64. Nchimi, A., Defawe, O., Brisbois, D., Broussaud, T.K.Y., Defraigne, J.O., Magotteaux, P., Massart, B., Serfaty, J.M., Houard, X., Michel, J.B., Sakalihasan, N.: MR imaging of iron phagocytosis in intraluminal thrombi of abdominal aortic aneurysms in humans. Radiology **254**, 973–981 (2010)
65. Pekkola, J., Kangasniemi, M.: Imaging of blood flow in cerebral arteries with dynamic helical computed tomography angiography (DHCTA) using a 64-Row CT scanner. Acta Radiol. **50**, 798–805 (2009)
66. Peterson, L.H., Jensen, R.E., Parnell, J.: Mechanical properties of arteries in vivo. Circ. Res. **8**, 622–639 (1960)
67. Powell, J.T., Brady, A.R., Brown, L.C., Forbes, J.F., Fowkes, F.G.R., Greenhalgh, R.M., Ruckley, C.V., Thompson, S.G.: Mortality results for randomised controlled trial of early elective surgery or ultrasonographic surveillance for small abdominal aortic aneurysms. Lancet **352**, 1649–1655 (1998)
68. Prager, R.W., Ijaz, U.Z., Gee, A.H., Treece, G.M.: Three-dimensional ultrasound imaging. J. Eng. **224**, 193–223 (2009)
69. Prince, M.R., Yucel, E.K., Kaufman, J.A., Harrison, D.C., Geller, S.C.: Dynamic gadolinium-enhanced three-dimensional abdominal MR arteriography. J. Magn. Reson. Imaging **3**, 877–881 (1993)
70. Raghavan, M.L., Vorp, D.A., Federle, M.P., Makaroun, M.S., Webster, M.W.: Wall stress distribution on three-dimensionally reconstructed models of human abdominal aortic aneurysm. J. Vasc. Surg. **31**, 760–769 (2000)
71. Rayz, V.L., Boussel, L., Acevedo-Bolton, G., Martin, A.J., Young, W.L., Lawton, M.T., Higashida, R., Saloner, D.: Numerical simulations of flow in cerebral aneurysms: comparison of CFD results and in vivo MRI measurements. J. Biomech. Eng. 130: article Number: 051011 (2008)
72. Reeps, C., Essler, M., Pelisek, J., Seidl, S., Eckstein, H.H., Krause, B.J.: Increased 18F-fluorodeoxyglucose uptake in abdominal aortic aneurysms in positron emission/computed tomography is associated with inflammation, aortic wall instability, and acute symptoms. J. Vasc. Surg. **48**, 417–423 (2008)
73. Rudd, J.H.F., Myers, K.S., Bansilal, S., Machac, J., Rafique, A., Farkouh, M., Fuster, V., Fayad, Z.A.: 18Fluorodeoxyglucose positron emission tomography Imaging of atherosclerotic plaque inflammation is highly reproducible—Implications for atherosclerosis therapy trials. J. Am. Coll. Cardiol. **50**, 892–896 (2007)
74. Saha, G.B., Macintyre, W.J., Go, R.T.: Radiopharmaceuticals for brain imaging. Semin. Nucl. Med. **24**, 324–349 (1994)
75. Schmieder, G.C., Stout, C.L., Stokes, G.K., Parent, F.N., Panneton, J.M.: Endoleak after endovascular aneurysm repair: duplex ultrasound imaging is better than computed tomography at determining the need for intervention. J. Vasc. Surg. **50**, 1012–1017 (2009). discussion 1017–1018
76. Schroeder, A., Kost, J., Barenholz, Y.: Ultrasound, liposomes, and drug delivery: principles for using ultrasound to control the release of drugs from liposomes. Chem. Phys. Lipids **162**, 1–16 (2009)
77. Schwope, R.B., Alper, H.J., Talenfeld, A.D., Cohen, E.I., Lookstein, R.A.: MR angiography for patient surveillance after endovascular repair of abdominal aortic aneurysms. Am. J. Roentgenol. **188**, W334–W340 (2007)

78. Sekhri, A.R., Lees, W.R., Adiseshiah, M.: Measurement of aortic compliance in abdominal aortic aneurysms before and after open and endoluminal repair: preliminary results. J. Endovasc. Ther. **11**, 472–482 (2004)
79. Shmitz, S.A., Couplan, S.E., Gust, R., et al.: Superparamagnetic iron oxide-enhanced MRI of atherosclerotic plaques in Watanabe hereditable rabbits. Invest. Radiol. **35**, 460–471 (2000)
80. Shojima, M., Oshima, M., Takagi, K., Torii, R., Hayakawa, M., Katada, K., Morita, A., Kirino, T.: Magnitude and role of wall shear stress on cerebral aneurysm—computational fluid dynamic study of 20 middle cerebral artery aneurysms. Stroke **35**, 2500–2505 (2004)
81. Sloan, M.A., Haley, E.C., Kassell, N.F., Henry, M.L., Stewart, S.R., Beskin, R.R., Sevilla, E.A., Tomer, J.C.: Sensitivity and specificity of transcranial Doppler ultrasonography in the diagnosis of vasospasm following subarachnoid hemorrhage. Neurology **39**, 1514–1518 (1989)
82. Struffert, T., Deuerling-Zheng, Y., Kloska, S., Engelhorn, T., Strother, C.M., Kalender, W.A., Köhrmann, M., Schwab, S., Doerfler, A.: Flat detector CT in the evaluation of brain parenchyma, intracranial vasculature, and cerebral blood volume: a pilot study in patients with acute symptoms of cerebral ischemia. Am. J. Neuroradiol. (available on line) (2010)
83. Steinman, D.A., Milner, J.S., Norley, C.J., Lownie, S.P., Holdsworth, D.W.: Image-based computational simulation of flow dynamics in a giant intracranial aneurysm. Am. J. Neuroradiol. **24**, 559–566 (2003)
84. Sun, Z.H.: Diagnostic value of color duplex ultrasonography in the follow-up of endovascular repair of abdominal aortic aneurysm. J. Vasc. Interv. Radiol. **17**, 759–764 (2006)
85. Tawakol, A., Migrino, R.Q., Bashian, G.G., Bedri, S., Vermylen, D., Cury, R.C., Yates, D., LaMuraglia, G.M., Furie, K., Houser, S., Gewirtz, H., Muller, J.E., Brady, T.J., Fischman, A.J.: In vivo F-18-fluorodeoxyglucose positron emission tomography imaging provides a noninvasive measure of carotid plaque inflammation in patients. J. Am. Coll. Cardiol. **48**, 1818–1824 (2006)
86. Teutelink, A., Rutten, A., Muhs, B.E., Olree, M., van Herwaarden, J.A., de Vos, A.M., et al.: Pilot study of dynamic cine CT angiography for the evaluation of abdominal aortic aneurysms: implications for endograft treatment. J. Endovasc. Ther. **13**, 139–144 (2006)
87. Teutelink, A., Muhs, B.E., Vincken, K.L., Bartels, L.W., Cornelissen, S.A., van Herwaarden, J.A., Prokop, M., Moll, F.L., Verhagen, H.J.M.: Use of dynamic computed tomography to evaluate pre- and postoperative aortic changes in AAA patients undergoing endovascular aneurysm repair. J. Endovasc. Ther. **14**, 44–49 (2007)
88. Thompson, M.M., Jones, L., Nasim, A., Sayers, R.D., Bell, P.R.F.: Angiogenesis in abdominal aortic aneurysms. Eur. J. Endovasc. Surg. **11**, 464–469 (1995)
89. Tippera, G., U-King-Ima, J.M., Priceb, S.J., et al.: Detection and evaluation of intracranial aneurysms with 16-row multislice CT angiography. Clin. Radiol. **60**, 565–572 (2005)
90. Truijers, M., Pol, J.A., SchultzeKool, L.J., van Sterkenburg, S.M., Fillinger, M.F., Blankensteijn, J.D.: Wall stress analysis in small asymptomatic, symptomatic and ruptured abdominal aortic aneurysms. Eur. J. Vasc. Endovasc. Surg. **33**, 401–407 (2007)
91. Truijers, M., Fütter, J.J., Takahashi, S., Heesakkers, R.A., Blankensteijn, J.D., Barentsz, J.O.: In vivo imaging of the aneurysm wall with MRI and a macrophage-specific contrast agent. Am. J. Roentgenol. **193**, W437–W441 (2009)
92. U.S. Preventive Services Task Force: Screening for abdominal aortic aneurysm: recommendation statement. Ann. Intern. Med. **142**, 198–202 (2005)
93. van der Vaart, M.G., Meerwaidt, R., Slart, R.H.J.A., van Dam, G.M., Tio, R.A., Zeebregts, C.J.: Application of PET/SPECT imaging in vascular disease. Eur. J. Vasc. Endovasc. Surg. **35**, 507–513 (2008)
94. van Herwaarden, J.A., Bartels, L.W., Muhs, B.E., Vincken, K.L., Lindeboom, M.Y.A., Teutelink, A., et al.: Dynamic magnetic resonance angiography of the aneurysm neck: conformational changes during the cardiac cycle with possible consequences for endograft sizing and future design. J. Vasc. Surg. **44**, 22–28 (2006)

95. van Herwaarden, J.A., Muhs, B.E., Vincken, K.L., Van Prehn, J., Teutelink, A., Bartels, L.W., et al.: Aortic compliance following EVAR and the influence of different endografts: determination using dynamic MRA. J. Endovasc. Ther. **13**, 406–414 (2006)
96. van Keulen, J.W., van Prehn, J., Prokop, M., Moll, F.L., van Herwaarden, J.A.: Dynamics of the aorta before and after endovascular aneurysm repair: a systematic review. Eur. J. Vasc. Endovasc. Surg. **38**, 586–596 (2009)
97. van Rooij, W.J., Sprengers, M.E., de Gast, A., Peluso, J.P.P., Sluzewski, M.: 3D rotational angiography: the new gold standard in the detection of additional intracranial aneurysms. Am. J. Neuroradiol. **29**, 976–979 (2008)
98. van 't Veer, M., Buth, J., Merkx, M., Tonino, P., van den Bosch, H., Pijls, N., van de Vosse, F.: Biomechanical properties of abdominal aortic aneurysms assessed by simultaneously measured pressure and volume changes in humans. J. Vasc. Surg. **48**, 1401–1407 (2008)
99. Vos, A.W., Wisselink, W., Marcus, J.T., Manoliu, R.A., Rauwerda, J.A.: Aortic aneurysm pulsatile wall motion imaged by cine MRI: a tool to evaluate efficacy of endovascular aneurysm repair? Eur. J. Vasc. Endovasc. Surg. **23**, 158–161 (2002)
100. Vos, A.W., Wisselink, W., Marcus, J.T., Vahl, A.C., Manoliu, R.A., Rauwerda, J.A.: Cine MRI assessment of aortic aneurysm dynamics before and after endovascular repair. J. Endovasc. Ther. **10**, 433–439 (2003)
101. Wardlaw, J.M., Cannon, J.C.: Color transcranial ''power'' Doppler ultrasound of intracranial aneurysms. J. Neurosurg. **84**, 459–461 (1996)
102. Wardlaw, J.M., Cannon, J., Statham, P.F.X., Price, R.: Does the size of intracranial aneurysms change with intracranial pressure? Observations based on color "power" transcranial Doppler ultrasound. J. Neurosurg. **88**, 846–850 (1998)
103. Weerakkody, R.A., Walsh, S.R., Cousins, C., Goldstone, K.E., Tang, T.Y., Gaunt, M.E.: Radiation exposure during endovascular aneurysm repair. Br. J. Surg. **95**, 699–702 (2008)
104. White, P.M., Wardlaw, J.M., Easton, V.: Can noninvasive imaging accurately depict intracranial aneurysms? A systematic review. Radiology **217**, 361–370 (2000)
105. White, P.M., Wardlaw, J.M., Teasdale, E., Sloss, S., Cannon, J., Easton, V.: Power transcranial Doppler ultrasound in the detection of intracranial aneurysms. Stroke **32**, 1291–1297 (2001)
106. White, P.M., Gilmour, J.N., Weir, N.W., Innes, B., Sellar, R.J.: AngioCT in the management of neurointerventional patients: a prospective, consecutive series with associated dosimetry and resolution data. Neuroradiology **50**, 321–330 (2008)
107. White, P.M., McPherson, R., Sellar, R.J.: The use of CT angiography in acute subarachnoid haemorrhage in Eastern Scotland. Scott. Med. J. **54**, 20–23 (2009)
108. White, P.M., Wardlaw, J.M.: Unruptured aneurysms: detection and management. Invited review. J. Neuroradiol. **30**, 336–350 (2003)
109. Whitaker, S.C.: Imaging of abdominal aortic aneurysm before and after endoluminal stent-graft repair. Eur. J. Radiol. **39**, 3–15 (2001)
110. Wilson, K., Bradbury, A., Whyman, M., Hoskins, P.R., Lee, A., Fowkes, G., McCollum, P., Ruckley, C.V.: Relationship between abdominal aortic aneurysm wall compliance and future risk of rupture: a preliminary analysis. Eur. J. Vasc. Endovasc. Surg. **15**, 478–482 (1998)
111. Wilson, K.A., Hoskins, P.R., Lee, A.J., Fowkes, F.G.R., Ruckley, C.V., Bradbury, A.W.: Ultrasonic measurement of abdominal aortic aneurysm wall compliance: a reproducibility study. J. Vasc. Surg. **31**, 507–513 (2000)
112. Wilson, K.A., Lindholt, J.S., Hoskins, P.R., Heickendorff, L., Vammen, S., Bradbury, A.W.: The relationship between abdominal aortic aneurysm distensibility and serum markers of elastin and collagen metabolism. Eur. J. Vasc. Endovasc. Surg. **21**, 175–178 (2001)
113. Wilson, K.A., Lee, A.J., Lee, A.J., Hoskins, P.R., Fowkes, F.G.R., Ruckley, C.V., Bradbury, A.W.: The relationship between aortic wall distensibility and rupture of infrarenal abdominal aortic aneurysms. J. Vasc. Surg. **37**, 112–117 (2003)

114. Wilson, K., Whyman, M., Hoskins, P.R., Lee, A.J., Bradbury, A.W., Fowkes, F.G.R., Ruckley, C.V.: Relationship between abdominal aortic aneurysm wall compliance, maximum diameter and growth rate. Cardiovasc. Surg. **7**, 208-213, (1999)
115. Wilson, W.R.W., Anderton, M., Schwalbe, E.C., et al.: Matrix metalloproteinases-8 and -9 are increased at the site of AAA rupture. Circulation **113**, 438–445 (2006)
116. Wintermark, M., Ko, N.U., Smith, W.S., Liu, S., Higashida, R.T., Dillon, W.P.: Vasospasm after subarachnoid hemorrhage: utility of perfusion CT and CT angiography on diagnosis and management. Am. J. Neuroradiol. **27**, 26–34 (2006)
117. Womersley, J.R.: Method for the calculation of velocity, rate of flow and viscous drag in arteries when the pressure gradient is known. J. Physiol. Lond. **127**, 553–563 (1955)
118. Xu, X.Y., Borghi, A., Nchimi, A., Leung, J., Gomez, P., Cheng, Z., Defraigne, J.O., Sakalihasan, N.: High levels of 18F-FDG uptake in aortic aneurysm wall are associated with high wall stress. Eur. J. Vasc. Endovasc. Surg. **39**, 295–301 (2010)
119. Yaghmai, V., Rohany, M., Shaibani, A., Huber, M., Soud, H., Russell, E.J., Walker, M.T.: Pulsatility imaging of saccular aneurysm model by 64-slice CT with dynamic multiscan technique. J. Vasc. Interv. Radiol. **18**, 785–788 (2007)

Biomechanics and Pathobiology of Aortic Aneurysms

Julie A. Phillippi, Salvatore Pasta and David A. Vorp

Abstract Biomechanical weakening of the aorta leads to aneurysm formation and/or dissection and total biomechanical failure results in rupture, which is often fatal. The most common aneurysm is the abdominal aortic aneurysm (AAA) whereas thoracic aortic aneurysms (TAA) involve the ascending or descending segments of the aorta. Biomechanical strength of the aorta is maintained in part via balance between the integrity of the aortic medial and adventitial extracellular matrix and the health of the mural cells. From a biomechanical perspective, aneurysms rupture or dissect when wall stresses locally exceed the wall strength. Pathobiologic mechanisms, pre-disposing disorders and variability of patient demographic characteristics can weaken the aortic wall while increased blood pressure and dilatation increase the stress acting on it, leading to further aneurysm expansion. Thoracic and abdominal aortic aneurysms arise from very different pathophysiologies that ultimately result in a final common outcome of matrix degeneration and biomechanical failure. Therefore, the patient-specific knowledge of both wall stress and wall strength distributions for a given aneurysm will greatly improve the ability to identify those aortic aneurysms that are at highest risk of rupture. Towards this end, the biomechanics of AAA has been studied extensively by many groups whereas TAA biomechanics has not been fully considered. This chapter articulates the state-of-the-art

J. A. Phillippi (✉)
Department of Cardiothoracic Surgery, University of Pittsburgh, Pittsburgh, Pennsylvania, USA
e-mail: phillippija@upmc.edu

S. Pasta
Department of Bioengineering, University of Pittsburgh, Pittsburgh, Pennsylvania, USA
e-mail: pastas@upmc.edu

D. A. Vorp
Departments of Bioengineering and Cardiothoracic Surgery, University of Pittsburgh, Pittsburgh, Pennsylvania, USA
e-mail: vorpda@upmc.edu

Stud Mechanobiol Tissue Eng Biomater (2011) 7: 67–118
DOI: 10.1007/8415_2011_84

Published Online: 16 June 2011

of aortic biomechanics, including the modeling of tensile strength and wall stress distributions and the biological mechanisms which influence them. The potential clinical utility of these biomechanical estimates in predicting AAA rupture is also discussed.

1 Introduction

Aortic aneurysm is a consequence of biomechanical weakening of the vessel that leads to a local enlargement of the aorta over years but can rupture suddenly, often without warning. Thus, aortic rupture is a catastrophic failure of the vessel that is responsible for high mortality and morbidity. The most common aneurysm is the abdominal aortic aneurysm (AAA) whereas thoracic aortic aneurysms (TAA) involve the ascending or descending segments of the aorta. Dissection is another pathology in which the aortic wall dissects, thereby permitting the blood to enter the wall and causing enlargement. Biomechanical strength of the aorta is maintained in part via balance between the integrity of the aortic medial and adventitial extracellular matrix (ECM) and the health of the mural cells; namely, the medial smooth muscle cells and adventitial fibroblasts. This state of balance is known as tissue homeostasis and is essential for proper functioning of the aorta.

The diagnosis of an aortic aneurysm occurs principally via radiologic imaging of the aorta. Current clinical practice dictates that replacement of the aneurysmal aorta should be performed when the aneurysm size reaches 5.0–5.5 cm in maximum diameter for both TAAs [1] and AAAs [2–4]. However, rupture or dissection can occur at diameters <5.0 mm in 0–23% cases (Table 1) [5–9]. This wide discrepancy in rupture or dissection occurrence may be due to variability in the pathology of the aortic disease, which can be influenced by patient demographics, co-morbidities or pre-disposing conditions such as connective tissue disorders. Furthermore, studies reporting rates of rupture are limited by lack of information regarding the diagnosis criteria and post-mortem studies [7, 10] and are likely unreliable due to diameter measurement discrepancies [5]. Clearly, a patient-specific approach would be desirable for more reliable rupture or dissection risk predictions to improve clinical management, and tailoring a prognosis and treatment course unique to the patient.

From a biomechanical perspective, aneurysms rupture or dissect when wall stresses locally exceeds the wall strength. Pathobiologic mechanisms, pre-disposing disorders and variability of patient demographic characteristics tend to weaken the aneurysm wall [11–13] while increased blood pressure and aneurysmal enlargement increase the stress acting on it [14, 15], leading to further aneurysmal expansion. This cycle continues as the aneurysm enlarges eventually ruptures or dissects, unless repaired or the patient succumbs to other causes.

The earliest predictions of wall stress used the law of Laplace (Eq. 1) as the theoretical basis for using the "maximum diameter" as the criterion for predicting AAA rupture potential. This law states that wall tension (T) increases in proportion

Table 1 Adapted from Darling et al. [7]; Vorp and Vande Geest [236]

Size, cm	N ruptured	N unruptured	Total	% Ruptured
≤5.0	34	231	265	12.8
>5.0	78	116	194	40.0
No size recorded	6	8	14	43.0
Total	118	355	473	24.9

to an increase in the intraluminal pressure (P), an increase in diameter D(R) or a decrease in wall thickness (t).

$$T = (P \times D)/2t) \tag{1}$$

The law of Laplace has two major fallacies for its application in biomechanics. First, the aorta is not a simple cylindrical or spherical shape of uniform radius of curvature for which the law of Laplace is valid. Rather, the geometry of the aortic wall is complex, containing a greater and lesser curvature, and these vary spatially [16, 17]. Therefore, the maximum diameter criterion to predict rupture risk ignores the important roles of vessel wall topography and geometry. This notion is supported by studies which demonstrated that stress on the wall is dependent on wall shape unique to each aneurysm [15, 16, 18]. Secondly, rupture predictions based on only the wall stress are not sufficient because a region of the aneurysmal wall that is under elevated wall stress may also have higher wall strength, thus equalizing its rupture potential. The wall strength will have spatial variance within a given aneurysm and also from patient-to-patient [19–21]. The rupture potential is highest in the location of peak ratio of wall stress to wall strength. However, risk analysis for rupture based upon the maximum diameter criterion is improved when a measurement of peak wall stress is also considered [22–24]. Ruptured AAAs demonstrated higher peak stress compared to diameter-matched electively repaired AAAs [12]. Clearly, the ability to noninvasively predict the locally acting wall stress and wall strength of an aortic aneurysm on a patient-specific basis will provide a more accurate diagnostic tool for evaluating the high risk of failure of an aortic aneurysm. Preliminary efforts toward this end have been reported [18, 25–28].

Although the in vivo wall strength distribution of an aneurysm can only be estimated [29], biomechanical experimental testing has been performed to assess ex vivo the biomechanical behavior of the aneurysmal and non-aneurysmal human aorta. Several studies have found that aneurysmal tissue is substantially weaker and stiffer than normal aorta [11, 12, 26, 29–31]. The aorta displays anisotropic biomechanical strength that is spatially dependent in both ascending [32–34] and abdominal [20, 31] segments. For both aneurysmal and non-aneurysmal aorta, the longitudinal segments were stiffer than segments measured in the circumferential orientation (Fig. 1) [32, 33]. Perhaps the most important piece of biomechanical data is the finding that tensile strength does not appear to be related to aneurysm diameter [33]. Hence, aortic diameter is not the only determinant of either wall strength or wall stress [15], and rupture risk prediction criteria that account solely for aortic diameters do not offer a complete picture of the true risk for

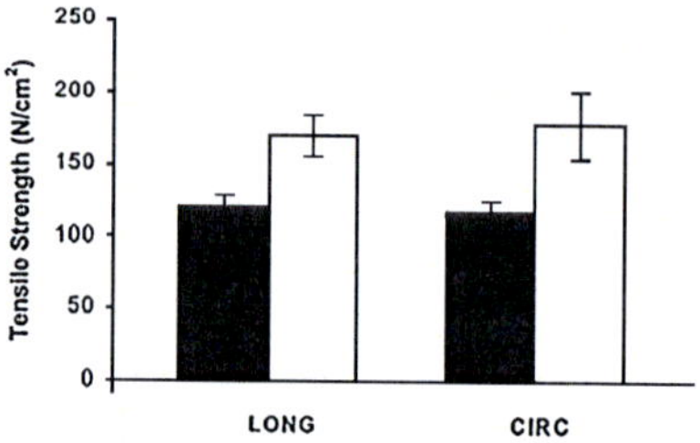

Fig. 1 Tensile strength of human thoracic aortic aneurysms (*black bars*) in the longitudinal (LONG) and circumferential (CIRC) orientations and non-aneurysmal controls (*white bars*). Bars present mean ±SEM. (Adapted from Fig. 2 of Vorp et al. [33])

aortic failure. Other criteria have been proposed for rupture prediction models such as rate of aneurysm growth [3, 35, 36], stiffness [37] intraluminal thrombus (ILT) thickness in AAA [38], wall tension [8] and peak AAA wall stress [22–24]. However, all these were empirical approaches and have their limitations and could potentially lead to sometimes fatal errors in the decision pertaining to clinical management of AAA and TAA. Using a concept of the Rupture Potential Index (RPI) proposed and preliminarily applied by Vorp et al. [26–28] demonstrated for AAA that this new index may hold promise for a patient-specific diagnostic assessment. The RPI, which is the ratio of acting wall stress to the wall strength at a particular point on the aneurysm wall, was found to be higher in AAA that ruptured compared to those that did not rupture. Currently, there is only one model to assess the RPI of AAAs that has been used recently [28, 39]. Indeed, significant differentiation of rupture risk by RPI was found for aneurysm diameter ranging between 55 and 75 mm of asymptomatic and symptomatic/ruptured AAAs where the maximum diameter criterion failed to adequately predict risk [28].

Thoracic and abdominal aortic aneurysms are thought to arise from different, very distinct pathophysiologies. However, both pathologies culminate to a final common pathway of matrix degeneration and biomechanical failure. Aneurysms of the thoracic aorta are generally non-inflammatory in nature and are related to connective tissue disorders such as Marfan, Ehlers-Danlos or Loeys-Dietz syndrome, the presence of a congenital defect such as the bicuspid aortic valve (BAV) or other familial aortic disease. TAAs and dissections can also arise from non-genetic pathologies including Takayasu's arteritis [40], giant cell arteritis [41], or tuberculous mycotic infection [42] and these conditions involve an inflammatory component. Aneurysms arising from these causes are less common than those of a genetic basis.

Matrix metalloproteinases (MMPs) are over-expressed in aortic aneurysms [43–46] specifically in TAAs [47–50] MMPs control degradation of elastin and collagen [51] and rupture or dissection occurs when the physiologic forces exerted on the vessel wall exceed its strength. Although AAAs exhibit marked inflammatory changes and associated atherosclerosis [52–59] there is a lack of evidence of an inflammatory component in TAAs [60–62]. This non-inflammatory phenomena is uniformly consistent with aortopathies of the ascending aorta, of which, key effector proteins inciting the phenotype have been identified for the Marfan (*fibrillin-1*) [63], Loeys-Dietz (*TGF-βR*) [64], and Ehlers-Danlos-vascular type

(*Type III collagen*) [65] syndromes and familial forms of TAAs, including mutations in genes that regulate SMC contractility (α-smooth muscle actin (SMA) (*ACTA2*) and smooth muscle myosin heavy chain (*SM-MHC*) [66, 67]). The presence of a BAV also predisposes the patient to development of TAA and/or dissection in addition to aortic stenosis or insufficiency. [44, 45, 68–72]. The incidence of BAV overlaps with *α-SMA* and *TGF-βR* mutations at a frequency of 2.5–3% [73] but no gene or pathway responsible for inciting TAAs in BAV patients has been identified.

Common histopathological evidence for many incidences of TAAs include non-inflammatory smooth muscle cell loss, fragmentation of the elastin fibers, mucoid degeneration, and accumulation of proteoglycans within areas of cell depletion. These phenomena are collectively termed "cystic medial degeneration" or CMD and are observed within the medial layer of the aortic wall [45, 48, 60–62, 74, 75]. Development of AAAs often includes a contribution from inflammatory cells as increased MMP production by macrophages and occurrence of ILT which are generally absent in TAAs. The differing pathobiologies involved in aneurysm formation in the thoracic and abdominal aortas likely results in variations in the mechanical stresses and wall weakening associated with each type of aneurysm with patient-to-patient nuances. Hence, the clinical recommendation for patient management will depend upon understanding, with patient specificity, the role of biomechanics in diagnosis, counseling, monitoring and treating patients at risk for and presenting with, aortic aneurysm and determining their relative risk for rupture.

This chapter will detail the relative biological mechanisms that contribute to aneurysm formation in both the thoracic aorta and abdominal aorta and their role in vessel wall biomechanics, which ultimately leads to aneurysm formation and potential rupture of the aorta through weakened strength and increased stress.

2 Biological Mechanisms that Impact Mechanical Properties of the Aneurysmal Thoracic Aorta

2.1 Marfan Syndrome

Marfan syndrome (MFS) is a hereditary connective tissue disorder that is autosomal dominant and manifests in several organs including those of the cardiovascular, ocular and skeletal systems. Patients with MFS display a definitive body habitus, namely a tall stature and wing span that exceeds the height, arachnodactyle, craniofacial defects, pectus excavatum or carinatum. Ectopia lentis may occur in up to 80% of patients, usually bilaterally. MFS has widespread clinical presentation but the most dangerous and often lethal manifestation of the disorder is acute aortic dissection.

Aortic dissection often occurs after a period of progressive aortic dilation and involves weakening of the vessel wall. In MFS, this weakening results at least in part due to degeneration of the minute architecture of the extracellular matrix (ECM). In 1991, mutations in the fibrillin-1 (fbn1) were identified as the cause of MFS [63]. Fibrillins are ECM proteins that comprise microfibrils, which are the key connectors between elastin and collagen to local cells in the microenvironment. These microfibrils are essential for not only the microarchitecture of the tissue, but also provide substantial biomechanical support. Similar to fibrillins, a related protein family, the latent TGF-β-binding proteins (LTBPs) participate in both structure of the matrix and also function as a signaling molecules to direct cell behavior. LTBPs directly interact with TGF-β to sequester the growth factor within the tissue microenvironment and thus regulate the bioavailability. When bound to LTBPS, TGF-β isoforms cannot interact with their receptors and therefore, signaling activity is attenuated. LTBPs retain TGFs within the matrix, keeping them at the ready to be released in their active state and capable of receptor binding. Fibrillins are similar in structure to LTBPs. Both are cysteine-rich with multiple epidermal growth factor (EGF)-like domains that bind calcium [76]. An 8-cysteine module, which is distinct only to fibrillins and LTBPs may be critical for microfibril assembly [77] and also be responsible for interactions with TGF-β [78–79].

Fibrillins interact with TGF-βs as the microfibrils of the ECM. These microfibrils self-assemble and interact with integrins on the cell surface and other matrix molecules within the microenvironment. Fibrillins can be viewed as a physical tether between local cells and their matrix and contain the known integrin-interacting sequence RGD to support cell adhesion [80–82]. The amino acid sequence RGD serves as code for cell adhesion to matrix molecules and provides a means for direct cell–matrix interactions. Such interactions are essential for integrity of the matrix and the biomechanical strength of the tissue. In the case of the aorta, the elastic laminae are connected to the endothelial cells on the intimal side and the SMCs of the medial layer. Fibrillin-based microfibrils provide this connectivity between matrix and cells which contributes to stability of the tissue. Mice lacking these connections displayed increased elastin degradation and alterations in SMC phenotype in an apparent attempted repair mechanism to sustain loss of matrix. These defenses also include an increase in the MMPs that are induced as part of the remodeling process, but unfortunately mediate further degeneration of the matrix and contribute to weakening of the vessel. When cells are no longer physically linked to their matrix, they cannot sense the ongoing needs of the matrix for homeostasis. Cell–matrix interactions are a constant dynamic dialogue and exchange of information. Loss of cell–matrix connections is itself a signal to rebuild because the cell senses a matrix deficit. However, loss of cell-ECM connections is also a trigger for apoptosis which is a self-initiated program of induced cell death. SMC apoptosis is a characteristic of aortic diseases and vessel weakening and in the case of MFS, may stem from improperly formed fibrillin-based microfibrils.

The fibrillin microfibrils elicit their structural contribution to matrix homeostasis by providing a template for tropoelastin deposition. The microfibril network provides elasticity in elastin-based and non-elastin tissues. Normal organization of

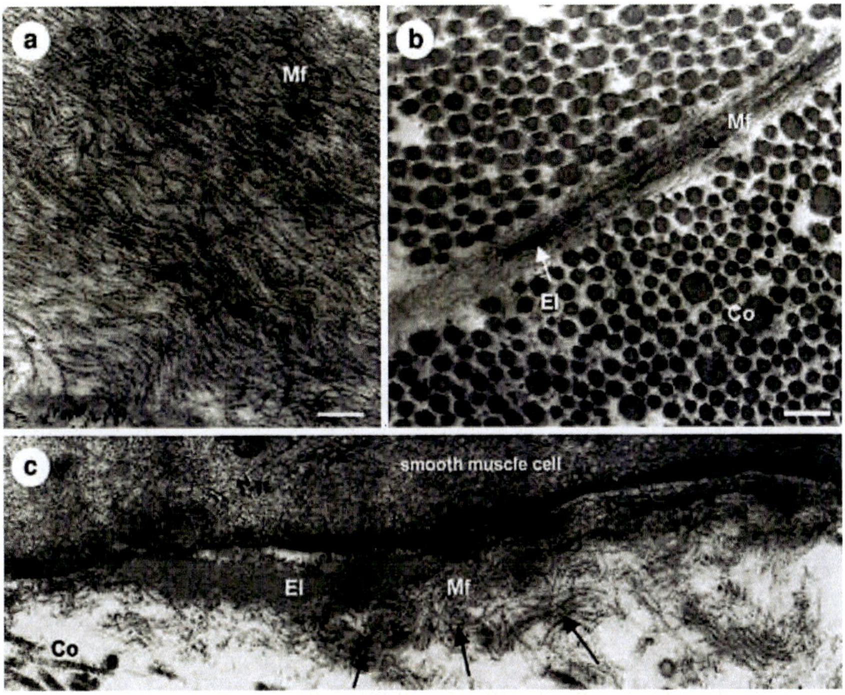

Fig. 2 Fibrillin microfibrils in (**a**) lobster aorta and mouse aorta (**b**, **c**). Mf=microfibrillar mantle; El=elastin; Co=collagen fibrils. Scale bar is 400 μm. (Reproduced from Fig. 1, Kielty et al. [83])

the microfibrils involves self-assembly of the fibrillin in moderately tight bundles of fibers that are oriented parallel to each other. The strength of the microfibrils is spatially dictated and controlled not only by the cells within the tissue, but also the forces upon which the tissue is subjected to. For this reason, microfibrils are often found within tissues that need to withstand considerable mechanical stress such as bone, and blood vessel, serving as anchors for various connective tissues (Fig. 2) [83]. In the MFS patient, mutations in fibrillin result in reductions in synthesis and deposition of fibrillin within the ECM and improper assembly of the microfibril bundles. These defects in fibrillin expression and organization are likely to impact the dual function of fibrillin as a regulator of TGF-β signaling and a biomechanical support for the aorta.

There are over 600 mutations of the Fbn1 gene. The majority of the mutations are point mutations, with some being nonsense or missense mutations. Substitutions of cysteine residues that alter the disulfide bonds within the EGF or 8-Cys modules are the most frequent mutations, while other common mutations involve the calcium binding region. These alterations affect the organization of the fibrillin into microfibril bundles and substantially impact their ability to bind TGF-β, interact with the cell surface and orient properly to guide elastin deposition in the matrix. These shortcomings will ultimately contribute to reduced and possible failure of vessel biomechanical strength and result in a lethal rupture of the ascending aorta in MFS patients. A mouse model of MFS, the Fbn1 knockout support the theory that fibrillin-rich microfibrils significantly contribute to the overall biomechanical strength of the aorta. Fbn1-/- mice die within one year of birth due to aortic dissection and rupture. Although elastin fibers appear normal at birth, aneurysm initiates within 2 weeks of birth and continues to enlarge with time [84]. There is elastin degradation and disorganization along with considerable loss of fibrillin-1 [85]. SMCs adopt a synthetic phenotype in attempt to repair the insufficient ECM. These phenomena progress to elastin breakdown initiating from the adventitia and moving inward to the media. Aortic dilatation, aneurysm, dissection and rupture then coalesce. Microfibrils work together with elastin to provide biomechanical support to the aorta [86]. Fibrillin deficiency is related to increased aortic stiffness and decreased compliance, possibly due to a compensatory role for collagen to endure the bulk of load bearing in the absence of adequate fibrillin and elastin. Fibrillin works together with elastin to regulate the pressure-diameter relationship of the aortic wall. The biomechanical function of fibrillin-1 microfibrils is dependent upon its proper expression, assembly and incorporation. Details of the molecular assembly of fibrillin 1 into complex microfibrils are not completely understood. However, recombinant approaches have allowed the investigation of various domains on fibrillin molecular assembly into the higher-order microfibrils. Inter-molecular cross-linking is known to be driven by tissue transglutaminase to form dimers with disulfide bonds and furin processing appears to accelerate but is not essential for this event. However, it remains to be seen if these biochemical manipulations occur in vivo to maintain aortic biomechanics. The vital role of fibrillin-1 in aortic homeostasis targets this protein as a focus of extensive research for eventual therapeutic intervention for the care of MFS patients.

Mutations in fibrillin that impair its TGF-β-binding role lead to increases in free TGF-β and excessive signaling of this pathway. This led investigators to hypothesize that TGF-β played a role in aortic dilation in MFS patients. TGF-β neutralizing antibodies prevent elastin breakdown in fbn-/- mice but such a therapy lacks clinical feasibility. Losartan, an angiotensin (Ang) II inhibitor also antagonizes TGF-β signaling. As a currently FDA-approved drug, losartan is widely successful medication used to treat hypertension. While blood pressure control is commonly sought in patients with MFS and aortic aneurysm, a means to attenuate or reverse matrix damage in the vessel wall and maintenance of biomechanical stability of the aorta to prevent rupture is necessary. Treatment of fbn-/- mice with losartan effectively prevented matrix degradation in the ascending aorta and mice were phenotypically and histologically indistinguishable from wild-type mice [85]. This treatment was also effective in preventing aortic root dilatation in children [87]. Losartan may prevent dilation through two mechanisms: (1) decrease in the absolute blood pressure or rate of change in blood pressure reduces the effect of hemodynamics on the already compromised aortic wall due to fibrillin mutations and/or (2) inhibition of TGF-β signaling through regulation of the Ang II type I receptor (AT1). How TGF-β signaling regulates matrix homeostasis remains unclear but may be due to altered SMC phenotype and synthesis of new matrix proteins in effort to remodel a matrix damaged by degradation. Furthermore, since TGF-βs are important for growth and development, it will be important to include tissue targeting strategies in any therapeutic drug design in order to limit the potential deleterious side effects of TGF-β on normal development and renal function [87, 88].

MFS patients present clinically with manifestations of this connective tissue disorder in several organ systems. A multitude of mutations in fibrillin have been identified that may contribute to the diseases. However, fundamentally, the microfibrils of the ECM are composed of fibrillin and when improperly synthesized, assembled or maintained, alterations in biomechanical stability and growth factor signaling through TGF-β are significantly altered. These mechanisms contribute to degradation of the ECM through elastolysis and SMC loss and culminate in reduced biomechanical strength, progressive weakening of the vessel wall, aortic aneurysm, dissection and the highly-fatal rupture. However, to our knowledge, no ex vivo mechanical testing has been conducted on the ascending aorta of the MFS patient population. There have been some reports that used noninvasive ultrasonography to demonstrate that MFS patients have increased stiffness of the infrarenal aorta and decreased strain, despite normal aortic diameters [89]. Ex vivo mechanical testing has been limited to mouse models of MFS. Increased stiffness was reported in carotid arteries and aortae of fbn1-/- mice and elastin and fibrillin were shown to provide the primary biomechanical strength of the tissue [86, 90]. Efforts to further understand the role of these biological mechanisms on the individual efforts of each matrix molecule on the overall biomechanical strength of the vessel itself should be pursued in depth in order to develop better treatments for patients with MFS.

2.2 *Loeys-Dietz Syndrome*

Loeys-Dietz syndrome (LDS) is an autosomal dominant disorder linked to mutations in the TGF-β receptors (TGFBR1 and TGFBR2) [64]. In addition to a pre-disposition for TAA, LDS patients exhibit other manifestations of this connective tissue disorder such as craniosynostosis, bifid uvula, cleft palate and transluscent skin [91]. The histopathologic observations of the aorta in LDS patients are consistent with those of other TAAs including CMD, increased collagen deposition, elastin fragmentation and lack of an inflammatory response [92]. Interestingly, the mutations in TGFBRs result in excessive TGF-β signaling rather than inhibited signaling. This leads to effects similar to those observed for MFS and underscores the importance of TGF-βs in regulating vascular tissue homeostasis. In LDS patients, contrary to expectation, increased collagen deposition was observed, accompanied by elevated connective tissue growth factor (CTGF) expression and nuclear Smad2 localization [64, 91] which are all clear indicators of enhanced TGF-β signaling. CTGF is directly regulated by TGF-β via the Smad2

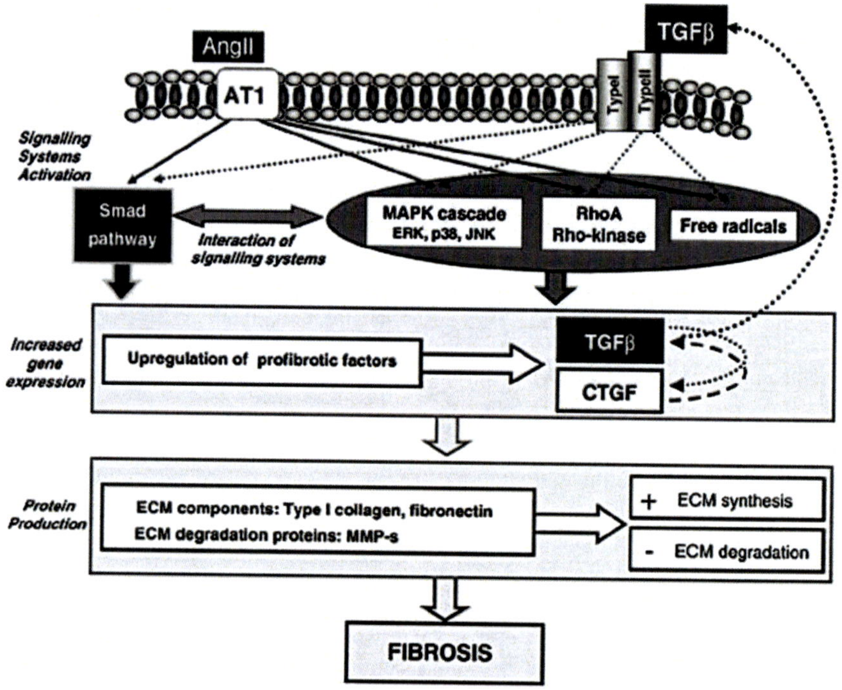

Fig. 3 Pathways of vascular fibrosis involving Ang II and TGF-β. (Figure 3 from Ruiz-Ortega et al. [94])

pathway and promotes proliferation and ECM production by fibroblasts and SMCs. The up-regulation of CTGF could propagate excessive TGF-β signaling via attenuation of the inhibitory Smad7, a key component of the TGF-β feedback loop [93]. CTGF has also been shown to regulate TGF-β responses and inhibition of CTGF can reduce TGF-β-induced collagen synthesis, while elevated levels of CTGF directly contribute to ECM accumulation (Fig. 3) [94]. As tissue homeostasis relies on a balance between organized matrix deposition and matrix degradation, TGF-β also elicits control over the remodeling process by activating inhibitors of tissue proteases such as plasminogen activator inhibitor-1 (PAI-1) and tissue inhibitors of the MMPs (TIMPs). These observations are counter-intuitive since aneurysm development is well known to include over-production of the ECM degenerative MMPs and calls into question the relative role of MMPs as instigators of aneurysm formation. Perhaps the causal agents lie upstream in the signaling cascade and initiate a faulty remodeling process which results in an improperly formed ECM, lacking sufficient biomechanical integrity. MMPs are essential for normal remodeling, evidenced in other tissues such as the skeletal system where ECM fragments are stimulators of new, organized matrix deposition. These apparently synergistic effects of TGF-β and CTGF create a fibrosis-like state of the aortic ECM in LDS patients, where biomechanical integrity is compromised. It has been well-established that both TGF-β and CTGF are potent fibrotic agents in many tissues; however the challenge that remains is devising an intervention at the molecular level to prevent these signaling events that cause impaired collagen deposition. CTGF, as a down-stream modulator of TGF-β-mediated fibrosis is an attractive target for therapeutic intervention to prevent collagen accumulation and matrix disorganization.

Fibrosis occurs as a result of tissue defense to an assault or injury and represents a poorly repaired tissue, lacking adequate biomechanical properties. Often there is excessive insoluble collagen with improperly formed cross-links or poorly organized into fibrils. More collagen does not translate to increased strength. However, unlike fibrosis, this accumulated collagen in TAA patients occurs in the absence of an inflammatory response. Contrary to MFS, there have been no studies measuring the biomechanical strength of the aorta in LDS patients. However, given the disruption of the TGF-β signaling cascade, it is reasonable to hypothesize, that dis-regulation of collagen synthesis and factors that regulate matrix remodeling such as MMPs and inhibitors of tissue proteases will significantly impact the biomechanical integrity of the matrix and the vessel wall as a whole. The direct contributions of each pathway or mechanism on the biomechanical strength will likely be determined through a combination of animal models and targeted interrogation of particular molecules in the signaling cascade, complemented by analysis of the biomechanical properties of the aneurysmal aorta. Within any given tissue niche, the cell senses the needs of the matrix and responds in kind to remodel, repair, or regenerate and disruptions upstream can have devastating effects that contribute to the pathogenesis of the connective tissues. To avoid surgical intervention, medical therapies should target correction of the upstream events (i.e., TGF-β signaling) to prevent the deleterious effects of an improperly organized and maintained ECM on aortic biomechanics.

In the LDS patient, medical intervention with Ang II receptor inhibitors to combat the effects of increased TGF-β signaling in LDS patients may show similar to promise as treated MFS patients, but to date, no conclusive efficacy studies have been performed.

2.3 *Ehlers-Danlos Syndrome*

In patients with Ehlers-Danlos syndrome (EDS), the predilection for vascular abnormalities is rupture or dissection of the first-order branch arteries but aortic dissection and rupture do also occur [70]. EDS patients show clinical manifestation of the disorder as hyperextensibility of the skin which is thin and translucent, hypermobility of joints, extensive bruising, rupture of the arteries, intestines and uterus. The leading cause of death for EDS patients is rupture or dissection of thoracic or abdominal medium-large vessels and this often occurs when the patient age is in the fifties. Clinical management is very difficult in these patients because the vascular tissue is extremely fragile and is not easily sutured. Therefore, surgical intervention often leads to further complications and increases the risk for aortic catastrophe [70]. Expert opinion promotes regular monitoring of EDS patients by CT or MR angiography of the vasculature but there are no well-established guidelines for elective intervention [70]. The EDS patient is extremely difficult to treat due to the high risk of operative treatment and the lack of understanding of the appropriate mechanism to target for medical intervention.

Of the at least 12 known types of EDS with various genetic bases, there are 6 types that are associated with thoracic aortic disease [95]. Over 70 mutations have been linked to EDS and patients displaying the vascular effects of EDS (type IV) are known to have a mutation in the gene encoding type III procollagen [65, 96]. Mutations in type V alpha 1 procollagen have also been reported for EDS patients with arterial rupture [97].

Type III collagen, along with Type I collagen is a predominant form of collagen in the aortic media and is synthesized by aortic SMCs [98, 99]. The effects of Col3a1 mutation has been difficult to analyze due to lack of a successful animal model. Mutations achieved through targeted ablation of Col3a1 showed aortic dissection and bruising of the skin, however only 5% of homozygous mice survived at weaning and most died within 48 h of birth. Recently, a new mouse model with spontaneous mutation of the Col3a1 gene was identified serendipitously whilst investigating the role of Mro in sexual development [100]. These mice died of aortic dissection of the descending thoracic and abdominal aorta with no evidence of inflammation. While there was no difference in overall collagen gene expression, histological analysis of collagen deposition showed an overall decrease in collagen content in the aortic adventitia. Transmission electron microscopy displayed disrupted collagen as well as elastin fibers (Fig. 4). The investigators also provided evidence of normal TGF-β signaling as well as lack of hypertension in col3a1-mutated mice. This model should prove useful in future investigations of the role of Type III collagen in the microarchitecture of the aorta. Since the adventitia has long been recognized as providing substantial biomechanical support for the aorta, disruption in Col31a may abrogate this role of the adventitia in EDS IV patients. Improperly formed collagen fibrils probably provide less strength against hemodynamic forces and lead to dissection, rather than aneurysm. The mouse model described by Smith et al. [100] also brings to light a putative role for Type III collagen in maintaining the layers of SMCs and the collagen/elastin matrix within the aorta. Evidence of shearing between layers was observed with infiltration of red blood cells. This is primary evidence for impending dissection. Therefore, the biomechanical contribution of Type III collagen may be to protect inter-layer strength and mutations in this key ECM protein may place the patient at risk for rupture by aortic dissection. While Type I collagen is thought to provide most of the tensile strength of the aorta and organize as thicker fibers, Type III collagen is found as thinner fibers in close association with elastin where it can participate in recoil capabilities [101]. A point mutation was identified in a patient with EDS IV, located in one of the Gly-X-Y backbone sequences of the alpha chain [102]. The mutation causes the Gly to become a Glu and given that Gly is required for the triple-helical structure of the collagen fiber, this mutation

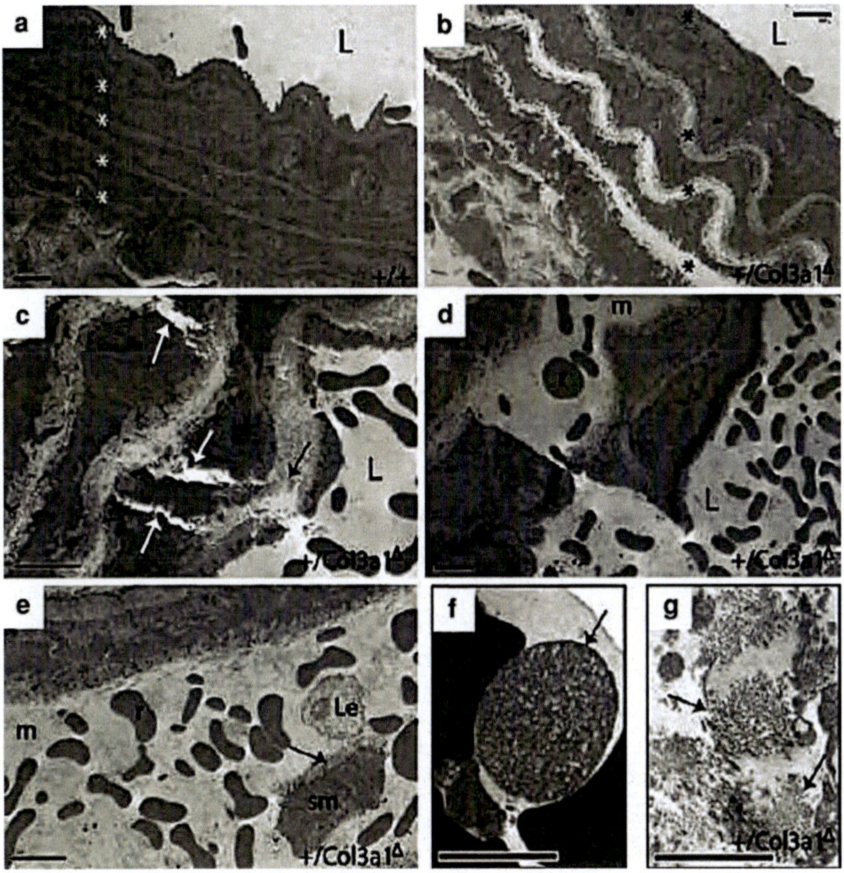

Fig. 4 Transmission electron microscopy of aortas from +/*Col3a1*$^{\Delta}$ mice display abnormalities in matrix architecture. **a** Normal +/+ mice show regular continuous elastin (*) while there is reduced density of elastin (*) in +/*Col3a1*$^{\Delta}$ mice (**b**). The elastin lamellae is discontinuous (*black arrows*) and dissection through the media layer is also noted in +/*Col3a1*$^{\Delta}$ mice (*white arrows*) (**c**). There is evidence of erythrocyte infiltration within the medial layer in (**d**) and (**e**) and the elastin lamellae (*arrow*). Collagen fiber abnormalities are evident as shown (**f**) and (**g**) (*arrows*). L=lumen; m=media; r=erythrocytes; sm=smooth muscle; Le=leukocyte. Scale bar=5 μm for **a**–**e** and 2 μm for **f**, **g**. (Figure 4 from Smith et al. [100])

de-stabilizes the collagen within the aorta and likely compromises its biomechanical integrity.

Collagen fiber strength is dependent on its cross-links formed between adjacent helical domains. The mechanical properties can be altered when cross-linking is modified either chemically or physically. The enzyme lysyl oxidase (Lox) catalyzes formation of lysine or hydroxylysine cross-links. Lox expression parallels that of Type I and III collagens [103–105]. Enzymatic activity of Lox is dependent on copper for the transfer of electrons to and from oxygen for delamination of peptidyl lysyl groups on collagen and elastin [106]. Furthermore, variations in

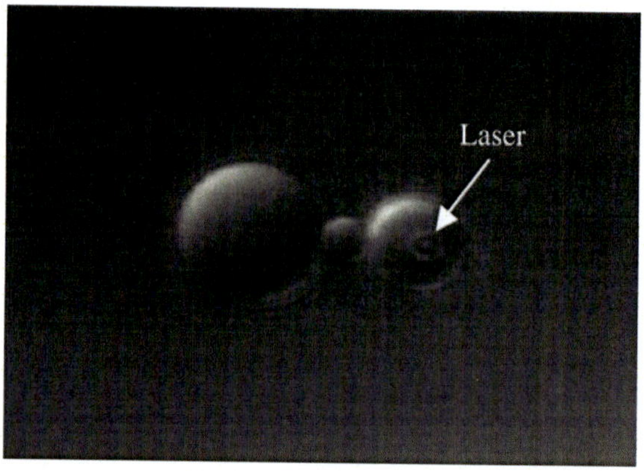

Fig. 5 Demonstrate of using optical tweezers to measure force of stretching a single collagen molecular. Such technology will be useful for characterizing the biomechanical properties of collagen fibrils. (Figure 1 from Sun et al. [110])

copper metabolism have long been linked to EDS type IX which has clinical manifestations including bladder diverticula, skin hyperelasticity and skeletal abnormalities [107]. In patients with impaired copper metabolism, a reduction in Lox expression and decreased collagen cross-linking in cultured skin fibroblasts was reported. Inactivation of Lox has also been linked to aortic aneurysm formation and death in a mouse model [108]. Mutations in Type III collagen that alter collagen fiber assembly may also negatively impact the cross-linking by Lox by directly altering the sites for cross-link formation which are critical to the biomechanical function of the collagen fibers.

There is also an established link between EDS patients and mutations in TGFBRs, accompanied by elevated TGF-β signaling, similar to LDS patients. It is also thought that reduced expression of Type III collagen may be compensated for by enhanced TGF-β in an ineffective attempt to rebuild a deficient matrix. It has also been proposed that cellular response to TGF-β within the aortic wall differ by cell type as reviewed recently by Jones et al. [109]. During the process of aneurysm formation or other forms of vascular injury, the relative proportion of SMCs, fibroblasts and myofibroblasts changes and the phenotype of SMCs is also altered toward the undifferentiated, synthetic state. It is important to understand the molecular nuances of TGF-β signaling in the normal and aneurysmal aorta in order to ascertain the role of these pathways on the overall tissue homeostasis. Balanced interplay between cells and matrix ultimately drives maintenance of adequate biomechanical strength to support the efforts of the heart to deliver blood throughout all body tissues. There are clues in the biochemical pathways, but the real knowledge of tissue homeostasis lies with biomechanical analysis of the tissue itself.

As with MFS and LDS patients, to date, the biomechanical strength of ascending aorta from EDS patients has not been investigated using ex vivo approaches. The work of Sun et al. [110] using optical tweezers offers another strategy to quantify the strength of individual collagen fibers from native tissue or those produced by culture cells in vitro (Fig. 5). These analyses could allow the interrogation of mutated forms of collagen produced recombinantly to determine the role of particular collagens and collagen structure on the biomechanical strength of the aorta and the causal role of mutations in EDS. Understanding the metabolism of collagen cross-linking and the impact of improper collagen assembly on the overall biomechanical strength of the aorta will improve the clinical management and treatment of patients with EDS who are at increased risk for complications during surgical intervention. Strategies to derive medicinal therapies for EDS patients might include focus on targeted repair of collagen cross-linking and/or assembly to improve the biomechanical strength of the aorta and prevent dissection or rupture in these patients.

2.4 Bicuspid Aortic Valve

BAV is the most common cardiac malformation and occurs in 1–2% of the general population with a gender bias of 2:1, males versus females [72]. The presence of a BAV pre-disposes the patient to formation of TAA and aortic dissection and usually requires careful surveillance and often surgical intervention for the associated co-morbidities of aortic insufficiency or stenosis in addition to ascending aortic replacement due to TAA or dissection. These phenomena that transpire in the BAV patients arise from an unknown pathway that is thought to stem from effects of altered hemodynamics through the BAV and/or due to an inherent defect in the wall of the ascending aorta which contributes to loss of biomechanical strength and leads to dilatation and aneurysm formation. The causal pathway of aneurysm formation in BAV patients appears to be distinct from patients with a morphologically normal aortic valve, known as tricuspid aortic valve (TAV). Within our clinical database, we confirm a remarkable bimodal age distribution for TAAs with a mean age of 51 for patients with BAV compared to a mean age of 68 for patients with TAV. This same bimodal age distribution was noted among patients with Stanford type A aortic dissection [111]. Strikingly, over 40% of patients undergoing replacement of the ascending aorta for aneurysm or dissection at large thoracic aortic surgical centers have BAV. BAV patients have uniformly larger diameter aortic roots and ascending aortas compared to age- and sex-matched controls [112]. The aortic roots of BAV patients, even in the absence of aneurysmal disease or valvular stenosis, have abnormal elasticity [113, 114]. This inherent aortopathy present with the BAV syndrome and the consequent aortic aneurysm formation greatly increases the risk of aortic dissection and sudden death in these patients [69].

Although linkage analysis shows evidence of genetic basis for BAV [70, 115, 116] and mutations in Notch 1 have been associated with calcification of the aortic valve and development of aortic stenosis [117], no gene or pathway to date has clearly defined a cause of the aortic pathologies occurring in patients with BAV. TAAs of the BAV patients display similar histopathology as those of connective tissue disorders discussed above. These included CMD, non-inflammatory SMC apoptosis, accumulation of proteoglycans, increased MMP activity and extensive elastin fragmentation. As with other TAAs, those occurring in BAV patients involve matrix degradation which results in a weakened biomechanical response, incapable of withstanding hemodynamic forces and as a result, aneurysm occurs and places the patient at increased risk for dissection and rupture.

Unlike the other connective tissue disorders, no gene has been identified for BAV, so there is considerably less known on the potential mechanism of TAA formation in BAV patients. The incidence of BAV overlaps with *α-SMA* and *TGF-βR* mutations at a frequency of 2.5–3% [73] but no gene or pathway responsible for inciting TAAs in BAV patients has been identified. A role for reactive oxygen species (ROS) in the development of TAAs in BAV patients has been reported by Phillippi and Gleason [118, 119] ROS have been implicated in the development of TAAs in non-BAV patients as well [120, 121], but the mechanism inciting TAAs remains elusive. Recent work from Phillippi and Gleason has focused on the role of the ascending aortic SMCs in maintenance of the aortic ECM. It was shown by gene array that several isoforms of a family of stress response proteins called metallothioneins (MTs) were substantially down-regulated in the ascending aorta of aneurysmal patients with BAV relative to normal non-aneurysmal patients and aneurysmal patients with TAV [119]. MTs are a superfamily of intracellular cysteine-rich proteins of low (<7 kD) molecular weight, with high metal binding and redox capabilities. MTs appear to be primarily stress response proteins and are transcriptionally activated under conditions of reactive oxygen species, hypoxia, UV irradiation and heavy metal exposure. The mechanism of action for MT has not been delineated. Many studies suggest that MT plays a role in the homeostasis of essential metals such as zinc [122] and cadmium (Cd) [123], detoxification of metals, and protection against oxidative stress [124–127]. MT-/- mice are non-lethal and studies have demonstrated that MT is not required for proper growth or development, however the absence of MT-1 or -2 increased Cd-induced lethality and hepatotoxicity, and overexpression of MTs provided protection from Cd-induced death [128]. MT has been implicated in the pathogenesis of multiple diseases including oxidative stress-induced cardiac dysfunction [129]. MT was also shown to rescue hypoxia-inducible factor 1-alpha (HIF-1α) activity and increase vascular endothelial growth factor (Vegf) expression in a mouse model of diabetic cardiomyopathy [130]. HIF-1α has been extensively characterized as a key player in processes of ECM degradation during neoangiogenesis in cancer progression [131–133] and in normal cells [134, 135]. Induction of Vegf in the presence of ROS was also dramatically reduced in primary SMCs from BAV patients [118]. However, MT has not yet been shown to play a direct role in ECM homeostasis. As part of the overall oxidative stress response and an upstream regulator of HIF-1α and Vegf, MT could elicit downstream effects that are proponents of ECM degradation.

In a mouse model of Marfan syndrome, $Fbn1^{C1039G/+}$ mice developed TAAs [136] with reduced expression of superoxide dismutase (SOD) and increased expression of iNOS, NADPH oxidase (NOX) and Xanthine oxidase (XO) [137]. Elevated levels of homocysteine and protein carbonyl content, accompanied by a decrease in total antioxidant capacity, were found in the serum of Marfan patients [121]. Homocysteine is thought to affect MMP/TIMP ratios by modulating Nitric oxide (NO) availability [138, 139], causing increased deposition of collagen and degradation of elastin leading ultimately to altered vessel biomechanics evidenced by reduced aortic flow velocity [140]. Oxidative stress has also been implicated in the pathogenesis of AAAs [141]. Increased expression of the NOX subunit $p22^{phox}$ in TAAs co-localized with MMP activity in the aortic media [120]. These data support a causal role for NOX activity in sub-clinical pathogenesis of BAV aortas. Mice deficient in Nox1 or NOX subunit $p47^{phox}$ were protected from angiotensin II-induced aneurysm and dissection, potentially through modulation of MMP/TIMP ratios [142, 143]. ROS appear to be associated with impaired vasomotor function [137].

A theoretical cause of TAAs is that the BAV elicits undue biomechanical stress on the aortic wall. This is substantiated by reports that altered hemodynamics cause aortic wall weakening and spatially dependent expression of matrix proteins [144–146] Evidence against a hemodynamic cause for TAAs in BAV patients arises from the clinical observation that BAV patients who have undergone aortic valve replacement prior to TAA development remain at increased risk for ongoing aortic dilatation [147] which is discordant with the degree of aortic stenosis or insufficiency [112, 114]. Patients with aneurysm or dissection of the ascending aorta, including those with BAV, exhibit evidence of ECM degeneration, such as elastin fragmentation and increased collagen deposition [45, 48, 60–62, 74, 75, 118] as do NOX-1-deficient animals [148, 149]. The increase in collagen matrix deposition [45, 47, 118, 150, 151] in BAV-TAAs as a putative result of ROS [142] may serve as a compensatory mechanism to counter the MMP/TIMP imbalance [152, 153]. Endothelial nitric oxide synthase (eNOS)-deficient mice demonstrate a 32–41% incidence of BAV [154, 155]. Polymorphisms of eNOS are linked to AAAs [156–158] and intracranial aneurysms [159–163] and reduced eNOS is observed in the aortic endothelium of BAV patients [164] and linked to experimental (elastase perfusion) [165] and age-related AAAs [166] but has not been characterized for TAAs. There is mounting evidence for a role of ROS in disrupted ECM homeostasis but a defined molecular mechanism has yet to be established. ROS is known to impact the SMC phenotype [167, 168], shifting them towards a de-differentiated state which increases ECM matrix synthesis [169] and is associated with TAAs of various origins [66, 67, 73, 170–173] and thus could modulate tissue biomechanical strength.

The BAV aorta is pathologically similar to those of the other connective tissue disorders but the mechanisms inciting the final common pathway of ECM degeneration and biomechanical failure appear to be distinct. Like LDS, collagen expression is elevated in TAAs of BAV patients [118]. Collagen is radically affected by the presence of ROS and the lucid SMC phenotype takes preference toward de-differentiation in TAAs [173] and is strongly influenced by oxidative stress [167, 168]. The inability of SMCs of BAV patients to adequately sense and cope with oxidative stress appears to negatively impact matrix deposition and orientation. Collagen fibers appear more disorganized in TAAs of BAV patients [174]. In agreement with the observations of increased collagen deposition and disorganization, preliminary results the authors' laboratories of ex vivo biomechanical testing of freshly-harvested TAA specimens demonstrated increased tensile strength and decreased delamination strength of BAV specimens when compared to aortas of healthy individuals [175–178].

These observations support a pivotal role of the SMCs in maintaining the aortic ECM. The aortic SMCs, presence of excess ROS, and the cells' ability to manage oxidative stress warrants further investigation pertaining to their role in overall ECM homeostasis and biomechanical integrity of the vessel wall. Improved treatment of BAV patients will likely come from drawing connections between SMC response to ROS in the form of cell viability and phenotypic modulations and the impact of these behaviors on ECM deposition, organization and biomechanical integrity.

2.5 *Other Familial Thoracic Aortic Aneurysms*

There are many instances of familial TAA where there is no connective tissue disorder present, nor incidence of BAV but there is heritability of aortic diseases within families. Linkage analysis has documented genetic loci that are implicated in familial aortic disease but no gene or pathway inciting ECM degeneration has been identified to date. In some patients, mutations in fibrillin-1 are similar to that of Marfan syndrome but lacking clinical manifestations in other organ systems [73, 179]. Expert surgical opinion suggests that these patients follow strict surveillance similar to that recommended for MFS patients [70]. The fundamental mechanism of failure is likely similar to that of other TAAs and include ECM degradation. As mutated genes and faulty pathways are identified in these patients, the impact of their dysfunction on ECM homeostasis and biomechanical integrity can be investigated.

3 Biological Mechanisms that Impact Mechanical Properties of the Aneurysmal Abdominal Aorta

During AAA formation and enlargement, the biomechanics of the aortic wall change dramatically [11, 26, 30, 31, 180–182] due to loss of ECM integrity which is spatially variable [21, 31, 183–185] (Fig. 6). Several studies demonstrated that the tensile strength decreases as the aneurysm expands [11, 26] and wall strength and stiffness are spatially variable [21, 29, 31]. The evidence supports the notion that local factors such as wall stress and/or hypoxia negatively influence the biomechanical strength of the vessel wall by promoting localized ECM degradation that is known to occur in AAAs [30, 186, 187]. Portions of AAAs were found to be stiffer, contain less elastin and collagen when compared to normal aorta. It is now well accepted that AAAs arise and progress up to failure when stress in the

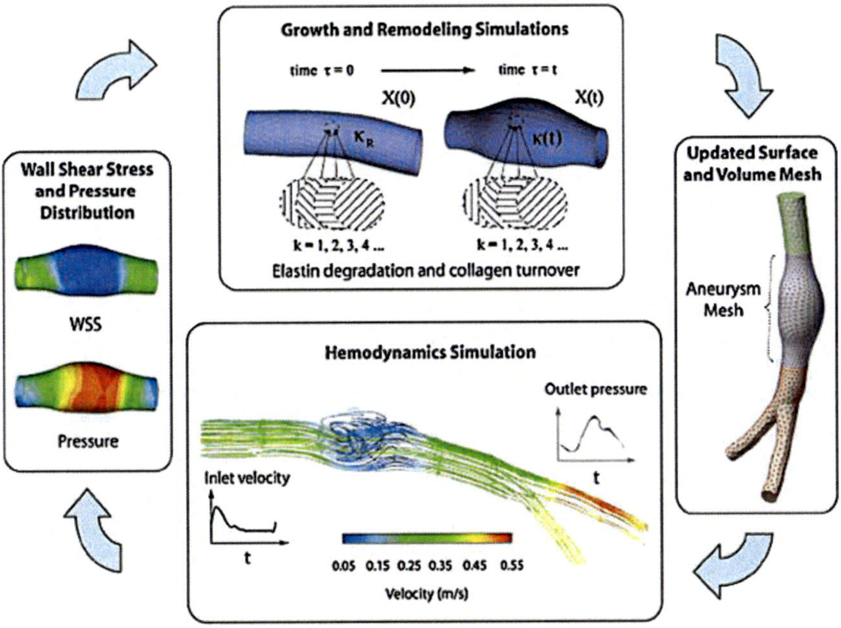

Fig. 6 Relationship between hemodynamics and growth and remodeling simulations of blood vessels. (Reproduced from Fig. 1, Sheidaei et al. [185])

aneurysm wall exceeds the strength of vessel wall. However, little is known of the biological mechanisms driving ECM degeneration in AAAs but two areas under close scrutiny are stress-mediated and strain-mediated wall weakening and hypoxia-mediated disruption of ECM homeostasis.

3.1 Stress/Strain-Mediated Formation of AAAs

Mechanical forces or deformations are considered to be paramount to maintenance of microstructure in normal tissue under physiological conditions. Elastin concentration was found to decrease with the diameter in the non-aneurysmal and aneurysmal aorta [188]. Variations in aortic wall stress are supported by the law of Laplace which indicates that vascular wall stress is directly proportional to its diameter. Hence, areas of dilatation may occur in response to spatial changes in wall stress [183]. The observation of focal saccular outpouchings, also known as "blebs" within the walls of AAAs are speculated to be sites of higher risk for rupture since these represent areas of stress concentration [189]. The latter is very important in practice because rupture of the material is commonly expected to start at the location of stress concentration. These blebs were also characterized by decreased collagen and elastin when compared to neighboring sites within the

AAA [190]. Mechanical forces are known to influence ECM production by SMCs [191, 192], endothelial cells [193] and other cells [194]. These combined data on spatial variability in ECM content and cell response to stress support the fact that stress concentration leads to disrupted ECM homeostasis, resulting in weakening of the aortic wall. In atherosclerosis, the occurrence of lesions at branch in artery is also correlated with the stress concentration. Similar mechanisms may contribute to formation and/or progression of TAA since ECM proteins were found to be differentially expressed on the greater and lesser curve of the ascending aorta [195].

Some of the major known contributors to ECM degradation are the MMPs. Elevated MMPs have been identified in both TAAs and AAAs and are influenced locally by stress [43, 153], aneurysm size [196] and growth rate [197]. MMPs, and other proteolytic enzymes such as tPA and urokinase plasminogen activator are enhanced by mechanical stress [153]. Macrophages also play a dominant role in increasing MMP production as part of the inflammatory response that occurs with AAA [21, 198]. Macrophages are centralized to areas of atherosclerotic plaques and elevated MMP production is observed in these zones of decreased mechanical strength [199, 200]. These studies suggest that macrophages may play a key role in reducing the strength of AAA via stress-mediated expression of proteolytic enzymes. The local SMCs and/or fibroblasts and especially the macrophages appear to "sense" the mechanical stimulation and respond by increasing production of proteolytic enzymes. Elevated levels of MMPs then work to degrade the ECM and thus weaken the biomechanical integrity of the aortic wall, leading to aneurysm formation and risk for rupture. The molecular signaling mechanisms by which mechanical stimulation promotes increased MMP expression remain elusive but likely involves a complex network of intracellular signaling pathways connected to cytoskeletal proteins that are indirectly tethered to the ECM. Some recent studies have targeted the Jnk and MAPK pathways as mediators of MMP regulation in AAA [201, 202].

Therefore, locally acting forces (and therefore wall stresses) play an important role in the pathophysiology of aneurysmal disease. Specifically, local ECM degeneration of the aneurysmal wall, which could result in a compromised structural integrity of the AAA wall, appears to coincide with areas of increased wall stress. These studies support the idea of spatial variation of wall strength that could result in increased risk for rupture in defined locations that vary patient-to-patient.

3.2 Hypoxia-Mediated Wall Weakening

Another powerful modulator of ECM deposition is hypoxia. As discussed in the TAA sections above, hypoxia and ROS disrupt the normal balance of ECM synthesis and degradation by directly influencing production of ECM protein and regulatory proteins. Hypoxia has been shown to both increase [203–205] and decrease [206, 207] production of collagen by SMCs. These observations suggest that local hypoxia contribute to aneurysm wall weakening through inhibition of ECM synthesis. Expression of MMPs as well as their release of elastase [208] and cytokines [209] is also greatly enhanced in macrophages under hypoxic conditions [210]. The hypoxic state may be caused by the presence of ILT [21, 211]. As a consequence, the mural cells within these zones likely respond to this environment. A greater number of macrophages were found within [212] and in the vicinity [198] of the ILT along with many other inflammatory cells when compared to adjacent locations within the AAA. These areas displayed reduced elastin and increased MMPs. These collective data point to a role for hypoxia in stimulating ECM degradation and thus, negatively influence the biomechanical integrity of the aorta wall, consequently leading to AAA formation in these patients.

3.3 Genetic Causes of Abdominal Aortic Aneurysm

There have been extensive efforts made to identify genetic causes of AAA and a genetic basis for aneurysm growth. A recent systematic review of the literature by Saratzis et al. summarized a majority of these studies as of the year 2011 [213]. Most of the genetic studies center on the matrix-regulating proteins (MMPs, TIMPs), factors related to the inflammatory (interleukins, C reactive protein) and wound healing responses (TGF-βs, platelet-activating factor, plasminogen activator inhibitor 1 (PAI-1), osteopontin and osteoprotegrin (OPG)). The majority of studies investigating the role of these proteins in AAA formation demonstrated alterations in gene or protein expression and identified gene polymorphisms causing these changes in expression. However, most analyses failed to demonstrate a direct association between gene polymorphisms and AAA formation or growth. A few polymorphisms were found to be associated with development (MMP-3, [214] MMP-9, [215] TIMP-1 and -2 [216–218] the β-chemokine receptors, CCR5 [219] and CCR2 [220] angiotensin converting enzyme [221] methylene tetra hydro folate reductase (MTHFR) [222–224] and potentially endothelial nitric oxide synthase (eNOS), although the sample size was relatively small [225]) or growth (PAI-1 [226], and an OPG-regulating factor called peroxisome proliferator-activated receptor gamma (PPARG) [227]).

Another recent review by Krishna et al. discussed the genetic and epigenetic causes of AAA [228] and was in agreement with Saratzis et al. for which there is a general lack of genetic polymorphisms strongly correlated with AAA. The "Aneurysm Consortium" has undertaken large Genome-Wide Association Studies (GWAS) to analyze variance in diseased patients and controls [229]. From this work, there was a strong correlation of chromosome 9p21.3 locus performed with respect to AAAs and was originally identified from seven different populations [230]. A more recent study demonstrated that this association remained despite removing other AAA risk factors such as smoking, coronary heart disease, family history and dyslipidemia [231]. This chromosomal region is thought to be non-coding DNA, consisting of repetitive DNA of retrotransposons which are transient genetic elements that move across the genome. Among these are a long interspersed nuclear elements 1 (LINE-1) which can invade a variety of genes and result in multiple phenotypic outcomes. A recent study evaluated several polymorphisms of 9p21.3 and their association with AAA. Only two polymorphisms were found to be significantly associated with two open reading frames (ORFs), providing evidence that LINE-1 elements may be involved in a genetic cause of AAA [232]. However, further larger studies are needed to confirm this association. The Aneurysm Consortium also identified a single nucleotide polymorphism (SNP) on 9p21 which was previously shown to be associated with heart disease and recently with AAA [230]. Neighboring genes to this particular SNP code for proteins related to cell proliferation, apoptosis and senescence [229] and could be related to SMC phenotype.

Gene products which have been shown to be altered in AAA and may or may not have had polymorphisms identified include MMP-2,3,7, 9, TIMP-3, inducible nitric oxide synthase (iNOS), eNOS, interferon-gamma (INF-γ), tumor necrosis factor (TBN)-α, intercellular adhesion molecule (ICAM)-1, IL-6, TGF-βRII, platelet-derived growth factor (PDGF)-A, estrogen receptor (ER)-α/β, c-fos, 5-lipoxygenase (5-LO), extracellular superoxide dismutase (EC-SOD), fatty acid desaturase (Fads)-2, MTHFR and p53 [228]. Several epigenetic changes such as histone modifications and DNA methylation can alter expression of these genes, and therefore cause the variation observed in AAA relative to normal patients. These epigenetic alterations can stem from known AAA-related risk factors such as smoking, ageing and inflammation. Among these, smoking is the most prevalent risk factor for AAA, and over 80% of patients developing AAA have a history of smoking [233].

4 Biomechanics of AAA and TAA

4.1 Methods for Measuring and Modeling Aortic Biomechanics

Aneurysm rupture can be seen as a mechanical failure of the degenerated aortic wall that occurs when the wall stress acting on the aneurysmal wall exceeds the wall strength. Therefore, the knowledge of both wall stress and wall strength distributions for a given aneurysm will greatly improve the ability to identify those aortic aneurysms that are at highest risk of rupture. Towards this end, the biomechanics of AAA has been studied extensively by many groups whereas TAA biomechanics has not been fully considered. This section provides the state-of-the-art of AAA biomechanics, including the modeling of tensile strength and wall stress distributions and factors which influence them. Then the potential clinical utility of these estimates in predicting AAA rupture will be also presented. A summary of recent advances in TAA biomechanics is also articulated.

4.2 Analysis of Biomechanical Strength and Wall Stress In vivo

4.2.1 AAA Wall Strength

Given the general recognition of the importance of wall mechanics in the natural history of AAAs (e.g., [183, 234–236]), there have been numerous studies of AAA biomechanical properties. Noninvasive diagnostic techniques such as ultrasound and computer tomography have long been used to understand AAAs wall mechanics. Indeed, early studies focused on gross measurement of structural stiffness [37, 237–239] or vessel compliance [240, 241], including the effect of ILT [242]. Although these investigations were useful for clinical correlations, the AAA wall strength was not evaluated by these approaches.

Vessel wall strength has not yet been measured in vivo due to the deficiency of current experimental techniques. However, researchers have easily assessed AAA wall strength performing ex vivo experimental testing on excised normal and pathological aortic tissue. Early ex vivo studies on the biomechanical properties of AAAs focused on understanding the effect of extracellular derangement observed in aneurysms [186, 187]. Reported uniaxial sub-failure tensile data for eight human AAAs and eight age-matched controls (all axially oriented specimens) revealed a stiffer AAA wall compared to that of the normal aorta [30]. Similar findings demonstrated that the aneurysmal aorta in the circumferential direction was stiffer than the axial direction, and that posterior portions were less stiff than anterior or lateral portions, suggesting a stiffness variation around the lesion [31]. These findings were corroborated by Vallabhaneni et al. [243] suggesting a correlation between strength and variation in MMP production, and Ragahvan et al. [244] reporting spatial variation of wall thickness and tensile strength in an excised AAA. Variation in AAA wall strength from point-to-point in a given aneurysm or from patient-to-patient support the fact that evaluation of AAA wall stress alone is insufficient to predict rupture and that AAA wall strength cannot be assumed to be constant on aneurysms. Vorp et al. widely investigated AAA wall mechanics, reporting measurements of AAA wall strength [26], and formulating both microstructure-based [11] and hyperelastic, continuum-mechanics-based models for the AAA wall mechanics [181]. In particular, the latter study reports uniaxial tensile tests on 69 freshly-excised human AAA specimens oriented along either the circumferential and longitudinal directions. They also quantified experimental data with a strain energy function of the form $W= \alpha(I_B-3) + \beta(I_B-3)^2$ where $I_B=tr\mathbf{B}=tr(\mathbf{F}\bullet\mathbf{F}^T)$ and $\mathbf{F}$ is the deformation gradient, and α and β are the model parameters indicative of the mechanical properties in longitudinal or circumferential directions (Fig. 7). This continuum-based constitutive model is appropriate for wall stress prediction [183] since it can be easily implemented in finite element analysis compared to microstructure-based models. To investigate the association of aortic wall weakening with AAA rupture, ex vivo tensile tests on freshly-excised wall tissue specimens from patient who suffered AAA rupture prior to surgery were compared to specimen tissues from elective repaired, asymptomatic

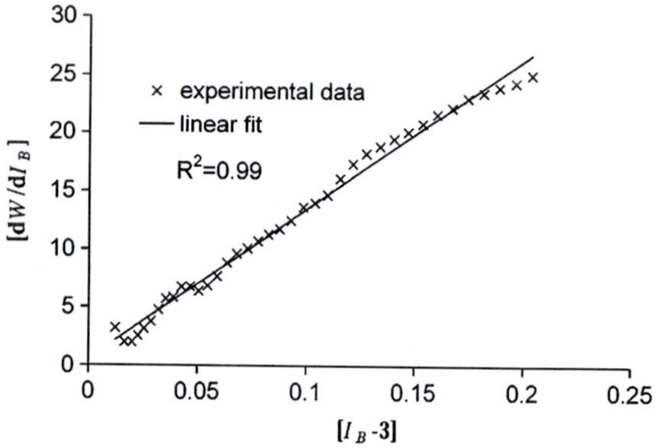

Fig. 7 The experimentally determined relationship between the first derivative of strain energy function and the first invariant of the stretch tensor for a representative AAA specimen. Linear regression resulted in a highly correlated fit (R^2=0.99), indicating a linear relationship between the two. (From Fig. 1, Raghavan and Vorp [14])

AAA [12]. The tensile strength of the ruptured AAA tissue was found to be significantly lower than that for the elective AAA tissue (Fig. 8). This data suggests that AAA rupture is associated with significant aneurysm wall weakening, supporting the idea that wall strength needs to be carefully considered on a patient-specific basis in order to accurately predict rupture potential of individual aneurysms.

Uniaxial testing is not sufficient for highlighting the multi-axial biomechanical response of vessel in vivo; therefore, normal and pathological abdominal aorta tissue specimens have been tested in a biaxial testing device [245] to obtain more appropriate constitutive relation of aortic tissue and to investigate any apparent anisotropy of the aorta. The high non-linear responses found on 26 AAA tissue samples and 8 aged-matched abdominal aorta samples were described well by a two-dimensional phenomenological Tong-Fung (1976) strain energy function. These results confirmed the general observation that AAAs are indeed stiffer and more anisotropic (i.e., stiffer in circumferential direction) than the normal aorta. Comparison between strain energy functions derived from uniaxial testing [14] and biaxial testing [246] provided evidence for remarkable differences in mechanical response of the tissue between the two models, demonstrating the importance of correct model choice in biomechanical simulations (Fig. 9).

There has been less attention to the influence of biomechanical properties of ILT that often lines AAAs. Di Martino [180] first reported uniaxial testing on 21 samples of thrombus obtained from 6 patients generating support for the idea that thrombus may be "mechanically protective" by providing stress shielding or a cushioning effect for AAA wall. The most comprehensive study of the biomechanical properties of ILT from human AAAs [247] revealed that the behavior is mildly non-linear over large strain, nearly isotropic [29] and inhomogeneous since portions of tissue from the luminal region were found to be stiffer and stronger when compared to middle region (Fig. 10).

Although these mentioned studies established that AAA biomechanical properties changes with aneurysm formation and that these properties are spatially variable, little progress has been made to provide methodology to noninvasively

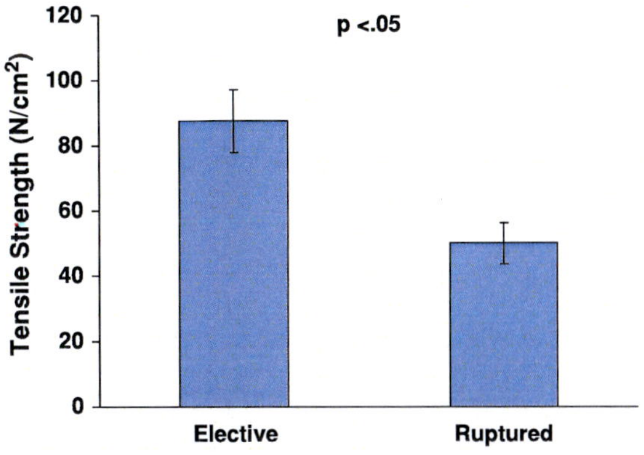

Fig. 8 Difference in tensile strength of ruptured and electively repaired AAA wall specimens. (From Fig. 10, Vorp [183])

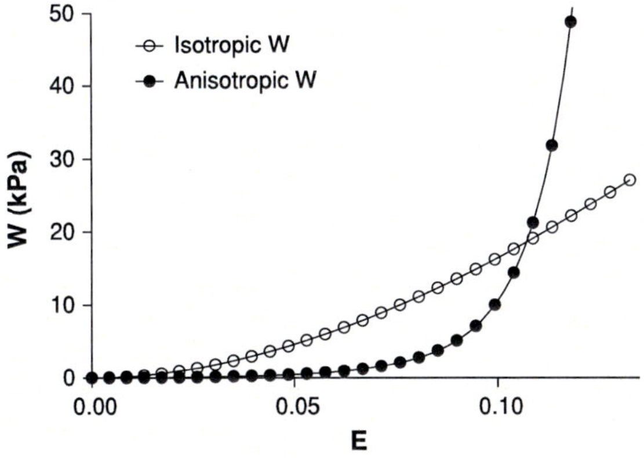

Fig. 9 The strain energy of the averaged isotropic relation [12] and the averaged anisotropic constitutive relation [218] for AAA versus equibiaxial strain (E). (From Fig. 2, Vorp [183])

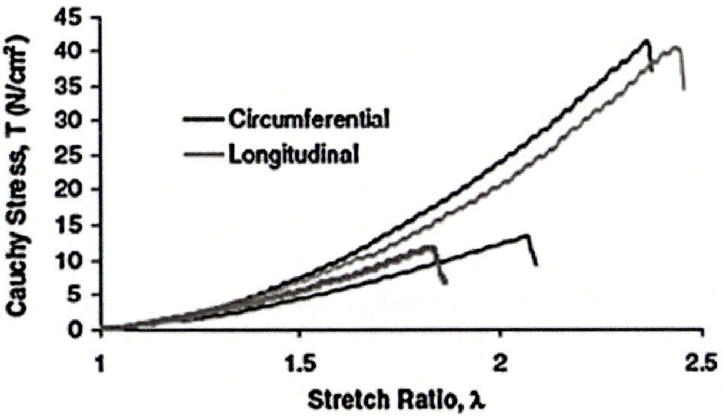

Fig. 10 Typical set of stress-stretch curves for ILT samples taken from the luminal region (*upper set of curves*) and the medial region (*lower set of curves*). All data were obtained from the same ILT (same patient). (From Fig. 3, Vorp [183])

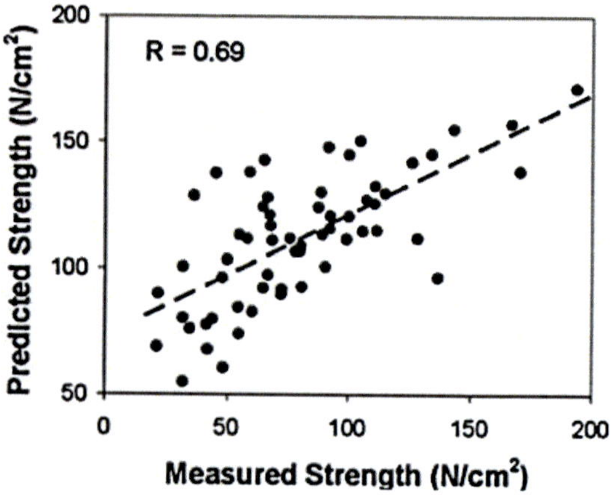

Fig. 11 Predicted versus measured strength for the statistical model of wall strength. The *dash line* represents the line of unity. (From Fig. 3A, Vande Geest et al. [20, 29])

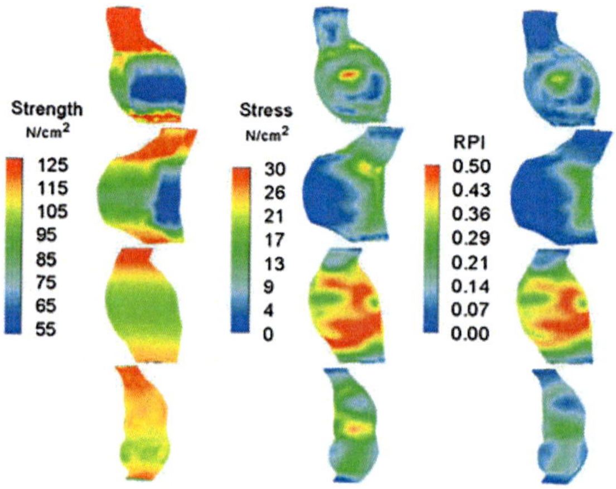

Fig. 12 Demonstrative application of the statistical model of wall strength (Eq. 3 in [20], left), wall stress (*middle*), and rupture potential index (*right*) for four representative AAAs. (From Fig. 4, Vande Geest et al. [20, 29])

predict the local wall strength distribution in a given AAA. A statistical-multifactorial mathematical model is currently the only model for assessing the AAA wall strength with a noninvasively approach. Specifically, multiple linear regression techniques were used to derive a statistical model for the prediction of AAA wall strength based on experimental measurements and measurable pertinent predictors, including local AAA diameter, local ILT thickness, patient age and gender, and patient family history of AAA disease [see the excellent work by Vande Geest [20] for more on this statistical model]. The predictability and example applications of such a model are shown in Figs. 11 and 12 [20]. In this study, the concept of RPI, which is defined as the locally acting wall stress divided by the local wall strength, was also introduced to define the propensity to rupture of a given AAA.

4.2.2 AAA Wall Stress

In recent years, there has been an increasing effort to develop appropriate finite element (FE) models to assess the wall stress in dictating the fate of aneurysm failure. Early stress analyses have assumed idealized geometries, and are based on law of Laplace [248–250], axisymmetric membrane theory [16] or linear elastic material behavior [251, 252]. Moreover, Inzoli et al. [251] were the first group to introduce an important, but controversial, concept: that ILT may act to reduce AAA wall stress by a shielding effect. However, these models are inappropriate to describe the inherent non-linearity of the aorta under large deformations and fail to demonstrate the likely importance of blebs which, in turn, cause stress concentration (Fig. 13). The best models available utilize patient-specific geometries and treat the aneurysmal wall as a non-linear elastic, isotropic, homogenous body [22, 23, 253] (Fig. 14). From these studies, the stress magnitude and distribution were found to be dependent on either the shape of the AAA bulge as well as aneurysm diameter. These findings strengthen the argument for which aneurysm with the same diameter may not necessarily have the same propensity to failure.

To accurately mimic the physiological mechanical behavior of the normal and pathological aorta, the proper utilization of the constitutive material models implemented in FE models needs to be evaluated. This was accomplished by Vorp et al. who investigated the influence of constitutive-model material parameters [13, 181] and tissue anisotropy [246] on wall stress prediction. They performed a parametrical analysis by changing the α and β material parameters of the constitutive material model described by Raghavan et al. [181], on several FE stress analyses of hypothetical 3D asymmetric AAAs. The maximum variation of material parameters led to less than 5% in the predicted wall stress, and this suggests that the differences in AAA wall stress from patient-to-patient are driven more by the differences in surface geometry than in material properties. In a similar way, FE analyses on two non-ruptured and one-ruptured AAAs modeled with both isotropic [14] and anisotropic [29] wall constitutive relation demonstrated that an anisotropic material model provides a better prediction of the stress distribution in AAAs compared to an isotropic model. The most advanced FE models also consider the presence of ILT contained in most AAAs. Indeed, the incorporation of thrombus into computational stress analysis models the profound influence on the magnitude and distribution of stress acting on the AAA wall, and this should be considered for accurate wall stress estimations (Fig. 15). Recently, the statistical model for predicting wall strength and rupture potential index proposed by Vande Geest et al. [182] were implemented in FE analysis [28] and compared with the maximum diameter criterion to evaluate their reliability in AAA rupture prediction. The strength model reported by Vande Geest was misused by the authors (the local normalized diameter that was used was not as specified by the original model), and it is unclear how this affected their results.

There has been increasing interest in fluid–structure interaction (FSI) modeling of AAAs, yet most studies assume the non-aneurysmal and aneurysmal aorta, and ILT as linear elastic, homogeneous, isotropic, nearly incompressible body

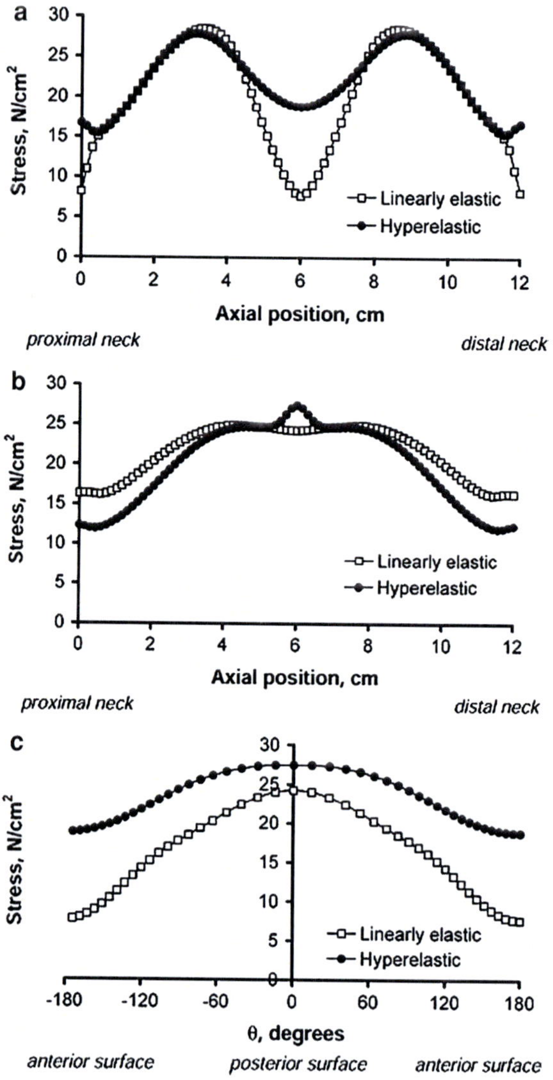

Fig. 13 Comparison of stresses computed for a 3D asymmetric AAA model using the hyperelastic constitutive model [12] with those using a linearized elasticity model along anterior surface (**a**), along posterior surface (**b**), and around mid-section (**c**). Note the substantial error involved with using the theory of linearized elasticity. (From Fig. 7, Raghavan and Vorp [14])

[254–256]. While these studies have been important to demonstrate that the pressure field in AAAs is relatively constant [257, 258], the effect of shear stress was neglected. Perhaps, the most important sophisticated FSI model to date is that by Rissland et al. [259], incorporating ILT, wall calcification, and anisotropic tissue properties. Experimental biaxial data of AAAs [29] were fitted with the Holzapfel orthotropic material formulations [260] which models vessel wall as fiber-reinforced composite material. Calcified plaques surrounding AAAs were assumed to behave as stiff isotropic materials whereas the ILT was modeled as linearly elastic. Later, Xenos et al. [39] validated this FSI methodology to assess

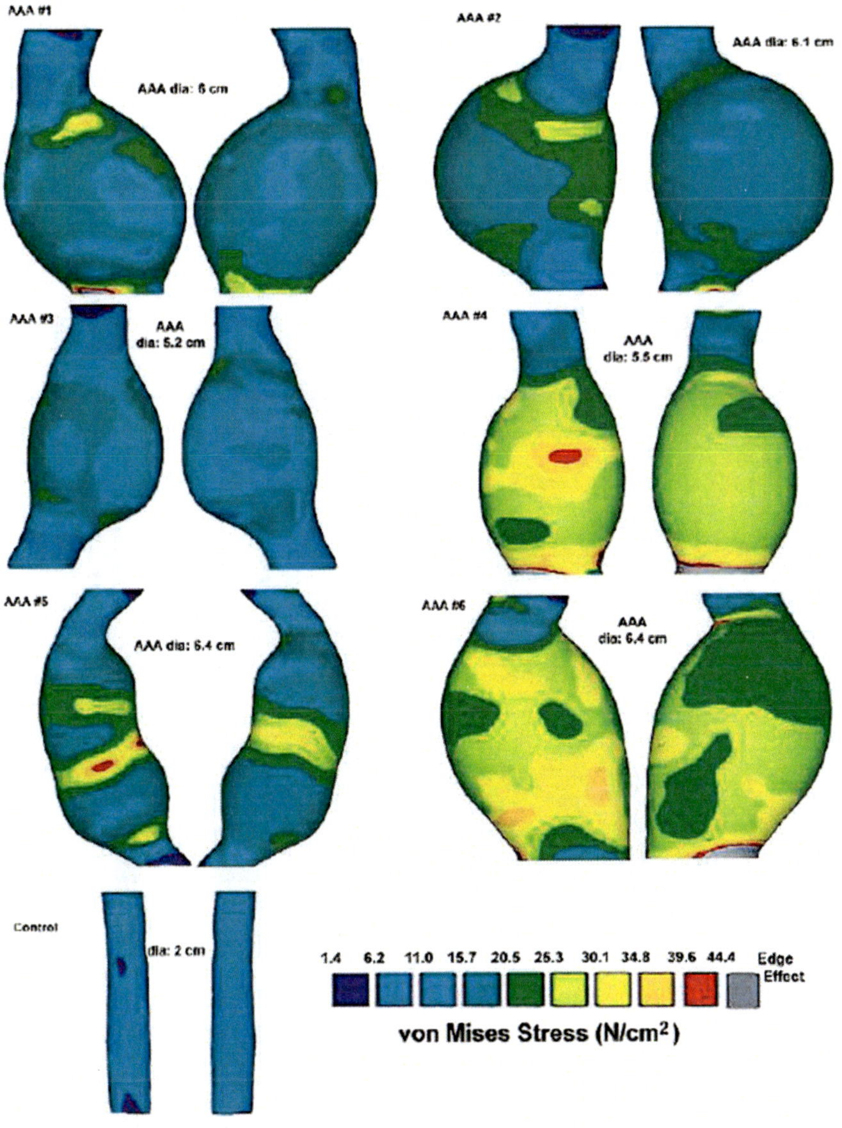

Fig. 14 The distribution of von Mises stress on the posterior and anterior abdominal aortic walls of the control subject and the patients with abdominal aortic aneurysms. Grey regions are those with artificially high stress concentrations because of edge effects. The individual color scales give the stress magnitude. In all cases, *blue* represents the lowest stress magnitude, and *red* represents the highest stress magnitude. Note the comparatively lower range of stress in the control aorta. (From Fig. 4, Raghavan et al. [181])

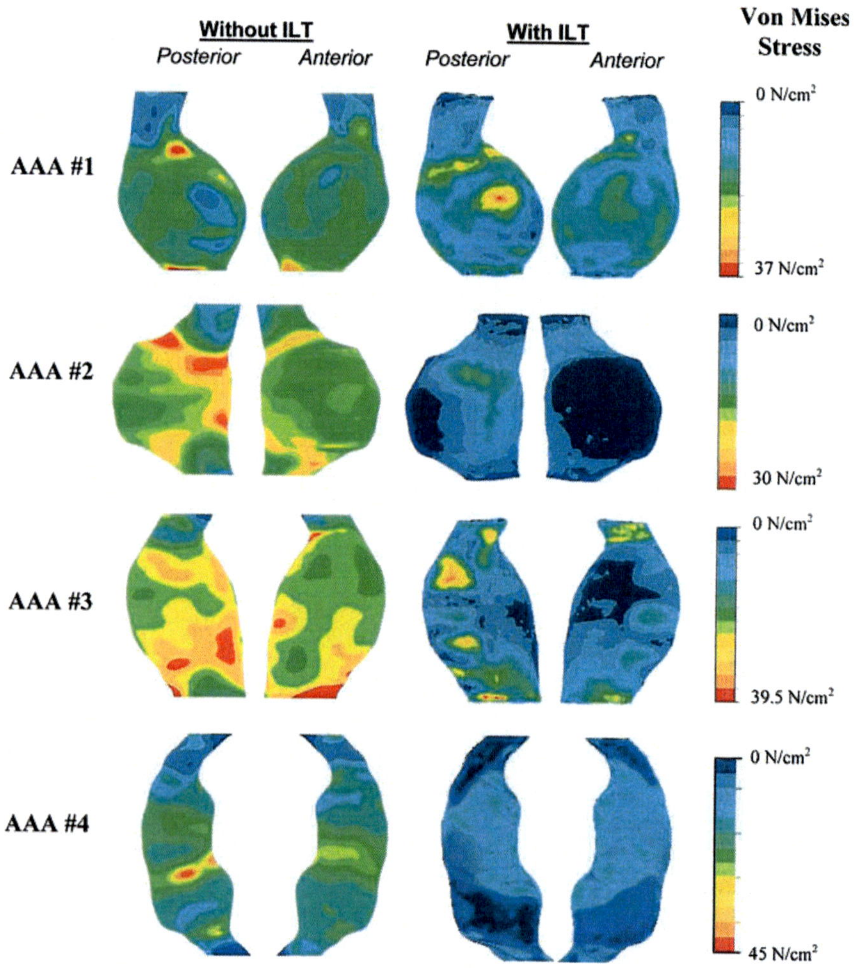

Fig. 15 Comparison of 3D wall stress distribution between AAA models with and without ILT. Individual color scales (*right*) indicate von Mises stress for each AAA. Both posterior and anterior views are shown for each case. (From Fig. 4, Wang et al. [19])

rupture propensity on cases of ruptured AAA using the wall strength model and RPI proposed by Vande Geest et al. [182] (Fig. 16).

These studies have demonstrated that computational models have potential to improve the predictions of rupture potential. It should be also noted that many investigators report the Von Mises stress to assess the maximum stress, inducing to AAA failure. However, Von Mises stress has its importance in the fracture mechanics of ductile materials, which yield due to excessive shear stresses. The aorta has a nonlinearly elastic behavior, not exhibiting yielding, where failure is governed by maximum normal stress [261–263]. Therefore, this issue of a most appropriate failure criterion must be addressed in future investigations.

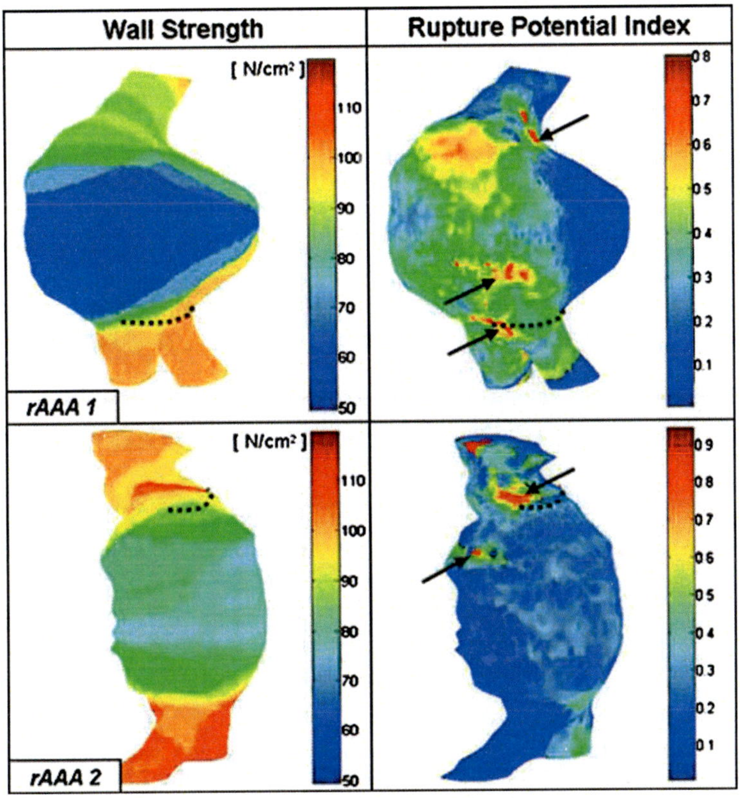

Fig. 16 Wall strength mapping and the corresponding rupture potential index, for the ruptured AAA cases. Multiple distinct regions of high RPI are observed, with at least one that coincides with the actual location of rupture. The RPI for both cases is close to unity indicating a very high risk of rupture, with a higher RPI (0.95) for the smaller rAAA. (from Fig. 6, Xenos et al. [39])

4.2.3 TAA Wall Strength and Wall Stress

Despite the aforementioned importance of wall strength in the mechanics underling the aortic failure, there has been less attention to that of thoracic aortic aneurysm. This pathology is also particular lethal although the relatively rare incidence. Similar to AAA, aneurysm involving the ascending and descending aortic segments can rupture. However, a more common pathology is dissection where an intimal tear, often circumferential in origin, permits blood to enter the aortic wall splitting the media and progressively separating the medial plane along the axial direction of the aorta. From a biomechanical viewpoint, the dissection represents a separation of the elastic layers of the degenerated aortic wall that occurs when the hemodynamic loads exerted on the aneurysmal wall exceed bonding forces that normally hold the layers together. Clinical hemodynamic disturbances that render the aorta susceptible to the onset of dissection are principally elevated maximum systolic and mean aortic blood pressure [264].

Early studies reported experimental data for normal descending and mid-thoracic aorta from uniaxial tensile tests [261, 265] and inflation tests [261, 266]. These studies indicated that wall strength is dependent on the anatomic site of the vessel and that tensile failure behavior is perhaps anisotropic. Early wall strength measurements of ascending thoracic aortic aneurysm (ATAA) revealed biomechanical characteristics similar to those of the abdominal aorta. Ex vivo studies by our laboratory on thirty aneurysmal aortic tissue specimens oriented in both the longitudinal and circumferential direction demonstrated that tensile strength in ATAA is 30% lower than non-aneurysmal ascending aorta, suggesting higher risk of rupture in patients with ATAA [33]. Additionally, tissue stiffness of ATAA wall was found to be higher than normal aorta and particularly, the longitudinal direction was stiffer than the axial direction. Later, Iliopoulos et al. [32] reported lower longitudinal tensile strength with aneurysm enlargement. The low wall strength in longitudinal direction found in these investigations may be responsible for aortic dissection in ATAAs.

As found for AAAs, the ATAA wall strength and stiffness vary spatially. Ex vivo tensile testing data revealed uniform ATAA tissue strength and stiffness along the circumferential direction whereas, in the longitudinal direction the anterior region was the weakest and least stiff part of the aneurysm, and this can be associated with the clinical evidences of preferential ATAA bulging in the anterior region [267]. Tensile wall strength data associated with abnormal valve morphology revealed that ATAA tissues excised from patients with bicuspid aortic valve are stiffer than those from those patients with tricuspid aortic valve [34]. However, these findings are limited by a relatively small sample size ($n = 6$). Only one study reported media splitting testing measurements to understand the dissection properties of thoracic aorta [268]. By infusing liquid into media, a bleb was created and thus the propagation of the liquid was monitored to explore the mechanics of dissection.

Although FE methodology has been applied extensively in the prediction of wall stress for AAAs, it has not been given the same degree of importance in the investigation of wall stress in ATAAs. Therefore, computational modeling of ATAAs is very limited and recent. In the earliest study, the ATAA was modeled as an axisymmetric dilation in a cylindrical aorta as an isotropic, nonlinear, hyperelastic material behavior [269]. As the aneurysm grows, the longitudinal stress in the bulb of the aneurysm is the only stress that was found to increase significantly and thus is likely to cause rupture or dissection. The influence of the aortic root movement on the aortic wall stress as related to aortic dissection was also investigated in a single-layered, linear elastic, isotropic, homogeneous ATAA model [270]. The aortic root motion determined a direct impact on the longitudinal stress that was found to increase critically in the ascending aorta above the sinotubular junction. This may explain why the intima tears circumferentially and aortic dissections occur more often in this orientation. FSI analyses of descending dissecting aneurysms were reported by several groups [271–273]. Recently, a FSI

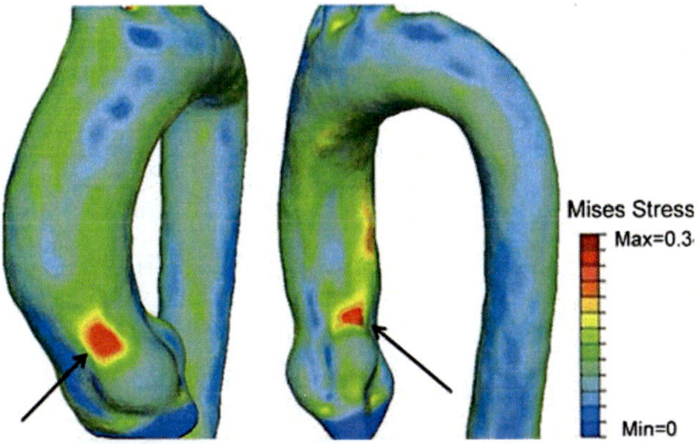

Fig. 17 Three-dimensional wall stress distribution for the normal ascending aorta. Stress in mega-pascals (MPa) is mapped, with the highest stress in *red* and lowest stress in *blue*. The *black arrows* indicate maxima of stress on the (**a**) convex and (**b**) concave side of the ascending aorta. (From Fig. 3, Nathan et al. [274])

analysis of an idealized three-layered descending aorta under turbulent pulsatile flow revealed that the media layer exhibited a larger wall stress compared with the intimal and adventitial layers, and thus the high stress due to elastic properties mismatch in the aortic wall is a risk factor for tearing in aortic dissections [272]. The most sophisticated FE model of an ATAA has been presented recently by Nathan et al. [274]. For the purpose of understanding the pathogenesis of aortic dissection, the authors developed a patient-specific model of an ATAA, though their assumptions of homogeneous, incompressible, isotropic, linear elastic properties and uniform thickness are limitations of this work (Figs. 17 and 18). Their results suggested that peak systolic pressure induces localized high wall stress above the sino-tubular junction and at the ostia of the left subclavian artery that may account for the development of type A and type B aortic dissection, respectively.

5 Using Aortic Biomechanical Data to Transform Clinical Care

All of the criteria that have been proposed to treat AAA and ATAA are based on empirical data with less emphasis on sound physical principles. The current gold standard for surgical intervention for a patient is based primarily on the maximum orthogonal diameter of the vessel, namely the “maximum diameter criterion”, which is 5.0–5.5 cm for both TAAs and AAAs. Nevertheless, this criterion can be improved as it has resulted in a rupture rate of 1% per year for patients under observation [2, 275]. Areas of peak wall stress were found to overlap with sites of rupture [39] and locations of deficient extracellular matrix [30, 186]. Efforts to

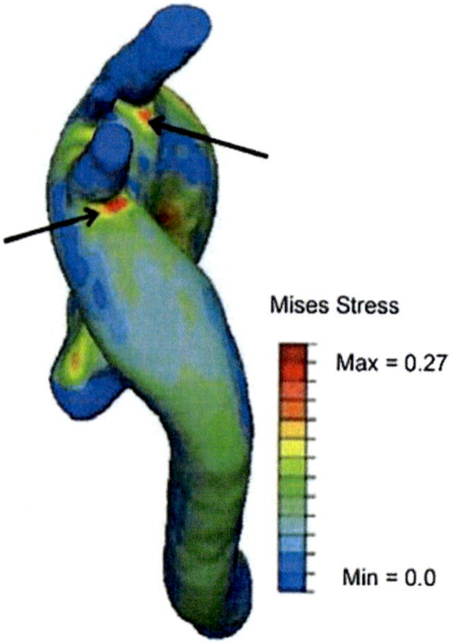

Fig. 18 Three-dimensional wall stress distribution for the normal thoracic aorta. Stress in mega-pascals (MPa) is mapped, with the highest stress in *red* and the lowest stress in *blue*. The *black arrows* indicate maxima of stress distal to left subclavian and innominate arteries. (From Fig. 4, Nathan et al. [274])

refine surgical recommendations and predict relative risk for rupture in a patient-specific manner have included consideration of wall stiffness [37], ILT thickness [38], wall tension [8], peak wall stress [22–24] and growth rate [276]. Serum biomarkers of elastolysis also offer promising non-invasive methods and correlated with disease progression but are not of sufficient predicative value. However, all of these approaches have their own limitations, and may lead to errors in decisions pertaining to clinical management of aortic aneurysm.

In this section, we have emphasized the importance of both the wall stress and the wall strength and that these physical parameters derived from failure mechanics principles may be translated in a novel improved criterion to identify those aneurysms that are at highest risk of rupture or dissection. Toward this end, the RPI, which is defined as the locally acting wall stress divided by the local wall strength, appears to be currently the most promising approach. This model relies on the relatively well-established methods for estimating wall strength stress and the evolving field on noninvasive techniques for modeling wall stress.

Despite the great efforts over the past 20–30 years in identifying not only the pathology of aneurysm, but also the significant advances in discovering genetic causes for several forms of ATAAs and the risk factors for AAA, the current risk prediction models are unreliable. Current improvements on this model which incorporate wall stress and wall strength are not yet ready for clinical translation, and the potential future success of these ideas is dependent on the development of novel noninvasive imaging technologies to measure wall strength. One promising technique is 4-dimensional MR that has recently been demonstrated to evaluate systolic flow patterns in the ascending aorta of aneurysmal patients and healthy volunteers [144]. This study shows in dramatic fashion the disrupted flow in BAV

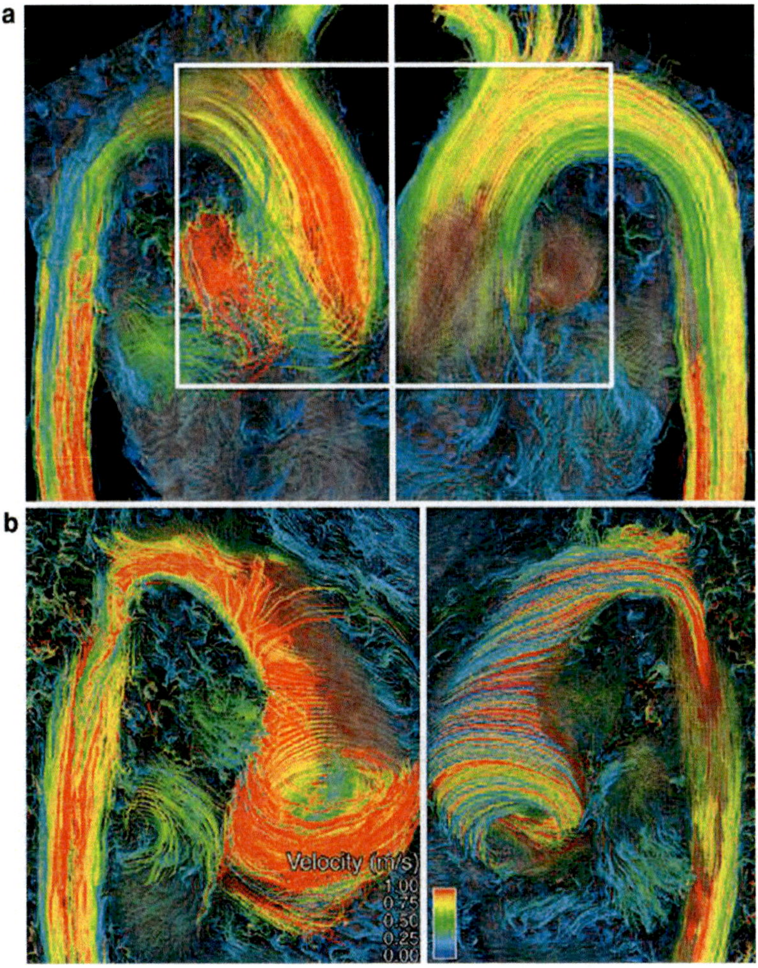

Fig. 19 Use of 4-dimensional flow magenetic resonance (MR) to characterize blood flow velocity in the human ascending aorta. **a** Smooth trajectory of normal systolic flow in a non-aneurysmal patient with tricuspid aortic valve. **b** Disrupted, turbulent and helical flow in a patient with an ascending aortic aneurysm and the abnormal bicuspid aortic valve. (Figure 1a and 2c, d from Hope et al. [144])

patients in a nested helical pattern and could be used to not only estimate local wall shear stress but also model wall strength (Fig. 19).

Our understanding of aortic diseases continues to advance as new partnerships between surgeons, biologists, engineers and mathematicians are established. The future of clinical care for aneurysmal patients lies at the intersection of these fields. In summary, the puzzle of how to care for patients with aneurysm will ideally incorporate knowledge of the biological pathways inciting matrix degradation, an understanding of the biomechanical contributions of the individual molecular entities comprising the matrix, the matrix as a whole and a thorough quantification of both wall strength and stress. The missing pieces lie in developing the enabling noninvasive technologies to measure wall stress and strain, refinement of the mathematical models and establishing links between the clinical manifestations and the biological mechanisms inciting them.

References

1. Ergin, M.A., Spielvogel, D., Apaydin, A., Lansman, S.L., McCullough, J.N., Galla, J.D., Griepp, R.B.: Surgical treatment of the dilated ascending aorta: when and how? Ann. Thorac. Surg. **67**(6), 1834–1839 (1999). discussion 1853–1856 PMID:10391320
2. Ashton, H.A., Buxton, M.J., Day, N.E., Kim, L.G., Marteau, T.M., Scott, R.A., Thompson, S.G., Walker, N.M.: The Multicentre Aneurysm Screening Study (MASS) into the effect of abdominal aortic aneurysm screening on mortality in men: a randomised controlled trial. Lancet **360**(9345), 1531–1539 (2002). PMID:12443589
3. Lederle, F.A., Johnson, G.R., Wilson, S.E., Ballard, D.J., Jordan W.D., Jr, Blebea, J., Littooy, F.N., Freischlag, J.A., Bandyk, D., Rapp, J.H., Salam, A.A.: Rupture rate of large abdominal aortic aneurysms in patients refusing or unfit for elective repair. JAMA **287**(22), 2968–2972 (2002). PMID:12052126
4. Powell, J.T., Brady, A.R.: Detection, management, and prospects for the medical treatment of small abdominal aortic aneurysms. Arterioscler. Thromb. Vasc. Biol. **24**(2), 241–245 (2004). PMID:14604835
5. Powell, J.T., Gotensparre, S.M., Sweeting, M.J., Brown, L.C., Fowkes, F.G., Thompson, S.G.: Rupture rates of small abdominal aortic aneurysms: a systematic review of the literature. Eur. J. Vasc. Endovasc. Surg. **41**(1), 2–10 (2011). PMID:20952216
6. Pape, L.A., Tsai, T.T., Isselbacher, E.M., Oh, J.K., O'Gara, P.T., Evangelista, A., Fattori, R., Meinhardt, G., Trimarchi, S., Bossone, E., Suzuki, T., Cooper, J.V., Froehlich, J.B., Nienaber, C.A., Eagle, K.A.: Aortic diameter >or = 5.5 cm is not a good predictor of type A aortic dissection: observations from the International Registry of Acute Aortic Dissection (IRAD). Circulation **116**(10), 1120–1127 (2007). PMID:17709637
7. Darling, R.C., Messina, C.R., Brewster, D.C., Ottinger, L.W.: Autopsy study of unoperated abdominal aortic aneurysms. The case for early resection. Circulation **56**(Suppl 3), II161–II164 (1977). PMID:884821
8. Hall, A.J., Busse, E.F., McCarville, D.J., Burgess, J.J.: Aortic wall tension as a predictive factor for abdominal aortic aneurysm rupture: improving the selection of patients for abdominal aortic aneurysm repair. Ann. Vasc. Surg. **14**(2), 152–157 (2000). PMID: 10742430
9. Knipp, B.S., Deeb, G.M., Prager, R.L., Williams, C.Y., Upchurch, G.R., Patel, H.J.: A contemporary analysis of outcomes for operative repair of type A aortic dissection in the United States (vol **142**, p 524, 2007). Surgery **143**(2), 301 (2008) PMID:ISI:000253172300033
10. Simao da Silva, E., Rodrigues, A.J., Magalhaes Castro de Tolosa, E., Rodrigues, C.J., Villas Boas do Prado, G., Nakamoto, J.C.: Morphology and diameter of infrarenal aortic aneurysms: a prospective autopsy study. Cardiovasc. Surg. **8**(7), 526–532 (2000). PMID:11068212
11. Raghavan, M.L., Webster, M.W., Vorp, D.A.: Ex vivo biomechanical behavior of abdominal aortic aneurysm: assessment using a new mathematical model. Ann. Biomed. Eng. **24**(5), 573–582 (1996). PMID:8886238
12. Di Martino, E.S., Bohra, A., Vande Geest, J.P., Gupta, N., Makaroun, M.S., Vorp, D.A.: Biomechanical properties of ruptured versus electively repaired abdominal aortic aneurysm wall tissue. J. Vasc. Surg. **43**(3), 570–576 (2006). PMID:ISI:000235848800027

13. Di Martino, E.S., Vorp, D.A.: Effect of variation in intraluminal thrombus constitutive properties on abdominal aortic aneurysm wall stress. Ann. Biomed. Eng. **31**(7), 804–809 (2003). PMID:12971613
14. Raghavan, M.L., Vorp, D.A.: Toward a biomechanical tool to evaluate rupture potential of abdominal aortic aneurysm: identification of a finite strain constitutive model and evaluation of its applicability. J. Biomech. **33**(4), 475–482 (2000). PMID:10768396
15. Vorp, D.A., Raghavan, M.L., Webster, M.W.: Mechanical wall stress in abdominal aortic aneurysm: influence of diameter and asymmetry. J. Vasc. Surg. **27**(4), 632–639 (1998). PMID:9576075
16. Elger, D.F., Blackketter, D.M., Budwig, R.S., Johansen, K.H.: The influence of shape on the stresses in model abdominal aortic aneurysms. J. Biomech. Eng. **118**(3), 326–332 (1996). PMID:8872254
17. Sacks, M.S., Vorp, D.A., Raghavan, M.L., Federle, M.P., Webster, M.W.: In vivo three-dimensional surface geometry of abdominal aortic aneurysms. Ann. Biomed. Eng. **27**(4), 469–479 (1999). PMID:10468231
18. Doyle, B.J., Callanan, A., Burke, P.E., Grace, P.A., Walsh, M.T., Vorp, D.A., McGloughlin, T.M.: Vessel asymmetry as an additional diagnostic tool in the assessment of abdominal aortic aneurysms. J. Vasc. Surg. **49**(2), 443–454 (2009). PMID:19028061
19. Wang, D.H., Makaroun, M.S., Webster, M.W., Vorp, D.A.: Effect of intraluminal thrombus on wall stress in patient-specific models of abdominal aortic aneurysm. J. Vasc. Surg. **36**(3), 598–604 (2002). PMID:12218961
20. Vande Geest, J.P., Wang, D.H., Wisniewski, S.R., Makaroun, M.S., Vorp, D.A.: Towards a noninvasive method for determination of patient-specific wall strength distribution in abdominal aortic aneurysms. Ann. Biomed. Eng. **34**(7), 1098–1106 (2006). PMID: 16786395
21. Vorp, D.A., Lee, P.C., Wang, D.H., Makaroun, M.S., Nemoto, E.M., Ogawa, S., Webster, M.W.: Association of intraluminal thrombus in abdominal aortic aneurysm with local hypoxia and wall weakening. J. Vasc. Surg. **34**(2), 291–299 (2001). PMID:11496282
22. Fillinger, M.F., Marra, S.P., Raghavan, M.L., Kennedy, F.E.: Prediction of rupture risk in abdominal aortic aneurysm during observation: wall stress versus diameter. J. Vasc. Surg. **37**(4), 724–732 (2003). PMID:12663969
23. Fillinger, M.F., Raghavan, M.L., Marra, S.P., Cronenwett, J.L., Kennedy, F.E.: In vivo analysis of mechanical wall stress and abdominal aortic aneurysm rupture risk. J. Vasc. Surg. **36**(3), 589–597 (2002). PMID:12218986
24. Venkatasubramaniam, A.K., Fagan, M.J., Mehta, T., Mylankal, K.J., Ray, B., Kuhan, G., Chetter, I.C., McCollum, P.T.: A comparative study of aortic wall stress using finite element analysis for ruptured and non-ruptured abdominal aortic aneurysms. Eur. J. Vasc. Endovasc. Surg. **28**(2), 168–176 (2004). PMID:15234698
25. Doyle, B.J., Cloonan, A.J., Walsh, M.T., Vorp, D.A., McGloughlin, T.M.: Identification of rupture locations in patient-specific abdominal aortic aneurysms using experimental and computational techniques. J. Biomech. **43**(7), 1408–1416 (2010). PMID:20152982
26. Vorp, D.A., Raghavan, M.L., Muluk, S.C., Makaroun, M.S., Steed, D.L., Shapiro, R., Webster, M.W.: Wall strength and stiffness of aneurysmal and nonaneurysmal abdominal aorta. Ann. N. Y. Acad. Sci. **800**, 274–276 (1996). PMID:8959012
27. Wang, D., Makaroun, M., Vorp, D.: Noninvasive estimation of wall strength distribution in human abdominal aortic aneurysm. In: Proceedings of ASME-IMECE (2001)
28. Maier, A., Gee, M.W., Reeps, C., Pongratz, J., Eckstein, H.H., Wall, W.A.: A comparison of diameter, wall stress, and rupture potential index for abdominal aortic aneurysm rupture risk prediction. Ann. Biomed. Eng. **38**(10), 3124–3134 (2010). PMID:ISI:000281895500008
29. Vande Geest, J.P., Sacks, M.S., Vorp, D.A.: The effects of aneurysm on the biaxial mechanical behavior of human abdominal aorta. J. Biomech. **39**(7), 1324–1334 (2006). PMID:15885699
30. He, C.M., Roach, M.R.: The composition and mechanical properties of abdominal aortic aneurysms. J. Vasc. Surg. **20**(1), 6–13 (1994). PMID:8028090

31. Thubrikar, M.J., Labrosse, M., Robicsek, F., Al-Soudi, J., Fowler, B.: Mechanical properties of abdominal aortic aneurysm wall. J. Med. Eng. Technol. **25**(4), 133–142 (2001). PMID: 11601439
32. Iliopoulos, D.C., Deveja, R.P., Kritharis, E.P., Perrea, D., Sionis, G.D., Toutouzas, K., Stefanadis, C., Sokolis, D.P.: Regional and directional variations in the mechanical properties of ascending thoracic aortic aneurysms. Med. Eng. Phys. **31**(1), 1–9 (2009). PMID:18434231
33. Vorp, D.A., Schiro, B.J., Ehrlich, M.P., Juvonen, T.S., Ergin, M.A., Griffith, B.P.: Effect of aneurysm on the tensile strength and biomechanical behavior of the ascending thoracic aorta. Ann. Thorac. Surg. **75**(4), 1210–1214 (2003). PMID:12683565
34. Duprey, A., Khanafer, K., Schlicht, M., Avril, S., Williams, D., Berguer, R.: In vitro characterisation of physiological and maximum elastic modulus of ascending thoracic aortic aneurysms using uniaxial tensile testing. Eur. J. Vasc. Endovasc. Surg. **39**(6), 700–707 (2010). PMID:ISI:000279522700008
35. Brown, P.M., Zelt, D.T., Sobolev, B.: The risk of rupture in untreated aneurysms: the impact of size, gender, and expansion rate. J. Vasc. Surg. **37**(2), 280–284 (2003). PMID:12563196
36. Hatakeyama, T., Shigematsu, H., Muto, T.: Risk factors for rupture of abdominal aortic aneurysm based on three-dimensional study. J. Vasc. Surg. **33**(3), 453–461 (2001). PMID: 11241112
37. Sonesson, B., Sandgren, T., Lanne, T.: Abdominal aortic aneurysm wall mechanics and their relation to risk of rupture. Eur. J. Vasc. Endovasc. Surg. **18**(6), 487–493 (1999). PMID:10637144
38. Stenbaek, J., Kalin, B., Swedenborg, J.: Growth of thrombus may be a better predictor of rupture than diameter in patients with abdominal aortic aneurysms. Eur. J. Vasc. Endovasc. Surg. **20**(5), 466–469 (2000). PMID:11112467
39. Xenos, M., Rambhia, S.H., Alemu, Y., Einav, S., Labropoulos, N., Tassiopoulos, A., Ricotta, J.J., Bluestein, D.: Patient-based abdominal aortic aneurysm rupture risk prediction with fluid structure interaction modeling. Ann. Biomed. Eng. **38**(11), 3323–3337 (2010). PMID:ISI:000282506400006
40. Yadav, M.K.: Takayasu arteritis: clinical and CT-angiography profile of 25 patients and a brief review of literature. Indian Heart J. **59**(6), 468–474 (2007). PMID:19151461
41. Cid, M.C., Garcia-Martinez, A., Lozano, E., Espigol-Frigole, G., Hernandez-Rodriguez, J.: Five clinical conundrums in the management of giant cell arteritis. Rheum. Dis. Clin. North Am. **33**(4), 819–834 (2007). vii PMID:18037119
42. Long, R., Guzman, R., Greenberg, H., Safneck, J., Hershfield, E.: Tuberculous mycotic aneurysm of the aorta: review of published medical and surgical experience. Chest **115**(2), 522–531 (1999). PMID:10027455
43. Boyum, J., Fellinger, E.K., Schmoker, J.D., Trombley, L., McPartland, K., Ittleman, F.P., Howard, A.B.: Matrix metalloproteinase activity in thoracic aortic aneurysms associated with bicuspid and tricuspid aortic valves. J. Thorac. Cardiovasc. Surg. **127**(3), 686–691 (2004). PMID:15001896
44. Fedak, P.W., de Sa, M.P., Verma, S., Nili, N., Kazemian, P., Butany, J., Strauss, B.H., Weisel, R.D., David, T.E.: Vascular matrix remodeling in patients with bicuspid aortic valve malformations: implications for aortic dilatation. J. Thorac. Cardiovasc. Surg. **126**(3), 797–806 (2003). PMID:14502156
45. Nataatmadja, M., West, M., West, J., Summers, K., Walker, P., Nagata, M., Watanabe, T.: Abnormal extracellular matrix protein transport associated with increased apoptosis of vascular smooth muscle cells in marfan syndrome and bicuspid aortic valve thoracic aortic aneurysm. Circulation **108**(Suppl 1), II329–II334 (2003). PMID:12970255
46. Ikonomidis, J.S., Jones, J.A., Barbour, J.R., Stroud, R.E., Clark, L.L., Kaplan, B.S., Zeeshan, A., Bavaria, J.E., Gorman 3rd, J.H., Spinale, F.G., Gorman, R.C.: Expression of matrix metalloproteinases and endogenous inhibitors within ascending aortic aneurysms of patients with bicuspid or tricuspid aortic valves. J. Thorac. Cardiovasc. Surg. **133**(4), 1028–1036 (2007). PMID:17382648

47. Lesauskaite, V., Tanganelli, P., Sassi, C., Neri, E., Diciolla, F., Ivanoviene, L., Epistolato, M.C., Lalinga, A.V., Alessandrini, C., Spina, D.: Smooth muscle cells of the media in the dilatative pathology of ascending thoracic aorta: morphology, immunoreactivity for osteopontin, matrix metalloproteinases, and their inhibitors. Hum. Pathol. **32**(9), 1003–1011 (2001). PMID:11567232
48. Segura, A.M., Luna, R.E., Horiba, K., Stetler-Stevenson, W.G., McAllister Jr, H.A, Willerson, J.T., Ferrans, V.J.: Immunohistochemistry of matrix metalloproteinases and their inhibitors in thoracic aortic aneurysms and aortic valves of patients with Marfan's syndrome. Circulation **98**(19 Supp), II331–II337 (1998). discussion II337–II338 PMID:9852923
49. Tamarina, N.A., McMillan, W.D., Shively, V.P., Pearce, W.H.: Expression of matrix metalloproteinases and their inhibitors in aneurysms and normal aorta. Surgery **122**(2), 264–271 (1997). discussion 271–272 PMID:9288131
50. LeMaire, S.A., Wang, X., Wilks, J.A., Carter, S.A., Wen, S., Won, T., Leonardelli, D., Anand, G., Conklin, L.D., Wang, X.L., Thompson, R.W., Coselli, J.S.: Matrix metalloproteinases in ascending aortic aneurysms: bicuspid versus trileaflet aortic valves. J. Surg. Res. **123**(1), 40–48 (2005). PMID:15652949
51. Jacob, M.P.: Extracellular matrix remodeling and matrix metalloproteinases in the vascular wall during aging and in pathological conditions. Biomed. Pharmacother. **57**(5–6), 195–202 (2003). PMID:12888254
52. Xiong, W., Zhao, Y., Prall, A., Greiner, T.C., Baxter, B.T.: Key roles of CD4+ T cells and IFN-gamma in the development of abdominal aortic aneurysms in a murine model. J. Immunol. **172**(4), 2607–2612 (2004). PMID:14764734
53. Pan, J.H., Lindholt, J.S., Sukhova, G.K., Baugh, J.A., Henneberg, E.W., Bucala, R., Donnelly, S.C., Libby, P., Metz, C., Shi, G.P.: Macrophage migration inhibitory factor is associated with aneurysmal expansion. J. Vasc. Surg. **37**(3), 628–635 (2003). PMID: 12618703
54. Satta, J., Mennander, A., Soini, Y.: Increased medial TUNEL-positive staining associated with apoptotic bodies is linked to smooth muscle cell diminution during evolution of abdominal aortic aneurysms. Ann. Vasc. Surg. **16**(4), 462–466 (2002). PMID:12132024
55. Treska, V., Kocova, J., Boudova, L., Neprasova, P., Topolcan, O., Pecen, L., Tonar, Z.: Inflammation in the wall of abdominal aortic aneurysm and its role in the symptomatology of aneurysm. Cytokines Cell Mol. Ther. **7**(3), 91–97 (2002). PMID:12850808
56. Bobryshev, Y.V., Lord, R.S., Parsson, H.: Immunophenotypic analysis of the aortic aneurysm wall suggests that vascular dendritic cells are involved in immune responses. Cardiovasc. Surg. **6**(3), 240–249 (1998). PMID:9705095
57. Curci, J.A., Liao, S., Huffman, M.D., Shapiro, S.D., Thompson, R.W.: Expression and localization of macrophage elastase (matrix metalloproteinase-12) in abdominal aortic aneurysms. J. Clin. Invest. **102**(11), 1900–1910 (1998). PMID:9835614
58. Anidjar, S., Dobrin, P.B., Eichorst, M., Graham, G.P., Chejfec, G.: Correlation of inflammatory infiltrate with the enlargement of experimental aortic aneurysms. J. Vasc. Surg. **16**(2), 139–147 (1992). PMID:1386635
59. Koch, A.E., Haines, G.K., Rizzo, R.J., Radosevich, J.A., Pope, R.M., Robinson, P.G., Pearce, W.H.: Human abdominal aortic aneurysms. Immunophenotypic analysis suggesting an immune-mediated response. Am. J. Pathol. **137**(5), 1199–1213 (1990). PMID:1700620
60. Schlatmann, T.J., Becker, A.E.: Pathogenesis of dissecting aneurysm of aorta. Comparative histopathologic study of significance of medial changes. Am. J. Cardiol. **39**(1), 21–26 (1977). PMID:831424
61. Marsalese, D.L., Moodie, D.S., Lytle, B.W., Cosgrove, D.M., Ratliff, N.B., Goormastic, M., Kovacs, A.: Cystic medial necrosis of the aorta in patients without Marfan's syndrome: surgical outcome and long-term follow-up. J. Am. Coll. Cardiol. **16**(1), 68–73 (1990). PMID:2358606
62. Coady, M.A., Rizzo, J.A., Goldstein, L.J., Elefteriades, J.A.: Natural history, pathogenesis, and etiology of thoracic aortic aneurysms and dissections. Cardiol. Clin. **17**(4), 615–635 (1999). vii PMID:10589336

63. Dietz, H.C., Cutting, G.R., Pyeritz, R.E., Maslen, C.L., Sakai, L.Y., Corson, G.M., Puffenberger, E.G., Hamosh, A., Nanthakumar, E.J., Curristin, S.M., et al.: Marfan syndrome caused by a recurrent de novo missense mutation in the fibrillin gene. Nature **352**(6333), 337–339 (1991). PMID:1852208
64. Loeys, B.L., Schwarze, U., Holm, T., Callewaert, B.L., Thomas, G.H., Pannu, H., De Backer, J.F., Oswald, G.L., Symoens, S., Manouvrier, S., Roberts, A.E., Faravelli, F., Greco, M.A., Pyeritz, R.E., Milewicz, D.M., Coucke, P.J., Cameron, D.E., Braverman, A.C., Byers, P.H., De Paepe, A.M., Dietz, H.C.: Aneurysm syndromes caused by mutations in the TGF-beta receptor. N. Engl. J. Med. **355**(8), 788–798 (2006). PMID:16928994
65. Narcisi, P., Richards, A.J., Ferguson, S.D., Pope, F.M.: A family with Ehlers-Danlos syndrome type III/articular hypermobility syndrome has a glycine 637 to serine substitution in type III collagen. Hum. Mol. Genet. **3**(9), 1617–1620 (1994). PMID:7833919
66. Zhu, L., Vranckx, R., Khau Van Kien, P., Lalande, A., Boisset, N., Mathieu, F., Wegman, M., Glancy, L., Gasc, J.M., Brunotte, F., Bruneval, P., Wolf, J.E., Michel, J.B., Jeunemaitre, X.: Mutations in myosin heavy chain 11 cause a syndrome associating thoracic aortic aneurysm/aortic dissection and patent ductus arteriosus. Nat. Genet. **38**(3), 343–349 (2006). PMID:16444274
67. Pannu, H., Tran-Fadulu, V., Papke, C.L., Scherer, S., Liu, Y., Presley, C., Guo, D., Estrera, A.L., Safi, H.J., Brasier, A.R., Vick, G.W., Marian, A.J., Raman, C.S., Buja, L.M., Milewicz, D.M.: MYH11 mutations result in a distinct vascular pathology driven by insulin-like growth factor 1 and angiotensin II. Hum. Mol. Genet. **16**(20), 2453–2462 (2007). PMID:17666408
68. Edwards, W.D., Leaf, D.S., Edwards, J.E.: Dissecting aortic aneurysm associated with congenital bicuspid aortic valve. Circulation **57**(5), 1022–1025 (1978). PMID:639201
69. Fedak, P.W., Verma, S., David, T.E., Leask, R.L., Weisel, R.D., Butany, J.: Clinical and pathophysiological implications of a bicuspid aortic valve. Circulation **106**(8), 900–904 (2002). PMID:12186790
70. Gleason, T.G.: Heritable disorders predisposing to aortic dissection. Semin. Thorac. Cardiovasc. Surg. **17**(3), 274–281 (2005). PMID:16253833
71. Roberts, W.C.: The congenitally bicuspid aortic valve. A study of 85 autopsy cases. Am. J. Cardiol. **26**(1), 72–83 (1970). PMID:5427836
72. Ward, C.: Clinical significance of the bicuspid aortic valve. Heart **83**(1), 81–85 (2000). PMID:10618341
73. Milewicz, D.M., Guo, D.C., Tran-Fadulu, V., Lafont, A.L., Papke, C.L., Inamoto, S., Kwartler, C.S., Pannu, H.: Genetic basis of thoracic aortic aneurysms and dissections: focus on smooth muscle cell contractile dysfunction. Annu. Rev. Genomics Hum. Genet. **9**, 283–302 (2008). PMID:18544034
74. Niitsuya, M., Kuwao, S., Sato, B., Kameya, T., Kikawada, R.: Histopathological study of aortic wall dissection. J. Cardiol. **21**(2), 445–452 (1991). PMID:1841931
75. Niwa, K., Perloff, J.K., Bhuta, S.M., Laks, H., Drinkwater, D.C., Child, J.S., Miner, P.D.: Structural abnormalities of great arterial walls in congenital heart disease: light and electron microscopic analyses. Circulation **103**(3), 393–400 (2001). PMID:11157691
76. Ramirez, F., Rifkin, D.B.: Cell signaling events: a view from the matrix. Matrix Biol. **22**(2), 101–107 (2003). PMID:12782137
77. Reinhardt, D.P., Gambee, J.E., Ono, R.N., Bachinger, H.P., Sakai, L.Y.: Initial steps in assembly of microfibrils. Formation of disulfide-cross-linked multimers containing fibrillin-1. J. Biol. Chem. **275**(3), 2205–2210 (2000). PMID:10636927
78. Gleizes, P.E., Beavis, R.C., Mazzieri, R., Shen, B., Rifkin, D.B.: Identification and characterization of an eight-cysteine repeat of the latent transforming growth factor-beta binding protein-1 that mediates bonding to the latent transforming growth factor-beta1. J. Biol. Chem. **271**(47), 29891–29896 (1996). PMID:8939931
79. Saharinen, J., Taipale, J., Keski-Oja, J.: Association of the small latent transforming growth factor-beta with an eight cysteine repeat of its binding protein LTBP-1. EMBO J. **15**(2), 245–253 (1996). PMID:8617200

80. D'Arrigo, C., Burl, S., Withers, A.P., Dobson, H., Black, C., Boxer, M.: TGF-beta1 binding protein-like modules of fibrillin-1 and -2 mediate integrin-dependent cell adhesion. Connect. Tissue Res. **37**(1–2), 29–51 (1998). PMID:9643645
81. Pfaff, M., Reinhardt, D.P., Sakai, L.Y., Timpl, R.: Cell adhesion and integrin binding to recombinant human fibrillin-1. FEBS Lett. **384**(3), 247–250 (1996). PMID:8617364
82. Sakamoto, H., Broekelmann, T., Cheresh, D.A., Ramirez, F., Rosenbloom, J., Mecham, R.P.: Cell-type specific recognition of RGD- and non-RGD-containing cell binding domains in fibrillin-1. J. Biol. Chem. **271**(9), 4916–4922 (1996). PMID:8617764
83. Kielty, C.M., Baldock, C., Lee, D., Rock, M.J., Ashworth, J.L., Shuttleworth, C.A.: Fibrillin: from microfibril assembly to biomechanical function. Philos. Trans. R. Soc. Lond. B Biol. Sci. **357**(1418), 207–217 (2002). PMID:11911778
84. Judge, D.P., Biery, N.J., Keene, D.R., Geubtner, J., Myers, L., Huso, D.L., Sakai, L.Y., Dietz, H.C.: Evidence for a critical contribution of haploinsufficiency in the complex pathogenesis of Marfan syndrome. J. Clin. Invest. **114**(2), 172–181 (2004). PMID:15254584
85. Habashi, J.P., Judge, D.P., Holm, T.M., Cohn, R.D., Loeys, B.L., Cooper, T.K., Myers, L., Klein, E.C., Liu, G., Calvi, C., Podowski, M., Neptune, E.R., Halushka, M.K., Bedja, D., Gabrielson, K., Rifkin, D.B., Carta, L., Ramirez, F., Huso, D.L., Dietz, H.C.: Losartan, an AT1 antagonist, prevents aortic aneurysm in a mouse model of Marfan syndrome. Science **312**(5770), 117–121 (2006). PMID:16601194
86. Carta, L., Wagenseil, J.E., Knutsen, R.H., Mariko, B., Faury, G., Davis, E.C., Starcher, B., Mecham, R.P., Ramirez, F.: Discrete contributions of elastic fiber components to arterial development and mechanical compliance. Arterioscler. Thromb. Vasc. Biol. **29**(12), 2083–2089 (2009). PMID:19850904
87. Brooke, B.S., Habashi, J.P., Judge, D.P., Patel, N., Loeys, B., Dietz 3rd, H.C.: Angiotensin II blockade and aortic-root dilation in Marfan's syndrome. N. Engl. J. Med. **358**(26), 2787–2795 (2008). PMID:18579813
88. Kalluri, R., Han, Y.: Targeting TGF-beta and the extracellular matrix in Marfan's syndrome. Dev. Cell **15**(1), 1–2 (2008). PMID:18606132
89. Sonesson, B., Hansen, F., Lanne, T.: Abnormal mechanical properties of the aorta in Marfan's syndrome. Eur J. Vasc. Surg. **8**(5), 595–601 (1994). PMID:7813727
90. Eberth, J.F., Taucer, A.I., Wilson, E., Humphrey, J.D.: Mechanics of carotid arteries in a mouse model of Marfan syndrome. Ann. Biomed. Eng. **37**(6), 1093–1104 (2009). PMID:19350391
91. Loeys, B.L., Chen, J., Neptune, E.R., Judge, D.P., Podowski, M., Holm, T., Meyers, J., Leitch, C.C., Katsanis, N., Sharifi, N., Xu, F.L., Myers, L.A., Spevak, P.J., Cameron, D.E., De Backer, J., Hellemans, J., Chen, Y., Davis, E.C., Webb, C.L., Kress, W., Coucke, P., Rifkin, D.B., De Paepe, A.M., Dietz, H.C.: A syndrome of altered cardiovascular, craniofacial, neurocognitive and skeletal development caused by mutations in TGFBR1 or TGFBR2. Nat. Genet. **37**(3), 275–281 (2005). PMID:15731757
92. Maleszewski, J.J., Miller, D.V., Lu, J., Dietz, H.C., Halushka, M.K.: Histopathologic findings in ascending aortas from individuals with Loeys-Dietz syndrome (LDS). Am. J. Surg. Pathol. **33**(2), 194–201 (2009). PMID:18852674
93. Wahab, N.A., Weston, B.S., Mason, R.M.: Modulation of the TGFbeta/Smad signaling pathway in mesangial cells by CTGF/CCN2. Exp. Cell Res. **307**(2), 305–314 (2005). PMID:15950619
94. Ruiz-Ortega, M., Rodriguez-Vita, J., Sanchez-Lopez, E., Carvajal, G., Egido, J.: TGF-beta signaling in vascular fibrosis. Cardiovasc. Res. **74**(2), 196–206 (2007). PMID:17376414
95. Beighton, P., De Paepe, A., Steinmann, B., Tsipouras, P., Wenstrup, R.J.: Ehlers-Danlos syndromes: revised nosology, Villefranche, 1997. Ehlers-Danlos National Foundation (USA) and Ehlers-Danlos Support Group (UK). Am. J. Med. Genet. 77(1):31–7 (1998) PMID:9557891
96. Pope, F.M., Narcisi, P., Nicholls, A.C., Germaine, D., Pals, G., Richards, A.J.: COL3A1 mutations cause variable clinical phenotypes including acrogeria and vascular rupture. Br. J. Dermatol. **135**(2), 163–181 (1996). PMID:8881656

97. Borck, G., Beighton, P., Wilhelm, C., Kohlhase, J., Kubisch, C.: Arterial rupture in classic Ehlers-Danlos syndrome with COL5A1 mutation. Am. J. Med. Genet. A **152A**(8), 2090–2093 (2010). PMID:20635400
98. Burke, J.M., Balian, G., Ross, R., Bornstein, P.: Synthesis of types I and III procollagen and collagen by monkey aortic smooth muscle cells in vitro. Biochemistry **16**(14), 3243–3249 (1977). PMID:407926
99. McCullagh, K.A., Balian, G.: Collagen characterisation and cell transformation in human atherosclerosis. Nature **258**(5530), 73–75 (1975). PMID:1186882
100. Smith, L.B., Hadoke, P.W., Dyer, E., Denvir, M.A., Brownstein, D., Miller, E., Nelson, N., Wells, S., Cheeseman, M., Greenfield, A.: Haploinsufficiency of the murine Col3a1 locus causes aortic dissection: a novel model of the vascular type of Ehlers Danlos Syndrome. Cardiovasc. Res. **90**, 182–190 (2010) PMID:21071432
101. McCullagh, K.G., Duance, V.C., Bishop, K.A.: The distribution of collagen types I, III and V (AB) in normal and atherosclerotic human aorta. J. Pathol. **130**(1), 45–55 (1980). PMID:6991657
102. Sadakata, R., Hatamochi, A., Kodama, K., Kaga, A., Yamaguchi, T., Soma, T., Usui, Y., Nagata, M., Ohtake, A., Hagiwara, K., Kanazawa, M.: Ehlers-Danlos syndrome type IV, vascular type, which demonstrated a novel point mutation in the COL3A1 gene. Intern. Med. **49**(16), 1797–1800 (2010). PMID:20720362
103. Kim, Y., Peyrol, S., So, C.K., Boyd, C.D., Csiszar, K.: Coexpression of the lysyl oxidase-like gene (LOXL) and the gene encoding type III procollagen in induced liver fibrosis. J. Cell Biochem. **72**(2), 181–188 (1999). PMID:10022501
104. Sharma, R., Kramer, J.A., Krawetz, S.A.: Lysyl oxidase, cellular senescence and tumor suppression. Biosci. Rep. **17**(4), 409–414 (1997). PMID:9367056
105. Sommer, P., Gleyzal, C., Raccurt, M., Delbourg, M., Serrar, M., Joazeiro, P., Peyrol, S., Kagan, H., Trackman, P.C., Grimaud, J.A.: Transient expression of lysyl oxidase by liver myofibroblasts in murine schistosomiasis. Lab. Invest. **69**(4), 460–470 (1993). PMID: 7901452
106. Rucker, R.B., Kosonen, T., Clegg, M.S., Mitchell, A.E., Rucker, B.R., Uriu-Hare, J.Y., Keen, C.L.: Copper, lysyl oxidase, and extracellular matrix protein cross-linking. Am. J. Clin. Nutr. **67**(5 Suppl), 996S–1002S (1998). PMID:9587142
107. Peltonen, L., Kuivaniemi, H., Palotie, A., Horn, N., Kaitila, I., Kivirikko, K.I.: Alterations in copper and collagen metabolism in the Menkes syndrome and a new subtype of the Ehlers-Danlos syndrome. Biochemistry **22**(26), 6156–6163 (1983). PMID:6140952
108. Maki, J.M., Rasanen, J., Tikkanen, H., Sormunen, R., Makikallio, K., Kivirikko, K.I., Soininen, R.: Inactivation of the lysyl oxidase gene Lox leads to aortic aneurysms, cardiovascular dysfunction, and perinatal death in mice. Circulation **106**(19), 2503–2509 (2002). PMID:12417550
109. Jones, J.A., Spinale, F.G., Ikonomidis, J.S.: Transforming growth factor-beta signaling in thoracic aortic aneurysm development: a paradox in pathogenesis. J. Vasc. Res. **46**(2), 119–137 (2009). PMID:18765947
110. Sun, Y.L., Luo, Z.P., Fertala, A., An, K.N.: Direct quantification of the flexibility of type I collagen monomer. Biochem. Biophys. Res. Commun. **295**(2), 382–386 (2002). PMID: 12150960
111. Januzzi, J.L., Isselbacher, E.M., Fattori, R., Cooper, J.V., Smith, D.E., Fang, J., Eagle, K.A., Mehta, R.H., Nienaber, C.A., Pape, L.A.: Characterizing the young patient with aortic dissection: results from the International Registry of Aortic Dissection (IRAD). J. Am. Coll. Cardiol. **43**(4), 665–669 (2004). PMID:14975480
112. Keane, M.G., Wiegers, S.E., Plappert, T., Pochettino, A., Bavaria, J.E., Sutton, M.G.: Bicuspid aortic valves are associated with aortic dilatation out of proportion to coexistent valvular lesions. Circulation **102**(19 Suppl 3), III35–III39 (2000). PMID:11082359
113. Nistri, S., Grande-Allen, J., Noale, M., Basso, C., Siviero, P., Maggi, S., Crepaldi, G., Thiene, G.: Aortic elasticity and size in bicuspid aortic valve syndrome. Eur. Heart J. **29**(4), 472–479 (2008). PMID:18096569

114. Nistri, S., Sorbo, M.D., Marin, M., Palisi, M., Scognamiglio, R., Thiene, G.: Aortic root dilatation in young men with normally functioning bicuspid aortic valves. Heart **82**(1), 19–22 (1999). PMID:10377302
115. Cripe, L., Andelfinger, G., Martin, L.J., Shooner, K., Benson, D.W.: Bicuspid aortic valve is heritable. J. Am. Coll. Cardiol. **44**(1), 138–143 (2004). PMID:15234422
116. Martin, L.J., Ramachandran, V., Cripe, L.H., Hinton, R.B., Andelfinger, G., Tabangin, M., Shooner, K., Keddache, M., Benson, D.W.: Evidence in favor of linkage to human chromosomal regions 18q, 5q and 13q for bicuspid aortic valvc and associated cardiovascular malformations. Hum. Genet. **121**(2), 275–284 (2007). PMID:17203300
117. Garg, V., Muth, A.N., Ransom, J.F., Schluterman, M.K., Barnes, R., King, I.N., Grossfeld, P.D., Srivastava, D.: Mutations in NOTCH1 cause aortic valve disease. Nature **437**(7056), 270–274 (2005). PMID:16025100
118. Phillippi, J.A., Eskay, M.A., Kubala, A.A., Pitt, B.R., Gleason, T.G.: Altered oxidative stress responses and increased type I collagen expression in bicuspid aortic valve patients. Ann. Thorac. Surg. **90**(6), 1893–1898 (2010). PMID:21095332
119. Phillippi, J.A., Klyachko, E.A., Kenny, J.P., Eskay, M.A., Gorman, R.C., Gleason, T.G.: Basal and oxidative stress-induced expression of metallothionein is decreased in ascending aortic aneurysms of bicuspid aortic valve patients. Circulation **119**(18), 2498–2506 (2009). PMID:ISI:000266011500011
120. Ejiri, J., Inoue, N., Tsukube, T., Munezane, T., Hino, Y., Kobayashi, S., Hirata, K., Kawashima, S., Imajoh-Ohmi, S., Hayashi, Y., Yokozaki, H., Okita, Y., Yokoyama, M.: Oxidative stress in the pathogenesis of thoracic aortic aneurysm: protective role of statin and angiotensin II type 1 receptor blocker. Cardiovasc. Res. **59**(4), 988–996 (2003). PMID: 14553839
121. Fiorillo, C., Becatti, M., Attanasio, M., Lucarini, L., Nassi, N., Evangelisti, L., Porciani, M.C., Nassi, P., Gensini, G.F., Abbate, R., Pepe, G.: Evidence for oxidative stress in plasma of patients with Marfan syndrome. Int. J. Cardiol. **145**, 544–546 (2010) PMID:20537738
122. Robbins, A.H., McRee, D.E., Williamson, M., Collett, S.A., Xuong, N.H., Furey, W.F., Wang, B.C., Stout, C.D.: Refined crystal structure of Cd, Zn metallothionein at 2.0 A resolution. J. Mol. Biol. **221**(4), 1269–1293 (1991). PMID:1942051
123. Arseniev, A., Schultze, P., Worgotter, E., Braun, W., Wagner, G., Vasak, M., Kagi, J.H., Wuthrich, K.: Three-dimensional structure of rabbit liver [Cd7] metallothionein-2a in aqueous solution determined by nuclear magnetic resonance. J. Mol. Biol. **201**(3), 637–657 (1988). PMID:3418714
124. Pitt, B.R., Schwarz, M., Woo, E.S., Yee, E., Wasserloos, K., Tran, S., Weng, W., Mannix, R.J., Watkins, S.A., Tyurina, Y.Y., Tyurin, V.A., Kagan, V.E., Lazo, J.S.: Overexpression of metallothionein decreases sensitivity of pulmonary endothelial cells to oxidant injury. Am. J. Physiol. **273**(4 Pt 1), L856–L865 (1997). PMID:9357862
125. Abel, J., de Ruiter, N.: Inhibition of hydroxyl-radical-generated DNA degradation by metallothionein. Toxicol. Lett. **47**(2), 191–196 (1989). PMID:2545017
126. Thornally, P., Vasak, M.: Possible role for metallothionein in protections against radiation-induced oxidative stress. Lineteics and mechanism of its reaction with superoxide hydroxyl radicals. Biochem. Biophys. Acta. **827**, 36–44 (1985)
127. Oh, S.H., Deagen, J.T., Whanger, P.D., Weswig, P.H.: Biological function of metallothionein. V. Its induction in rats by various stresses. Am. J. Physiol. **234**(3), E282–E285 (1978). PMID:629343
128. Klaassen, C.D., Liu, J.: Metallothionein transgenic and knock-out mouse models in the study of cadmium toxicity. J. Toxicol. Sci. **23**(Suppl 2), 97–102 (1998). PMID:9760441
129. Fang, C.X., Doser, T.A., Yang, X., Sreejayan, N., Ren, J.: Metallothionein antagonizes aging-induced cardiac contractile dysfunction: role of PTP1B, insulin receptor tyrosine phosphorylation and Akt. Aging Cell **5**(2), 177–185 (2006). PMID:16626396
130. Feng, W., Wang, Y., Cai, L., Kang, Y.J.: Metallothionein rescues hypoxia-inducible factor-1 transcriptional activity in cardiomyocytes under diabetic conditions. Biochem. Biophys. Res. Commun. **360**(1), 286–289 (2007). PMID:17586470

131. Surazynski, A., Donald, S.P., Cooper, S.K., Whiteside, M.A., Salnikow, K., Liu, Y., Phang, J.M.: Extracellular matrix and HIF-1 signaling: the role of prolidase. Int. J. Cancer **122**(6), 1435–1440 (2008). PMID:17999410
132. Kurban, G., Duplan, E., Ramlal, N., Hudon, V., Sado, Y., Ninomiya, Y., Pause, A.: Collagen matrix assembly is driven by the interaction of von Hippel-Lindau tumor suppressor protein with hydroxylated collagen IV alpha 2. Oncogene **27**(7), 1004–1012 (2008). PMID:17700531
133. Elstner, A., Holtkamp, N., von Deimling, A.: Involvement of Hif-1 in desferrioxamine-induced invasion of glioblastoma cells. Clin. Exp. Metastasis **24**(1), 57–66 (2007). PMID:17357815
134. Norman, J.T., Orphanides, C., Garcia, P., Fine, L.G.: Hypoxia-induced changes in extracellular matrix metabolism in renal cells. Exp. Nephrol. **7**(5–6), 463–469 (1999). PMID:10559644
135. Kivela, R., Kyrolainen, H., Selanne, H., Komi, P.V., Kainulainen, H., Vihko, V.: A single bout of exercise with high mechanical loading induces the expression of Cyr61/CCN1 and CTGF/CCN2 in human skeletal muscle. J. Appl. Physiol. **103**(4), 1395–1401 (2007). PMID:17673559
136. Pereira, L., Lee, S.Y., Gayraud, B., Andrikopoulos, K., Shapiro, S.D., Bunton, T., Biery, N.J., Dietz, H.C., Sakai, L.Y., Ramirez, F.: Pathogenetic sequence for aneurysm revealed in mice underexpressing fibrillin-1. Proc Natl Acad Sci USA **96**(7), 3819–3823 (1999). PMID:10097121
137. Yang, H.H., van Breemen, C., Chung, A.W.: Vasomotor dysfunction in the thoracic aorta of Marfan syndrome is associated with accumulation of oxidative stress. Vascul. Pharmacol. **52**(1–2), 37–45 (2010). PMID:19879959
138. Radomski, A., Sawicki, G., Olson, D.M., Radomski, M.W.: The role of nitric oxide and metalloproteinases in the pathogenesis of hyperoxia-induced lung injury in newborn rats. Br. J. Pharmacol. **125**(7), 1455–1462 (1998). PMID:9884073
139. Tyagi, S.C., Matsubara, L., Weber, K.T.: Direct extraction and estimation of collagenase(s) activity by zymography in microquantities of rat myocardium and uterus. Clin. Biochem. **26**(3), 191–198 (1993). PMID:8330388
140. Steed, M.M., Tyagi, N., Sen, U., Schuschke, D.A., Joshua, I.G., Tyagi, S.C.: Functional consequences of the collagen/elastin switch in vascular remodeling in hyperhomocysteinemic wild-type, eNOS-/-, and iNOS-/- mice. Am. J. Physiol. Lung Cell. Mol. Physiol. **299**(3), L301–L311 (2010). PMID:20581102
141. McCormick, M.L., Gavrila, D., Weintraub, N.L.: Role of oxidative stress in the pathogenesis of abdominal aortic aneurysms. Arterioscler. Thromb. Vasc. Biol. **27**(3), 461–469 (2007). PMID:17218601
142. Gavazzi, G., Deffert, C., Trocme, C., Schappi, M., Herrmann, F.R., Krause, K.H.: NOX1 deficiency protects from aortic dissection in response to angiotensin II. Hypertension **50**(1), 189–196 (2007). PMID:17502491
143. Thomas, M., Gavrila, D., McCormick, M.L., Miller Jr, F.J., Daugherty, A., Cassis, L.A., Dellsperger, K.C., Weintraub, N.L.: Deletion of p47phox attenuates angiotensin II-induced abdominal aortic aneurysm formation in apolipoprotein E-deficient mice. Circulation **114**(5), 404–413 (2006). PMID:16864727
144. Hope, M.D., Hope, T.A., Meadows, A.K., Ordovas, K.G., Urbania, T.H., Alley, M.T., Higgins, C.B.: Bicuspid aortic valve: four-dimensional MR evaluation of ascending aortic systolic flow patterns. Radiology **255**(1), 53–61 (2010). PMID:20308444
145. Bauer, M., Siniawski, H., Pasic, M., Schaumann, B., Hetzer, R.: Different hemodynamic stress of the ascending aorta wall in patients with bicuspid and tricuspid aortic valve. J. Card. Surg. **21**(3), 218–220 (2006). PMID:16684044
146. Nkomo, V.T., Enriquez-Sarano, M., Ammash, N.M., Melton 3rd, L.J., Bailey, K.R., Desjardins, V., Horn, R.A., Tajik, A.J.: Bicuspid aortic valve associated with aortic dilatation: a community-based study. Arterioscler. Thromb. Vasc. Biol. **23**(2), 351–356 (2003). PMID:12588783

147. Yasuda, H., Nakatani, S., Stugaard, M., Tsujita-Kuroda, Y., Bando, K., Kobayashi, J., Yamagishi, M., Kitakaze, M., Kitamura, S., Miyatake, K.: Failure to prevent progressive dilation of ascending aorta by aortic valve replacement in patients with bicuspid aortic valve: comparison with tricuspid aortic valve. Circulation **108**(Suppl 1), II291–II294 (2003). PMID:12970248
148. Wang, X., LeMaire, S.A., Chen, L., Shen, Y.H., Gan, Y., Bartsch, H., Carter, S.A., Utama, B., Ou, H., Coselli, J.S., Wang, X.L.: Increased collagen deposition and elevated expression of connective tissue growth factor in human thoracic aortic dissection. Circulation **114**(1 Suppl), I200–I205 (2006). PMID:16820572
149. Whittle, M.A., Hasleton, P.S., Anderson, J.C., Gibbs, A.C.: Collagen in dissecting aneurysms of the human thoracic aorta. Increased collagen content and decreased collagen concentration may be predisposing factors in dissecting aneurysms. Am. J. Cardiovasc. Pathol. **3**(4), 311–319 (1990). PMID:2129573
150. Choudhury, N., Bouchot, O., Rouleau, L., Tremblay, D., Cartier, R., Butany, J., Mongrain, R., Leask, R.L.: Local mechanical and structural properties of healthy and diseased human ascending aorta tissue. Cardiovasc. Pathol. **18**(2), 83–91 (2009). PMID:18402840
151. Absi, T.S., Sundt 3rd, T.M., Tung, W.S., Moon, M., Lee, J.K., Damiano Jr, R.R., Thompson, R.W.: Altered patterns of gene expression distinguishing ascending aortic aneurysms from abdominal aortic aneurysms: complementary DNA expression profiling in the molecular characterization of aortic disease. J. Thorac. Cardiovasc. Surg. **126**(2), 344–357 (2003). discission 357 PMID:12928630
152. Daugherty, A., Cassis, L.A.: Mouse models of abdominal aortic aneurysms. Arterioscler. Thromb. Vasc. Biol. **24**(3), 429–434 (2004). PMID:14739119
153. Kadoglou, N.P., Liapis, C.D.: Matrix metalloproteinases: contribution to pathogenesis, diagnosis, surveillance and treatment of abdominal aortic aneurysms. Curr. Med. Res. Opin. **20**(4), 419–432 (2004). PMID:15119978
154. Fernandez, B., Duran, A.C., Fernandez-Gallego, T., Fernandez, M.C., Such, M., Arque, J.M., Sans-Coma, V.: Bicuspid aortic valves with different spatial orientations of the leaflets are distinct etiological entities. J. Am. Coll. Cardiol. **54**(24), 2312–2318 (2009). PMID: 19958967
155. Lee, T.C., Zhao, Y.D., Courtman, D.W., Stewart, D.J.: Abnormal aortic valve development in mice lacking endothelial nitric oxide synthase. Circulation **101**(20), 2345–2348 (2000). PMID:10821808
156. Atli, F.H., Manduz, S., Katrancioglu, N., Ozum, U., Disli, O.M., Atahan, E., Ozdemir, O., Dogan, K., Berkan, O.: eNOS G894T polymorphism and abdominal aortic aneurysms. Angiology **61**, 125–130 (2010) PMID:19638352
157. Fatini, C., Sofi, F., Sticchi, E., Bolli, P., Sestini, I., Falciani, M., Azas, L., Pratesi, G.: eNOS G894T polymorphism as a mild predisposing factor for abdominal aortic aneurysm. J. Vasc. Surg. **42**(3), 415–419 (2005). PMID:16171581
158. Sandrim, V.C.: Regarding "eNOS G894T polymorphism as a mild predisposing factor for abdominal aortic aneurysm". J. Vasc. Surg. **43**(5), 1079 (2006). PMID:16678714
159. Ko, N.U., Rajendran, P., Kim, H., Rutkowski, M., Pawlikowska, L., Kwok, P.Y., Higashida, R.T., Lawton, M.T., Smith, W.S., Zaroff, J.G., Young, W.L.: Endothelial nitric oxide synthase polymorphism (-786T->C) and increased risk of angiographic vasospasm after aneurysmal subarachnoid hemorrhage. Stroke **39**(4), 1103–1108 (2008). PMID:18309169
160. Krischek, B., Kasuya, H., Akagawa, H., Tajima, A., Narita, A., Onda, H., Hori, T., Inoue, I.: Using endothelial nitric oxide synthase gene polymorphisms to identify intracranial aneurysms more prone to rupture in Japanese patients. J. Neurosurg. **105**(5), 717–722 (2006). PMID:17121133
161. Ozum, U., Bolat, N., Gul, E., Ozdemir, O.: Endothelial nitric oxide synthase gene [G894T] polymorphism as a possible risk factor in aneurysmal subarachnoid haemorrhage. Acta Neurochir. (Wien) **150**(1), 57–61 (2008). discussion 62 PMID:18046500
162. Starke, R.M., Kim, G.H., Komotar, R.J., Hickman, Z.L., Black, E.M., Rosales, M.B., Kellner, C.P., Hahn, D.K., Otten, M.L., Edwards, J., Wang, T., Russo, J.J., Mayer, S.A.,

Connolly Jr, E.S.: Endothelial nitric oxide synthase gene single-nucleotide polymorphism predicts cerebral vasospasm after aneurysmal subarachnoid hemorrhage. J. Cereb. Blood Flow Metab. **28**(6), 1204–1211 (2008). PMID:18319732

163. Tamura, T., Jamous, M.A., Kitazato, K.T., Yagi, K., Tada, Y., Uno, M., Nagahiro, S.: Endothelial damage due to impaired nitric oxide bioavailability triggers cerebral aneurysm formation in female rats. J. Hypertens. **27**(6), 1284–1292 (2009). PMID:19307983
164. Aicher, D., Urbich, C., Zeiher, A., Dimmeler, S., Schafers, H.J.: Endothelial nitric oxide synthase in bicuspid aortic valve disease. Ann. Thorac. Surg. **83**(4), 1290–1294 (2007). PMID:17383329
165. Lee, J.K., Borhani, M., Ennis, T.L., Upchurch Jr, G.R., Thompson, R.W.: Experimental abdominal aortic aneurysms in mice lacking expression of inducible nitric oxide synthase. Arterioscler. Thromb. Vasc. Biol. **21**(9), 1393–1401 (2001). PMID:11557662
166. Pimiento, J.M., Maloney, S.P., Tang, P.C., Muto, A., Westvik, T.S., Fitzgerald, T.N., Fancher, T.T., Tellides, G., Dardik, A.: Endothelial nitric oxide synthase stimulates aneurysm growth in aged mice. J. Vasc. Res. **45**(3), 251–258 (2008). PMID:18182824
167. Hultgardh-Nilsson, A., Lovdahl, C., Blomgren, K., Kallin, B., Thyberg, J.: Expression of phenotype- and proliferation-related genes in rat aortic smooth muscle cells in rimary culture. Cardiovasc. Res. **34**(2), 418–430 (1997). PMID:9205557
168. Thyberg, J.: Differentiated properties and proliferation of arterial smooth muscle cells in culture. Int. Rev. Cytol. **169**, 183–265 (1996). PMID:8843655
169. Rensen, S.S., Doevendans, P.A., van Eys, G.J.: Regulation and characteristics of vascular smooth muscle cell phenotypic diversity. Neth. Heart J. **15**(3), 100–108 (2007). PMID: 17612668
170. Bunton, T.E., Biery, N.J., Myers, L., Gayraud, B., Ramirez, F., Dietz, H.C.: Phenotypic alteration of vascular smooth muscle cells precedes elastolysis in a mouse model of Marfan syndrome. Circ. Res. **88**(1), 37–43 (2001). PMID:11139471
171. Guo, D.C., Pannu, H., Tran-Fadulu, V., Papke, C.L., Yu, R.K., Avidan, N., Bourgeois, S., Estrera, A.L., Safi, H.J., Sparks, E., Amor, D., Ades, L., McConnell, V., Willoughby, C.E., Abuelo, D., Willing, M., Lewis, R.A., Kim, D.H., Scherer, S., Tung, P.P., Ahn, C., Buja, L.M., Raman, C.S., Shete, S.S., Milewicz, D.M.: Mutations in smooth muscle alpha-actin (ACTA2) lead to thoracic aortic aneurysms and dissections. Nat. Genet. **39**(12), 1488–1493 (2007). PMID:17994018
172. Morisaki, H., Akutsu, K., Ogino, H., Kondo, N., Yamanaka, I., Tsutsumi, Y., Yoshimuta, T., Okajima, T., Matsuda, H., Minatoya, K., Sasaki, H., Tanaka, H., Ishibashi-Ueda, H., Morisaki, T.: Mutation of ACTA2 gene as an important cause of familial and nonfamilial nonsyndromatic thoracic aortic aneurysm and/or dissection (TAAD). Hum. Mutat. **30**(10), 1406–1411 (2009). PMID:19639654
173. Phillippi, J.A., Eskay, M.A., Kubala, A.A., Gleason, T.G.: Smooth Muscle Cell Phenotype Alterations in Patients With Bicuspid Aortic Valve. American Society for Cell Biology, Philadelphia (2010)
174. Phillippi, J.A., Kubala, A.A., Eskay, M.A., Chew, D.W., Gaitan, D., Hempel, J., Vorp, D.A., Watkins, S.C., Gleason, T.G.: Disrupted Collagen Homeostasis in Thoracic Aortic Aneurysms in Patients with Bicuspid Aortic Valve. International Society of Applied Cardiovascular Biology, Cambridge (2010)
175. Pasta, S., Phillippi, J.A., Gleason, T.G., Vorp, D.A.: Dissection properties of aneurysmal and nonaneurysmal human ascending thoracic aorta: preliminary results. In: ASME Summer Bioengineering Conference, Naples, FL (2010)
176. Pasta, S., Phillippi, J.A., Watkins, S.C., Gleason, T.G., Vorp, D.A.: The effect of aneurysm on the delamination strength in human ascending thoracic aorta. Biomedical Engineering Society Annual Meeting, Austin, TX (2010)
177. Pichamuthu, J.E., Phillippi, J.A., Gleason, T.G., Vorp, D.A.: Ascending aortic aneurysm biomechanical properties are variable depending on aortic valve morphology. Aortic Symposium, NY (2010)

178. Pichamuthu, J.E., Phillippi, J.A., Gleason, T.G., Vorp, D.A.: Association of wall strength of ascending thoracic aneurysms with different aortic valve morphologies. Biomedical Engineering Society, Austin, TX (2010)
179. Hayward, C., Brock, D.J.: Fibrillin-1 mutations in Marfan syndrome and other type-1 fibrillinopathies. Hum. Mutat. **10**(6), 415–423 (1997). PMID:9401003
180. Di Martino, E., Mantero, S., Inzoli, F., Melissano, G., Astore, D., Chiesa, R., Fumero, R.: Biomechanics of abdominal aortic aneurysm in the presence of endoluminal thrombus: experimental characterisation and structural static computational analysis. Eur. J. Vasc. Endovasc. Surg. **15**(4), 290–299 (1998). PMID:9610340
181. Raghavan, M.L., Vorp, D.A., Federle, M.P., Makaroun, M.S., Webster, M.W.: Wall stress distribution on three-dimensionally reconstructed models of human abdominal aortic aneurysm. J. Vasc. Surg. **31**(4), 760–769 (2000). PMID:10753284
182. Van de Geest, J.P., Di Martino, E.S., Bohra, A., Makaroun, M.S., Vorp, D.A.: A biomechanics-based rupture potential index for abdominal aortic aneurysm risk assessment: demonstrative application. Ann. N. Y. Acad. Sci. **1085**, 11–21 (2006). PMID:17182918
183. Vorp, D.A.: Biomechanics of abdominal aortic aneurysm. J. Biomech. **40**(9), 1887–1902 (2007). PMID:17254589
184. Xenos, M., Alemu, Y., Zamfir, D., Einav, S., Ricotta, J.J., Labropoulos, N., Tassiopoulos, A., Bluestein, D.: The effect of angulation in abdominal aortic aneurysms: fluid-structure interaction simulations of idealized geometries. Med. Biol. Eng. Comput. **48**, 1175–1190 (2010) PMID:21088917 DOI 10.1007/s11517-010-0714-y
185. Sheidaei, A., Hunley, S.C., Zeinali-Davarani, S., Raguin, L.G., Baek, S.: Simulation of abdominal aortic aneurysm growth with updating hemodynamic loads using a realistic geometry. Med. Eng. Phys. **33**(1), 80–88 (2011) PMID:20961796
186. Dobrin, P.B.: Pathophysiology and pathogenesis of aortic aneurysms. Current concepts. Surg. Clin. North Am. **69**(4), 687–703 (1989). PMID:2665139
187. Sumner, D.S., Hokanson, D.E., Strandness Jr, D.E.: Stress-strain characteristics and collagen-elastin content of abdominal aortic aneurysms. Surg. Gynecol. Obstet. **130**(3), 459–466 (1970). PMID:5413429
188. Sakalihasan, N., Heyeres, A., Nusgens, B.V., Limet, R., Lapiere, C.M.: Modifications of the extracellular-matrix of aneurysmal abdominal aortas as a function of their size. Eur. J. Vasc. Surg. **7**(6), 633–637 (1993). PMID:ISI:A1993MG43300007
189. Hunter, G.C., Leong, S.C., Yu, G.S., McIntyre, K.E., Bernhard, V.M.: Aortic blebs: possible site of aneurysm rupture. J. Vasc. Surg. **10**(1), 93–99 (1989). PMID:2746804
190. Hunter, G.C., Smyth, S.H., Aguirre, M.L., Baxter, B.T., Bull, D.A., King, D.D., Wang, Y.P., Hall, K.A., Putnam, C.W.: Incidence and histologic characteristics of blebs in patients with abdominal aortic aneurysms. J. Vasc. Surg. **24**(1), 93–101 (1996). PMID:8691533
191. Kapadia, S.R., Oral, H., Lee, J., Nakano, M., Taffet, G.E., Mann, D.L.: Hemodynamic regulation of tumor necrosis factor-alpha gene and protein expression in adult feline myocardium. Circ. Res. **81**(2), 187–195 (1997). PMID:9242179
192. Leung, D.Y., Glagov, S., Mathews, M.B.: Cyclic stretching stimulates synthesis of matrix components by arterial smooth muscle cells in vitro. Science **191**(4226), 475–477 (1976). PMID:128820
193. Sumpio, B.E., Banes, A.J., Link, G.W., Iba, T.: Modulation of endothelial cell phenotype by cyclic stretch: inhibition of collagen production. J. Surg. Res. **48**(5), 415–420 (1990). PMID:2352417
194. Howard, P.S., Kucich, U., Taliwal, R., Korostoff, J.M.: Mechanical forces alter extracellular matrix synthesis by human periodontal ligament fibroblasts. J. Periodontal Res. **33**(8), 500–508 (1998). PMID:9879524
195. Cotrufo, M., Della Corte, A., De Santo, L.S., Quarto, C., De Feo, M., Romano, G., Amarelli, C., Scardone, M., Di Meglio, F., Guerra, G., Scarano, M., Vitale, S., Castaldo, C., Montagnani, S.: Different patterns of extracellular matrix protein expression in the convexity and the concavity of the dilated aorta with bicuspid aortic valve: preliminary results. J. Thorac. Cardiovasc. Surg. **130**(2), 504–511 (2005). PMID:16077420

196. McMillan, W.D., Tamarina, N.A., Cipollone, M., Johnson, D.A., Parker, M.A., Pearce, W.H.: Size matters: the relationship between MMP-9 expression and aortic diameter. Circulation **96**(7), 2228–2232 (1997). PMID:ISI:A1997YA09800020
197. Speelman, L., Hellenthal, F.A., Pulinx, B., Bosboom, E.M.H., Breeuwer, M., van Sambeek, M.R., de Vosse, F.N.V., Jacobs, M.J., Wodzig, W.K.W.H., Schurink, G.W.H.: The influence of wall stress on AAA growth and biomarkers. Eur. J. Vasc. Endovasc. Surg. **39**(4), 410–416 (2010). PMID:ISI:000277816300005
198. Kazi, M., Thyberg, J., Religa, P., Roy, J., Eriksson, P., Hedin, U., Swedenborg, J.: Influence of intraluminal thrombus on structural and cellular composition of abdominal aortic aneurysm wall. J. Vasc. Surg. **38**(6), 1283–1292 (2003). PMID:14681629
199. Lee, R.T., Schoen, F.J., Loree, H.M., Lark, M.W., Libby, P.: Circumferential stress and matrix metalloproteinase 1 in human coronary atherosclerosis. Implications for plaque rupture. Arterioscler. Thromb. Vasc. Biol. **16**(8), 1070–1073 (1996). PMID:8696948
200. Lendon, C.L., Davies, M.J., Born, G.V., Richardson, P.D.: Atherosclerotic plaque caps are locally weakened when macrophages density is increased. Atherosclerosis **87**(1), 87–90 (1991). PMID:1872926
201. Yoshimura, K., Aoki, H., Ikeda, Y., Furutani, A., Hamano, K., Matsuzaki, M.: Identification of c-Jun N-terminal kinase as a therapeutic target for abdominal aortic aneurysm. Ann. N. Y. Acad. Sci. **1085**, 403–406 (2006). PMID:17182964
202. Fernandez-Varo, G., Morales-Ruiz, M., Ros, J., Tugues, S., Munoz-Luque, J., Casals, G., Arroyo, V., Rodes, J., Jimenez, W.: Impaired extracellular matrix degradation in aortic vessels of cirrhotic rats. J. Hepatol. **46**(3), 440–446 (2007). PMID:17156884
203. Gong, L.M., Du, J.B., Shi, L., Shi, Y., Tang, C.S.: Effects of endogenous carbon monoxide on collagen synthesis in pulmonary artery in rats under hypoxia. Life Sci. **74**(10), 1225–1241 (2004). PMID:14697406
204. Horino, Y., Takahashi, S., Miura, T., Takahashi, Y.: Prolonged hypoxia accelerates the posttranscriptional process of collagen synthesis in cultured fibroblasts. Life Sci. **71**(26), 3031–3045 (2002). PMID:12408871
205. Patel, R., Cardneau, J.D., Colles, S.M., Graham, L.M.: Synthetic smooth muscle cell phenotype is associated with increased nicotinamide adenine dinucleotide phosphate oxidase activity: effect on collagen secretion. J. Vasc. Surg. **43**(2), 364–371 (2006). PMID:16476616
206. Levene, C.I., Kapoor, R., Heale, G.: The effect of hypoxia on the synthesis of collagen and glycosaminoglycans by cultured pig aortic endothelium. Atherosclerosis **44**(3), 327–337 (1982). PMID:6816251
207. Pietila, K., Jaakkola, O.: Effect of hypoxia on the synthesis of glycosaminoglycans and collagen by rabbit aortic smooth muscle cells in culture. Atherosclerosis **50**(2), 183–190 (1984). PMID:6712771
208. Campbell, E.J., Wald, M.S.: Hypoxic injury to human alveolar macrophages accelerates release of previously bound neutrophil elastase—implications for lung connective-tissue injury including pulmonary-emphysema. Am. Rev. Respir. Dis. **127**(5), 631–635 (1983). PMID:ISI:A1983QP96200021
209. Scannell, G., Waxman, K., Kaml, G.J., Ioli, G., Gatanaga, T., Yamamoto, R., Granger, G.A.: Hypoxia induces a human macrophage cell-line to release tumor-necrosis-factor-alpha and its soluble receptors in vitro. J. Surg. Res. **54**(4), 281–285 (1993). PMID: ISI:A1993LM66700003
210. Burke, B., Giannoudis, A., Corke, K.P., Gill, D., Wells, M., Ziegler-Heitbrock, L., Lewis, C.E.: Hypoxia-induced gene expression in human macrophages—implications for ischemic tissues and hypoxia-regulated gene therapy. Am. J. Pathol. **163**(4), 1233–1243 (2003). PMID:ISI:000185517500003
211. Vorp, D.A., Wang, D.H., Webster, M.W.: Federspiel WJ effect of intraluminal thrombus thickness and bulge diameter on the oxygen diffusion in abdominal aortic aneurysm. J. Biomech. Eng. **120**(5), 579–583 (1998). PMID:10412434

212. Adolph, R., Vorp, D.A., Steed, D.L., Webster, M.W., Kameneva, M.V., Watkins, S.C.: Cellular content and permeability of intraluminal thrombus in abdominal aortic aneurysm. J. Vasc. Surg. **25**(5), 916–926 (1997). PMID:9152321
213. Saratzis, A., Abbas, A.A., Kiskinis, D., Melas, N., Saratzis, N., Kitas, G.D.: Abdominal aortic aneurysm: a review of the genetic basis. Angiology **62**(1), 18–32 (2011). PMID: 20566578
214. Yoon, S., Tromp, G., Vongpunsawad, S., Ronkainen, A., Juvonen, T., Kuivaniemi, H.: Genetic analysis of MMP3, MMP9, and PAI-1 in Finnish patients with abdominal aortic or intracranial aneurysms. Biochem. Biophys. Res. Commun. **265**(2), 563–568 (1999). PMID: 10558909
215. Jones, G.T., Phillips, V.L., Harris, E.L., Rossaak, J.I., van Rij, A.M.: Functional matrix metalloproteinase-9 polymorphism (C-1562T) associated with abdominal aortic aneurysm. J. Vasc. Surg. **38**(6), 1363–1367 (2003). PMID:14681642
216. Hinterseher, I., Bergert, H., Kuhlisch, E., Bloomenthal, A., Pilarsky, C., Ockert, D., Schellong, S., Saeger, H.D., Krex, D.: Matrix metalloproteinase 2 polymorphisms in a caucasian population with abdominal aortic aneurysm. J. Surg. Res. **133**(2), 121–128 (2006). PMID:16458924
217. Ogata, T., Shibamura, H., Tromp, G., Sinha, M., Goddard, K.A., Sakalihasan, N., Limet, R., MacKean, G.L., Arthur, C., Sueda, T., Land, S., Kuivaniemi, H.: Genetic analysis of polymorphisms in biologically relevant candidate genes in patients with abdominal aortic aneurysms. J. Vasc. Surg. **41**(6), 1036–1042 (2005). PMID:15944607
218. Wang, X., Tromp, G., Cole, C.W., Verloes, A., Sakalihasan, N., Yoon, S., Kuivaniemi, H.: Analysis of coding sequences for tissue inhibitor of metalloproteinases 1 (TIMP1) and 2 (TIMP2) in patients with aneurysms. Matrix Biol. **18**(2), 121–124 (1999). PMID:10372551
219. Ghilardi, G., Biondi, M.L., Battaglioli, L., Zambon, A., Guagnellini, E., Scorza, R.: Genetic risk factor characterizes abdominal aortic aneurysm from arterial occlusive disease in human beings: CCR5 Delta 32 deletion. J. Vasc. Surg. **40**(5), 995–1000 (2004). PMID: 15557916
220. Katrancioglu, N., Manduz, S., Karahan, O., Yilmaz, M.B., Sezgin, I., Bagci, G., Berkan, O.: The role of the CCR2 gene polymorphism in abdominal aortic aneurysms. Angiology **62**(2), 140–143 (2011). PMID:21220372
221. Pola, R., Gaetani, E., Santoliquido, A., Gerardino, L., Cattani, P., Serricchio, M., Tondi, P., Flore, R., Grande, M., Carbonin, P., Fadda, G., Pola, P.: Abdominal aortic aneurysm in normotensive patients: association with angiotensin-converting enzyme gene polymorphism. Eur. J. Vasc. Endovasc. Surg. **21**(5), 445–449 (2001). PMID:11352521
222. Brunelli, T., Prisco, D., Fedi, S., Rogolino, A., Farsi, A., Marcucci, R., Giusti, B., Pratesi, C., Pulli, R., Gensini, G.F., Abbate, R., Pepe, G.: High prevalence of mild hyperhomocysteinemia in patients with abdominal aortic aneurysm. J. Vasc. Surg. **32**(3), 531–536 (2000). PMID:10957660
223. Ferrara, F., Novo, S., Grimaudo, S., Raimondi, F., Meli, F., Amato, C., Amodeo, G., Lo Presti, R., Caimi, G.: Methylenetetrahydrofolate reductase mutation in subjects with abdominal aortic aneurysm subdivided for age. Clin. Hemorheol. Microcirc. **34**(3), 421–426 (2006). PMID:16614466
224. Sofi, F., Marcucci, R., Giusti, B., Pratesi, G., Lari, B., Sestini, I., Lo Sapio, P., Pulli, R., Pratesi, C., Abbate, R., Gensini, G.F.: High levels of homocysteine, lipoprotein (a) and plasminogen activator inhibitor-1 are present in patients with abdominal aortic aneurysm. Thromb. Haemost. **94**(5), 1094–1098 (2005). PMID:16363254
225. Atli, F.H., Manduz, S., Katrancioglu, N., Ozum, U., Disli, O.M., Atahan, E., Ozdemir, O., Dogan, K., Berkan, O.: eNOS G894T polymorphism and abdominal aortic aneurysms. Angiology **61**(2), 125–130 (2010). PMID:19638352
226. Jones, K., Powell, J., Brown, L., Greenhalgh, R., Jormsjo, S., Eriksson, P.: The influence of 4G/5G polymorphism in the plasminogen activator inhibitor-1 gene promoter on the incidence, growth and operative risk of abdominal aortic aneurysm. Eur. J. Vasc. Endovasc. Surg. **23**(5), 421–425 (2002). PMID:12027469

227. Moran, C.S., Clancy, P., Biros, E., Blanco-Martin, B., McCaskie, P., Palmer, L.J., Coomans, D., Norman, P.E., Golledge, J.: Association of PPARgamma allelic variation, osteoprotegerin and abdominal aortic aneurysm. Clin. Endocrinol. (Oxf) **72**(1), 128–132 (2010). PMID: 19438902
228. Krishna, S.M., Dear, A.E., Norman, P.E., Golledge, J.: Genetic and epigenetic mechanisms and their possible role in abdominal aortic aneurysm. Atherosclerosis **212**(1), 16–29 (2010). PMID:20347091
229. Genome Wide Association Studies: Identifying the genes that determine the risk of abdominal aortic aneurysm. Eur. J. Vasc. Endovasc. Surg. **36**(4), 395–396 PMID:18621558 (2008)
230. Helgadottir, A., Thorleifsson, G., Magnusson, K.P., Gretarsdottir, S., Steinthorsdottir, V., Manolescu, A., Jones, G.T., Rinkel, G.J., Blankensteijn, J.D., Ronkainen, A., Jaaskelainen, J.E., Kyo, Y., Lenk, G.M., Sakalihasan, N., Kostulas, K., Gottsater, A., Flex, A., Stefansson, H., Hansen, T., Andersen, G., Weinsheimer, S., Borch-Johnsen, K., Jorgensen, T., Shah, S.H., Quyyumi, A.A., Granger, C.B., Reilly, M.P., Austin, H., Levey, A.I., Vaccarino, V., Palsdottir, E., Walters, G.B., onsdottir, T., Snorradottir, S., Magnusdottir, D., Gudmundsson, G., Ferrell, R.E., Sveinbjornsdottir, S., Hernesniemi, J., Niemela, M., Limet, R., Andersen, K., Sigurdsson, G., Benediktsson, R., Verhoeven, E.L., Teijink, J.A., Grobbee, D.E., Rader, D.J., Collier, D.A., Pedersen, O., Pola, R., Hillert, J., Lindblad, B., Valdimarsson, E.M., Magnadottir, H.B., Wijmenga, C., Tromp, G., Baas, A.F., Ruigrok, Y.M., van Rij, A.M., Kuivaniemi, H., Powell, J.T., Matthiasson, S.E., Gulcher, J.R., Thorgeirsson, G., Kong, A., Thorsteinsdottir, U., Stefansson, K.: The same sequence variant on 9p21 associates with myocardial infarction, abdominal aortic aneurysm and intracranial aneurysm. Nat. Genet. **40**(2), 217–224 (2008). PMID:18176561
231. Bown, M.J., Braund, P.S., Thompson, J., London, N.J., Samani, N.J., Sayers, R.D.: Association between the coronary artery disease risk locus on chromosome 9p21.3 and abdominal aortic aneurysm. Circ. Cardiovasc. Genet. **1**(1), 39–42 (2008). PMID:20031540
232. Biros, E., Cooper, M., Palmer, L.J., Walker, P.J., Norman, P.E., Golledge, J.: Association of an allele on chromosome 9 and abdominal aortic aneurysm. Atherosclerosis **212**(2), 539–542 (2010). PMID:20605023
233. Annambhotla, S., Bourgeois, S., Wang, X., Lin, P.H., Yao, Q., Chen, C.: Recent advances in molecular mechanisms of abdominal aortic aneurysm formation. World J. Surg. **32**(6), 976–986 (2008). PMID:18259804
234. Alexander, J.J.: The pathobiology of aortic aneurysms. J. Surg. Res. **117**(1), 163–175 (2004). PMID:ISI:000220182100021
235. Fillinger, M.F., Racusin, J., Baker, R.K., Cronenwett, J.L., Teutelink, A., Schermerhorn, M.L., Zwolak, R.M., Powell, R.J., Walsh, D.B., Rzucidlo, E.M.: Anatomic characteristics of ruptured abdominal aortic aneurysm on conventional CT scans: implications for rupture risk. J. Vasc. Surg. **39**(6), 1243–1252 (2004). PMID:ISI:000222018600016
236. Vorp, D.A., Vande Geest, J.P.: Biomechanical determinants of abdominal aortic aneurysm rupture. Arterioscler. Thromb. Vasc. Biol. **25**(8), 1558–1566 (2005). PMID:16055757
237. Sonesson, B., Hansen, F., Lanne, T.: Abdominal aortic aneurysm: a general defect in the vasculature with focal manifestations in the abdominal aorta? J. Vasc. Surg. **26**(2), 247–254 (1997). PMID:9279311
238. Wilson, K., Whyman, M., Hoskins, P., Lee, A.J., Bradbury, A.W., Fowkes, F.G., Ruckley, C.V.: The relationship between abdominal aortic aneurysm wall compliance, maximum diameter and growth rate. Cardiovasc. Surg. **7**(2), 208–213 (1999). PMID:10353673
239. Wilson, K.A., Lindholt, J.S., Hoskins, P.R., Heickendorff, L., Vammen, S., Bradbury, A.W.: The relationship between abdominal aortic aneurysm distensibility and serum markers of elastin and collagen metabolism. Eur. J. Vasc. Endovasc. Surg. **21**(2), 175–178 (2001). PMID:11237793
240. MacSweeney, S.T., Young, G., Greenhalgh, R.M., Powell, J.T.: Mechanical properties of the aneurysmal aorta. Br. J. Surg. **79**(12), 1281–1284 (1992). PMID:1486417

241. Lanne, T., Sonesson, B., Bergqvist, D., Bengtsson, H., Gustafsson, D.: Diameter and compliance in the male human abdominal aorta: influence of age and aortic aneurysm. Eur. J. Vasc. Surg. **6**(2), 178–184 (1992). PMID:1572458
242. Vorp, D.A., Mandarino, W.A., Webster, M.W., Gorcsan 3rd, J.: Potential influence of intraluminal thrombus on abdominal aortic aneurysm as assessed by a new non-invasive method. Cardiovasc. Surg. **4**(6), 732–739 (1996). PMID:9013001
243. Vallabhaneni, S.R., Gilling-Smith, G.L., How, T.V., Carter, S.D., Brennan, J.A., Harris, P.L.: Heterogeneity of tensile strength and matrix metalloproteinase activity in the wall of abdominal aortic aneurysms. J. Endovasc. Ther. **11**(4), 494–502 (2004). PMID:ISI: 000223415300019
244. Raghavan, M.L., Kratzberg, J., DaSilva, E.S.: Heterogeneous, variable wall-thickness modeling of a ruptured abdominal aortic aneurysm. In: 2004 ASME International Mechanical Engineering Conference, Anaheim (CA) (2004)
245. Sacks, M.S.: Biaxial mechanical evaluation of planar biological materials. J. Elast. **61**(1–3), 199–246 (2000). PMID:ISI:000169741700009
246. Van de Geest, J.P., Sacks, M.S., Vorp, D.A.: Age dependency of the biaxial biomechanical behavior of human abdominal aorta. J. Biomech. Eng. **126**(6), 815–822 (2004). PMID: 15796340
247. Wang, D.H., Makaroun, M., Webster, M.W., Vorp, D.A.: Mechanical properties and microstructure of intraluminal thrombus from abdominal aortic aneurysm. J. Biomech. Eng. **123**(6), 536–539 (2001). PMID:11783723
248. Stringfellow, M.M., Lawrence, P.F., Stringfellow, R.G.: The influence of aorta aneurysm geometry upon stress in the aneurysm wall. J. Surg. Res. **42**(4), 425–433 (1987). PMID:ISI: A1987H239900014
249. Marston, W.A., Criado, E., Baird, C.A., Keagy, B.A.: Reduction of aneurysm pressure and wall stress after endovascular repair of abdominal aortic aneurysm in a canine model. Ann. Vasc. Surg. **10**(2), 166–173 (1996). PMID:ISI:A1996UK57200010
250. Mcgiffin, D.C., Mcgiffin, P.B., Galbraith, A.J., Cross, R.B.: Aortic-wall stress profile after repair of coarctation of the aorta—is it related to subsequent true aneurysm formation. J. Thorac. Cardiovasc. Surg. **104**(4), 924–931 (1992). PMID:ISI:A1992JT19800011
251. Inzoli, F., Boschetti, F., Zappa, M., Longo, T., Fumero, R.: Biomechanical factors in abdominal aortic aneurysm rupture. Eur J. Vasc. Surg. **7**(6), 667–674 (1993). PMID:8270069
252. Mower, W.R., Baraff, L.J., Sneyd, J.: Stress distributions in vascular aneurysms—factors affecting risk of aneurysm rupture. J. Surg. Res. **55**(2), 155–161 (1993). PMID:ISI: A1993MC02400007
253. Raghavan, M.L., Fillinger, M.F., Marra, S.P., Naegelein, B.P., Kennedy, F.E.: Automated methodology for determination of stress distribution in human abdominal aortic aneurysm. J. Biomech. Eng.-Trans. ASME **127**(5), 868–871 (2005). PMID:ISI:000232193200018
254. Di Martino, E.S., Guadagni, G., Fumero, A., Ballerini, G., Spirito, R., Biglioli, P., Redaelli, A.: Fluid-structure interaction within realistic three-dimensional models of the aneurysmatic aorta as a guidance to assess the risk of rupture of the aneurysm. Med. Eng. Phys. **23**(9), 647–655 (2001). PMID:ISI:000173836400005
255. Scotti, C.M., Shkolnik, A.D., Muluk, S.C., Finol, E.A.: Fluid-structure interaction in abdominal aortic aneurysms: effects of asymmetry and wall thickness. Biomed. Eng. Online **4**, 64 (2005). PMID:16271141
256. Finol, E.A., Amon, C.H.: Flow-induced wall shear stress in abdominal aortic aneurysms: part II–pulsatile flow hemodynamics. Comput. Methods Biomech. Biomed. Eng. **5**(4), 319–328 (2002). PMID:12186711
257. Asbury, C.L., Ruberti, J.W., Bluth, E.I., Peattie, R.A.: Experimental investigation of steady flow in rigid models of abdominal aortic aneurysms. Ann. Biomed. Eng. **23**(1), 29–39 (1995). PMID:7762880
258. Peattie, R.A., Riehle, T.J., Bluth, E.I.: Pulsatile flow in fusiform models of abdominal aortic aneurysms: flow fields, velocity patterns and flow-induced wall stresses. J. Biomech. Eng.-Trans. ASME **126**(4), 438–446 (2004). PMID:ISI:000224668800006

259. Rissland, P., Alemu, Y., Einav, S., Ricotta, J., Bluestein, D.: Abdominal aortic aneurysm risk of rupture: patient-specific FSI simulations using anisotropic model. J. Biomech. Eng. **131**(3), 031001 (2009). PMID:19154060
260. Holzapfel, G.A., Gasser, T.C., Stadler, M.: A structural model for the viscoelastic behavior of arterial walls: continuum formulation and finite element analysis. Eur. J. Mech. A-Solids **21**(3), 441–463 (2002). PMID:ISI:000176078400007
261. Mohan, D., Melvin, J.W.: Failure properties of passive human aortic tissue.2. Biaxial tension tests. J. Biomech. **16**(1), 31 (1983). PMID:ISI:A1983QE15900004
262. Lee, M.C., Haut, R.C.: Strain rate effects on tensile failure properties of the common carotid-artery and jugular veins of ferrets. J. Biomech. **25**(8), 925–927 (1992). PMID: ISI:A1992JD39100011
263. Purslow, P.P.: Positional variations in fracture toughness, stiffness and strength of descending thoracic pig aorta. J. Biomech. **16**(11), 947–953 (1983)
264. Rajagopal, K., Bridges, C., Rajagopal, K.R.: Towards an understanding of the mechanics underlying aortic dissection. Biomech. Model. Mechanobiol. **6**(5), 345–359 (2007)
265. Adham, M., Gournier, J.P., Favre, J.P., De La Roche, E., Ducerf, C., Baulieux, J., Barral, X., Pouyet, M.: Mechanical characteristics of fresh and frozen human descending thoracic aorta. J. Surg. Res. **32**(1), 32–34 (1996). PMID:8806470
266. Groenink, M., Langerak, S.E., Vanbavel, E., van der Wall, E.E., Mulder, B.J., van der Wal, A.C., Spaan, J.A.: The influence of aging and aortic stiffness on permanent dilation and breaking stress of the thoracic descending aorta. Cardiovasc. Res. **43**(2), 471–480 (1999). PMID:10536677
267. Iliopoulos, D.C., Kritharis, E.P., Giagini, A.T., Papadodima, S.A., Sokolis, D.P.: Ascending thoracic aortic aneurysms are associated with compositional remodeling and vessel stiffening but not weakening in age-matched subjects. J. Thorac. Cardiovasc. Surg. **137**(1), 101–109 (2009). PMID:19154911
268. Tiessen, I.M., Roach, M.R.: Factors in the initiation and propagation of aortic dissections in human autopsy aortas. J. Biomech. Eng.-Trans. ASME **115**(1), 123–125 (1993). PMID:ISI:A1993KQ01400019
269. Thubrikar, M.J., Agali, P., Robicsek, F.: Wall stress as a possible mechanism for the development of transverse intimal tears in aortic dissections. J. Med. Eng. Technol. **23**(4), 127–134 (1999). PMID:ISI:000082863900002
270. Beller, C.J., Labrosse, M.R., Thubrikar, M.J., Robicsek, F.: Role of aortic root motion in the pathogenesis of aortic dissection. Circulation **109**(6), 763–769 (2004). PMID:ISI: 000189005500013
271. Borghi, A., Wood, N.B., Mohiaddin, R.H., Xu, X.Y.: Fluid-solid interaction simulation of flow and stress pattern in thoracoabdominal aneurysms: a patient-specific study. J. Fluids Struct. **24**(2), 270–280 (2008). PMID:ISI:000254747600007
272. Khanafer, K., Berguer, R.: Fluid-structure interaction analysis of turbulent pulsatile flow within a layered aortic wall as related to aortic dissection. J. Biomech. **42**(16), 2642–2648 (2009). PMID:ISI:000273135200003
273. Tse, K.M., Chiu, P., Lee, H.P., Ho, P.: Investigation of hemodynamics in the development of dissecting aneurysm within patient-specific dissecting aneurismal aortas using computational fluid dynamics (CFD) simulations. J. Biomech. **44**(5), 827–836 (2011). PMID:21256491
274. Nathan, D.P., Xu, C., Gorman III, J.H., Fairman, R.M., Bavaria, J.E., Gorman, R.C., Chandran, K.B., Jackson, B.M.: Pathogenesis of acute aortic dissection: a finite element stress analysis. Ann. Thorac. Surg. **91**(2), 458–463 (2011)
275. Lederle, F.A., Wilson, S.E., Johnson, G.R., Reinke, D.B., Littooy, F.N., Acher, C.W., Ballard, D.J., Messina, L.M., Gordon, I.L., Chute, E.P., Krupski, W.C., Busuttil, S.J., Barone, G.W., Sparks, S., Graham, L.M., Rapp, J.H., Makaroun, M.S., Moneta, G.L., Cambria, R.A., Makhoul, R.G., Eton, D., Ansel, H.J., Freischlag, J.A., Bandyk, D.: Immediate repair compared with surveillance of small abdominal aortic aneurysms. N. Engl. J. Med. **346**(19), 1437–1444 (2002). PMID:12000813
276. Li, Z.Y.: Computed wall stress may predict the growth of abdominal aortic aneurysm. Conf. Proc. IEEE Eng. Med. Biol. Soc. **1**, 2626–2629 (2010). PMID:21096184

Computer-Aided Diagnosis of Abdominal Aortic Aneurysms

Barry J. Doyle and Timothy M. McGloughlin

Abstract Computer-aided diagnosis (CAD) systems have been used in several areas of medicine for the last number of years. A typical CAD system interprets medical images and provides guidance for the clinician. The concept of CAD in the assessment of abdominal aortic aneurysm (AAA) has been around for several years, however, the technique is gaining momentum as of late. Computer modeling of AAAs is becoming more prevalent with several novel approaches of CAD reported over the past number of years. CAD is possible through computer-aided detection (CADe) and computer-aided quantification (CADq) techniques that work together to return usable quantities aimed at helping identify AAAs that may be at risk of rupture. This chapter examines some recent developments within the area of CAD for AAAs, in particular the use of peak wall stress, and also asymmetry and the finite element analysis rupture index. All three tools provide additional data to the clinician through the CAD system and help complement the use of maximum diameter in identifying high-risk AAAs.

1 Background

Abdominal aortic aneurysms (AAAs) are notoriously asymptomatic and often referred to as a "silent killer". Patients frequently present at hospital with abdominal and/or back pain, where examination reveals the cause of the pain to be

B. J. Doyle (✉) · T. M. McGloughlin
Department of Mechanical, Aeronautical and Biomedical Engineering,
Materials and Surface Science Institute, Centre for Applied Biomedical
Engineering Research (CABER), University of Limerick,
MSSI Building, MSG-013-022, Limerick, Ireland
e-mail: Barry.Doyle@ul.ie

Stud Mechanobiol Tissue Eng Biomater (2011) 7: 119–138
DOI: 10.1007/8415_2011_70

Published Online: 24 April 2011

a pulsating mass deep within the abdomen, in other words, an AAA. Ultrasound is the preferred method to diagnose AAAs, primarily because the tool is cost effective and measurements are reproducible within a range of 6 mm [16]. Screening involves the use of ultrasonography to detect AAA and the implementation of these programs is becoming increasingly common. It has been recommended that people over the age of 60–65 years, in particular men, should be screened for AAA, with the recommended age reducing to 50–55 when there is a history of aneurysmal disease in the family. AAA screening programs are becoming more widespread in the UK with many private institutions providing screening. The UK National Health Service (NHS) recently announced that a full screening program will be made available throughout the UK, but is unlikely to become widely available until 2013 [38]. According to the US Preventative Services Task Force [55], the potential benefit of screening for AAA among women over the age of 65 is low because of the number of age-related deaths in this population. The majority of AAA related deaths occur in women over the age of 80, and as there are many competing health risks at this age, any benefit of screening would be minimal [55].

Currently, the trend in determining the severity of an AAA is to use the maximum diameter criterion [4, 22]. Patients with an AAA that has a maximum diameter greater than 5–5.5 cm are deemed a high rupture risk and are usually recommended for surgical repair [26]. In the case of smaller AAAs where the diameter <5 cm, the preferred approach is often careful and frequent observation using either ultrasonography or computed tomography (CT). AAA growth rate is also used as an additional parameter with AAAs that exceed growth rates of 1 cm/year deemed a high rupture risk. Recent research however, has questioned the suitability of surgical repair based on the maximum diameter criterion alone [9, 10, 13, 18, 19, 25, 27, 41, 46, 57–59]. Although the diameter-criterion can be justified, as the rupture risk for an AAA is clearly related to its maximum diameter [3, 18], surgical decision-making using solely this parameter may in fact lead to unnecessary AAA repairs and also exclude certain cases (AAA < 5 cm) from surgical repair [4, 5, 18, 40]. Nicholls et al. [40] reported that 10–24% of ruptured AAAs were less than 5 cm in diameter. Darling et al. [5] also reported that of 473 non-repaired AAAs examined from autopsy reports, there were 118 cases of rupture, 13% of which were less than 5 cm in diameter. They also showed that 60% of the AAAs greater than 5 cm (including 54% of those AAAs between 7.1 and 10 cm) never experienced rupture. Vorp et al. [62] later concluded from the findings of Darling et al. [5] that if the maximum diameter criterion were followed for the 473 subjects, only 7% (34/473) of cases would have succumbed to rupture prior to surgical intervention as the diameter was less than 5 cm, with 25% (116/473) of cases possibly undergoing unnecessary surgery since these AAAs may never have ruptured.

Alternative approaches to rupture assessment have been recently reported. The majority of these methods involve the numerical analysis of AAAs using the common engineering technique of the finite element method (FEM) to determine the wall stress distributions. Studies have reported that these stress distributions correlate to the overall geometry of the AAA rather than to the maximum diameter [9, 45, 63]. It is also known that wall stress alone does not completely dictate failure as an AAA will rupture when the local wall stress exceeds the local wall strength. Therefore, rupture risk should consider the patient-specific wall stress along with the patient-specific wall strength. A non-invasive method of approximating patient-specific wall strength was recently reported by Vande Geest et al. [57–59], with more traditional approaches to strength determination via tensile testing performed by others [6, 42, 44, 52, 68]. Newly proposed AAA rupture risk parameters include: AAA wall stress [18, 19, 60]; AAA expansion rate [23]; degree of asymmetry [9]; presence of intraluminal thrombus (ILT)[1] [65]; a rupture potential index (RPI) [57, 58]; a finite element analysis rupture index (FEARI) [10, 13]; biomechanical factors coupled with computer analysis [25]; growth of ILT [50]; geometrical parameters of the AAA [9, 21]; and also a method of determining AAA growth and rupture based on mathematical models [61, 66]. Based on this recent work by both our group and others, it is believed that improved AAA rupture risk parameters are necessary and could have major clinical relevance.

As computational modeling, and computers in general, become a vital tool for clinicians in all aspects of healthcare, the role of CAD systems will become more prevalent. This chapter will briefly describe the use of CAD for AAAs and also present the preliminary validation of computational modeling for AAA rupture-prediction.

2 Computer-Aided Diagnosis (CAD)

Computer modeling has become an important tool in the clinical work flow when treating several types of disorders as there is significant potential to improve diagnosis, optimise clinical treatment by predicting outcomes, and inform the design of surgical training practices [39]. This approach is becoming more prevalent in AAA treatment nowadays [11] as clinicians become more aware of the ease in which patient-specific modeling can be applied to the disease. CAD can be applied to AAAs to identify those aneurysms that are a high rupture risk, and also indicate AAAs that may be relatively safe from rupture, thus preventing the trauma and cost associated with surgical treatment. CAD can be divided into a sequence of Computer-Aided Detection (CADe) and Computer-Aided Quantification (CADq), whereby CADe entails the detection, medical imaging and three-dimensional (3D) reconstruction of the model, and CADq encompasses some of the recently proposed parameters that aim to improve the determination of rupture risk. These terms have been adapted from their regular use in radiology, in particular, the use of CAD systems in mammography [17], to suit the current application of AAAs.

[1] ILT consists of a fibrin structure incorporated with blood cells, platelets, blood proteins and cellular debris, and are found in most AAAs.

The impact computers have had on medical imaging over recent years is indisputable, as without computers, imaging modalities such as CT or magnetic resonance imaging (MRI) would not even exist. However, the interpretation of medical images is still a human task and this is set to continue for many years as false positives are hard to rule out when relying on computers alone. There is immense research, in particular within the area of cancer research, to help reduce false positive readings and allow image interpretation to become an automated process. Whilst it is unlikely that AAA diagnosis will ever become completely automated, the role of computers must not be under-estimated. CADe systems will allow the AAA to be properly detected and reconstructed, whereas CADq systems allow an estimate of rupture risk to be determined. This process is semi-automated and requires significant user input, with some attempting to almost remove the user input by designing software to perform the CADe and CADq elements (A4 Clinics^{TM}, VASCOPS GmbH, Austria). The use of finite element analysis (FEA) to predict rupture locations in AAAs has recently been validated in vitro [12, 14] with some preliminary results of in vivo validation described later on in this chapter.

3 Computer-Aided Detection (CADe)

When concerned with AAAs, CADe refers to the imaging of the aneurysm and the subsequent 3D reconstruction. Initial examination of a patient suspected of AAA is usually performed via ultrasound. Ultrasound can identify AAAs and allow maximum diameter measurements, however, if any further examination of the disease is required, CT, or in some cases MRI, must be employed. CT capabilities vary from institution to institution, with the performance potential of CAD systems heavily dependent on the initial image quality. Poor pixel resolution of the scans will result in poor image interpretation, poor reconstructions and ultimately, poor diagnosis. For the purposes of this chapter, CADe will entail the use of the commercially available software Mimics (Materialise NV, Belgium) which has been used in several reports in the literature [7–11, 13, 30, 31, 67]. The process of reconstruction using Mimics has been reported in-depth previously [8] but in essence relies on thresholding and segmentation of the CT image. Thresholding is based on the pixel intensity, often measured in Hounsfield Units (HU), and allows the user to identify regions of interest, in this case the AAA. Segmenting the AAA from the remainder of the image is then possible and can be applied to every scan in the series. Mimics employs the *marching cubes* and *marching squares* algorithms [29] to produce a triangle mesh by computing isosurfaces from discrete data. From connecting the patches from all the cubes on the isosurface boundary, a surface representation can be obtained. This procedure, as shown in Fig. 1, results in 3D models of exquisite detail, which can be further refined and smoothed by the user if desired.

The majority of AAAs contain intraluminal thrombus. In computational models, this thrombus has been shown to significantly alter the biomechanics of the

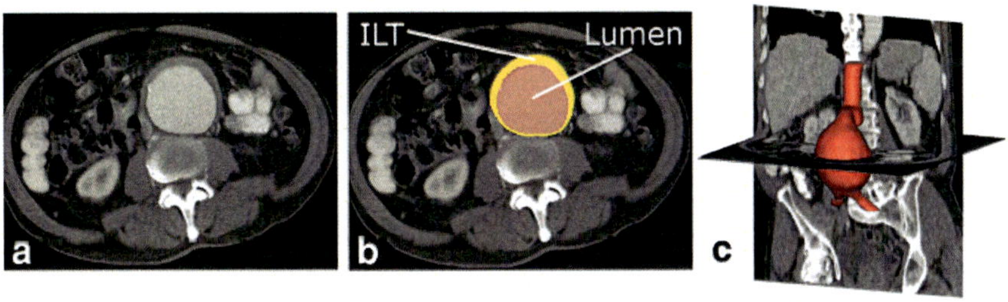

Fig. 1 Procedure of 3D reconstruction from CT scan. **a** Typical CT scan of an AAA through the maximum diameter region. **b** Image after thresholding and segmentation of the region of interest with the lumen shown in *yellow* and the intraluminal thrombus (ILT) shown in *blue*. **c** Resulting 3D reconstruction of the AAA

aneurysm, yet, it has been reported that in vivo pressure transmission through the thrombus can vary considerably from patient to patient [56] and may not act as the "mechanical buffer" that it is often described as. Regardless of the on-going debate over the role of the ILT, from both clinical [47] and engineering [1] viewpoints, the structure must be included in numerical models if rupture-prediction estimates are desired. In 2001, Wang and colleagues [64] mechanically characterised the ILT using 50 specimens harvested from 14 patients. The resulting population-mean material model is possibly the most employed throughout the literature and is applied to all FEA computations in this chapter.

Calcifications are also a common feature of AAAs. Calcified deposits primarily occur within the intima and intima-media interface, but can also occur within the ILT. The role of calcifications in numerical analyses is still somewhat under debate and is strongly dependent on the modeling approach. Speelman et al. [49] modeled calcifications by assigning modified material properties to regions of the diseased AAA wall, whereas Li et al. [28] included the structures as separate entities. More recently, Maier et al. [31] examined several different approaches to calcification inclusion and concluded that it is doubtful that rupture risk will increase by including calcified deposits in numerical models as they can act as load bearing structures. Inclusion of calcifications are therefore omitted from the CAD aspects of this chapter, but could be easily implemented in future studies should the role of these deposits be clarified.

The cohort included for examination in this present study consists of both electively-repaired (n = 42) and ruptured (n = 10) AAA cases, of which the general details and AAA characteristics are shown in Table 1.

4 Computer-Aided Quantification (CADq)

Computer-Aided Quantification (CADq) can take several forms in the context of AAA assessment. The current clinical standard used to quantify AAA rupture-threat is to measure the maximum diameter and, if possible the growth rate, of the

Table 1 Patient details and AAA characteristics of study cohort (Mean ± SD)

	Repaired	Ruptured	*P*
n	42	10	–
Male/female	34/8	7/3	–
Age	71.9 ± 6.4	69.1 ± 6.0	0.205
Diameter (mm)	64.3 ± 12.7	81.7 ± 12.5	0.0003
AAA volume (cm^3)	228.2 ± 119.6	428.8 ± 120.8	0.015
% ILT	50.9 ± 20.1	39.5 ± 14.8	0.057
Surface area (cm^2)	209.3 ± 73.7	317.1 ± 101.8	0.009
Length (mm)	111 ± 16	131 ± 25	0.03
Diameter/length	0.58 ± 0.09	0.63 ± 0.09	0.138
ROD[a]	2.05 ± 0.45	2.26 ± 0.53	0.268

[a] Ratio of maximum diameter to proximal neck diameter

aneurysm. However, as discussed earlier, there is growing concern over the use of these parameters in all AAA cases. Small AAAs can have similar, if not higher wall stress than much larger AAAs, and vice versa. Recent research suggests that more than size alone may contribute to rupture risk, however, reports [5] show that larger AAAs are more likely to rupture than smaller AAAs and therefore clinicians will seldom, if ever, recommend surveillance for large AAAs. Thus, biomechanical analysis of AAAs may only be applicable for small to medium-sized AAAs.

Over recent years, several laboratories have aimed at developing more robust rupture parameters than size alone. The Law of Laplace relates wall stress and diameter in a linear fashion and can estimate the wall stress exhibited on a cylinder. However, the Law fails to approximate the wall stress in more complex shapes and is therefore not suitable for AAAs. The use of computational techniques such as the finite element method (FEM) have allowed the wall stresses in these complex 3D AAA structures to be estimated and also provides an indication of the likely rupture location. For a detailed introduction and background into the FEM the reader can is referred to Zienkiewicz et al. [69]. Numerical results obtained using the FEM indicate the maximum wall stress in the model and usually generate a contour plot showing the distribution of wall stresses in the aneurysm. FEM uses the 3D geometry created from the medical images, converts the geometry into a series of elements connected with nodes, applies user-defined material properties and boundary conditions to the model, and then solves the problem to determine the stresses and strains within the structure. This process is illustrated in Fig. 2. For the purposes of the numerical analyses presented throughout this chapter the ILT was modeled using the material characteristics developed by Wang et al. [64] and the diseased AAA wall assigned the material model proposed by [41]. All analyses were performed using the commercially available finite element code ABAQUS v6.9 (SIMULIA, RI, USA). A static internal pressure of 120 mmHg (16 kPa) was applied to the luminal surface of all models and each AAA was rigidly constrained at the proximal and distal regions to represent the tethering to the remainder of the aorta. These are the standard boundary conditions used throughout the literature when analysing AAAs using

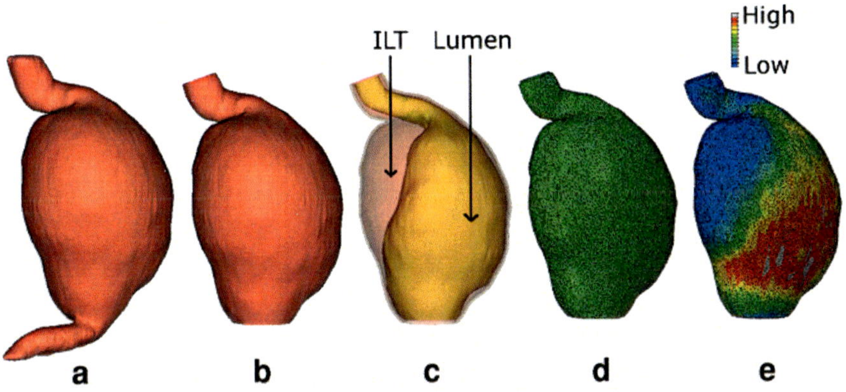

Fig. 2 **a** Complete 3D reconstruction of the AAA from above the mesenteric arteries to below the iliac bifurcation. **b** For the FE simulations, only the region from below the renal arteries to immediately above the iliac bifurcation is examined. **c** Each 3D reconstruction consists of the lumen region and the ILT. An artificial 1.5 mm uniform wall is then created around the ILT. **d** The 3D model is meshed and the appropriate boundary conditions applied. **e** Resulting von Mises stress contours on the model. In this particular case the large volume of ILT (86% of the total AAA volume) is shielding the diseased outer AAA wall from the force of the blood pressure resulting in very low wall stress in regions of thick ILT. All models are shown from the *right*

the FEM. Smoothing is necessary on 3D reconstructions of AAAs and the geometries used in the work presented here was smoothed according to a previous study [7] whereby all models are smoothed so that unwanted surface artefacts are removed without a significant loss in the accuracy of the structure. Other methods of smoothing have also been described [35, 36] and the process is currently very much user-dependent. There is no defined standard with respect to smoothing of AAAs and therefore care must be taken when generating reconstructions, with results analysed under several different levels of smoothing to assess the direct influence on 3D models.

CADq techniques using the FEM have allowed several novel approaches to AAA assessment to be proposed. Peak wall stress, the FEARI and asymmetry, and the severity parameter (SP) [25] have all been recently suggested as possible indicators of rupture risk. This chapter, however, will further describe the use of peak wall stress, FEARI and asymmetry within the context of AAA rupture-prediction.

4.1 Peak Wall Stress

Since the FEM was first applied to the problem of AAA in the late 1980's by Stringfellow et al. [51] the method has become the primary tool used to predict the stress exerted on the diseased AAA wall. In this first study [51] the aneurysm geometry was idealised as either 2D cylindrical or spherical. However, this work paved the way for others to investigate AAA disease using the FEM.

In the early 1990's, several others performed wall stress estimations using the FEM [15, 24, 37], again using idealised 2D geometries. These early studies were far from close approximations of the realistic wall stress experienced in vivo, yet they did introduce some important and still debatable topics that may influence wall stress. ILT was first included in numerical analyses by Inzoli et al. [24] in which they concluded that not only does wall stress increase with diameter (as it certainly does when employing idealised geometries) but also that the presence of ILT may significantly reduce wall stress, by up to 30% in their study. The inclusion of ILT into numerical analyses is routine nowadays as it is generally accepted that it does influence wall stress and is certainly important. On the other hand, however, some clinical studies [47, 56] have indicated that the ILT may not be as protective as generally believed.

One of the first reports to couple FE analyses together with patient-specific AAA 3D reconstructions was performed by Raghavan et al. [43] where, six AAA cases and one healthy control were examined, with very positive results. Fillinger et al. [18, 19] then furthered this work to show that peak wall stress may be superior to diameter in assessing rupture risk of patient-specific AAAs. These studies used large cohorts (n = 48 and n = 103, respectively) to determine statistical significance of results and concluded that not only is peak wall stress significantly higher in those cases that ruptured [18], but that peak wall stress seems superior to diameter in differentiating patients under observation who will experience catastrophic outcome [19]. Venkatasubramaniam et al. [60] concluded similar results in a smaller study group (n = 27). Peak wall stress appears to be an effective method of differentiating small AAAs that may be at risk of rupture also. Truijers et al. [54] reported that in a cohort of thirty small AAAs (diameter < 55 mm) that peak wall stress was significantly higher in ruptured cases compared to the repaired group.

Peak wall stress in the cohort examined for this chapter, showed that peak wall stress was on average (± standard deviation) 56% significantly higher in the ruptured cases compared to the repaired group (0.89 ± 0.35 vs. 0.57 ± 0.23 MPa, $P = 0.018$), as shown in Fig. 3.

Peak wall stress appears to be repeatedly higher in ruptured AAAs compared to similarly-sized repaired AAAs and may be a very useful tool in helping with the decision-making process. However, to comprehensively determine the risk of rupture, one must account for the strength of the AAA wall also. Areas of high wall stress may be relatively safe from rupture due to the strength of the wall at that location, with wall strength known to vary significantly from region to region [44, 52]. Therefore, biomechanics-based rupture parameters have also been suggested, that may be even more superior than peak wall stress alone.

4.2 FEARI

The FEARI uses a ratio of wall stress to strength to assess the rupture threat. Wall stress can be easily estimated using the FEM, however, in vivo wall strength is

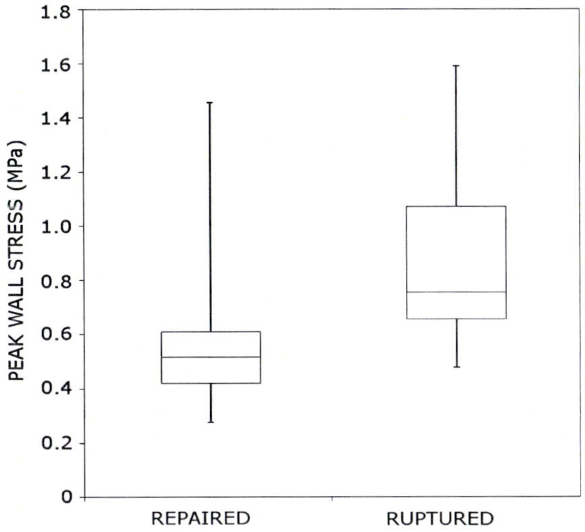

Fig. 3 Box and whisker plot of the peak wall stress results for the electively-repaired and ruptured cohort. The box plot represents the median, 25th and 75th percentiles with the whiskers representing the maximum and minimum values observed

significantly more difficult to obtain. There have been several reports published whereby the mechanical behaviour of AAA tissue has been presented [6, 42, 44, 52, 59]. By compiling the results from three large studies [42, 44, 52] and further analysing the location-dependent ultimate tensile strength (UTS), strength values based on a cohort of 69 patients and 149 tissue samples could be developed for each of the key regions. This wall strength is then used in conjunction with the peak wall stress measurement to assess the likelihood of rupture.

The FEARI method [10, 13] was applied to the study group of this chapter. 3D reconstructions were performed as before and wall stress computations determined using ABAQUS v6.9. All models were constrained proximally and distally to represent tethering to the remainder of the aorta, and a static uniform pressure loading of 120 mmHg was applied to the luminal surface. This standardised internal blood pressure allows the effect of geometry to be examined, as the variability of patient-specific blood pressure would add another uncertainty to the model. All models included the ILT and a uniform wall thickness of 1.5 mm. Fig. 4 illustrates the methodology behind the FEARI. Once the location of peak wall stress is determined, the stress is related to wall strength at that location. FEARI provides a rupture risk based on the ratio of stress to strength, with values larger than 1 indicating failure.

The results of the FEARI assessment (Fig. 5) indicate that the ruptured AAA cohort had a higher mean FEARI than the repaired group (1.03 ± 0.42 vs. 0.65 ± 0.3, $P = 0.019$). Maximum diameters were also 26% higher in the ruptured group (81.7 ± 12.5 vs. 64.7 ± 12.3 mm, $P = 0.0003$). The relationships between FEARI and maximum diameter in both the repaired and ruptured cohorts are insignificant (repaired, $P = 0.072$ and ruptured, $P = 0.174$). FEARI is possible through the mechanical testing of excised tissue in order to average population-mean UTS values for each of the primary locations in the AAA. The work to

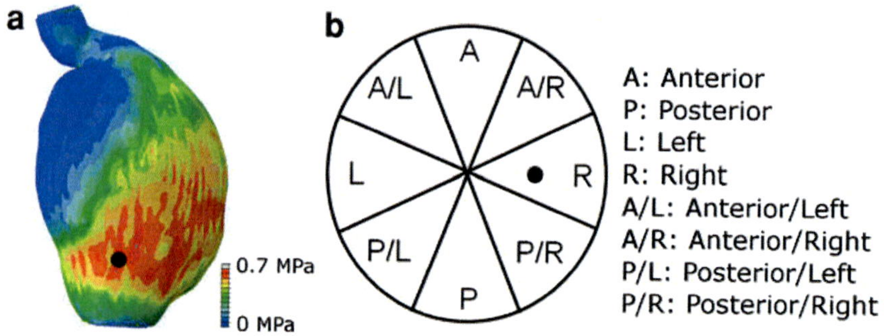

Fig. 4 FEARI methodology. **a** Peak wall stress quantity and location is determined using the FEM. **b** The location of peak wall stress corresponds to a distinct region, which in turn is assigned a wall strength value. FEARI then divides the peak wall strength by the wall strength at the location of peak stress to return a rupture index

Fig. 5 FEARI results for the repaired and ruptured groups. The box plot represents the median, 25th and 75th percentiles with the whiskers representing the maximum and minimum values observed

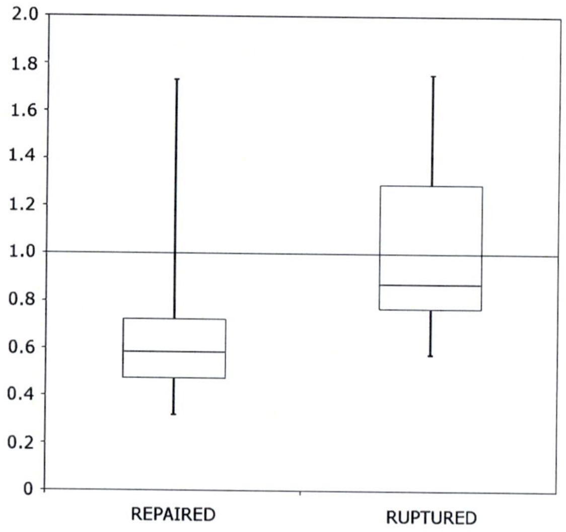

date using the FEARI model is based on the previous tests of others and therefore the approach can be improved. A large multi-centre study involved in the excision and testing of tissue would help strengthen the FEARI wall strength data, thus strengthening the applicability of the tool.

It is also possible to statistically estimate patient-specific wall strength thanks to the work of Vande Geest et al. [58]. This approach allows a point-wise wall strength estimation which can be coupled with the point-wise wall stress estimations from numerical analyses to create the RPI as discussed in Chap. 3.

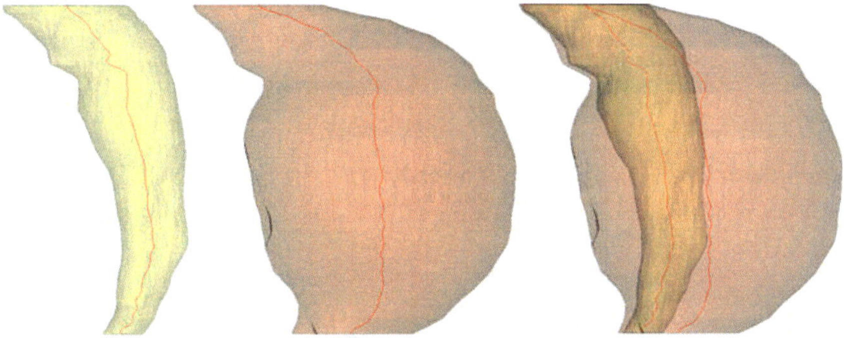

Fig. 6 Comparison of the centrelines generated from the lumen geometry (**a**) and the complete AAA volume, including ILT (**b**), and the visual difference in both centrelines (**c**)

4.3 Asymmetry

In 2009, a method of measuring asymmetry in the anterior-posterior plane of AAAs was reported [9] with this asymmetry correlated to both peak and posterior wall stress. This previous publication had some significant limitations, however, and work has continued since to address these shortcomings. Firstly, anterior-posterior asymmetry is essentially 2D and although the majority of AAAs naturally bulge outward in this direction due to the constraint of the spine, this measurement excludes certain AAAs. A new method of measurement has been recently developed by the authors whereby the asymmetry is measured in 3D therefore, all AAAs can be investigated. Another major limitation of this earlier report was the omission of the ILT. For any meaningful numerical analyses, ILT must be incorporated into the model. In the latest work, the ILT is modeled using the material model of Wang et al. [64] as mentioned previously. Within the finite element analysis, the use of shell elements have been replaced with 3D solid stress elements which are deemed a more accurate modeling approach compared to shell elements [7, 34], in particular when analysing these complex 3D simulations.

Therefore, for this aspect of the CADq, the 3D asymmetry of the study group used throughout this chapter was measured. When determining the asymmetry for AAA models in the presence of ILT, two centrelines are generated, one for the lumen and one for the luminal surface of AAA wall (Fig. 6). The method of measurement of 3D asymmetry remains similar to that of 2D asymmetry, in that the centreline is connected with an axis and the asymmetry defined as the distance from this axis to the centreline. However, as AAA centrelines are tortuous and travel through three dimensions, the distance from the centreline is now measured in three dimensions also. This updated measurement technique can be seen in Fig. 7. A further refinement to the tool can be made by smoothing the centrelines generated within the reconstruction software, as slight changes in geometry can result in large changes in the centreline.

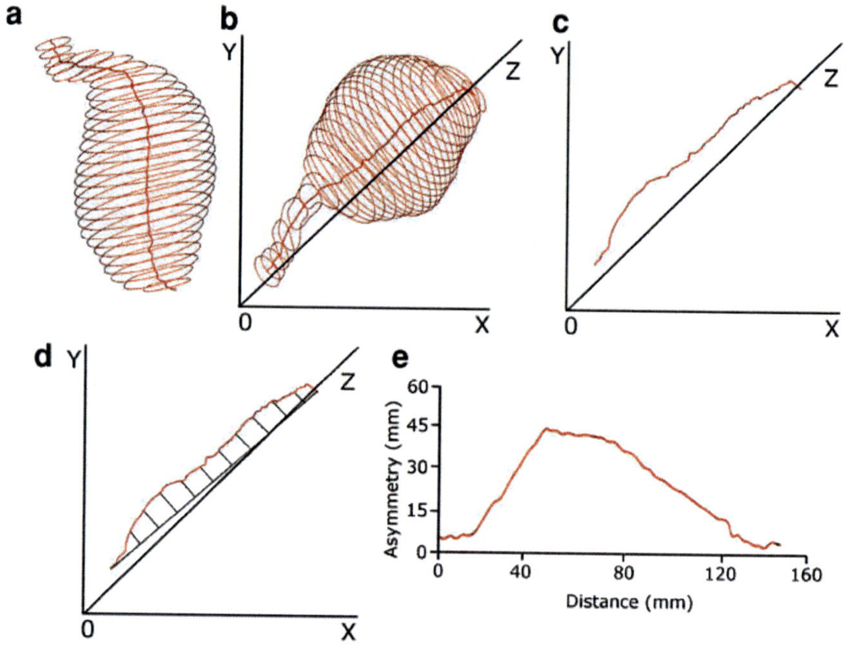

Fig. 7 Process of determining 3D asymmetry. **a** Polylines and centrelines are generated from the 3D reconstruction and **b** then plotted in 3D using our custom-built program. The program then automatically **c** isolates the centreline and **d** connects the ends of the centreline. The distance from this connecting axis to the centreline is then measured through three dimensions before **e** the program then automatically graphs the asymmetry along the length of the AAA

The CADq asymmetry results (Fig. 8) observed in this study show that the mean lumen asymmetry is 33% higher for the ruptured group, yet the difference was not significant ($P = 0.1$). The AAA wall asymmetry was only 3% higher in the ruptured group. By correlating the maximum wall stress to both the maximum diameter and the lumen asymmetry, it was noted that the lumen asymmetry is more significantly correlated to maximum wall stress for the repaired ($P = 0.002$) and ruptured ($P = 0.033$) groups compared to diameter for the same groups (repaired, $P = 0.032$; ruptured, $P = 0.174$).

Asymmetry as a CADq tool can be used as an additional technique to identify high-risk AAAs where diameter may fail. By comparing similarly-sized AAAs from the repaired and ruptured groups, it was seen that asymmetry may be useful alongside diameter. The AAAs compared in Fig. 9a were both large 11 cm AAAs, where one had ruptured and one was electively-repaired. The diameters of both AAAs were comparable but the resulting peak wall stress of the ruptured aneurysm was 157% higher than the repaired case (1.13 vs. 0.44 MPa) and the maximum measured lumen asymmetry was 125% higher in the ruptured AAA (45 vs.

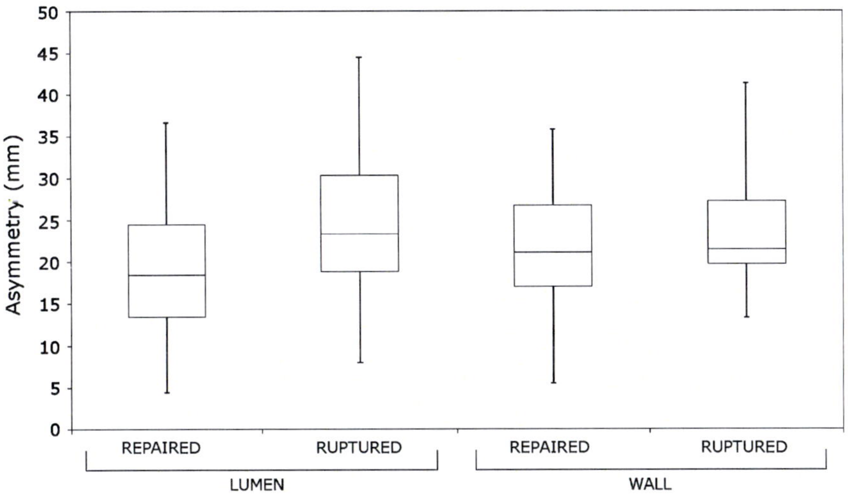

Fig. 8 Asymmetry results for both the lumen centrelines and the AAA wall centrelines for the repaired and ruptured groups. The box plot represents the median, 25th and 75th percentiles with the whiskers representing the maximum and minimum values observed

20 mm). The use of asymmetry also applies in some smaller AAAs. In the two aneurysms shown in Fig. 9b, both of which are 6.5 cm in maximum transverse diameter, the results show that the peak wall stress is 184% higher in the ruptured case (1.39 vs. 0.49 MPa) and the maximum lumen asymmetry is 157% higher in the ruptured case (18 vs. 7 mm). However, care should be taken when interpreting these preliminary results. The results presented in Fig. 8 show there can be a large distribution in asymmetry results and the measurement is not necessarily higher in ruptured cases. Increasing the numbers of cases examined may help reduce the distribution of results as there are only 10 patients in the ruptured cohort and only 42 in the repaired group.

5 Validation of CAD

Validation of numerical tools in the context of AAA rupture-prediction is difficult to achieve [33], with the benefits and challenges of patient-specific risk assessment well documented [2]. In vitro validation of numerical modeling was recently performed [14] where patient-specific silicone rubber AAA models were manufactured and then ruptured. High-speed photography was used to capture the event of rupture, with FEA used to validate the rupture locations. Excellent agreement in the experimental and computational results was observed with FEA accurately predicting the location of rupture in 90% of the models examined. The 10% that did not correlate were found to contain flaws within the wall as a result of the

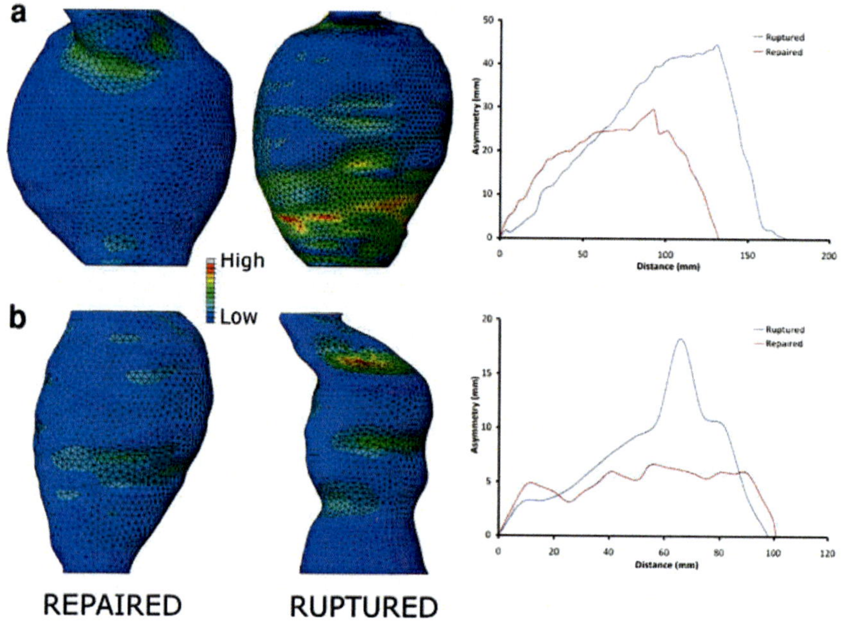

Fig. 9 **a** Comparison of an 11 cm repaired AAA and an 11 cm ruptured case. Measured lumen asymmetry for both cases is presented on the right. Both peak wall stress (1.13 vs. 0.44 MPa) and lumen asymmetry (45 vs. 20 mm) were higher in the ruptured case, whereas diameter did not differentiate. **b** Comparison of a 6.5 cm repaired AAA and a 6.5 cm ruptured case. Measured lumen asymmetry for both cases is presented on the right. Both peak wall stress (1.39 vs. 0.49 MPa) and lumen asymmetry (18 vs. 7 mm) were again higher in the ruptured case, with diameter identifying them both as high-risk (diameters >5.5 cm). Models are not shown to scale

manufacturing process, thus altering the burst site. In vivo validation however, is significantly more challenging.

However, recently a 73 year old male presented himself to a local vascular outpatients department with intermittent claudication. The clinician examined the patient with ultrasound and observed a 7.5 cm AAA (Fig. 10a). CT imaging was subsequently performed (Fig. 10b) and endovascular aneurysm repair (EVAR) planned. Exactly one week after CT imaging, the patient collapsed at a pre-operative assessment clinic complaining of acute abdominal pain and nausea. The patient was transferred to the HSE Midwestern Regional Hospital, Limerick and was hypotensive (80/40 mmHg) on arrival. The clinicians immediately performed a laparotomy and open surgical repair of the leaking infrarenal AAA. During the procedure the clinicians sketched (Fig. 11a) and later recorded the location of rupture on the CT image (Fig. 11b). The patient fully recovered from the operation and was discharged nine days later with no complications.

The CT dataset of this case was imported into Mimics and analysed by the author using the techniques outlined throughout this chapter. The model was

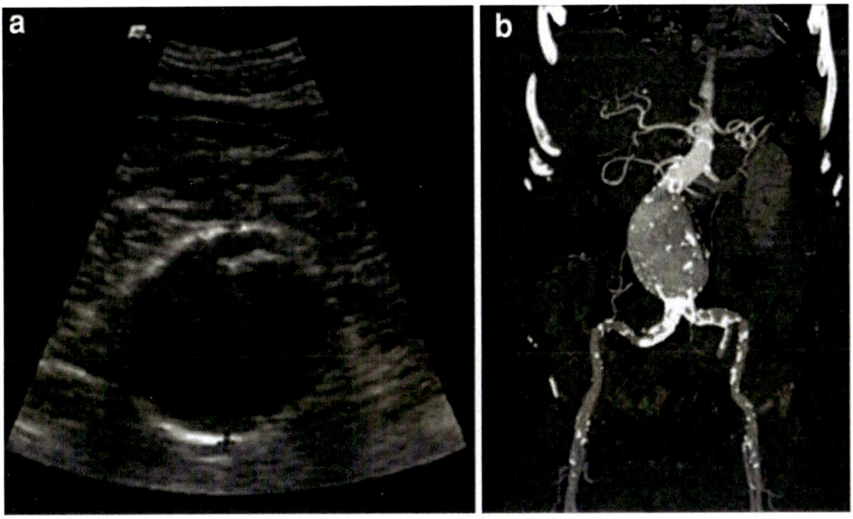

Fig. 10 **a** Ultrasound image showing the patient's 7.5 cm AAA and **b** CT image showing the infrarenal AAA from the anterior view

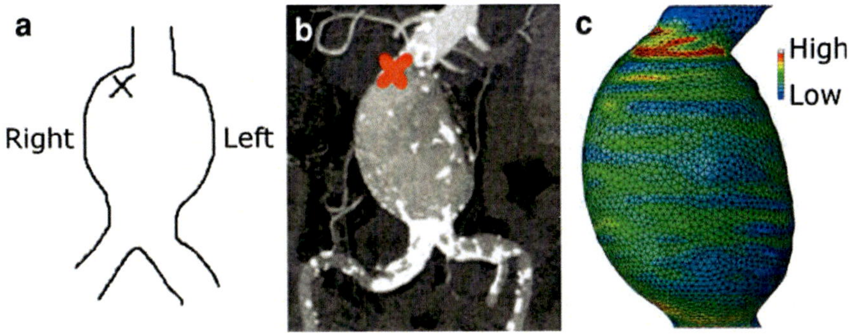

Fig. 11 Comparison of clinical rupture locations with the FEA-predicted location of peak wall stress. **a** Intraoperative sketch of AAA rupture site. **b** Clinicians then recorded the rupture site on the CT image. **c** Resulting FEA-predicted wall stress distribution (Peak wall stress = 0.719 MPa)

examined with finite element analysis using a uniform wall thickness of 1.5 mm and the standard AAA wall [41] and ILT [64] material properties. The exact location of rupture was withheld from the author to ensure there was no bias in the modeling approach. Wall stress results (Fig. 11c) showed that the AAA experienced high wall stress ($\sim$0.7 MPa) located at the anterior-right proximal inflection region of the AAA sac. An excellent correlation was observed when the location of FEA-predicted peak stress was compared with the actual rupture site (Fig. 11). According to our FEARI model this particular case presented an 85%

chance of rupture (FEARI = 0.85) based on a peak wall stress of 0.719 MPa located at the anterior-right region (wall strength = 0.847 MPa).

6 Conclusion

Numerical modeling of AAAs is becoming more prevalent with several laboratories working towards improving AAA risk-prediction. To date, these efforts remain essentially un-validated and the brief description of the validation procedure presented here is one of the few reports of correlation between numerical and clinical rupture sites [54]. However, regardless of the efforts underway and the regular improvements in numerical modeling, possibly the biggest barrier to CAD becoming a clinical tool is the difficulty in translating the approach to the clinic. Neal and Kerckhoffs [39] recently reported on the current progress of patient-specific modeling, concluding that the incorporation of patient-specific modeling into the workflow of the clinician will require, among other things, regulatory approval by the relevant bodies, for example, the Food and Drug Administration (FDA). The techniques presented in this chapter are relatively easy to translate to the clinic as results of peak wall stress, FEARI and asymmetry can be obtained within approximately 3 h of CT imaging and therefore these tools would not detract from the treatment course of the patient. Wall stress has been repeatedly shown to be higher in symptomatic/ruptured AAAs compared to electively-repaired cases [13, 18–20, 30, 54, 60] and the results reported here agree with these previous studies. However, further refinement of CAD, in particular CADe of the AAA wall is vital if patient-specific analysis is desired. The aortic wall thickness remains elusive from current CT and MR imaging. A method of detecting the wall thickness was reported by Martufi et al. [32] however the technique is yet to be fully adopted within the numerical models of others. Assuming a uniform wall thickness in numerical models may be a gross assumption as it is known that the wall thickness influences the wall stress [48, 53, 60], yet in the CAD validation presented in Sect. 5 and in the work of Truijers et al. [54], a uniform wall thickness was employed and the location of rupture was still predicted. A similar finding was observed in studies on an idealised AAA model [12] where the wall thickness was varied according to the measured wall thickness in a silicone idealised AAA analogue. Although wall thickness does alter the numerical quantities of wall stress, it may not significantly influence the stress patterns to the same extent. In this previous report the overall trend in wall stress distribution remained similar regardless of wall thickness, that is, wall stress was higher at the proximal and distal inflection regions than that found in the area of maximum diameter.

It is understood that it may be unlikely these tools will ever directly replace the use of maximum diameter as clinicians will always feel that large AAAs represent a rupture-threat and should be repaired. It is the small to medium-sized AAAs however, that could be examined using these alternative diagnostic tools, and someday may prove to be a useful adjunct to maximum diameter.

References

1. Bluestein, D., Dumont, K., De Beule, M., Ricotta, J., Impellizzeri, P., Verhegghe, B., Verdonck, P.: Intraluminal thrombus and risk of rupture in patient-specific abdominal aortic —FSI modeling. Comput. Methods Biomech Biomed Eng. **12**, 73–81 (2009)
2. Breeuwer, M., de Putter, S., Kose, U., Speelman, L., Visser, K., Gerritsen, F., Hoogeveen, R., Krams, R., van den Bosch, H., Buth, J., Gunther, T., Wolters, B., van Dam, E., van de Vosse, F.: Towards patient-specific risk assessment of abdominal aortic aneurysm. Med. Biol. Eng. Comput. **46**, 1085–1095 (2008)
3. Conway, K.P, Byrne, J., Townsend, M., Lane, IF.: Prognosis of patients turned down for conventional abdominal aortic aneurysm repair in the endovascular and sonographic era: Szilagyi revisted? J. Vasc. Surg. **33**, 752–757 (2001)
4. Cronenwett, J.L., Murphy, T.F., Zelenock, G.B., Whitehouse, Jr W.M., Lindenauer S.M., Graham, L.M., Quint, L.E., Silver T.M., Stanley, J.C.: Actuarial analysis of variables associated with rupture of small abdominal aortic aneurysms. Surgery **98**, 472–483 (1985)
5. Darling, R.C., Messina, C.R., Brewster D.C., Ottinger L.W.: Autopsy study of unoperated abdominal aortic aneurysms. The case for early resection. Circulation **56**, 161–164 (1977)
6. Di Martino, E.S., Bohra, A., Vande Geest, J.P., Gupta, N., Makaroun M.S., Vorp, D.A.: Biomechanical properties of ruptured versus electively repaired abdominal aortic aneurysm wall tissue. J. Vasc. Surg. **43**, 570–576 (2006)
7. Doyle, B.J., Callanan, A., McGloughlin, T.M.: A comparison of modelling techniques for computing wall stress in abdominal aortic aneurysms. Biomed. Eng. Online **6**, 38 (2007)
8. Doyle, B.J., Morris, L.G., Callanan, A., Kelly, P., Vorp, D.A., McGloughlin, T.M.: 3D reconstruction and manufacture of real abdominal aortic aneurysms: From CT scan to silicone model. J. Biomech. Eng. **130**, 034501 (2008)
9. Doyle, B.J., Callanan, A., Burke, P.E., Grace, P.A., Walsh, M.T., Vorp, D.A., McGloughlin, T.M.: Vessel asymmetry as an additional tool in the assessment of abdominal aortic aneurysms. J. Vasc. Surg. **49**, 443–454 (2009)
10. Doyle, B.J., Callanan, A., Walsh M.T., Grace P.A., McGloughlin, T.M.: A finite element analysis rupture index (FEARI) as an additional tool for abdominal aortic aneurysm rupture prediction. Vasc. Dis. Prev. **6**, 114–121 (2009)
11. Doyle, B.J., Grace, P.A., Kavanagh, E.G., Burke, P.E., Wallis, F., Walsh, M.T., McGloughlin, T.M.: Improved assessment and treatment of abdominal aortic aneurysms: The use of 3D reconstructions as a surgical guidance tool in endovascular repair. Ir. J. Med. Sci. **178**, 321–328 (2009)
12. Doyle, B.J., Corbett, T.J., Callanan, A., Walsh, M.T., Vorp, D.A., McGloughlin, T.M.: An experimental and numerical comparison of the rupture locations of an abdominal aortic aneurysm. J. Endovasc. Ther. **16**, 322–335 (2009)
13. Doyle, B.J., Coyle, P., Kavanagh, E.G., Grace, P.A., McGloughlin, T.M.: A finite element analysis rupture index (FEARI) assessment of electively repaired and symptomatic/ruptured abdominal aortic aneurysms. IFMBE Proc. **31**, 883–886 (2010)
14. Doyle, B.J., Cloonan, A.J., Walsh, M.T., Vorp, D.A., McGloughlin, T.M.: Identification of rupture locations in patient-specific abdominal aortic aneurysms using experimental and computational techniques. J. Biomech. **43**, 1408–1416 (2010)
15. Elger, D.F., Blackletter, D.M., Budwig, R.S., Johansen, K.H.: The influence of shape on the stresses in model abdominal aortic aneurysms. J. Biomech. Eng. **118**, 326–332 (1996)

16. Ernst, C.B.: Abdominal aortic aneurysm. N. Eng. J. Med. **328**, 1167–1172 (1993)
17. Fenton, J.J., Taplin, S.H., Carney, P.A., Abraham, L., Sickles, E.A., D'Orsi, C., Berns, E.A., Cutter, G., Hendrick, E., Barlow, W.E., Elmore, J.G.: Influence of computer-aided detection on performance of screening mammography. N. Eng. J. Med. **356**, 1399–1409 (2007)
18. Fillinger, M.F., Raghavan, M.L., Marra, S.P., Cronenwett, J.L., Kennedy F.E.: In vivo analysis of mechanical wall stress and abdominal aortic aneurysm rupture risk. J. Vasc. Surg. **36**, 589–597 (2002)
19. Fillinger, M.F., Marra, S.P., Raghavan, M.L., Kennedy, F.E.: Prediction of rupture risk in abdominal aortic aneurysm during observation: wall stress versus diameter. J. Vasc. Surg. **37**, 724–732 (2003)
20. Gasser, T.C., Auer, M., Labruto, F., Swedenborg, J., Roy, J.: Biomechanical rupture risk assessment of abdominal aortic aneurysms: model complexity versus predictability of finite element simulations. Eur. J. Vasc. Endovasc. Surg. (2010). doi:10.1016/j.ejvs.2010.04.003
21. Giannoglu, G., Giannakoulas, G., Soulis, J., Chatzizisis, Y., Perdikides, T., Melas, N., Parcharidis, G., Louridas, G.: Predicting the risk of rupture of abdominal aortic aneurysms by utilizing various geometrical parameters: revisiting the diameter criterion. Angiology **57**, 487–494 (2006)
22. Glimaker, H., Holmberg, L., Elvin, A., Nybacka, O., Almgren. B., Bjorck, C.G., Eriksson, I.: Natural history of patients with abdominal aortic aneurysm. Eur. J. Vasc. Surg. **5**, 125–130 (1991)
23. Hirose, Y., Takamiya, M.: Growth curve of ruptured aortic aneurysm. J. Cardiovasc. Surg. **39**, 9–13 (1998)
24. Inzoli, F., Boschetti, F., Zappa, M., Longo, T., Fumero, R.: Biomechanical factors in abdominal aortic aneurysm rupture. Eur. J. Vasc. Surg. 7, 667–674 (1993)
25. Kleinstreuer, C., Li, Z.: Analysis and computer program for rupture risk prediction of abdominal aortic aneurysms. Biomed. Eng. Online **5**, 19 (2006)
26. Lederle, F.A., Johnson, G.R., Wilson, S.E., Ballard, D.J., Jordan Jr ,W.D., Blebea, J., Littooy, F.N., Freischlag, J.A., Bandyk, D., Rapp, J.H., Salam, A.A.: Rupture rate of large abdominal aortic aneurysms in patients refusing or unfit for elective repair. JAMA **287**, 2968–2972 (2002)
27. Leung, J.H., Wright, A.R., Cheshire, N., Crane, J., Thom, S.A., Hughes, A.D., Xu Y Fluid structure interaction of patient specific abdominal aortic aneurysms: a comparison with solid stress models. Biomed. Eng. Online **5**, 33 (2006)
28. Li, Z.Y., U-King-Im, J., Tang, T.Y., Soh, E., See, T.C., Gillard, J.H.: Impact of calcification and intraluminal thrombus on the computed wall stresses of abdominal aortic aneurysm. J. Vasc. Surg. **47**, 928–935 (2008)
29. Lorensen, W.E., Cline, H.E.: Marching cubes: a high resolution 3D surface construction algorithm. Comp. Graphics **21**, 163–169 (1987)
30. Maier, A., Gee, M.W., Reeps, C., Pongratz, J., Eckstein, H.H., Wall, W.A.: A comparison of diameter, wall stress, and rupture potential index for abdominal aortic aneurysm rupture risk prediction. Ann. Biomed. Eng. **38**, 3124–3134 (2010)
31. Maier, A., Gee, M.W., Reeps, C., Eckstein, H.H., Wall, W.A.: Impact of calcifications on patient-specific wall stress analysis of abdominal aortic aneurysms. Biomech. Model. Mechanobiol. **9**, 511–521 (2010)
32. Martufi, G., DiMartino, E.S., Amon, C.H., Muluk, S.C., Finol, E.A.: Three-dimensional geometrical characterization of abdominal aortic aneurysms: image-based wall thickness distribution. J. Biomech. Eng. **131**, 061015 (2009)
33. McGloughlin, T.M., Doyle, B.J.: New approaches to abdominal aortic aneurysm rupture risk assessment: engineering insights with clinical gain. Arterioscler. Thromb. Vasc. Biol. **30**, 1687–1694 (2010)
34. Meyer, C.A., Guivier-Curien, C., Moore, J.E.: Trans-thrombus blood pressure effects in abdominal aortic aneurysms. J. Biomech. Eng. **132**, 071005 (2010)

35. Moore, J.A., Steinman, D.A., Ethier, C.R.: Computational blood flow modeling: errors associated with reconstructing finite element models from magnetic resonance images. J. Biomech. **31**, 179–184 (1998)
36. Morris, L., Delassus, P., Callanan, A., Walsh, M., Wallis, F., Grace, P., McGloughlin, T.: 3D numerical simulation of blood flow through models of the human aorta. J. Biomech. Eng. **127**, 767–775 (2005)
37. Mower, W.R., Baraff, L.J., Sneyd, J.: Stress distributions in vascular aneurysms: factors affecting risk of aneurysm rupture. J. Surg. Res. **55**, 155–161 (1993)
38. National Health Service. National Screening Program for Abdominal Aortic Aneurysm [online] available: http://aaa.screening.nhs.uk (2009). Accessed 9 Feb 2009
39. Neal, M.L., Kerckhoffs, R.: Current progress in patient-specific modeling. Brief Bioinform. **11**, 111–126 (2009)
40. Nicholls, S.C., Gardner, J.B., Meissner, M.H., Johansen, H.K: Rupture in small abdominal aortic aneurysms. J. Vasc. Surg. **28**, 884–888 (1998)
41. Raghavan, M.L., Vorp, D.A.: Toward a biomechanical tool to evaluate rupture potential of abdominal aortic aneurysm: identification of a finite strain constitutive model and evaluation of its applicability. J. Vasc. Surg. **33**, 475–482 (2000)
42. Raghavan, M.L., Webster, M.W., Vorp, D.A.: Ex vivo biomechanical behaviour of abdominal aortic aneurysm: assessment using a new mathematical model. Ann. Biomed. Eng. **24**, 573–582 (1996)
43. Raghavan, M.L., Vorp, D.A., Federle, M.P., Makaroun, M.S., Webster, M.W.: Wall stress distribution on three-dimensionally reconstructed models of human abdominal aortic aneurysm. J. Vasc. Surg. **31**, 760–769 (2000)
44. Raghavan, M.L., Kratzberg, J., de Tolosa, E.M.C., Hanaoka, M.M., Walter, P., da Silva, E.S.: Regional distribution of wall thickness and failure properties of human abdominal aortic aneurysm. J. Biomech. **39**, 3010–3016 (2006)
45. Sacks, M.S, Vorp, D.A, Raghavan, M.L., Federle, M.P, Webster, M.W.: In vivo three-dimensional surface geometry of abdominal aortic aneurysms. Ann. Biomed. Eng. **27**, 469–479 (1999)
46. Sayers, R.D.: Aortic aneurysms, inflammatory pathways and nitric oxide. Ann. Royal Col. Surg. Eng. **84**, 239–246 (2002)
47. Schurink, G.W.H., van Baalen, J.M., Visser, M.J.T., van Bockel, J.H.: Thrombus within an aortic aneurysm does not reduce pressure on the aneurysmal wall. J. Vasc. Surg. **31**, 501–506 (2000)
48. Scotti, C.M., Shkolnik, A.D., Muluk, S.C., Finol, E.: Fluid-structure interaction in abdominal aortic aneurysms: effect of asymmetry and wall thickness. Biomed. Eng. Online **4**, 64 (2005)
49. Speelman, L., Bohra, A, Bosboom, E.M.H., Schurink, G.W.H., van de Vosse, F.N., Makaroun, M.S., Vorp, D.A.: Effects of wall calcifications in patient-specific wall stress analyses of abdominal aortic aneurysms. J. Biomech. Eng. **129**, 1–5 (2007)
50. Stenbaek, J., Kalin, B., Swedenborg, J.: Growth of thrombus may be a better predictor of rupture than diameter in patients with abdominal aortic aneurysms. Eur J Vasc Endovasc Surg **20**, 466–469 (2000)
51. Stringfellow, M.M., Lawrence, P.F., Stringfellow, R.G.: The influence of aorta geometry upon stress in the aneurysm wall. J. Surg. Res. **42**, 425–433 (1987)
52. Thubrikar, M.J., Labrosse, M., Robicsek, F., Al-Soudi, J., Fowler, B.: Mechanical properties of abdominal aortic aneurysm wall. J. Med. Eng. Tech. 25:133–142 (2001)
53. Thubrikar M.J, Al-Soudi, J., Robicsek, F.: Wall stress studies of abdominal aortic aneurysm in a clinical model. Ann. Vasc. Surg. **15**, 355–366 (2001)
54. Truijers M., Pol, J.A., SchultzeKool, L.J., van Sterkenburg, S.M., Fillinger, M.F., Blankensteijn, J.D.: Wall stress analysis in small asymptomatic, symptomatic and ruptured abdominal aortic aneurysms. Eur. J. Vasc. Endovasc. Surg. **33**, 401–407 (2007)
55. United States Preventative Services Task Force. Screening for abdominal aortic aneurysm: recommendation statement. Ann. Int. Med. **142**, 198–202 (2005)

56. Vallabhaneni, S.R., Gilling-Smith, G.L., Brennan, J.A., Heyes, R.R., Hunt, J.A., How, T.V., Harris, P.L.: Can intrasac pressure monitoring reliably predict failure of endovascular aneurysm repair? J. Endovasc. Ther. **10**, 524–530 (2003)
57. Vande Geest, J.P., Di Martino, E.S., Bohra, A., Makaroun, M.S., Vorp, D.A.: A biomechanics-based rupture potential index for abdominal aortic aneurysm risk assessment. Ann. NY Acad. Sci. **1085**, 11–21 (2006)
58. Vande Geest, J.P., Wang, D.H.J., Wisniewski, S.R., Makaroun, M.S., Vorp, D.A.: Towards a non-invasive method for determination of patient-specific wall strength distribution in abdominal aortic aneurysms. Ann. Biomed. Eng. **34**, 1098–1106 (2006)
59. Vande Geest, J.P., Sacks, M.S, Vorp, D.A.: The effects of aneurysm on the biaxial mechanical behaviour of human abdominal aorta. J. Biomech. **39**, 1324–1334 (2006)
60. Venkatasubramaniam, A.K., Fagan, M.J., Mehta, T., Mylankal K.J., Ray, B., Kuhan, G., Chetter, I.C., McCollum, P.T.: A comparative study of aortic wall stress using finite element analysis for ruptured and non-ruptured abdominal aortic aneurysms. Eur. J. Vasc. Endovasc. Surg. **28**, 168–176 (2004)
61. Volokh, K.Y., Vorp, D.A.: A model of growth and rupture of abdominal aortic aneurysm. J. Biomech. **41**, 1015–1021 (2008)
62. Vorp, DA.: Biomechanics of abdominal aortic aneurysm. J. Biomech. **40**, 1887–1902 (2008)
63. Vorp, D.A., Raghavan, M.L., Webster, M.W.: Mechanical wall stress in abdominal aortic aneurysm: influence of diameter and asymmetry. J. Vasc. Surg. **27**, 632–639 (1998)
64. Wang, D.H.J., Makaroun, M.S., Webster, M.W., Vorp, D.A.: Mechanical properties and microstructure of intraluminal thrombus from abdominal aortic aneurysm. J. Biomech. Eng. **123**, 536–539 (2001)
65. Wang, D.H.J., Makaroun, M.S., Webster, M.W., Vorp, D.A.: Effect of intraluminal thrombus on wall stress in patient-specific models of abdominal aortic aneurysm. J. Vasc. Surg. **36**, 598–604 (2002)
66. Watton, P., Hill, N., Heil, M.: A mathematical model for the growth of the abdominal aortic aneurysm. Biomech. Model. Mechanobiol. **3**(2), 98–113 (2004)
67. Wilarusmee, C., Suvikrom, J., Suthakorn, J., Lertsithichai, P., Sitthiseriprapip, K., Proprom, N., Kittur, D.S.: Three-dimensional aortic aneurysm model and endovascular repair: an educational tool for surgical trainees. Int. J. Angiol. **17**, 129–133 (2008)
68. Xiong, J., Wang, S.M., Zhou, W., Wu, J.G.: Measurement and analysis of ultimate mechanical properties, stress-strain curve fit, and elastic modulus formula of human abdominal aortic aneurysm and nonaneurysmal abdominal aorta. J. Vasc. Surg. **48**, 189–195 (2008)
69. Zienkiewicz, O.C., Taylor, R.L., Zhu, J.Z.: The finite element method: its basis and fundamentals, 6th edn. Elsevier Butterworth-Heinemann, UK (2005)

Mechanical Properties of AAA Tissue

Madhavan L. Raghavan and Erasmo Simão da Silva

Abstract Aneurysms of the infrarenal abdominal aorta involve 100–800% growth of the aortic tissue. While the precise pathogenesis of AAA is poorly understood, it should be expected that this change to the structure of the aortic wall will significantly impact its mechanical behavior. It is only reasonable to postulate that a better understanding of the mechanical behavior of AAA tissue will aid in our understanding of its causes and assessing its severity. Even for developing ways to treat it—currently done with implants deployed surgically or endovascularly—understanding the mechanical properties of the AAA wall will enhance our ability to design implants that can stay in place and/or protect the aneurysm wall from blood pressure. This chapter will review the state of literature on the mechanical properties of AAA tissue, the challenges faced, recent advances and future directions.

1 Introduction

Aneurysms are most commonly found in the infrarenal abdominal aorta—abdominal aortic aneurysm (AAA)—where the wall undergoes growths of 100–800%. While the precise pathogenesis of AAA is poorly understood, it should

M. L. Raghavan (✉)
Department of Biomedical Engineering, University of Iowa, 1136 Seamans Center, Iowa City IA 52242, USA
e-mail: ml-ragahvan@uiowa.edu

E. S. da Silva
Department of Surgery, University of São Paulo, São Paulo, SP, Brazil

Stud Mechanobiol Tissue Eng Biomater (2011) 7: 139–162
DOI: 10.1007/8415_2011_71

Published Online: 31 July 2011

be expected that this change to the structure of the aortic wall will significantly impact its mechanical behavior. This chapter will review the state of literature on the mechanical properties of AAA tissue, the challenges faced, recent advances and future directions.

In order to understand the biomechanical issues in the etiology and treatment of abdominal aortic aneurysms, it is important to understand the structure of the aortic wall. The aortic wall is composed of three layers: intima, media and adventitia. The biomechanical properties of the tissue are predominated by the structure of the media and adventitia and these are affected by natural issues such as aging and gender, or pathologic process such as, smoking, hypertension, atherosclerosis or aneurysms and dissections. The aortic media contains elastin, collagen, smooth muscle cells and non-fibrous matrix. Light and electron microscopic studies indicate that the structural components in the media are arranged in an orderly fashion. Concentric fibrillar elastin lamellae are connected by intricate network of elastin fibrils with interspersed collagen fibers and smooth muscle cells [1]. Elastin and collagen—connective tissue elements—found in the arterial media and adventitia, provide passive stiffness and strength to the aortic wall. Different studies have shown that this structural arrangement is very complex and varies according to the aortic segment (thoracic or abdominal) [2]. Several hypotheses have been postulated to explain the predilection of aneurysms to the infrarenal aorta, including flow characteristics, anatomic features of the media lamella, and mural ischemia. Although this localization suggests a focal process, abdominal aortic aneurysms are associated with generalized dilation and elongation of other major arteries at remote sites such popliteal arteries. It is conceivable therefore that even the non-dilated segment (aneurysm neck) is similarly affected [3–6]. In abdominal aortic aneurysm pathogenesis, the structural alterations of aortic wall leading to progressive dilatation and eventual ruptures are thought to be the degradation of matrix proteins, decrease of smooth muscle cells, inflammatory infiltration and neovascularization [7]. The striking feature among these is elastin degradation and excessive collagen deposition resulting in likely large changes in their mechanical properties. Two issues are worth noting. First, these modifications to aortic tissue need not be homogeneous within an aneurysm. Indeed, it can be quite heterogeneous and account for a large variability in wall thickness and mechanical properties. Second, pathological processes may be occurring in undilated regions. The information provided in this chapter should be understood within that context.

The AAA is made up of many entities, some of which are known to be specifically relevant in the mechanical sense—these are the aneurysmal aortic tissue itself, the large amounts of thrombus seen inside the AAA, and regionally localized entities such as the calcifications on/of the wall tissue and non-calcified atherosclerotic plaque lining the wall (see Fig. 1). This chapter will focus on the AAA wall itself rather than calcified tissue or intraluminal thrombus (see important review by Vorp [8] and Vorp and Vande Geest [9] for a discussion of these properties).

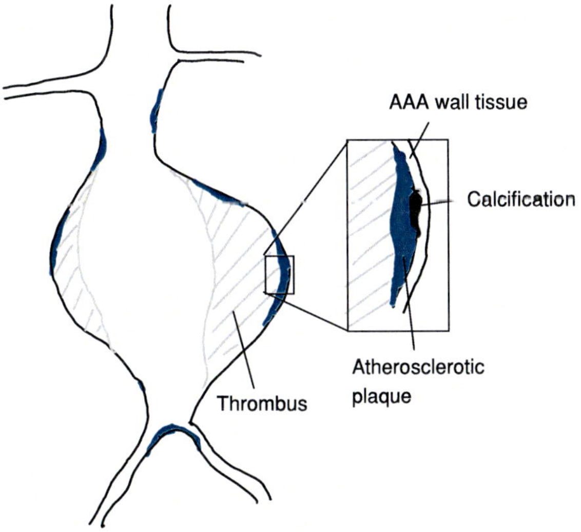

Fig. 1 Distinct structural components in the AAA wall

1.1 Clinical Significance of Understanding Mechanical Properties of AAA

Three questions are of clinical importance in AAA—what are the causes leading up to aneurysm development, how to diagnose its severity to assess its risk of rupture and how to treat it. Irrespective of events leading up to it, rupture itself is a mechanical event—the failure of the aortic wall tissue. It is therefore only reasonable to postulate that a better understanding of the mechanical behavior of the AAA tissue will aid in our understanding of its causes and assessing its severity. Even for developing ways to treat it—currently done with implants deployed surgically or endovascularly—understanding the mechanical properties of the AAA wall will enhance our ability to design implants that can stay in place and/or protect the aneurysm wall from blood pressure. Gaining an understanding of the mechanical properties of the AAA tissue therefore, is of clinical significance.

1.2 Approaches Used in Inferring Mechanical Properties of AAA

Mechanical properties of solid entities capture the relationship between forces and constraints on these entities and their resulting motion. Gaining a clear understanding of this relationship, especially for biological entities involves many complexities that do not necessarily present themselves in the field of classical engineering mechanics mainly because of limitations posed by the unique biological environment. For a synthetic material, the classical approach would be to start with specimens of the solid entities and subject them to controlled loads while measuring the resulting motion (load-controlled tests) or vice versa (motion-controlled tests). But because biological entities are part of a larger living system that is complex and poorly understood, isolating them for testing may introduce as-yet unknown changes to their behavior affecting the results of such tests. On the other hand, one may reason that within the biological environment itself, such mechanical tests are being conducted in vivo and one could leverage these 'tests' to extract the mechanical properties. That is, in vivo, the entity may be subject to forces and constraints and therefore undergo deformation in its natural physiological environment (for example, the aorta is undergoing motion because of pulsating blood pressure), which may themselves be the 'experimental' data required to estimate their properties. But in this latter paradigm, the challenges are in our ability to accurately determine the true force and the displacement distribution in the biological entity, not to mention, ascertaining its stress-free configuration. The use of animal models, where perhaps control and measurability are greater, is an alternative, but remains a poor one for disease-affected tissues mainly because of legitimate questions on the reliability of such animal models to be surrogates of human disease states, especially when it comes to mechanical properties of the diseased entities. Therefore, on studying mechanical properties of biological entities, there is no one entirely reliable approach, but rather, different approaches that offer advantages and disadvantages. In the case of ascertaining mechanical properties of AAA, studies have used all the above approaches—ex vivo/in vitro testing and in vivo 'tests'. The following sections will discuss many of these approaches and reported results in the literature.

1.3 Quantities Used in Inferring Mechanical Properties of AAA

The relationship between force and motion in AAA tissue may be characterized by multiple measures that capture different aspects of that relationship with multiple mathematical definitions for a given measure, reflective of the general lack of consensus on this topic.

2 In Vivo Distensibility of AAA

Measures of the motion of the wall under pulsatile blood pressure provide a first look at its in vivo behavior. These measures usually capture the relationship between the pressure pulse and resulting change in size. There are many mathematical definitions of distensibility however, because of differences in definition of 'size', whether or not size change is normalized and if/how the definition is reconciled with the notion of elastic modulus. In 1992, Lanne et al. [10] quantified in vivo distensibility of the AAA wall using the pressure-strain elastic modulus

(E_p) and what they termed, the stiffness parameter (β) based on the following definitions:

$$E_p = D_d \times (P_s - P_d)/(D_s - D_d) \quad (1)$$

$$\beta = D_d \times \text{Ln}\,(P_s/P_d)/(D_s - D_d) \quad (2)$$

where, D is diameter, P is pressure and the subscripts, s and d refer to systolic and diastolic time points—which were typically assumed to be maximum and minimum values respectively. By these definitions, both E_p and β would drop with rising distensibility of the AAA wall. They used ultrasound phase-locked echo-tracking system and reported that E_p was higher on average and more widely dispersed in aneurysmal abdominal aorta ($N = 37$) compared to the non-aneurysmal aorta group ($N = 76$). In 1992 as well, MacSweeney et al. [11] also reported that E_p was higher in aneurysmal abdominal aorta compared to controls. In 1996, Vorp et al. [12] performed ultrasound imaging of nine AAA subjects to study the wall distensibility, which they quantified using compliance defined as,

$$C = (A_{max} - A_{min})/[A_{max}(P_{max} - P_{min})] \quad (3)$$

where, A is cross sectional area measured from ultrasound image, P is blood pressure, with subscripts, max and min referring to the maximum and minimum values. In addition to the compliance of the AAA wall, they also calculated the compliance of the lumen (recollect from Fig. 1 that the wall and the lumen are not the same owing to presence of large amounts of thrombus). They reported that the compliance of the wall was roughly half that of the lumen. They further noted that the thrombus cross sectional area was nearly constant during the cardiac cycle indicating its incompressible nature. In 1999, Wilson et al. [13] evaluated E_p and β in 60 AAA subjects who were also followed untreated for a median 21 months. They noted that while E_p (not β) correlated with mean arterial pressure, neither E_p nor β correlated with growth rate. While these previous studies used ultrasound imaging, Long et al. [14, 15] employed tissue Doppler imaging of 56 AAA patients and evaluated numerous measures of wall distensibility—essentially, variants of one or more of the above-defined, E_p, β or C. They report that segmental compliance (which is essentially, C without normalizing to D_d) correlated to AAA size, while E_p and β did not. More recently, Ganten et al. [16] used Time-resolved ECG-gated CT imaging data from 67 patients to calculate what they termed, distensibility, but identical in definition to compliance C reported by Vorp et al. [12] They found that the compliance of AAA did not differ between small (<5 cm; $N = 44$) and large (>5 cm; $N = 23$) lesions.

The above studies represent the evolution of reports on the topic of in vivo AAA distensibility. They provide valuable information on the in vivo motion of the AAA. There are some limitations to these measures and measurement techniques. For one, they provide what may be thought of as extrinsic information on the AAA behavior, not necessarily intrinsic properties of the AAA wall tissue. This is because, the motion of the wall due to pressure is affected not just by the caliber of the wall tissue, but also by size and other boundary conditions that are not yet fully understood. In other words, distensibility represents a lumped measure of the AAA, rather than specific to AAA tissue. Further, their definitions and physical meanings are unclear and varying leading to confusion in the field. For example, the physical distinction between E_p and β pose difficulties in explaining findings where one is different between two groups but not the other [13]. However, it is worth noting that in vivo distensibility measures do have some key merits in their favor. For one, the fact that it is in vivo data implies that the AAA structure is being studied undisturbed (i.e., it is not isolated from its surrounding environment). Consequently, there is high fidelity in the conditions under which measurements were made. Further, in many cases, a lumped extrinsic measure is precisely what is needed for practical interpretations. For these reasons, the above studies continue to be valuable in our understandings of AAA wall mechanical properties. Perhaps, there is a need to standardize the definition of distensibility in order to develop a consistent and physically meaningful definition for this important measure, so result from different reports may be compared and utilized in the clinical setting and for further research.

3 Ex Vivo Elastic Properties of AAA Tissue

Unlike in vivo estimations, ex vivo testing of harvested AAA tissue samples may be performed under controlled displacement and force conditions allowing for the ability to characterize intrinsic properties of the tissue itself rather than having to settle for a measure that lumps tissue caliber with other aspects. Such control permits the estimations of more precise measures of mechanical properties with independent physical meanings. Some of these mechanical properties include measures of stiffness, failure stress and strain.

3.1 Uniaxial Tests

Uniaxial extension testing is the simplest and most common of ex vivo elastic property testing methods. Here, a rectangular, planar sample is subjected to extension along its length at a constant displacement rate in a tester while the force is recorded during extension. Typically, these tests are conducted until failure of the specimen. From the resulting force-extension data, some elastic properties may be inferred.

Sumner et al. [17] and Drangova et al. [18] were the first known to have reported on ex vivo tests of human AAA harvested from cadavers. They reported that stiffness estimates based on pressure-inflation tests were found to be greater in aneurysmal aorta compared to non-aneurysmal aorta. They explained this as being due to reduced elastin and increased collagen content. He and Roach [19] obtained rectangular specimen strips during surgical resection of eight AAA patients and subjected them to uniaxial extension tests up to a pre-defined maximum load rather than until failure. The recorded force-extension data was converted to stress–strain data assuming incompressibility. They showed that the stress–strain behavior of AAA tissue was non-linear, as is often the case with biological soft tissues. They then fit their stress–strain data to the exponential functional form,

$$\sigma = \mathrm{a}e^{\mathrm{b}\varepsilon} \tag{4}$$

where, σ is stress, ε is strain and a and b are material parameters. Later, in two reports, Vorp et al. and Raghavan et al. [20, 21] reported on uniaxial extension testing of AAA specimen strips harvested from the anterior midsection of AAA during surgical resection in 69 patients. The specimens were extended until failure. In most cases, the rectangular specimens' length was in the axial orientation, but in a small population, they were oriented circumferentially. They reported a novel mathematical model to estimate physically meaningful measures such as stiffness. The data was fit to a modified form of Michaelis-Menton model:

$$\varepsilon = [\,\mathrm{K} + \mathrm{A}/(\mathrm{B} + \sigma)\,]\sigma \tag{5}$$

where, K, A and B are material parameters. The model is based on the premise that the non-linearity of the stress–strain curve is a result of collagen recruitment into load bearing, a phenomenon proposed by Roach and Burton in 1957 [22]. The elastic (pre-yield) portion of the stress–strain curve was thought to transition through three phases:

- Phase 1: an initial low-strain phase where load is borne only by the taut elastin fibers. This will result in a linear stress–strain curve whose slope maybe thought of the stiffness of elastin in the tissue (E_E);
- Phase 2: an intermediate phase at medium strains when a gradually increasing number of collagen fibers that were previously tortuous start to straighten out and contribute to load bearing. This will cause a gradual increase in the slope of the stress–strain curve and hence its apparent non-linearity;
- Phase 3: a final phase at strains approaching the yield point (phase 3), when all collagen fibers have been recruited into load bearing. With increasing strain, the slope will remain constant and a linear stress–strain curve results. The slope of the stress–strain curve in this phase will be indicative of the combined stiffness of elastin and collagen in the tissue ($E_E + E_C$).

They ignored the contribution to stiffness from other structural components such as smooth muscle cells (the aorta is an elastic artery) and ground substance as being negligible. They then derived a relationship between the parameters of the model and E_E and E_C by exploiting the functional form of their model.

For phase 1, as $\sigma \rightarrow 0$, $(\mathrm{B} + \sigma) \approx \mathrm{B}$ and from Eq. 5:

$$\varepsilon|_{\text{phase 1}} = \left(K + \frac{A}{B}\right)\sigma \tag{6}$$

Inverting to the more typical stress as a function of strain,

$$\sigma|_{\text{phase1}} = \frac{1}{\left(K + \frac{A}{B}\right)}\varepsilon \tag{7}$$

Hence, Eq. 5 reduces to a linear relationship between σ and ε in phase 1. Since E_E is the slope of the stress strain curve at low strains, we note from Eq. 7 that

$$E_E = \left.\frac{d\sigma}{d\varepsilon}\right|_{\sigma\to 0} = \frac{1}{\left(K + \frac{A}{B}\right)} \tag{8}$$

Let us assume a priori that B $\ll \sigma_y$ during phase 3, so that (B + σ) $\approx \sigma$ and

$$\varepsilon|_{\text{phase 3}} = K\sigma + A \tag{9}$$

or

$$\sigma|_{\text{phase 3}} = \left(\frac{1}{K}\right)\varepsilon - \frac{A}{K} \tag{10}$$

Again the model reduces to a linear relationship between σ and ε in phase 3, where it has been postulated to be so. Since the constant slope in phase 3 is defined as E_E+E_C, we note from Eq. 10 that

$$E_E + E_C = \left.\frac{d\sigma}{d\varepsilon}\right|_{\sigma\to\sigma_y} = \frac{1}{K} \tag{11}$$

The contribution to tissue stiffness by the fully recruited collagen fibers can be found by subtracting Eq. 8 from Eq. 11:

$$E_C = \frac{1}{K} - \frac{1}{K + \frac{A}{B}} = \frac{A}{K(A + KB)} \tag{12}$$

Further, a direct physical meaning for the model parameter A may be ascribed by simply noting that it is the strain intercept for the linear fit to phase 3 (see Eq. 9). Higher A is thus indicative of a highly extensible tissue and therefore a measure of the strain required for complete collagen recruitment into load bearing. Additional details on this model may be found in the publication.

Thus, from model parameters, K, A and B, they were able to infer three independent physical meanings, namely, the stiffness contribution from elastin (E_E), stiffness contribution from collagen (E_C), and strain needed for complete collagen recruitment (the model parameter A, itself). They report that E_E and E_C were not significantly different between longitudinally oriented AAA specimens ($N = 45$) and longitudinally oriented normal aortic specimens ($N = 7$) that were harvested from healthy organ donors. The recruitment parameter A, however, was significantly lower (half) for aneurysmal versus control specimens indicating that collagen fibers get recruited at lower strains into load bearing in aneurysms which may be causing them to behave stiffer. It is in failure properties that they noted large differences between normal and aneurysmal tissue with both the yield stress (stress at which the slope of stress–strain curve starts to drops) and ultimate stress (maximum stress in the stress–strain curve) being significantly lower than controls. One limitation of the study was that the control group population was significantly younger than the AAA group (47 vs. 69 years). They note that the differences in failure properties held up even when they employed an age-correction equation for failure strength borrowed from another report [23]. Subsequently, this group used the stress–strain data to fit it to a more robust finite strain constitutive model and estimated its material parameters [24]. The strain energy density function was defined as,

$$W = \alpha(I_c - 3) + \beta(I_c - 3)^2 \tag{13}$$

where, α and β are the material parameters. For their study population, α and β were found to be 17.4 ± 1.5 and 188.1 ± 37.2 N/cm^2 (mean $\pm$ SEM). Details on the model and results may be found in their report [24].

3.2 Tissue Properties and Matrix Metalloproteinases

Later, Vallabhaneni et al. [25] performed uniaxial extension tests on 124 strips of AAA tissue excised from 24 patients and 60 normal abdominal aortic tissue strips from 6 organ donors. The aneurysm specimens were elliptical sections from the anterior portion of the AAA excised during surgical resection. Remarkably, unlike previous studies, they obtained an average of 6 rectangular test strips per AAA. They also biopsied small strips from these specimens and assayed for MMP-2 and MMP-9. The estimated mechanical property measures were elastic modulus, ultimate strength, ultimate strain (strain corresponding to ultimate stress), MMP-2 and MMP-9 levels. Given the non-linearity of the stress–strain curve, the elastic modulus was calculated as the slope at physiological stress. It was unclear from their report how they ascertained physiological stress in the aortic wall of each of their subjects. They noted that none of these parameters were statistically significant in the differences between aneurysmal and non-aneurysmal specimens mainly because of significant intra-subject variations rather than even inter-subject variations. They further note that there was little correlation between MMP levels and mechanical properties.

3.3 Regional Variations in Mechanical Properties

In order to rigorously study the regional heterogeneity of mechanical properties within AAA tissue, Thubrikar et al. [26] obtained five whole unruptured AAA specimens during surgical resection. They then cut about four axially oriented and two circumferentially oriented strips per AAA and subjected them to uniaxial tests until failure. They fit their data to the exponential functional form reported by

Table 1 Patient demographics, morphology and test details for the study population. All images are from the anterior view and to scale

	AAA #1	AAA #2	AAA #3	AAA #4
Sex	Male	Male	Male	Female
Age (years)	55	54	77	88
Days in storage	86 (frozen)	0	83 (frozen)	34 (frozen)
AAA Dia (cm)	4.5	4.0	7.5	8.0
AAA height (cm)	12.0	15.5	13.8	14.0
Cause of death	Cancer	Myocardial infarction	COPD	AAA rupture
Morphology				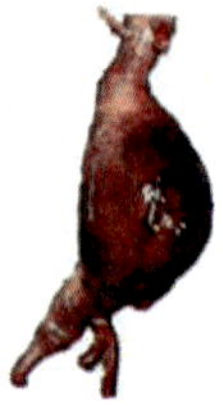

He and Roach [19]. They also measured thickness at about 10 sites in the AAA wall and the yield stress from their test data. They did not report any consistent pattern in the variation of any of these properties. The key finding from their study was that there is indeed significant regional variations in wall thickness and material properties (characterized by a and b of the exponential model mentioned earlier and the yield stress).

Following their study, the authors (of this chapter) also performed a similar study to assess the level and nature of regional variations in AAA tissue properties [27]. Three unruptured and one ruptured AAA were excised from cadavers submitted to necropsy for the elucidation of the cause of death at the Service for Confirmation of Deaths at São Paulo University School of Medicine. Wall thickness was measured at multiple points along the cut edge, approximately every 1.5 cm and recorded on the digital image. Multiple longitudinally oriented rectangular strips of specimens were cut all around the circumference of the AAA using an 8 mm wide cutting die. Wall thickness was continuously measured as the specimen strips were cut and measurement locations marked on the digital images. Localized characteristics such as blisters, calcified plaque and rupture were noted when present within or close to a specimen strip. Table 1 shows the patient demographics, and AAA morphology for the study population.

The specimen strips were then immersed in saline and refrigerated at 4°C for immediate mechanical testing. Uniaxial extension testing of rectangular specimen strips was performed. From the force-extension data, the failure tension (T_f) and failure stress (S_f) associated with the peak load were calculated (see Fig. 2).

The elastic portion of the stress strain curve was fit to a previously reported finite elastic material model [24] and the material parameters, α and β were estimated. For a quantitative assessment of regional distribution of thickness and mechanical properties, all locations and associated properties were mapped to

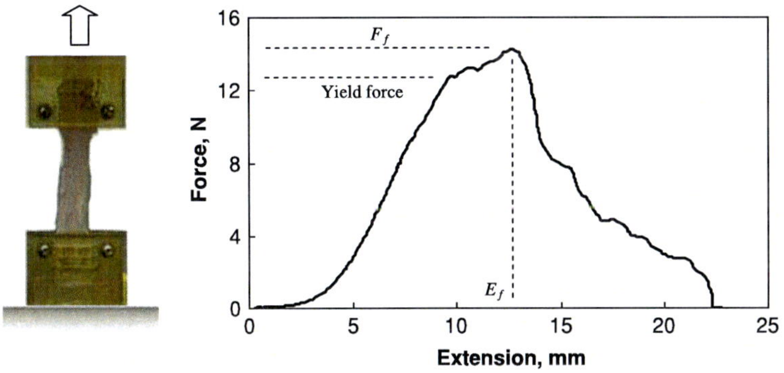

Fig. 2 Illustration of a typical uniaxial extension test and a resulting force-extension data. F_f is the maximum force in the data, E_f is the extension at F_f and yield force is the force at which the slope of the curve starts to decrease

cylindrical coordinates and the AAA were mapped into geographical regions. Circumferentially, it was subdivided into anterior, left, posterior and right regions, each covering a fourth of 360°. Longitudinally, it was divided into proximal neck, aneurysmal region and the distal neck. In addition, regions were classified as bulged or unbulged based on visual observation. Figure 3 schematically illustrates various classifications. For quantitative assessment of wall thickness, measurement points were first separated into those at the site of calcified or non-calcified atherosclerotic plaque (plaque group) and those with no evidence of plaque (non-plaque group) based on visual observation and touch.

The aneurysms studied showed significant regional variation but little difference between the small (#1 and 2) and large (#3 and 4) AAA. In all three unruptured aneurysms, the average wall thickness followed a similar pattern reaching a minimum at the posterior-right region (see Fig. 4). This is the site of the vertebral column, but the implications remain unclear. In the ruptured aneurysm alone, there were two minimas—the thinnest one at the anterior rupture site and a second at the posterior-right region. In three of four AAA, sites with atherosclerotic plaque were noted.

On the finite elastic properties, α and β also they note significant regional variation as documented in Tables 2 and 3, but the AAA-wide average values were found to be remarkably close to that from Raghavan et al.'s previous report [24] (mean α and β were 14 and 188 N/cm^2).

On failure properties, two measures were used—the failure tension, T_f and the failure stress, S_f. The former is the failure load/width while the latter is failure load/area. S_f is essentially T_f normalized for the specimen thickness. Because thickness is not an experimentally controlled variable (unlike specimen width), and because the ability of the AAA wall tissue to resist rupture is likely the lumped effect of both S_f and wall thickness, T_f may be closer to true tissue caliber than the traditional failure strength, S_f. T_f (and so also S_f) showed regional variations in addition to inter-subject variations as seen in Fig. 5.

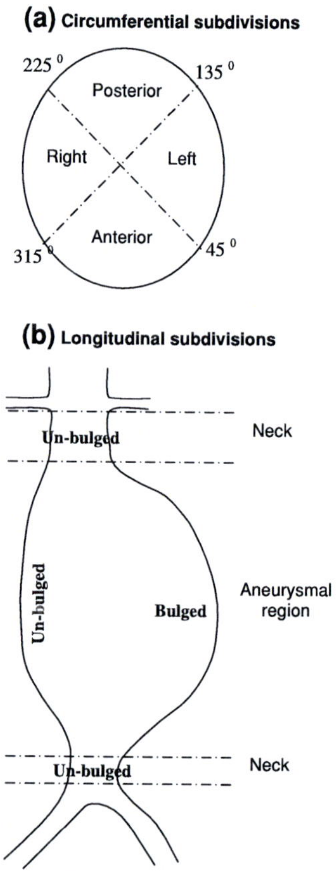

Fig. 3 Schematic illustration of geographical subdivisions along the **a** circumferential and **b** longitudinal axes. Examples of bulged versus un-bulged categorizations are illustrated. Notice that even areas within the aneurysmal region may be classified as unbulged

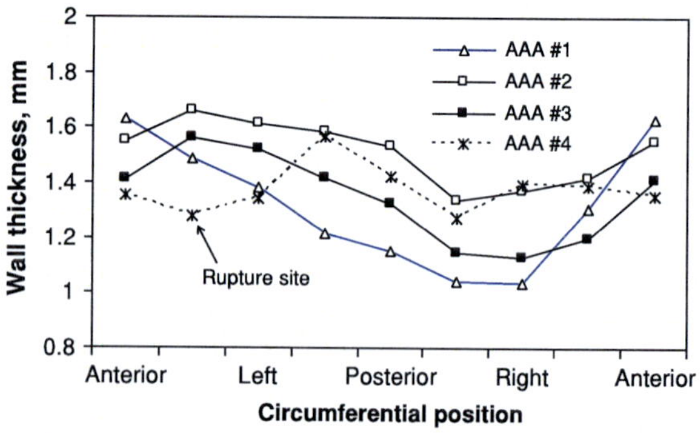

Fig. 4 Circumferential variation of wall thickness. Data points shown are the mean wall thickness for every 45 degree groupings along the circumferential direction. Thickness drop consistently between the posterior and right regions for all AAA. The ruptured AAA alone has a minimum along its rupture site

In the only ruptured AAA studied, a progressive reduction in thickness within a 1 cm radius around the rupture site was observed. Oddly, the specimen closest to the rupture site (which was about 1 cm from it) was only the third weakest, behind two strips closest to a blister at another location. On average however, the ruptured

Table 2 Material parameter, α in N/cm^2 for various geographical groups given as mean ± SD [number of specimen strips]

ID	Circumferential location				Longitudinal location		Dilation		All
	Anterior	Left	Posterior	Right	Neck	Aneurysmal region	Bulged	Unbulged	
AAA #1	21.5 ± 21.9 [2]	10.3 ± 8.7 [2]	15.6 ± 13.8 [3]	15.0 ± 0 [1]	20.6 ± 23.2 [2]	14.1 ± 9.6 [6]	9.6 ± 4.8 [3]	19.4 ± 14.4 [5]	15.7 ± 12.3 [8]
AAA #2	2.4 ± 3.4 [2]	14.2 ± 14.5 [4]	- [0]	12.1 ± 15.7 [2]	8.9 ± 14.7 [5]	13.8 ± 9.2 [3]	14.0 ± 13.1 [2]	9.6 ± 13.2 [6]	10.7 ± 12.4 [8]
AAA #3	22.2 ± 25.8 [4]	17.5 ± 12.6 [3]	12.4 ± 4.2 [3]	6.8 ± 8.0 [5]	- [0]	14.2 ± 15.1 [15]	11.7 ± 17.6 [9]	18.0 ± 10.7 [6]	14.2 ± 15.1 [15]
AAA #4	4.7 ± 2.4 [4]	2.7 ± 4.1 [6]	4.6 ± 3.9 [4]	9.1 ± 5.4 [3]	- [0]	4.8 ± 4.2 [17]	4.5 ± 4.3 [9]	5.1 ± 4.5 [8]	4.8 ± 4.2 [17]
All	13.0 ± 17.8 [12]	9.7 ± 10.9 [15]	10.2 ± 8.7 [10]	9.2 ± 8.0 [11]	12.2 ± 16.3 [7]	10.2 ± 11.1 [41]	8.8 ± 11.9 [23]	12.1 ± 11.8 [25]	10.5 ± 11.8 [48]

Table 3 Material parameter, β in N/cm^2 for various geographical groups given as mean ± SD [number of specimen strips]

ID	Circumferential location				Longitudinal location		Dilation		All
	Anterior	Left	Posterior	Right	Neck	Aneurysmal region	Bulged	Unbulged	
AAA #1	200.9 ± 19.9 [2]	199.1 ± 129.5 [2]	46.7 ± 14.4 [3]	239.2 ± 0 [1]	147.2 ± 56.1 [2]	147.5 ± 113.5 [6]	172.4 ± 95.4 [3]	132.4 ± 107.6 [5]	147.4 ± 98.2 [8]
AAA #2	126.5 ± 95.6 [2]	280.1 ± 263.0 [4]	- [0]	63.5 ± 60.6 [2]	243.9 ± 242.0 [5]	93.6 ± 90.0 [3]	107.4 ± 122.7 [2]	214.3 ± 228.3 [6]	187.5 ± 204.5 [8]
AAA #3	124.8 ± 110.9 [4]	48.5 ± 14.9 [3]	215.9 ± 113.8 [3]	372.4 ± 186.0 [5]	- [0]	210.3 ± 177.5 [15]	289.5 ± 183.8 [9]	91.5 ± 77.0 [6]	210.3 ± 177.5 [15]
AAA #4	201.2 ± 99.2 [4]	272.9 ± 173.7 [6]	193.6 ± 104.5 [4]	142.8 ± 71.7 [3]	- [0]	214.4 ± 127.9 [17]	214.8 ± 127.0 [9]	214.0 ± 137.7 [8]	214.4 ± 127.9 [17]
All	163.2 ± 92.0 [12]	220.1 ± 188.2 [15]	156.2 ± 111.2 [10]	241.5 ± 181.8 [11]	216.3 ± 204.5 [7]	194.3 ± 144.7 [41]	229.1 ± 151.7 [23]	168.4 ± 150.1 [25]	197.5 ± 152.4 [48]

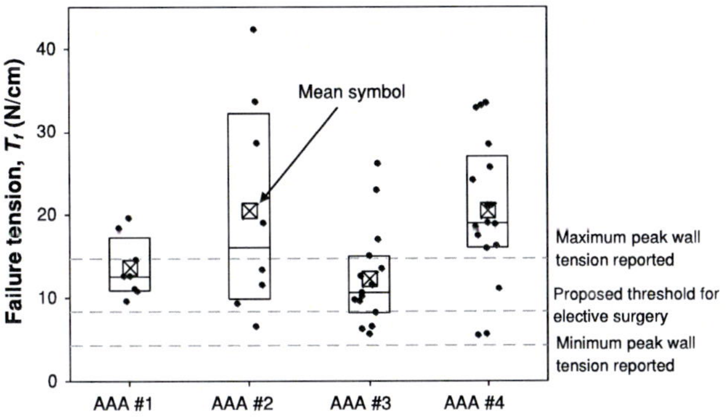

Fig. 5 Box plots of failure tension, T_f in the study population. The box indicates the 25th percentile, median and 75th percentile within each AAA. The mean value for each specimen is also shown. The *solid circles* refer to T_f for individual specimen strips. The *dotted lines* represent the range of FEA-based estimates of pressure-induced peak wall tension in patient population reported by Fillinger et al. [28] to help place failure tension values within the context of rupture

AAA was not weaker than any of the unruptured AAA. Perhaps the greatest limitation in their study was the freezing of AAA specimens for storage prior to testing.

3.4 Mechanical Properties of Ruptured Versus Unruptured AAA

Di Martino et al. [29] reported on a study where they compared mechanical properties of ruptured versus unruptured AAA. 26 specimen strips were obtained from 16 unruptured AAA and 13 strips from 9 ruptured AAA, all from the anterior midsection during surgical resection. Upon uniaxial extension testing until failure, they found that the ruptured AAA had a lower ultimate stress compared to the unruptured AAA with statistical significance (54 ± 6 vs. 82 ± 9 N/cm^2; $P < .04$). The elastic property was quantified using two measures of elastic modulus—the maximum slope of the stress–strain curve and the slope at physiological stress. They did not find a difference in the elastic moduli between ruptured and unruptured AAA.

In a study [30] that was similar to our earlier AAA heterogeneity study, but employing more rigor in approach [31], the authors harvested 9 unruptured and 4 ruptured whole AAA from cadavers. This time, specimen strips were studied histologically, histochemically and subjected to uniaxial extension tests until failure. Unlike the earlier study, all specimens were tested within 24 h of harvest and within 48 h of death. After removal of intraluminal thrombus, the AAA was longitudinally cut open on its anterior surface and multiple rectangular strips (typical dimensions: 4×0.4 cm) were cut from all around the AAA resulting in 5–15 specimens per subject within the aneurysmal portion and where possible, from the undilated neck regions. Specimen strip locations were marked on the AAA images. A small piece of each strip was also cut and stored for histology and histochemistry. The test strips were clamped with custom grips and attached to a

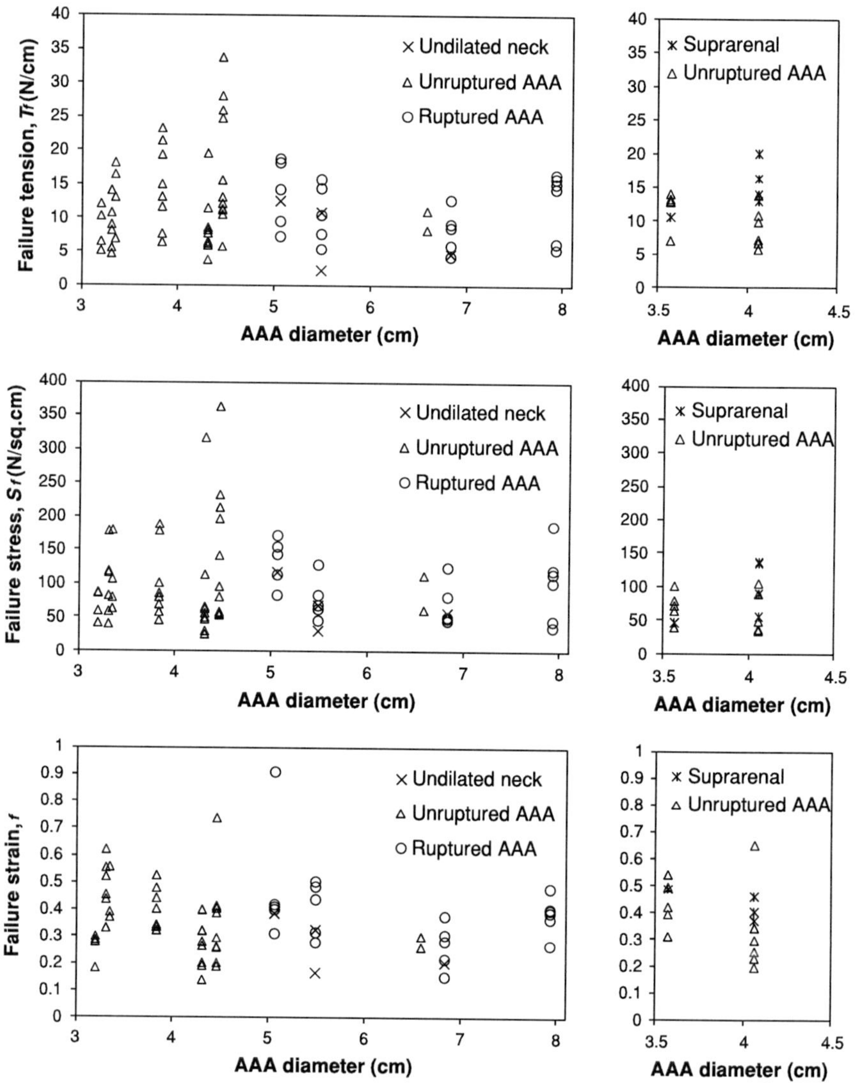

Fig. 6 Regional and inter-subject variations in measured mechanical properties of AAA harvested from autopsies

uniaxial extension tester (Instron Spec 2,200) equipped with a tissue bath, a 5 lb load cell and a displacement sensor. Strips were subjected to uniaxial extension at 20%/min strain rate until failure. The mechanical properties used earlier were determined, namely, the failure tension—T_f, the failure stress—S_f, and the failure strain—ε_f.

Histological study involved selective staining of tissue specimens from which the cross sectional area covered by elastin (Verhoeff's stain), collagen (trichrome stain), smooth muscle cells (*T*-Masson stain) and fat (clear areas) was recorded. The total medial thickness was also recorded. For histochemical analysis, the monoclonal antibodies against metalloproteinase MMP-9, C-reactive protein and inflammatory cells were used. For these, a nominal scale was used: 0—absent, 1 (scarce and focal positivity), 2 (multifocal positivity) and 3 (diffuse positivity).

Upon discarding test strips that failed at the clamps (~20% of 145 tests), an average of six test strips per subject from the infrarenal aneurysmal aortic region were obtained. The median failure properties were quite similar between those from ruptured and unruptured AAA: $T_f = 10.4$ versus 11.1 N/cm; $S_f = 83.4$ versus 80.1 N/cm^2; $\varepsilon_f = 0.39$ versus 0.34 N/cm^2; wall thickness = 1.46 versus 1.51 mm; % Collagen = 47 versus 64% respectively. Histochemical analysis was performed on one to two tissue pieces each from 4 ruptured and 6 unruptured AAA. For those AAA where multiple pieces were studied, the results were averaged. Student *t*-tests did not reveal a statistically significant difference for inflammatory cells (average score = 2.00 versus 1.58; $p = 0.49$), C-reactive protein (average score =2.1 versus 1.8; $p = 0.53$) or MMP-9 (average score = 2.1 versus 2.4; $p = 0.57$) levels between the ruptured ($N = 4$) and unruptured ($N = 7$) groups respectively (see Figs. 6 and 7). There was a remarkable similarity in failure properties between specimen strips cut from the posterior and anterior wall surface. For instance, anterior versus posterior averages were: $T_f = 11.1$ versus 11.4 N/cm and % Collagen = 48 versus 50% within the ruptured group; $T_f = 12.9$ versus 12.0 N/cm and % Collagen = 62 versus 65% within the unruptured group (all longitudinally orientated strips). Lumping longitudinal and circumferential specimens and ruptured and unruptured groups does not alter this general observation for all the properties (T_f, S_f, ε_f, thickness and % Collagen). These findings are in contrast to the finding of a difference in S_f by Di Martino et al. using specimen strips from the anterior midsection harvested during surgical resection. Further studies will likely shed more light on this important issue.

In the four ruptured AAA, primary rupture sites were on the lateral quadrants (left or right) with one rupture extending into the posterior (see Fig. 8). Rupture lengths ranged from 1 to 6 cm. Remarkably, all ruptures had a longitudinal orientation and appeared to follow a meridional geodesic line (shortest line on a surface).

In conclusion, the authors note that AAA-wide average failure tension was not significantly different between the ruptured ($N = 4$) and unruptured ($N = 7$) groups (inter-AAA avg of intra-AAA avg $T_f = 11.2$ vs. 11.6 N/cm; $p = 0.82$). This is also the case for failure strain, failure stress, wall thickness and % cross sectional area covered by collagen. The results are not consistent with the notion that ruptured aneurysms are globally weaker than unruptured ones in this small study population. Of course, the authors concede that despite being the largest study of its nature, the sample population is still too small to draw any conclusions.

In summary, uniaxial extension tests on harvested AAA tissue samples have provided us a first look at the elastic behavior of aneurysmal aortic tissue and its failure properties. By isolating the tissue for controlled testing, these studies were able to focus on the intrinsic tissue behavior rather than lumping it with geometry and boundary conditions as with the case of in vivo distensibility characterizations. There

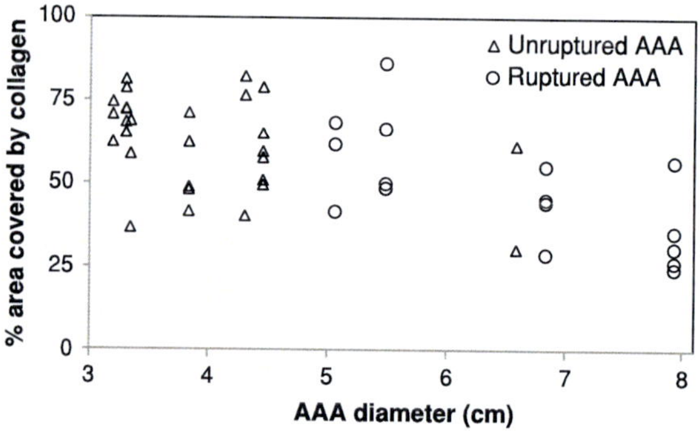

Fig. 7 Regional and inter-subject variations in measured collagen content defined as % area of histological section covered by collagen (trichrome stained)

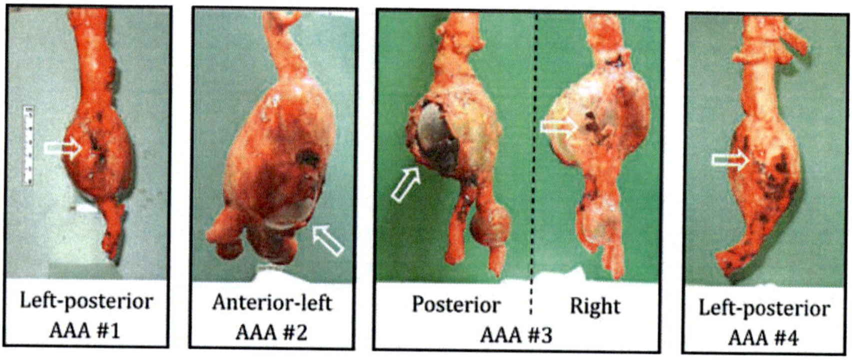

Fig. 8 Photographs of ruptured AAA in the study population. The *white arrows* point to the rupture sites. In two cases, the rupture site had sutures that were removed prior to testing. All images are to the same scale (5 cm ruler on *left*)

are however, limitations to this testing method. Chief among them is precisely the fact that they were isolated from their physiological environment inducing hitherto unknown changes to their behavior. They tend to provide passive properties devoid of the role played by SMCs. Fortunately, the aorta is a conducting artery and therefore SMC role is likely low to begin with. A second concern is the cutting of the specimen strip from the whole AAA. This too may induce variants to their behavior, chief among them, possible secretion of enzymes that may induce proteolysis prior to testing. A third limitation is that uniaxial extension induces uniaxial stress which is a deviation from the physiological loading conditions where, tissue is conceivably experiencing a planar biaxial tensile stress (axial and circumferential components) and a transmural compressive stress (radial component). And finally, uniaxial extension testing is not well suited for studying AAA wall tissue anisotropy—that is, the orientational differences in properties (axial versus circumferential). Testing individual strips from different orientations do provide us valuable information on anisotropy [32], but this requires the use of different specimens from different locations introducing heterogeneity as another variable that will not be controlled for. So differences in measured properties cannot be definitively attributed to orientation alone. Biaxial properties may indeed be ascertained from just uniaxial testing of a single strip, but only if the reduction in specimen width during extension is measured during testing [33]. This measurement can be rather imprecise. Perhaps for this or other reasons, all of the reported studies on uniaxial testing of AAA tissue have not measured width changes, but rather—where needed—merely assume isotropy (and incompressibility) to estimate the width and thickness changes in specimens during uniaxial extension. Despite these and other limitations, uniaxial extension tests form the first order information about AAA tissue and are therefore valuable in the information they provide.

3.5 Stochastic Models for Predicting Failure Properties

Failure properties measured using ex vivo mechanical testing studies of harvested specimens provide us with some understanding of how these properties vary in the study population. But the ultimate goal should be to predict these properties in an intact aneurysm so that they may better guide our interpretation of the criticality of AAA wall stress estimates. In this context two important efforts have been reported [34, 35]. Vande Geest et al. [35] took uniaxial data of 60 specimens from anterior AAA of 29 patients and developed a stochastic model relating failure strength (S_f) to various measurable factors. They reasoned that intraluminal thickness (ILT), patient age (AGE), AAA size (SIZE), regionally varying diameter of the aneurysm (normalized to neck diameter—NORD), family history of AAA (HIST), history of smoking (SMK), patient gender (SEX) and the method of measuring ILT thickness (METHOD) were all factors affecting strength. Based on a linear mixed-effects modeling, they built the following stochastic model:

$$\begin{aligned}\text{STRENGTH} = \beta_0 + [\beta_1 \times \text{ILT}] + [\beta_2 \times \text{AGE}] + [\beta_3 \times \text{SIZE}] + [\beta_4 \times \text{NORD}] \\ + [\beta_5 \times \text{HIST}] + [\beta_6 \times \text{SMK}] + [\beta_7 \times \text{SEX}] + [\beta_8 \times \text{METHOD}] + \varepsilon\end{aligned} \tag{14}$$

where, β_i are the statistical coefficients determined using their data from 29 patients. Subsequently, they tested the validity of the developed model on an independent validation set of data of 21 specimens from anterior AAA of 9 AAA patients and found reasonable agreement. This group then leveraged this predictive model in order to calculate the wall stress to strength ratio in the AAA of patients that they termed the rupture potential index (RPI) [36].

Doyle et al. [34] took an alternative approach for estimating strength in intact AAA by averaging the results reported by Thubrikar et al. [26] and Raghavan et al. [27] where they both directly measured properties covering all major regions of the AAA—anterior, posterior, lateral, distal, proximal and midsection, albeit in a total of 9 patients. Doyle et al. [34] combined these to arrive at mean values for ultimate stress in eight circumferentially subdivided regions—anterior, anterior-left, left, posterior/left, posterior, posterior/right, right, and anterior/right in that order. They then used these estimates to compare with finite element wall estimates in order to calculate a stress–strength ratio that they termed Finite element analysis rupture index (FEARI)./Section2>

3.6 Other Forms of Testing AAA Tissue

As noted in the above section, uniaxial testing has its limitations, key among them is the fact that the stress distribution does not conform fully to physiological conditions. The next best alternative is planar biaxial extension testing. In 2006, Vande Geest et al. [37] harvested 26 aneurysmal and 8 age-matched non-aneurysmal aortic tissue samples from surgical resection patients and organ donors respectively. Square specimens were cut and held to the movable cross heads of the biaxial extension test apparatus [38] on four edges by sutures. The specimen was then subjected to force-controlled testing with varying prescribed forces between the two orthogonal directions. A CCD camera was used to track the displacement of markers forming a 5 × 5 mm square placed on the specimen. Figure 9 shows representative stress-stretch data reported.

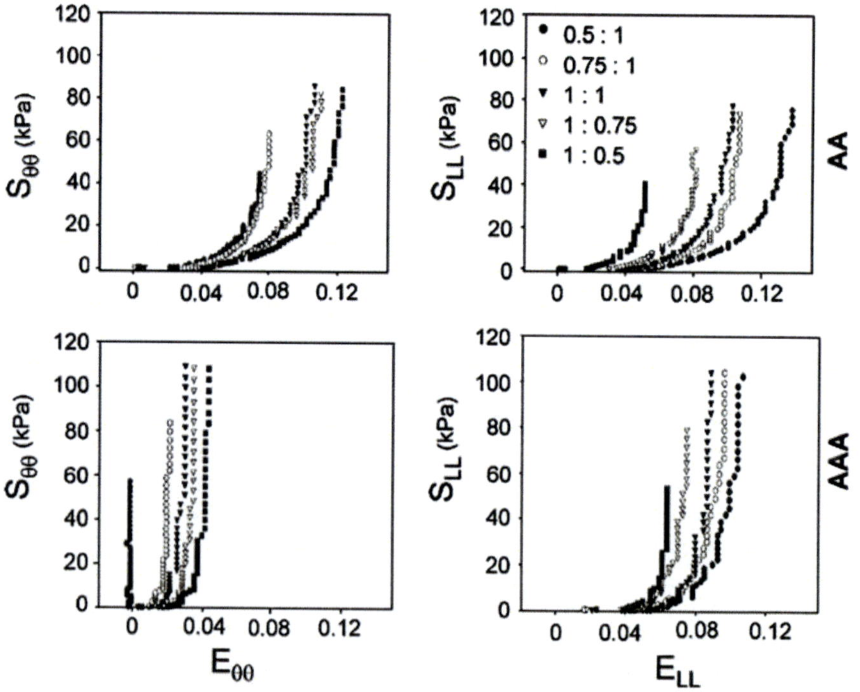

Fig. 9 Representative stress–strain plots for non-aneurysmal (*top*) and aneurysmal (*bottom*) aortic tissue under biaxial extension testing reported by Vande Geest et al. S denotes the Second Piola-Kirchoff stress, subscripts θ and L refer to the axial and circumferential orientations, and the ratios denoting the prescribed ratio of force between the θ and L directions during the test

To interpret this important data and extract constitutive model form and parameters, they adopted a methodical approach to guide their choice of a functional form reported by this group earlier [39]. They found that a constitutive functional form used earlier by Choi and Vito [40] would best suit their experimental data:

$$W = b_0\left[\exp\left(\tfrac{1}{2}b_1 E_{\theta\theta}^2\right) + \exp\left(\tfrac{1}{2}b_2 E_{LL}^2\right) + \exp\left(\tfrac{1}{2}b_3 E_{\theta\theta}E_{LL}\right) - 3\right] \quad (15)$$

where, E is the Green-strain tensor and b_0, b_1, b_2 and b_3 are the material parameters. The fits of above model to experimental data were found to be quite good (R^2 of greater than 0.9). They reported population average values for parameters, b_0, b_1, b_2, and b_3 to be 0.14 kPa, 477, 416, 408 for AAA tissue versus 0.32 kPa, 141.1, 143.3, 127.9 for non-aneurysmal aorta respectively. To interpret anisotropy, they calculated the sqrt(b_1/b_2)–the higher it is, greater the circumferential stiffness compared to axial stiffness—and found this to be higher for aneurysmal aorta indicative of preferential circumferential stiffening in aneurysms. Marra et al. [41] also reported on preliminary biaxial extension testing of aneurysmal tissue samples, but did not derive/characterize the data to extract constitutive model parameters.

4 Recent Advances

In the study of mechanical properties of AAA tissue, much remains unknown. Elastic behavior is just one aspect of AAA tissue behavior. Other aspects include, viscoelasticity, active behavior, plasticity and damage behavior, but little information is available on these important topics relevant to AAA pathogenesis. Even within the realm of elastic behavior, challenges abound. For one, uniaxial and even biaxial test methods do not completely conform to the physiological loading conditions. Biaxial testing does improve over uniaxial, but has the limitation of not providing important information on failure properties. An intriguing alternative to either of these methods might be inflation testing such as that reported by Marra et al. [42] for porcine aortic tissue. Here, they held a circular tissue specimen between two annular ring-like clamps and pressurized one side while recording the radius of curvature formed by inflation of the flat specimen in the other direction. They noted that all specimens tore in the circumferential direction. The pressure, curvature (and hence stretch ratio) data was curve fit to the Holzapfel model [43] and material parameters extracted. They did not report similar effort on human AAA tissue, but the test method proposed remains a useful alternative because it offers two key advantages over biaxial extension: (1) the loading conditions are more physiological (tensile axial and circumferential stress, compressive transmural stress), and (2) they can help elucidate failure properties.

A second classic challenge has been the dilemma between in vivo lumped 'property' estimation and controlled ex vivo testing. In this context, one recent advancement might be to leverage dynamic scanning modalities to obtain the 3D deformation of the AAA under pulse pressure and then use numerical methods to extract the regional variations of the intrinsic tissue mechanical parameters. This latter approach would in one swipe, make the classic dilemma moot because it would be the best of both worlds—using in vivo physiological 'test' data while extracting intrinsic material parameters, even if regionally varying. There are challenges though in this approach: (1) how well do we understand the boundary conditions on the AAA (vertebral constraints, external pressure etc.), (2) how to handle initial conditions (residual stress, axial tethering traction), (3) how well can we extract point-to-point correspondence between any two configurations during motion, and (4) how well can we handle the computational cost of optimizing for so many parameter values (especially if heterogeneity is included) while arriving at a unique global solution. Some work in this area has been recently reported for cerebral aneurysm [44, 45] and may well be applicable to aortic aneurysms as well. Recently, Tierney et al. [46] regarding the feasibility of using acoustic radiation force impulse (ARFI) imaging to extract valuable information on the elastic caliber of the abdominal aortic aneurysm wall.

And finally, while the above limitations are methodological, perhaps the greatest challenge may be in application of these methods in a large enough study population for the results to be representative and meaningful. Understandably due to practical considerations in terms of availability of data and specimens, most studies have been confined to small study populations restricting their universal applicability. Future studies should strive for larger study populations while leveraging the best of methodologies for improving our understanding of the mechanical properties of AAA tissue.

References

1. Wolinsky, H., Glagov, S.: Structural Basis for the Static Mechanical Properties of the Aortic Media. Circ. Res. **14**, 400–413 (1964)
2. Dingemans, K.P., et al.: Extracellular matrix of the human aortic media: an ultrastructural histochemical and immunohistochemical study of the adult aortic media. Anat. Rec. **258**(1), 1–14 (2000)
3. Baxter, B.T., et al.: Abdominal aortic aneurysms are associated with altered matrix proteins of the nonaneurysmal aortic segments. J. Vasc. Surg. **19**(5), 797–802 (1994) (discussion 803)
4. van Laake, L.W., et al.: Systemic dilation diathesis in patients with abdominal aortic aneurysms: a role for matrix metalloproteinase-9?. Eur. J. Vasc. Endovasc. Surg. **29**(4), 371–377 (2005)
5. Diehm, N., et al.: Aortic neck dilatation after endovascular abdominal aortic aneurysm repair: a word of caution. J. Vasc. Surg. **47**(4), 886–892 (2008)
6. Diehm, N., et al.: Severe structural damage of the seemingly non-diseased infrarenal aortic aneurysm neck. J. Vasc. Surg. **48**(2), 425–434 (2008)
7. Holmes, D.R., et al.: Medial neovascularization in abdominal aortic aneurysms: a histopathologic marker of aneurysmal degeneration with pathophysiologic implications. J. Vasc. Surg. **21**(5), 761–371 (1995) (discussion 771–772)

8. Vorp, D.A.: Biomechanics of abdominal aortic aneurysm. J. Biomech. **40**(9), 1887–1902 (2007)
9. Vorp, D.A., Vande Geest, J.P.: Biomechanical determinants of abdominal aortic aneurysm rupture. Arterioscler. Thromb. Vasc. Biol. **25**(8), 1558–1566 (2005)
10. Lanne, T., et al.: Diameter and compliance in the male human abdominal aorta: influence of age and aortic aneurysm. Eur. J. Vasc. Surg. **6**(2), 178–184 (1992)
11. MacSweeney, S.T., et al.: Mechanical properties of the aneurysmal aorta. Br. J. Surg. **79**(12), 1281–1284 (1992)
12. Vorp, D.A., et al.: Potential influence of intraluminal thrombus on abdominal aortic aneurysm as assessed by a new non-invasive method. Cardiovasc. Surg. **4**(6), 732–739 (1996)
13. Wilson, K., et al.: The relationship between abdominal aortic aneurysm wall compliance, maximum diameter and growth rate. Cardiovasc. Surg. **7**(2), 208–213 (1999)
14. Long, A., et al.: Compliance of abdominal aortic aneurysms: evaluation of tissue Doppler imaging. Ultrasound Med. Biol. **30**(9), 1099–1108 (2004)
15. Long, A., et al.: Compliance of abdominal aortic aneurysms evaluated by tissue Doppler imaging: correlation with aneurysm size. J. Vasc. Surg. **42**(1), 18–26 (2005)
16. Ganten, M.K., et al.: Quantification of aortic distensibility in abdominal aortic aneurysm using ECG-gated multi-detector computed tomography. Eur. Radiol. **18**(5), 966–973 (2008)
17. Sumner, D.S., Hokanson, D.E., Strandness Jr., D.E.: Stress-strain characteristics and collagen-elastin content of abdominal aortic aneurysms. Surg. Gynecol. Obstet. **130**(3), 459–466 (1970)
18. Drangova, M., et al.: Elasticity and geometry measurements of vascular specimens using a high-resolution laboratory CT scanner. Physiol. Meas. **14**(3), 277–290 (1993)
19. He, C.M., Roach, M.R.: The composition and mechanical properties of abdominal aortic aneurysms. J. Vasc. Surg. **20**(1), 6–13 (1994)
20. Raghavan, M.L., Webster M.W., Vorp, D.A.: Ex vivo biomechanical behavior of abdominal aortic aneurysm: assessment using a new mathematical model. Ann. Biomed. Eng. **24**(5), 573–582 (1996)
21. Vorp, D.A., et al.: Wall strength and stiffness of aneurysmal and nonaneurysmal abdominal aorta. Ann. N. Y. Acad. Sci. **800**, 274–276 (1996)
22. Roach, M.R., Burton, A.C.: The reason for the shape of the distensibility curves of arteries. Can. J. Biochem. Physiol. **35**(8), 681–690 (1957)
23. Sherebrin, M.H., Hegney, J.E., Roach, M.R.: Effects of age on the anisotropy of the descending human thoracic aorta determined by uniaxial tensile testing and digestion by NaOH under load. Can. J. Physiol. Pharmacol. **67**(8), 871–878 (1989)
24. Raghavan, M.L., Vorp, D.A.: Toward a biomechanical tool to evaluate rupture potential of abdominal aortic aneurysm: identification of a finite strain constitutive model and evaluation of its applicability. J. Biomech. **33**(4), 475–482 (2000)
25. Vallabhaneni, S.R., et al.: Heterogeneity of tensile strength and matrix metalloproteinase activity in the wall of abdominal aortic aneurysms. J. Endovasc. Ther. **11**(4), 494–502 (2004)
26. Thubrikar, M.J., et al.: Mechanical properties of abdominal aortic aneurysm wall. J. Med. Eng. Technol. **25**(4), 133–142 (2001)
27. Raghavan, M.L., et al.: Regional distribution of wall thickness and failure properties of human abdominal aortic aneurysm. J. Biomech. **39**(16), 3010–3016 (2006)
28. Fillinger, M.F., et al.: Prediction of rupture risk in abdominal aortic aneurysm during observation: wall stress versus diameter. J. Vasc. Surg. **37**(4), 724–732 (2003)
29. Di Martino, E.S., et al.: Biomechanical properties of ruptured versus electively repaired abdominal aortic aneurysm wall tissue. J. Vasc. Surg. **43**(3), 570–576 (2006) (discussion 576)
30. Raghavan, M.L., Hanaoka, M.M., Kratzberg, J.A., Higuchi, M.D., da Silva, E.S.: Biomechanical failure properties and microstructural content of ruptured and unruptured abdominal aortic aneurysms. J. Biomech. (in press)
31. Raghavan, M.L., et al.: Failure properties of ruptured and unruptured abdominal aortic aneurysms. In: Proceedings of Summer Bioengineering Conference (SBC2009-204060). Lake Tahoe, CA, USA (2009)

32. Holzapfel, G.A.: Determination of material models for arterial walls from uniaxial extension tests and histological structure. J. Theor. Biol. **238**(2), 290–302 (2006)
33. Holzapfel, G.A., Sommer, G., Regitnig, P.: Anisotropic mechanical properties of tissue components in human atherosclerotic plaques. J. Biomech. Eng. **126**(5), 657–665 (2004)
34. Doyle, B.J., et al.: A finite element analysis rupture index (FEARI) as an additional tool for abdominal aortic aneurysm rupture prediction. Vasc. Dis. Prev. **6**, 114–121 (2009)
35. Vande Geest, J.P., et al.: Towards a noninvasive method for determination of patient-specific wall strength distribution in abdominal aortic aneurysms. Ann. Biomed. Eng. **34**(7), 1098–1106 (2006)
36. Vande Geest, J.P., et al.: A biomechanics-based rupture potential index for abdominal aortic aneurysm risk assessment: demonstrative application. Ann. N. Y. Acad. Sci. **1085**, 11–21 (2006)
37. Vande Geest, J.P., Sacks, M.S., Vorp, D.A.: The effects of aneurysm on the biaxial mechanical behavior of human abdominal aorta. J. Biomech. **39**(7), 1324–1334 (2006)
38. Sacks, M.S.: Biaxial mechanical evaluation of planar biological materials. J. Elast. **61**(1–3), 199–246 (2000)
39. Vande Geest, J.P., Sacks, M.S., Vorp, D.A.: Age dependency of the biaxial biomechanical behavior of human abdominal aorta. J. Biomech. Eng. **126**(6), 815–822 (2004)
40. Choi, H.S., Vito, R.P.: Two-dimensional stress-strain relationship for canine pericardium. J. Biomech. Eng. **112**(2), 153–159 (1990)
41. Marra, S.P., Kennedy, F.E., Fillinger, M.F.: Mechanical properties characterization of abdominal aortic aneurysm tissue using biaxial testing. In: Proceedings of IMECE2002 ASME International Mechanical Engineering Congress & Exposition (IMECE2002-32779). New Orleans, LA (2002)
42. Marra, S.P., et al.: Elastic and rupture properties of porcine aortic tissue measured using inflation testing. Cardiovasc. Eng. **6**(4), 123–131 (2006)
43. Holzapfel, G.A., Gasser, T.C., Ogden, R.W.: A new constitutive framework for arterial wall mechanics and a comparative study of material models. J. Elast. **61**, 48 (2006)
44. Balocco, S., et al.: Feasibility of estimating regional mechanical properties of cerebral aneurysms in vivo. Med. Phys. **37**(4), 1689–1706 (2010)
45. Zhao, X., Raghavan, M.L., Lu, J.: Identifying heterogeneous anisotropic properties in cerebral aneurysms: a pointwise approach. Biomech. Model. Mechanobiol. **7**, 477–489 (2008)
46. Tierney, A.P., Callanan, A., McGloughlin, T.M.: In vivo feasibility case study for evaluating abdominal aortic aneurysm tissue properties and rupture potential using acoustic radiation force impulse imaging. J. Mech. Behav. Biomed. Mater. **4**(3), 507–513 (2011)

Fluid–Structure Interaction in Healthy, Diseased and Endovascularly Treated Abdominal Aortic Aneurysms

David S. Molony, Stephen Broderick, Anthony Callanan, Tim M. McGloughlin and Michael T. Walsh

Abstract Abdominal aortic aneurysms are irreversible dilations of the infrarenal aorta. If left untreated the aneurysm may continue to grow until eventually rupturing. Endovascular aneurysm repair (EVAR) is an established method for the treatment of abdominal aortic aneurysms (AAAs). Complications arising from this treatment include endoleaks and graft migration. Computational methods such as FEA, CFD and FSI can be used to investigate both the disease manifestation and its treatment. FSI is a particularly useful tool for the investigation of EVAR as both the fluid forces acting on the graft and stresses on the aneurysm wall are of interest. This work investigates the stresses and haemodynamics in healthy, diseased and treated aneurysms through the use of FSI. Higher stresses and more disturbed haemodynamics are seen in aneurysms than in a healthy aorta. The insertion of a stent-graft significantly reduces the aneurysm wall stress and redistributes it. The stent-graft is subject to large haemodynamic forces which can cause migration of the device. These forces do not necessarily act primarily in a caudal direction, hence resulting in a non-caudal migration. The inclusion of patient-specific data such as patient-specific pressure and graft oversize was investigated. These were found to have a large effect on the accuracy of the results and in future best efforts should be made to include as much patient-specific data as possible in numerical models.

D. S. Molony · S. Broderick · A. Callanan · T. M. McGloughlin · M. T. Walsh (✉)
Centre for Applied Biomedical Engineering Research (CABER), Materials and Surface Science Institute and the Department of Mechanical & Aeronautical Engineering, University of Limerick, Limerick, Ireland
e-mail: michael.walsh@ul.ie

Stud Mechanobiol Tissue Eng Biomater (2011) 7: 163–179

DOI: 10.1007/8415_2011_85

Published Online: 16 June 2011

1 Introduction

Abdominal aortic aneurysm (AAA) is defined as a dilation of the infrarenal aorta of 1.5 times the normal size of the artery. The aneurysm is normally treated when it reaches the clinically regarded maximum diameter threshold of 5 or 5.5 cm. Treatment is either by open repair (OR) or endovascular aneurysm repair (EVAR). EVAR involves the use of a device called a stent-graft and this treatment has become commonplace in the last decade due to the minimal invasiveness of the deployment technique. These stent-grafts provide a conduit for the blood flow and shield the weakened aneurysm wall from the pulsatile blood pressure.

This technique is still associated with problems, which can lead to failure of the device, and often results in conversion to open repair (OR). These problems are primarily but not entirely associated with endoleaks, graft migration and graft occlusions. Endoleaks occur when blood enters the aneurysm sac, resulting in re-pressurization of the sac. Increased aneurysm sac pressure results in increased stress on the aneurysm wall [17] and is strongly associated with increasing aneurysm diameter after EVAR [4]. In a similar fashion a dilation of the proximal neck of the aneurysm can occur after EVAR. The mechanism of neck dilation is not fully understood but may be attributable to oversizing of the stent-graft causing increased stress on the aneurysm neck. Certainly there is an initial dilation of the aneurysm neck in the first year following implantation [27], but there is differing opinions as to whether the degree of oversize further dilates the neck during follow-ups [3, 29]. It should be noted that most neck enlargement studies are considered to have been flawed through poor methodology [31]. The other main complication associated with stent-graft treatment is graft migration. Graft migration occurs when the graft is dislodged by either 5 or 10 mm from its initial deployment location. Part of the reason for graft migration is the haemodynamic force created by the pressure of the blood flow. This works to dislodge the graft and may explain the cases of late graft migration. Other important factors also contributing to migration are lack of adequate proximal fixation length [11] and excessive oversizing [14].

Computational tools have become a useful technique in analysing biomedical and in this case vascular afflictions. These numerical methods can provide detailed information about wall stresses and fluid flow. In particular, fluid–structure interaction (FSI) is of major benefit as both graft and aneurysm wall stresses as well as fluid forces can be determined. FSI also has the added benefit of a more accurate representation of the in vivo environment than standard finite element analysis (FEA) or computational fluid dynamics (CFD). In standard FEA the pressure gradient across the region of interest is neglected and a constant static pressure assumed, while in CFD studies the wall boundary is assumed to be rigid. In reality the artery wall is deforming due to the pulsatile blood pressure which can be represented by FSI. FSI has been shown to results in large differences compared to standard FEA [28], though this has been disputed by others [8, 15].

Despite EVAR being a problem that is very suitable for FSI simulation, there has been limited research in this area. A large body of the work has focused on the drag or migration forces experienced by the graft, while only one set of publications has dealt with the coupled problem. The drag force acting on a graft can be determined analytically using control volume analysis or numerically with CFD by summing the pressure and viscous forces acting on a fluid surface. It has been shown that neck angulation, iliac angulation, neck diameter and neck-iliac area ration are the primary contributors to increased drag force magnitude [22]; [16, 18]. Increased pressure in the aneurysm sac results in a reduction of the drag force acting on a graft due to the decreased pressure gradient across the graft wall [12]. Previous FSI investigations in 3D representative models of EVAR incorporating stagnant blood in the aneurysm sac have shown that the pressure in the aneurysm sac is significantly reduced after EVAR, though the stagnant blood will normally thrombose hours or days after the procedure [16, 18].

In this work, patient-specific cases of a healthy aorta, abdominal aortic aneurysm and EVAR treated aneurysm are investigated. First, details on the geometry reconstruction techniques as well as the FSI methodology are presented in Sect. 2. Results of the simulations for the normal aorta are shown in Sect. 3. Simulation results for the AAA are shown in Sect. 4. The effect of EVAR is shown in Sect. 5 as well as the influence of including patient-specific data. Discussion of the results and their clinical relevance is presented in Sect. 6.

2 Numerical Methods

2.1 Image Segmentation and Model Construction

All segmentations were performed using Mimics 12.0 (Materialise, Belgium). CT scan images of a healthy patient, patient with an aneurysm and a patient who had undergone EVAR were obtained. Both the aneurysm and EVAR cases had contrast enhanced regions where there was blood flow. This region could be semi-automatically segmented, while for the EVAR cases the aneurysm wall was manually segmented. The stents in the EVAR scans appear with a lot of noise, preventing their reconstruction. As a result of this the stent-graft was assumed to consist only of the region visible with contrast dye. The region between the stent-graft and the aneurysm wall was assumed to be fully solid intraluminal thrombus (ILT). This region was segmented by subtracting a stereolithography (stl) file of the stent-graft from a stl of the inner aneurysm region. Polylines were created along the edges of the segmented regions and were then offset in order to create a wall thickness. The artery/aneurysm wall was uniformly offset by 1.5 mm for all models, this is a standard value that is often assumed [15, 28]. The stent-graft wall was offset by a value of 0.2 mm, again this is a commonly used value in literature [16, 18]; [20]. An overview of the reconstruction process is illustrated in Fig. 1.

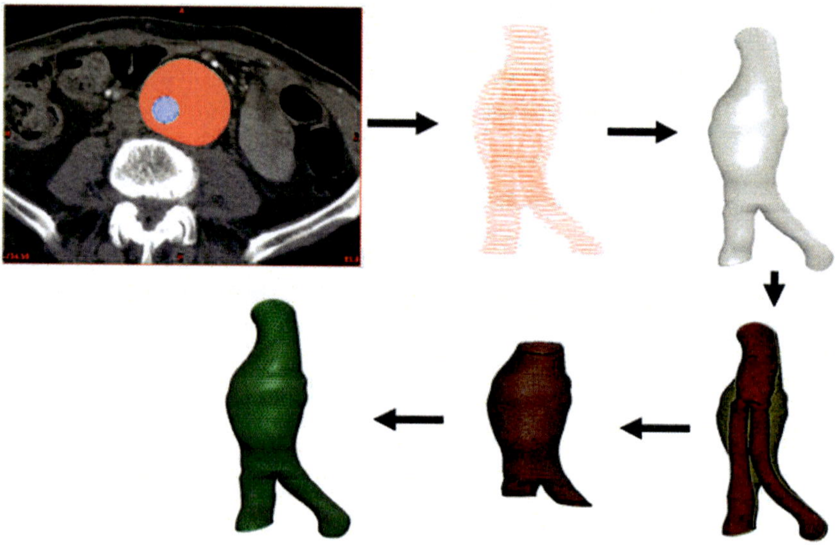

Fig. 1 Reconstruction of EVAR patient geometry—from CT scan to mesh. First, both the stent-graft lumen and aneurysm are segmented semi-automatically and manually respectively (1), polylines of these models are exported (2) and a CAD model is built (3), stls of the graft domain and inner aneurysm domain are subtracted from each other (4) leaving the ILT region behind (5). Finally meshed aneurysm (shown), graft and ILT regions are assembled (6)

2.2 *Fluid–Solid Coupling*

The primary technique for solving FSI problems is the Arbitrary Lagrangian–Eulerian (ALE) method. Other techniques that exist are, for example, the coupled momentum method (CMM) [7] or the immersed boundary method [5]. In the ALE method the nodes of the mesh may be moved in an arbitrarily specified way. Using the ALE method solvers can either be monolithic or partitioned. Monolithic solvers solve the fluid and solid governing equations together while partitioned solvers use separate fluid and structural solvers and exchange information between. A partitioned approach is used here with the structural solver Abaqus coupled with the fluid solver Fluent. The interface and data exchange between both programmes is provided by a third party software MpCCI (Fraunhofer SCAI, Germany). As a partitioned coupling method is chosen time stepping between the codes is achieved explicitly. Implicit time stepping (exchange of data during each time step) between partitioned solvers is a more stable approach but can be difficult to implement in commercial software. Explicit time stepping can be unstable particularly in domains with a long geometry and when small time steps are used. In order to provide a more stable solution sub-cycling of the fluid solver was used here, with typically 5 fluid time-steps taken per solid time step. The benefits of using a partitioned approach are the ability to use the best in-field solid and fluid solvers (and their advanced capabilities) and being able to use non-conforming meshes for the fluid and solid domain.

For the explicit approach used here, at the end of each time step Abaqus will send the deformed nodal co-ordinates to Fluent while Fluent will send the pressure to Abaqus. Fluent will then update the fluid mesh based on the deformed nodal co-ordinates obtained from Abaqus. Both softwares then proceed with their individual time steps before exchanging data again.

2.3 Governing Equations

The governing equations for fluid flow are the continuity equation (1) and the momentum equation (2)

$$\nabla \cdot \mathbf{v} = 0 \tag{1}$$

$$\rho \frac{\partial \mathbf{v}}{\partial t} + (\mathbf{v} - \mathbf{v}_g) \cdot \nabla \mathbf{v} = -\nabla p + \mu \nabla^2 \mathbf{v} + f_b \tag{2}$$

Where $\mathbf{v}$ is the fluid velocity vector, p is the pressure ρ is the density and μ is the dynamic viscosity, f_b are the body forces. Because of the ALE formulation of the equation an extra term $\mathbf{v}_g$ to account for the movement of the grid is necessary. For the solid mechanics large deformations are considered.

2.4 Mesh Independence Study

Mesh independence was completed separately for the solid and fluid domains. Peak stress was chosen as the criteria for the solid domain independence, specifically that the peak stress did not differ by more than 2% between successive meshes. In the fluid domain the peak velocity as well as the drag force for the post-operative cases was chosen as the criteria for grid independence. A similar 2% difference between successive meshes was also used. Pulse cycle independence was achieved during the third cardiac cycle.

3 Normal Aorta Biomechanics

CT image data of a normal aorta was obtained and reconstructed based on the methods described earlier. FSI simulations were performed based on the techniques also described earlier. Figure 2 illustrates velocity and pressure waveforms from a healthy patient that were used as boundary conditions [19]. Specifically a velocity inlet was used at the inlet and pressure outlet at the outlets. The behaviour of the artery wall was accounted for by fitting a strain energy function to the

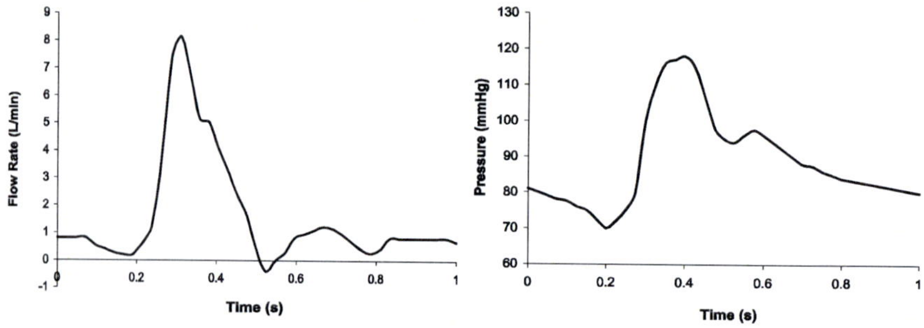

Fig. 2 Flow rate (*left*) and pressure (*right*) waveforms

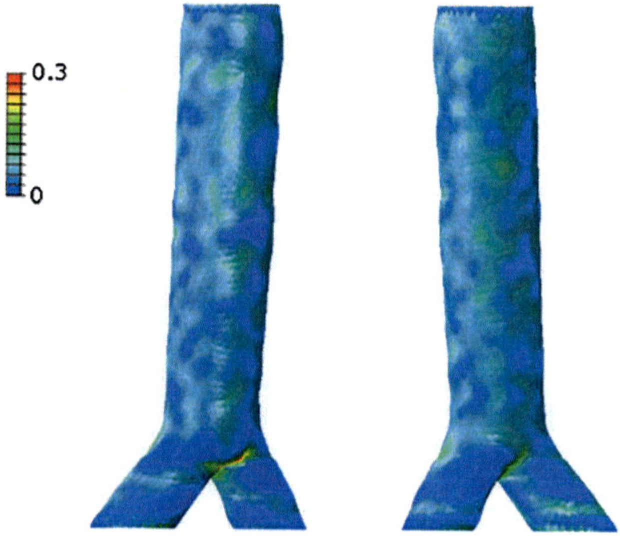

Fig. 3 Von Mises stress (MPa) on the healthy aorta at the time of maximum stress ($t = 0.43$). *Anterior view* (*left*); *Posterior view* (*right*)

stress–strain data of 7 patients with normal aortas from Raghavan et al., [25]. The inlet and outlets were constrained in all degrees-of-freedom.

3.1 Wall Stress

Anterior and posterior views of the von Mises stress on the normal aorta wall at the time of maximum stress ($t = 0.43$) are shown in Fig. 3. In this case the peak stress occurs at the aortic bifurcation. The stress in this location is normally not of physiological interest, whereas, the stresses of interest on the aorta wall all fall below 0.25 MPa. Maximum wall stress did not occur at the time of peak pressure,

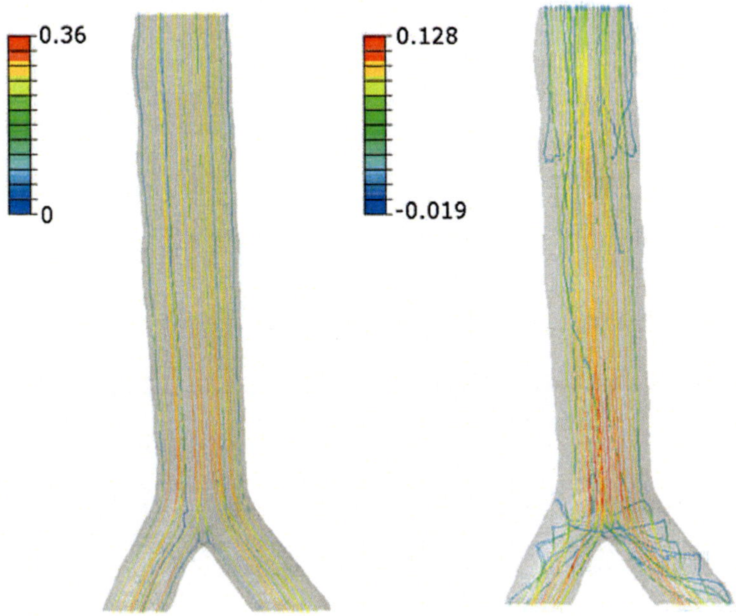

Fig. 4 Velocity streamlines (m/s) in the healthy aorta during maximum velocity ($t = 0.295$ s) and maximum deceleration ($t = 0.45$ s)

this is due to FSI effects creating a non-uniform pressure acting on the aorta wall. Due to the near symmetric shape of the healthy aorta there are negligible differences in wall stress on the anterior and posterior wall.

3.2 *Velocity Streamlines*

Streamlines through the aorta at maximum velocity and maximum deceleration are shown in Fig. 4. Unsurprisingly little flow disturbance can be seen in the straight tapering healthy geometry, particularly at the time of peak velocity. However, it can be seen that there is some swirling of the flow in the iliac leg region during fluid deceleration. This can be expected as blood flow is much more unstable during flow deceleration [23].

4 Aneurysm Biomechanics

A patient suffering from AAA was selected from our database so as to provide an example of the biomechanics of diseased dilated aortas. Again, the same velocity and pressure waveforms from Mills et al., [19] were used as boundary profiles at

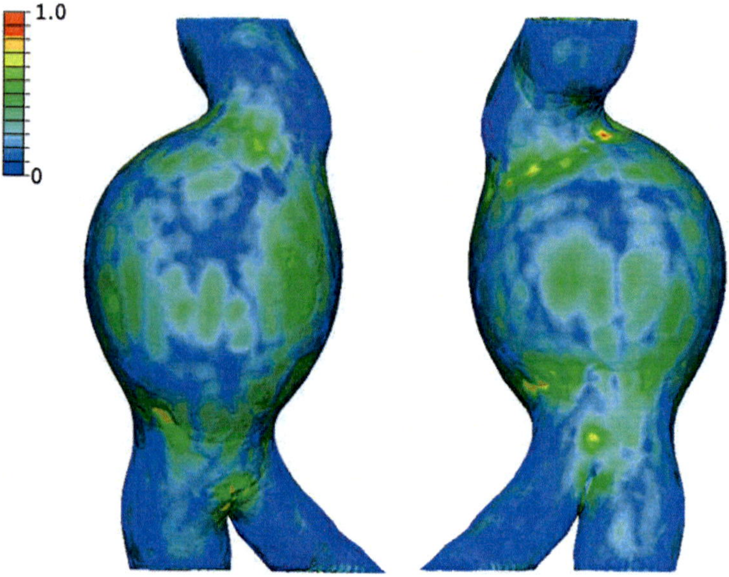

Fig. 5 Von Mises stress on aneurysm wall at the time of maximum stress ($t = 0.43$ s). Anterior view (*left*); Posterior view (*right*)

the inlet and outlets respectively (Fig. 5). The aneurysm wall was described by a 2 term strain-energy density function proposed by Raghavan and Vorp [26]. The inlet and outlets were constrained in all degrees-of-freedom.

4.1 Wall Stress

The peak pressure was located on the posterior wall of the patient just below the proximal neck. As this patient did not contain ILT this may have resulted in higher wall stress as ILT can act as a buffer. Similar regions of high stress can be seen on both the anterior and posterior walls.

4.2 Velocity Streamlines

Velocity streamlines are shown in Fig. 6 at maximum velocity and during deceleration of the flow. During maximum velocity the blood flows into the aneurysm sac and there is little presence of disturbances such as vortices and recirculations except for a small portion of helical flow as the blood flows down the aneurysm. Later, as the blood is decelerating the flow is more unstable and this is reflected by vortices in the sac causing strong recirculations.

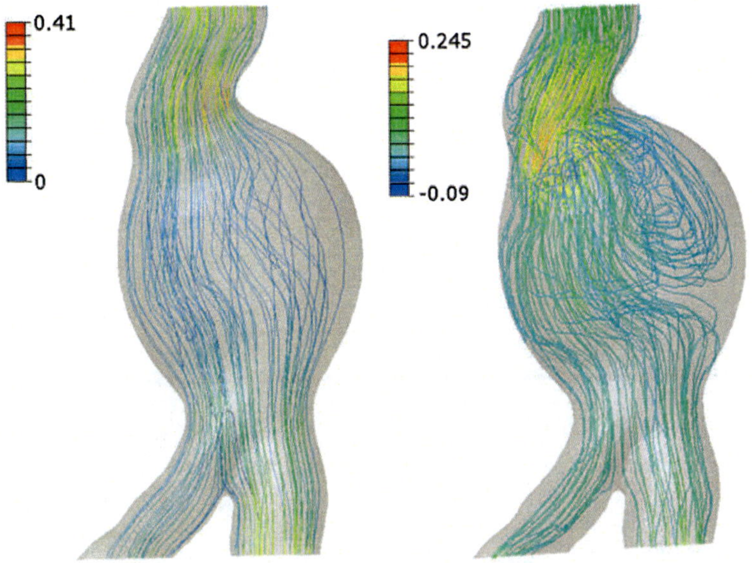

Fig. 6 Velocity streamlines (m/s) in the aneurysm during maximum velocity ($t = 0.295$ s) and maximum deceleration ($t = 0.45$ s)

5 Post-Operative Biomechanics

Similarly to the normal and diseased cases a patient who had undergone EVAR treatment for AAA was selected from our database. The CT images were obtained from a 6 month follow-up examination for the patient. The stent-graft was assumed to be linear elastic with a Young's Modulus of 10 MPa [16, 18]. The aneurysm wall was again prescribed the coefficients described by Ragahavan and Vorp [26], while the ILT was similarly described [33].

Initially oversize of the stent-graft was neglected and the graft was assumed not to be exerting a normal stress on the aneurysm neck. Also, the same velocity and pressure profiles as used in the normal aorta and aneurysm cases were initially used as boundary conditions. The influence of these assumptions is addressed later as patient-specific pressure and oversize are considered.

5.1 Wall Stress

The implantation of a stent-graft changes the stress pattern one would expect to see in a AAA. Instead of high stresses in the sac region there are high stresses in the aneurysm neck as the stent-graft is in direct contact with the neck. As there is now

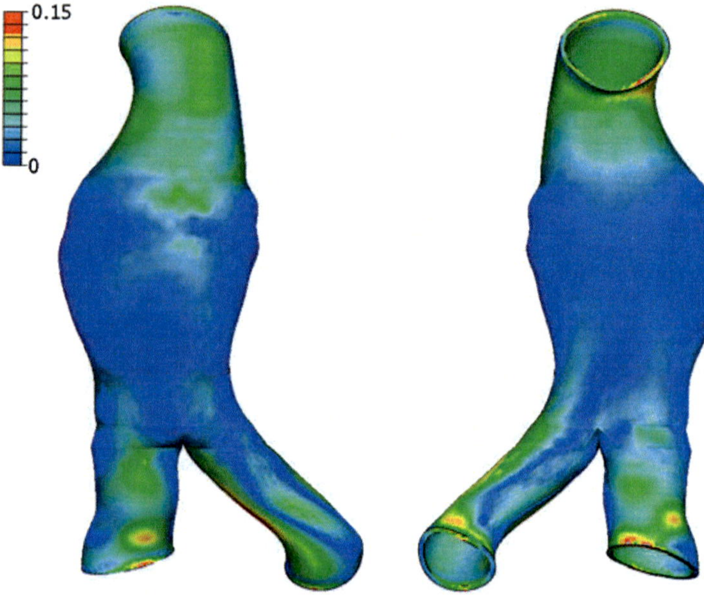

Fig. 7 Von Mises wall stress in EVAR model at time of maximum wall stress ($t = 0.43$ s). Anterior view (*left*); Posterior view (*right*)

a large amount of material between the blood pressure and sac wall (stent-graft and ILT) stresses are greatly reduced in this region (Fig. 7).

5.2 Velocity Streamlines

From Fig. 8 it can be seen that the blood flow is quite streamlined through the stent-graft. Due to the smooth geometry the blood flows into the iliacs with no disturbance during peak velocity. Similarly to the other models shown the flow is more helical during the deceleration phase.

5.3 Stent-Graft Drag Forces

The drag force which can lead to migration of the graft from its anchoring location was determined numerically. It was calculated by summing the pressure and viscous force components acting on the graft wall. The drag force and its components are plotted in Fig. 9. It can clearly be seen that the y-component is the largest contributor to the resultant force. As a result of this large component the resultant drag force acts primarily in an anterior direction. Components in the

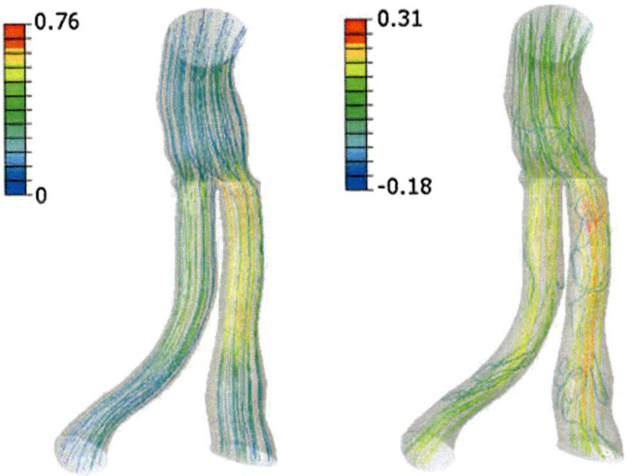

Fig. 8 Velocity streamlines (m/s) in the EVAR model during maximum velocity ($t = 0.295$ s) and maximum deceleration ($t = 0.45$ s)

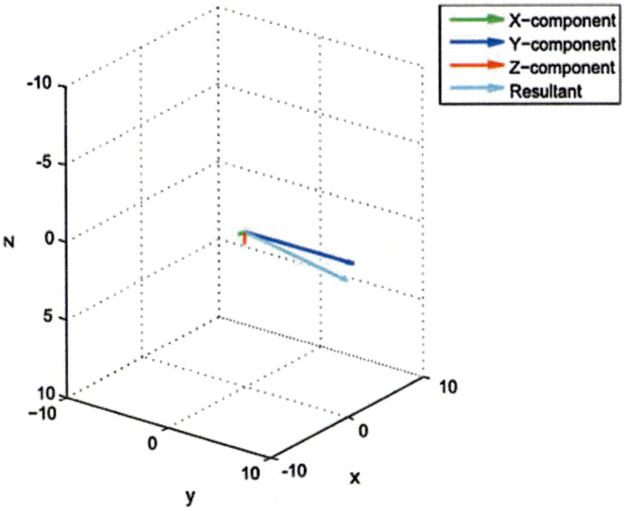

Fig. 9 Resultant drag force and the *x*, *y* and *z* components

lateral (x) and caudal (z) directions are almost negligible. The generation of this force can be attributed to the large anterior-posterior angulation of the patient. Angulation of the inlet and outlets have been shown to have a large influence on drag force magnitude, that the majority of aneurysms have an anterior-caudal acting drag force [21] and that this may cause a lateral dislodgement of the stent-graft as well as caudal movement [6].

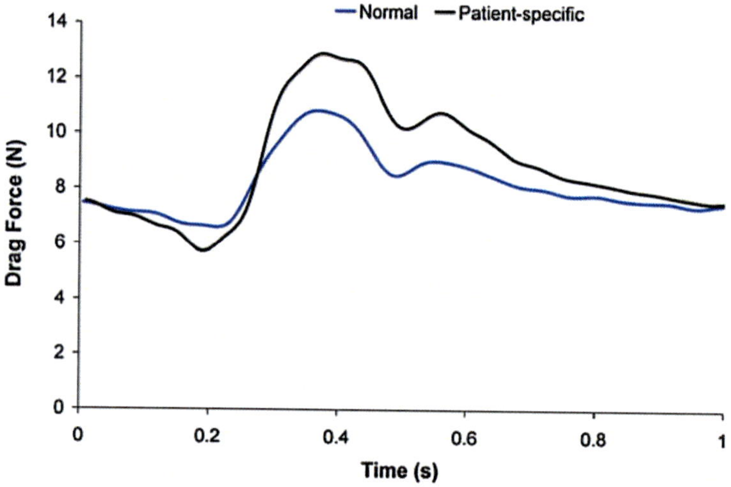

Fig. 10 Drag force acting on stent-graft due to a normal pressure waveform and a patient-specific pressure waveform

5.4 *Influence of Patient-Specific Pressure*

Systolic and diastolic blood pressures were recorded as part of post-operative patient surveillance. Patient blood pressure may vary greatly depending on medication so the largest recorded systolic value (141 mmHg) was used as well as the diastolic value recorded at the same time (67 mmHg). The maximum and minimum pressures of the previously used healthy waveform [19] were then scaled to reflect the newly recorded maximum and minimum. The drag forces for the normal and patient-specific pressure were then determined and can be seen in Fig. 10. With the use of the patient-specific waveform the peak drag force rises from 10.8 to 12.9 N. This under-prediction by using a normal waveform indicates the importance of including patient-specific pressure data. Information about pulse duration would also be of great value as patients with shorter pulse durations will be subject to more peak force hits.

Figure 11 shows the difference in aneurysm wall stress for both the normal and patient-specific pressure waveforms. Though the differences do not appear to be significantly different, higher wall stresses can be seen in the patient-specific waveform case. This is solely attributable to the higher pressure in the patient-specific waveform.

5.5 *Influence of Oversizing*

The degree of oversize was determined by the difference between the graft diameter and the outer aortic wall diameter (measuring adventitia to adventitia) as described by Sternbergh et al. [29]. A pre-stress was then applied to the regions of

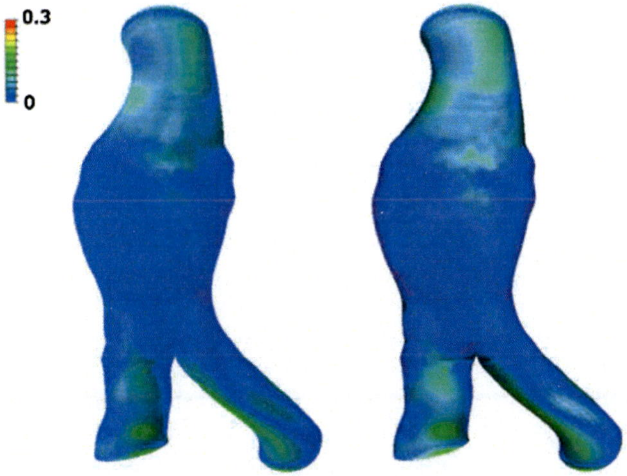

Fig. 11 Von Mises stress on aneurysm showing the influence of patient-specific pressures with normal waveform (*left*) and patient-specific waveform (*right*) at the time of maximum wall stress ($t = 0.43$)

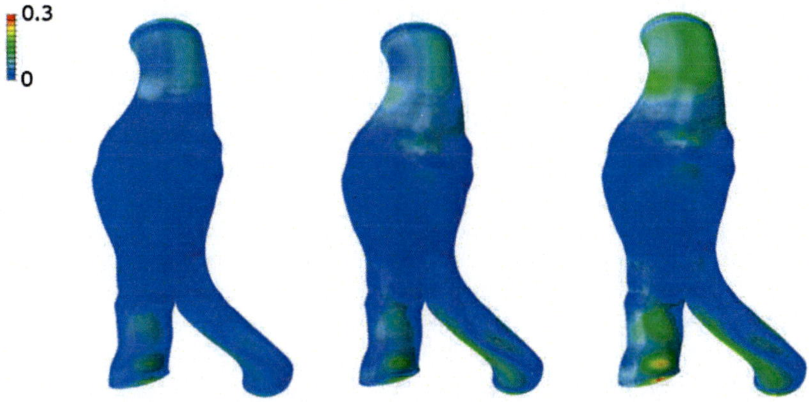

Fig. 12 Von Mises stress (MPa) on post-op models showing the influence of graft oversizing on aneurysm wall stress at the time of maximum wall stress ($t = 0.43$). Pre-stress only (*left*); No pre-stress at peak pressure (*middle*); Pre-stress and at peak pressure (*right*)

the aneurysm neck and iliac legs that were under stress due to the graft oversize. Based on the graft specification and CT scan aneurysm measurements it was found that the graft neck was oversized by 10%, the left iliac by 2.5% and the right iliac by 6%. As the pre-stress is applied only to the neck and iliac regions the increase in wall stress can be seen in this region, though a small elevation in the aneurysm sac was also noted (Fig. 12). A more uniform stress on the neck appears to be generated due to the pre-stress. While, even if just the pre-stress is included the stress is brought close to that of the case with pre-stress at peak pressure.

6 Discussion

The development of AAA results in elevated wall stresses and disturbed haemodynamics [34]. The examples here show the differences in healthy to diseased cases, namely in the presence of elevated wall stresses and greater flow disturbances. Normal healthy aorta geometry does not vary greatly from patient to patient and as a result wall stresses would not expect to vary greatly from the value of 0.3 MPa found here (at the bifurcation). On the other hand it has been shown that there are large variations in aneurysm geometries and as a result large differences in wall stress [9]. Unsurprisingly, in the example shown in this work the aneurysm had 3 times greater wall stress than the normal aorta. This increase can be attributed to the larger size, larger curvature and stiffer wall of the aneurysm model.

Previously it has been shown how the implantation of a stent-graft can improve haemodynamics and remove vortices from the flow [20]. In the example presented here streamlined blood flow can be seen in both the normal aorta and the stent-graft. This illustrates the benefits of a stent-graft in restoring blood flow close to the pre-op scenario. In the aneurysm the flow is of a more disturbed nature. During deceleration more disturbed flow can be seen in each model. Biasseti et al. [1] showed that blood flow differs significantly from fusiform to saccular aneurysms with large flow reversals in fusiform aneurysms, while the blood is almost at rest in saccular aneurysms. The aneurysm in this case was saccular in nature. As a result regions of, or almost of, zero flow are commonplace particularly during flow deceleration. Saccular aneurysms such as the example here would expect to have more disturbed flow, depending on the degree of ILT formation, due to the creation of pockets which the blood will flow into.

The patient in this study did not suffer from severe hypertension and hence there was not a large change in the stress due to the patient-specific waveform not differing greatly from the standard waveform. In patients suffering from severe hypertension drag forces would expect to be significantly higher. If elevated aneurysm neck stresses contribute to post-op neck dilation then these stresses may be under-predicted with the use of a generic pressure waveform. For the estimation of wall stress in the aneurysm neck the degree of oversize is crucial information if an accurate value is to be determined. As the specification of each implanted graft is known a pressure can be applied to give a reasonable representation of the normal stress acting on the aneurysm neck and iliacs. In the model presented here a large increase in wall stress was observed for the inclusion of graft oversizing (Fig. 12). The oversize applied in this case was 10% and would not be considered excessive by clinical standards. Indeed, oversizing of between 20 and 30% are not uncommon and would result in even greater neck and iliac stresses.

Based on the results seen here further advancement in patient-specific data would create more accurate patient-specific models. One of the major limitations of this study and most traditional FSI studies are in the implementation of boundary conditions. Most FSI studies of AAA use velocity inlet and pressure outlet boundary profiles from either Mills et al., [19] or Olufsen [24]. More accurate outlet boundary conditions can be obtained by representing the downstream vasculature with one-dimensional or impedance models [32]. Unfortunately, these types of boundary conditions can be difficult to implement in commercial softwares. An alternative is to record pressures in vivo. Recently Torii et al., [30] obtained patient-specific pressure profiles using a pressure probe, this is an additional and more invasive technique than acquiring velocity or flow rate data which can be acquired using Doppler ultrasound or phase contrast magnetic resonance imaging (PC-MRI). Nevertheless, it is a significant improvement on the method presented here of superimposing the measured patient pressure on the waveform from Mills et al. Though this method should give an accurate representation of the patients systolic and diastolic pressures, the pressures between these points are unknown. Information regarding the duration of the aortic pulses is also unknown.

Another source of potential inaccuracy is the assumption of zero stress in the initial aorta/aneurysm geometry. While efforts have been made to account for this [10] by finding the pre-stressed aneurysm geometry, this has not been extended to finding the pre-stressed aneurysm geometry after a stent-graft has been implanted. The assumption of linear elasticity for the stent-graft is also a limitation of the work. Nitinol is a shape memory alloy and behaves very differently under differing degrees of strain while ePTFE does not behave linearly [2]. Incorporation of some of these aspects has previously been achieved [13] but has not been extended to a patient-specific case. This can be primarily attributed to the difficulty in stent reconstruction, as they appear with considerable noise in CT images.

7 Conclusions

AAA results in elevated wall stresses and disturbed haemodynamics. The implantation of a stent-graft reduces stresses and improves haemodynamics to levels similar to healthy cases. EVAR treatment may create problems due to drag forces acting on the graft and high neck stresses. Accurate patient-specific data is crucial for the prediction of the stresses and forces acting on aneurysms and stent-grafts and this data has a large bearing on the predicted results. Vascular modelling that aims to predict disease formation or device performance should use as much patient data as is possibly available.

References

1. Biasseti, J., Gasser, T.C., Auer, M., Hedin, U., Labruto, F.: Hemodynamics of the normal aorta compared to fusiform and saccular abdominal aortic aneurysms with emphasis on a potential thrombus formation. Ann. Biomed. Eng. **38**(2), 380–390 (2010)

2. Catanese III, J., Cooke, D., Maas, C., Pruitt, L.: Mechanical properties of medical grad expanded polytetraflouroethylene: The effects of intermodal distance, density, and displacement rate. J. Biomed. Mat. Res. **48**(2), 187–192 (1999)
3. Conners III, M.S., Sternbergh, W.C., Carter, G., Tonnessen, B.H., Yoselevitz, M., Money, S.R.: Endograft migration one to four years after endovascular abdominal aortic aneurysm repair with the Aneurx device: A cautionary note. J. Vasc. Surg. **36**(3), 476–484 (2002)
4. Corriere, M.A., Feurer, I.D., Becker, S.Y., Dattilo, J.B., Passman, M.A., Guzman, R.J., Naslund, T.C.: Endoleak following endovascular abdominal aortic aneurysm repair. Ann. Surg. **239**, 800–807 (2004)
5. De Hart, J., Peters, G.W.M., Schreurs, P.J.G., Baaijens, F.P.T.: A three-dimensional computational analysis of fluid-structure interaction in the aortic valve. J. Biomech. **36**, 103–112 (2003)
6. Figueroa, C.A., Taylor, C.A., Yeh, V., Chiou, A.J., Gorrepati, M.L., Zarins, C.K.: Preliminary 3D computational analysis of the relationship between aortic displacement force and direction of endograft movement. J. Vasc. Surg. **51**(6), 1488–1497 (2010)
7. Figueroa, C.A., Vignon-Clementel, I.E., Jansen, K.C., Hughes, T.J.R., Taylor, C.A.: A coupled momentum method for modelling blood flow in three-dimensional deformable arteries. Comp. Meth. Appl. Mech. Eng. **195**, 5685–5706 (2006)
8. Fraser, K.H., Li, M.X., Lee, W.T., Easson, W.J., Hoskins, P.R.: Fluid-structure interaction in axially symmetric models of abdominal aortic aneurysms. Proc. Inst. Mech. Eng. H **223**(2), 195–209 (2009)
9. Gasser, T.C., Auer, M., Labruto, F., Swedenborg, J., Roy, J.: Biomechanical rupture risk assessment of abdominal aortic aneurysms: Model complexity versus predictability of finite element simulations. Euro. J. Vasc. Endovasc. Surg. **40**, 176–185 (2010)
10. Gee, M.W., Reeps, C., Eckstein, H.H., Wall, W.A.: Prestressing in finite deformation abdominal aortic aneurysm simulation. J. Biomech. **42**(11), 1732–1739 (2009)
11. Greenber, R.K., Turc, A., Haulon, S., Srivastava, S.D., Sarac, T.P., O'Hara, P.J., Lyden, S.P., Ouriel, K.: Stent-graft migration: A reappraisal of analysis methods and proposed revised definition. J. Endovasc. Ther. **11**, 353–363 (2004)
12. Howell, B.A., Kim, T., Cheer, A., Dwyer, H., Saloner, D., Chuter, T.A.M.: Computational fluid dynamics within bifurcated abdominal aortic stent-grafts. J. Endovasc. Ther. **14**(2), 138–143 (2007)
13. Kleinstreuer, C., Li, Z., Basciano, C.A., Seelecke, S., Farber, M.A.: Computational mechanics of Nitinol stent-grafts. J. Biomech. **41**, 2370–2378 (2008)
14. Kratzberg, J.A., Golzarian, J., Raghavan, M.L.: Role of graft oversizing in the fixation strength of barbed endovascular grafts. J. Vasc. Surg. **49**(6), 1453–1553 (2009)
15. Leung, J.H., Wright, A.R., Cheshire, N., Crane, J., Thom, S.A., Hughes, A.D., Xu, Y.: Fluid structure interaction of patient specific abdominal aortic aneurysms: a comparison with solid stress models. Biomed. Eng. Online **5**, 33 (2006)
16. Li, Z., Kleinstreuer, C.: Fluid structure interaction effects on sac-blood pressure and wall stress in a stented aneurysm. J. Biomech. Eng. **127**, 662–671 (2005)
17. Li, Z., Kleinstreuer, C.: Effects of major endoleaks on a stented abdominal aortic aneurysm. J. Biomech. Eng. **128**, 697–706 (2006)
18. Li, Z., Kleinstreuer, K.: Analysis of biomechanical factors affecting stent graft migration in an abdominal aortic aneurysm model. J. Biomech. **39**(12), 2264–2273 (2005)
19. Mills, C.J., Gabe, I.T., Gault, J.H., Mason, D.T., Ross, J., Braunwuld, E., Shillingford, J.P.: Pressure flow relationship and vascular impedance in man. Card. Res. **4**(4), 405–417 (1970)
20. Molony, D.S., Callanan, A., Kavanagh, E.G., Walsh, M.T., McGloughlin, T.: Fluid-structure interaction of a patient-specific abdominal aortic aneurysm treated with an endovascular stent-graft. Biomed. Eng. Online **8**, 24 (2009)
21. Molony, D.S., Kavanagh, E.G., Walsh, M.T., McGloughlin, T.M.: A computational study of the magnitude and direction of migration forces in patient-specific abdominal aortic aneurysm stent-grafts. Euro. J. Vasc. Endovasc. Surg. **40**(3), 332–339 (2010)

22. Morris, L., Delassus, P., Walsh, M., McGloughlin, T.: A mathematical model to predict the in vivo distraction forces acting on bifurcated stent grafts used in endovascular treatment of abdominal aortic aneurysm (AAA). J. Biomech. **37**(7), 1087–1095 (2004)
23. Morris, L., Delassus, P., Callanan, A., Walsh, M., Wallis, F., Grace, P., McGloughlin, T.: 3-D numerical simulation of blood flow through models of the human aorta. J. Biomech. Eng. **127**(5), 767–775 (2005)
24. Olufsen, M.S.: Structured tree outflow condition for blood flow in larger systemic arteries. Am. J. Physiol. Heart Circ. Physiol. **276**, 257–268 (1999)
25. Raghavan, M.L., Webster, M.W., Vorp, D.A.: Ex vivo biomechanical behaviour of abdominal aortic aneurysm: assessment using a new mathematical model. Ann. Biomed. Eng. **24**(5), 573–582 (1996)
26. Raghavan, M.L., Vorp, D.A.: Toward a biomechanical tool to evaluate rupture potential of abdominal aortic aneurysm: Identification of a finite strain constitutive model and evaluation of its applicability. J. Biomech. **33**(4), 475–482 (2000)
27. Sampaio, S.M., Panneton, J.M., Mozes, G., Andrews, J.C., Noel, A.A., Kalra, M., Bower, T.C., Cherry, K.J., Sullivan, T.M., Gloviczki, P.: Aortic neck dilation after endovascular abdominal aortic aneurysm repair: Should oversizing be blamed? Ann. Vasc. Surg. **20**(3), 338–345 (2006)
28. Scotti, C.M., Finol, E.A.: Compliant biomechanics of abdominal aortic aneurysms: A fluid structure interaction study. Comp. Struct. **85**, 1097–1113 (2007)
29. Sternbergh, W.C., Money, S.R., Greenberg, R.K., Investigators, Zenith: Influence of endograft oversizing on device migration, endoleak, aneurysm shrinkage and aortic neck dilation: Results from the Zenith multicenter trial. J. Vasc. Surg. **39**(1), 220–226 (2004)
30. Torii, R., Wood, N.B., Hadjiloizou, N., Dowsey, A.W., Wright, A.R., Hughes, A.D., Davies, J., Francis, D.P., Mayet, J., Yang, G., Thom, S.A., Xu, Y.: Fluid-structure interaction analysis of a patient-specific right coronary artery with physiological velocity and pressure waveforms. Comm. Num. Meth. Eng. **25**, 565–580 (2009)
31. Van Prehn, J., van Herwaarden, J.A., Vincken, K.L., Verhagen, H.J., Moll, F.L., Bartels, L.W.: Asymmetric aortic expansion of the aneurysm neck analysis and visualization of shape changes with electrocardiogram-gated magnetic resonance imaging. J. Vasc. Surg. **49**(6), 1395–1402 (2009)
32. Vignon-Clementel, I.E., Figueroa, C.A., Jansen, K.E., Taylor, C.A.: Outflow boundary conditions for three-dimensional finite element modelling of blood flow and pressure in arteries. Comp. Meth. Appl. Mech. Eng. **195**, 3776–3796 (2006)
33. Wang, D.H.J., Makaroun, M., Webster, M.W., Vorp, D.A.: Mechanical properties and microstructure of intraluminal thrombus from abdominal aortic aneurysm. J. Biomech. Eng. **123**(6), 536–539 (2001)
34. Xenos, M., Rambhia, S.H., Alemu, Y., Einav, S., Labropoulos, N., Tassiopoulos, A., Ricotta, J.J., Bluestein, D. Patient-based abdominal aortic aneurysm risk prediction with fluid structure interaction modelling, Ann. Biomed. Eng. doi: 10.1007/s10439-010-0094-3.

Biomechanical Aspects of Abdominal Aortic Aneurysm (AAA) and its Risk of Rupture: Fluid Structure Interaction (FSI) Studies

M. Xenos and D. Bluestein

Abstract A series of numerical studies to elucidate the effect of various biomechanical parameters on AAA disease progression and its risk of rupture is presented, highlighting a methodology based on biomechanical considerations for a reliable patient specific prediction of AAA risk of rupture. Fluid structure interaction simulations of normal aortas, non-ruptured, and contained ruptured AAA (rAAA) were conducted in patient specific geometries reconstructed from CT scans. The models included the arterial wall, intraluminal thrombus (ILT), and calcifications. Parametric studies of the effects of iliac bifurcation and neck angulation, and of hypertension were also conducted. Advanced constitutive material models included wall anisotropy and structural strength of the aorta. Our studies demonstrate that the anisotropic simulations showed higher peak wall stresses as compared to isotropic ones, indicating that the latter may underestimate the AAA risk of rupture. The ILT appeared to provide a cushioning effect reducing the stresses, while small calcifications appeared to weaken the wall and contribute to the rupture risk. The location of the maximal wall stresses and rupture potential index (RPI) in ruptured AAA simulations overlapped the actual rupture region, demonstrating the predictive capabilities of the methodology. Parametric studies of the AAA iliac bifurcation and neck angulation in idealized streamlined AAA geometries indicated a trend of increasing peak wall stresses with increasing iliacs and neck angulation, with the appearance of two local minima in the iliac bifurcation angle peak stresses pointing to the possibility of an adaptation response in an attempt to achieve a temporary reduction of the peak wall stresses during the disease progression. Such geometric parametric dependence can be further used to

M. Xenos · D. Bluestein (✉)
Department of Biomedical Engineering,
Stony Brook University, Stony Brook, NY, USA
e-mail: danny.bluestein@sunysb.edu

Stud Mechanobiol Tissue Eng Biomater (2011) 7: 181–220

DOI: 10.1007/8415_2011_72

Published Online: 24 April 2011

augment AAA diagnostics. This may help clinicians estimate the AAA risk of rupture to establish whether a risky surgical intervention is warranted.

1 Introduction

1.1 Abdominal Aortic Aneurysm (AAA) Disease: Prevalence and Mortality Rates

Aneurysmal dilation of the aorta occurs in 2–4% of males over the age of 65 in the western world. The disease risk is increased in patients who have evidence of coronary, carotid, or peripheral vascular disease, a history of smoking, or a family history of aneurysmal disease [121]. Recently, routine ultrasound surveillance screening for abdominal aortic aneurysm has been recommended for males over 65 years of age and selected high-risk females [47]. The major complication of aneurysmal disease is rupture, which is associated with mortality rates of 50–75% [6]. Prophylactic intervention to prevent aneurysm rupture is recommended for patients whose annual risk of rupture exceeds the risk of operation (2–5%, in patients with acceptable cardiovascular risk). Due to the lack of more rigorous diagnostic tools, many of the AAA surgical procedures may be deemed unnecessary, while some patients are under-treated with possibly devastating consequences. This trend is expected to keep rising due to the aging population [80].

1.2 Relationship of AAA Size and Other Factors to Rupture

Rupture risk is generally correlated to maximal AAA transverse diameter; consequently, this parameter has been used to determine the need for intervention. Current recommendations, based on prospective studies, indicate that AAA should be repaired when the maximal diameter exceeds 5.0–5.5 cm [47, 61]. However, the risk of rupture is multifactorial, and an absolute correlation between size and risk of rupture is impossible to obtain [39]. Important variables influencing rupture risk include the AAA configuration (fusiform versus saccular), size of the normal adjacent aorta, vessel tortuosity, and the presence or absence of thrombus and calcium. The ability to estimate the rate of aneurysm progression is also of great clinical importance. This estimate influences the frequency of follow up as well as decisions regarding surgical intervention. Initial AAA size is a major determinant of progression, with average AAA expansion rates of about 10% diameter per year [75, 121]. However, individual patient risk factors, vessel angulation, and hemodynamic forces undoubtedly influence this process. It has been established in the literature that smaller diameter aneurysms can rupture while larger ones can remain intact [17, 46, 61, 84]. Additionally, the mortality rate for aneurysm rupture is independent of size and exceeds 50% [92]. Some aneurysms contain intraluminal thrombus (ILT). Whether or not the ILT offers protection against rupture is subject to an ongoing investigation [20, 76, 89, 103, 106]. The degree of wall calcification in both aneurysmal and adjacent normal aorta varies widely. There are no data that correlate this distribution with rupture risk. When matched for age, gender, and diameter, ruptured AAAs tend to be less tortuous, yet have greater cross-sectional diameter asymmetry [26]. When diameter asymmetry is associated with low aortic tortuosity, the larger diameter on axial sections more accurately reflects rupture risk, and when diameter asymmetry is associated with moderate or severe tortuosity, the smaller diameter on axial sections more accurately reflects rupture risk. Smoking is significantly associated with rupture, even when controlling for gender and AAA anatomy [26].

1.3 Small Aneurysms can Rupture

Several clinical studies emphasized the fact that large aneurysms have larger mortality rate because of they are more prone to rupture, but small diameter aneurysms can also rupture [10, 13, 61]. Accordingly, physicians follow patients with small (4.0–5.5 cm) AAAs with serial imaging studies to document growth rate and absolute size. This helps determine the need for intervention. It is generally accepted that even a small aneurysm that shows rapid growth (>5 mm in 6 months) should be operated on because of an imputed increased risk of rupture. Within the cohort of patients examined, women have a threefold higher risk of aneurysm rupture than men. Effective control of blood pressure and cessation of smoking are likely to diminish the risk of rupture. Once rupture occurs, mortality rates of small (<5 cm) and large (>5 cm) aneurysms are similar, 70% for small AAAs versus 66% for large AAAs [56]. No significant differences are seen between the patients with small and large ruptured aneurysms with respect to the prevalence rates of hypertension (60 versus 50%) or of cardiac disease (20 versus 22%). In the [27] study, the smallest ruptured AAA was 4.8 cm, but this aneurysm had a stress equivalent to the average electively repaired 6.3 cm AAA. Recent FSI simulations conducted in models reconstructed from patients with ruptured AAA indicated similar results, whereas smaller ruptured AAA exhibited higher stresses than larger ruptured AAA [117, 118].

1.4 Mechanisms of AAA Wall Weakening

AAA development is multifactorial. A major mechanism postulated for AAA formation focuses on inflammatory processes where macrophage recruitment leads to MMP production and elastase release. The biomechanical changes associated with enzymatic degradation of structural proteins suggests that AAA expansion is primarily related to elastolysis, whereas rupture involves failure of the remaining collagen mesh [21]. Elastolysis promotes AAA expansion, increasing the wall tension according to Laplace's Law but does not necessarily weaken the aortic wall. There is clear evidence that inflammation, with macrophage infiltration and matrix digestion by matrix metalloproteinases (MMP) plays an important role in this process. There are two mechanisms that specifically weaken the AAA wall and increase the possibility of AAA rupture. The first one is based on the stress-mediated wall weakening and the second one is the hypoxia-mediated wall weakening. The wall strength distribution and the tissue stiffness within any AAA is spatially variable [87, 88, 102]. These observations support the hypothesis that local factors like the biomechanical wall and hypoxia could promote tissue degeneration in AAA. Aneurysmal portions of the aorta were found stiffer and contained less elastin and collagen than normal aorta [30]. A decreasing quadratic relationship was found between elastin concentration and diameter for normal aortas and for pathological increasing diameter [73]. Focal saccular outpouchings or "blebs", found within the walls of some AAAs may represent sites of potential rupture [40]. Further studies showed that these saccular outpounchings result in focal stress concentration [44, 105]. Increased matrix metalloproteinase MMP-9 mRNA in AAA correlated with increasing aneurysm size [52]. Further studies show that the degree of elastolysis increases in AAA as it enlarges [72].

Hypoxia may cause AAA wall weakening by upsetting the normal balance between extracellular matrix synthesis and degradation, by mediating cytokines activity such as tumor necrosis factor-α [22, 48, 60, 74]. Kazi et al. showed that AAA wall adjacent to ILT was thinner, and contained more macrophages and inflammatory cells than AAA wall adjacent to no ILT [43]. It was also shown by the Vorp group that macrophages are present in the ILT and are exposed to a hypoxic environment [1, 107]. This clearly indicates that the local hypoxic environment could lead to a decrease of the overall structural integrity of the wall and eventually to rupture.

1.5 Mechanical Properties of AAA Using Experimental Clinical Approaches

AAA rupture is a mechanical failure of the wall tissue due to elevated stresses developing within the AAA wall as a result of the interaction between the pathological AAA hemodynamics and the compromised integrity of the wall. Currently, the means of measuring in vivo the wall stresses or tissue strength of an aneurysm are extremely limited. The use of noninvasive techniques like ultrasound and computed tomography (CT) for the detection and treatment of AAAs have recently been expanded for estimating the mechanical behavior of the AAA wall and ILT, primarily with the recording of compliance measurements. It has been shown with the use of ultrasound tracking system that there is a decrease in compliance in patients with AAA as compared with control subjects [45]. Using M-mode ultrasonography an increase in the pressure-strain elastic modulus was reported in patients with AAA [51]. Automated ultrasonographic measurements of the aortic wall and ILT showed that the compliance of the AAA wall is decreased [101]. Several studies suggest that the ILT has a cushioning effect [41, 109, 118]. The relationship between compliance, maximum diameter and growth rate was studied, concluding that large aneurysms tend to be less compliant and thus stiffer compared to healthy subjects [112]. Recently the same group showed that gender, hypertension, a decrease in elastic modulus and a larger maximum diameter are significant parameters in AAA rupture [113].

1.6 AAA Wall Stress and Strength

There is a significant decrease in the tensile strength of AAA compared to nonaneurysmal tissue [64]. Larger AAAs have a higher risk of rupture [5, 49], suggesting that the failure strength of AAA wall reduces progressively as a AAA enlarges [96]. Earliest predictions for AAA wall stress used the law of Laplace in simple geometries [23, 41, 54]. These simplified geometries were initially used by many researches and provided useful information on the general factors influencing AAA wall stress [16, 23, 54, 55]. However, excluding the geometric complexity of the AAA wall can lead to over or underestimation of the wall stresses [71]. Linear elasticity was initially applied for describing the wall material properties of the AAA [16, 23, 54, 55]. However the AAA wall tissue is characterized by hyperelastic/viscoelastic material properties and significant anisotropy. The use of simplified geometries or linearized elasticity for AAA wall properties could lead to inconsistent results that may underestimate the wall stresses [66]. The rupture of AAA is not only a function of wall stress but wall strength. Thus, determination of the strength of the local tissue is of importance. Wilson et al. [111, 113, 114] distinguished between two major types of aneurysms, the ones that 'remodel' while growing (elastic aneurysms) and adjust to the new stress levels, and the ones that do not remodel (stiff aneurysms). AAA distensibility (pressure strain elastic modulus [Ep] and stiffness [β]) were measured at 6 months with an ultrasound scan-based echo-tracking technique [113].

1.7 Tensile Testing and Modeling of AAA Wall and ILT

Early ex vivo studies of the biomechanical properties of AAA were focused on the understanding of the extracellular matrix and its basic properties like the wall stiffness. A stiffer and less distensible uniaxial response was observed in AAA tissue specimens as compared with normal tissue, along with an associated decrease in volume fraction of elastin [30]. A decrease in AAA wall strength of specimens tested uniaxially was also reported [62, 104]. Biomechanical properties, e.g., yield stress (the critical stress value that is required for the rupture of the aortic tissue to occur) and its regional distribution within the aneurysmal wall [86–89] were observed, clearly indicating that spatial variations in wall stresses and strengths must be taken into account. Uniaxial testing of abdominal aortic tissue specimens was performed [63], and the mechanical behavior of the tissue was modeled using non-linear hyperelastic wall mechanical properties following the isotropic Mooney–Rivlin formulation [53]. Tension controlled biaxial testing of AAA wall specimens was used to carefully characterize the stress–strain relationship of the tissue [29]. This experimental data was incorporated into advanced anistropic models for the wall behavior using the innovative fluid structure interaction (FSI) approach for modeling AAA wall stress and deformation recently performed by our group [66].

A high degree of anisotropy in mechanical response was observed between the circumferential and longitudinal directions. The uniaxial tensile properties of ILT were measured and used for modeling it as a linear isotropic material [16], and a uniaxial tensile test was performed to develop a hyperelastic isotropic constitutive relation for the ILT [108]. It indicated the heterogeneity of the ILT-composed of three distinct layers with different mechanical properties. More recent work has shown that the luminal layer of ILT behaves as an isotropic material [16, 94, 108].

1.8 Image Segmentation and Reconstruction Tools

With the advances in graphics techniques for surface rendering, the potential for attaining useful information from graphics in the medical field has emerged. 2D images obtained in parallel planes from imaging modalities like CT and magnetic resonance imaging (MRI) scans can be reconstructed for visualizing the 3D morphology of diseased organs. Angiographic approaches utilize multisliced CT to obtain 3D reconstructions of the volumetric data. Several techniques have been developed, such as the maximum intensity projection (MIP), shaded surface display (SSD), volumetric rendering etc. [2, 11, 42, 59]. Visualization toolkit (VTK) and the insight toolkit (ITK) are a few examples of packages developed for performing registration and segmentation based on ITK and VTK libraries [120]. These open-source toolkits have active development community that includes laboratories, institutions and universities from around the world [3, 4, 50, 82, 110]. While these tools provide a qualitative technique for diagnosing pathologic conditions, there is also a need to quantify pertaining parameters. Despite the high resolution of these imaging techniques 3D representations alone do not provide blood flow patterns that develop within the lumen or the wall stresses that act on the wall and develop within it due to the hemodynamics. Accordingly, several steps still need to be developed in order to complete a quantitative framework for better diagnostics and predictive capabilities for estimating AAA risk of rupture.

1.9 Computational Models of AAA

The improving resolution of clinical imaging modalities and the advanced 3D reconstruction techniques provide qualitative and partially quantitative technique for diagnosing pathologies. However, they do not provide quantitative information of the luminal blood flow patterns and their complex interaction with the blood vessel wall, and the resultant hemodynamically driven wall stresses that may lead to the wall failure. In order to help the clinician arrive at an informed decision regarding the need for a surgical intervention, it is essential to map and quantify the stresses developing within the aneurysmal wall. This will offer a much better diagnosis of the patient's true condition. However, several steps still need to be developed for a framework that will lead to a successful quantitative prediction for hemodynamics in the vessels and the influence on the vessel wall.

Following the limitations of measuring the wall stresses or tissue strength of AAA in vivo, computational models offer a robust alternative to calculate such properties, and thus help establish the risk of rupture. Using 3D CT reconstructions Fillinger et al. [25] have shown that computational models using a static uniform internal pressure on a AAA can be 12% more accurate and 13% more sensitive than using maximum diameter as a sole predictor for risk of rupture. Further, they have studied the impact of wall stress as related to diameter on aneurysm rupture and progression [24, 90]. Idealized geometries to measure the effects of variations in wall thickness and the effects of including the ILT showing that variations in wall thickness generate higher stresses and increased risk of rupture than models that use a uniform wall thickness [77–79]. Di Martino el al. [16, 18–20] have shown that the presence of ILT can significantly reduce the stress on the wall.

1.10 FSI Approach

In recent years FSI simulations, in which the dynamic interaction between the AAA hemodynamics and wall deformation is modeled, were conducted by our group and others to simulate the biomechanical behavior of the AAA walls [8, 19, 25, 58, 77, 99]. Those were universally based on the isotropic assumption, in which the directional ambiguities associated with abdominal aortic tissue mechanical response to stresses-which may play a major role in the behavior of the tissue under elevated stresses, cannot be resolved. Recently our group performed FSI numerical studies where patient specific 3D geometries were reconstructed from CT scans in AAA of differing configurations, both with and without ILT [8, 66]. Complex flow trajectories within the AAA lumen indicated a putative mechanism for the formation and growth of the ILT. The resulting magnitude and location of the peak wall stresses was dependent on the shape of the AAA. Our data suggest that while thrombus does not significantly change the location of maximal stress in the aneurysm, the presence of thrombus within the AAA may reduce some of the stress on the wall. Accordingly, inclusion of ILT and calcifications in stress analysis of AAA is important and will likely increase the accuracy of predicting the risk of AAA rupture.

A proper model for the accurate prediction of stresses developing within the AAA wall requires detailed information, including (i) patient specific AAA geometry (ii) blood flow parameters and flow patterns (including flow rates and pressures at the various phases of the cardiac pulse wall thickness and its variability), and (iii) appropriate material models that can describe the mechanical response of AAA tissue. Recently, our group [66] fitted experimental data of AAA wall specimens [29] to an exponential strain energy orthotropic material model [66, 100], further applying a similar model to the orthotropic material model of Holzapfel et al. [34] which models the tissue as a fiber-reinforced composite material with the fibers corresponding to the collagenous component of the material. This material formulation was previously successfully applied to various arterial walls, e.g., aorta, coronaries, and carotid arteries [31–38].

2 Patient-Based FSI Methodology

A methodology for modeling the biomechanics of AAAs that may eventually lead to its rupture is presented, based on advanced material models and AAA geometries reconstructed from patient specific CT scans. The computational FSI approach is presented and discussed vis-à-vis its capabilities for predicting the AAA rupture potential and its location. AAA components and the biomechanical manifestations of its pathological features such as the ILT and calcifications are discussed, while describing aneurysmal disease progression from normal aortas, through AAA, to rupture (rAAA). Parametric studies of iliac bifurcation and AAA neck angulation are discussed in lieu of the role they may play in an adaptation response during the disease progression, as well as effects of hypertension.

The patient specific AAA FSI methodology follows several steps, schematically described in Fig. 1 [117]. AAA geometries were obtained from CT Angiograms as part of the routine evaluation of patients with AAA. The images were acquired using abdominal CT with intravenous contrast, which is the standard of care for AAA disease patients. Thin slice CT scans (CT angiograms) were obtained with high resolution CT scanner (0.625 mm slice) as part of the routine evaluation of patients. Computerized reconstruction using Medical Metrx Software (MMS Inc, NH, USA) was performed using a GE Lightspeed VCT 64-slice scanner. Abdominal CT scans were also acquired from patients with normal aortas in order to serve as controls and characterize aneurysmal disease progression.

The ability of the models to predict AAA rupture was studied by reconstructing retrospectively ruptured aneurysms in several cases of patients who arrived at the Stony Brook university hospital emergency room (ER) with contained ruptured AAA (rAAA). In all patients informed consent was obtained retrospectively. The protocol was approved by Stony Brook University Institutional Review Boards (IRB) Committees on Research Involving Human Subjects.

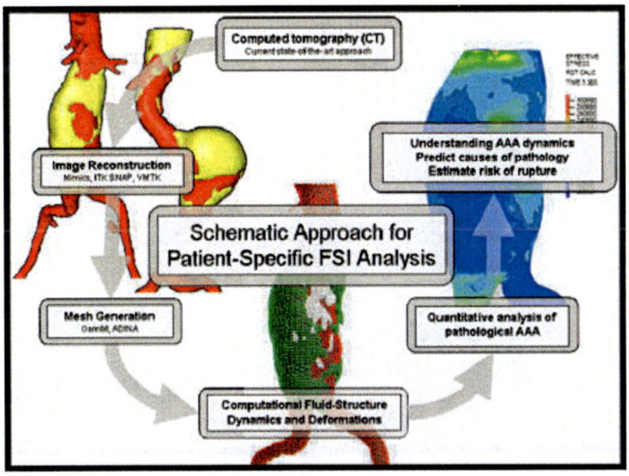

Fig. 1 Schematic diagram of patient specific fluid structure interaction (FSI) approach shows the main steps through which conventional patient CT data is used for estimating risk of rupture in AAAs. (i) Acquiring CT images from AAA patients. (ii) Reconstruction of abdominal aorta from below the iliac bifurcation till above the AAA neck. (iii) Discretization of the geometry to generate a computational mesh for FSI analysis. (iv) The FSI results provides the surgeon with quantitative mapping of wall stress and strength distribution to assess the AAAs risk of rupture

The coordinate points and dimensions of patient specific AAA geometry were extracted from the CT scans, including anatomical details of the vessel wall, lumen, ILT, and calcifications (Ca). For both the normal and pathological abdominal aortas, boundaries of the reconstructions were established from just below the renal arteries branching until 4.0 cm distal to the iliac bifurcation. Other smaller arteries, such as the gonadal and lumbar/spinal arteries, the inferior mesenteric artery and the median sacral artery were neglected due to lack of information about the blood flow exiting these arteries and their marginal effect on AAA hemodynamics. A uniform thickness of 2 mm was assigned to the AAA wall, as used in previous studies of patient specific geometries [14, 58, 95, 115]. The parallel plane scans were translated into 3D images using image reconstruction software packages (Mimics, Materialise, Leuven, Belgium; MMS, Medical Metrx Solutions Inc., West Lebanon, NH). Reconstruction of the anatomical details of the vessel included the artery wall, lumen, ILT, and calcifications by using different layered masks. In a normal abdominal aorta, the outer wall is a reinforced fibrous material, which surrounds the blood carrying lumen. Due to arterial wall degradation and weakening during aneurysmal disease progression, vessel remodeling ensues and the ballooning of the outer wall become prominent. This remodeling may also result from the combined effects of hardened calcified spots embedded in the wall and the presence of a soft, fatty intraluminal thrombus. The reconstructed surfaces were converted into numerical meshes using a grid generation package (Gambit, Ansys Inc, NH, USA). The three-dimensional geometry was finally discretized into smaller triangular finite elements for further CFD analysis or FSI simulations using the software package ADINA (ADINA R&D, Inc., Watertown, MA). The fluid domain was modeled with Newtonian

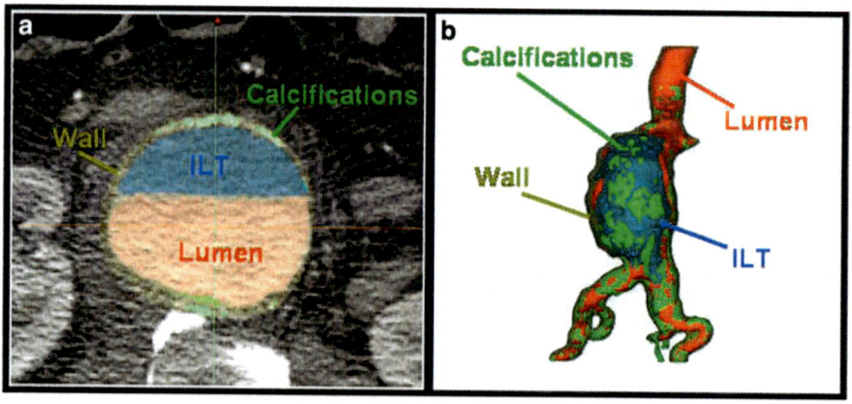

Fig. 2 Three-dimensional image reconstruction from patient CT data is performed using masks for each two-dimensional slice. **a** Masks are shown in a two dimensional CT image slice. The outer wall is shown in *yellow*, ILT in *blue*, calcifications in *green*, and lumen in *red*. **b** Three-dimensional rendering of the pathological ruptured structure is performed with further iterative smoothing techniques for the segmented surfaces

blood properties. The solid domain in our formulation was composed of three components, the vessel wall, ILT, and calcifications. Modeling the solid vessel structure involved mathematical description of the fiber orientation using orthotropic material formulation [34, 70], which was further customized to fit published experimental results in AAA specimens [8, 66, 95]. In the following section, we describe the details of the image reconstruction procedure with some representative ruptured AAA cases.

3 Image Reconstruction of AAA Geometries

The steps that are involved in the reconstruction of patient based AAA models and their adaptation to FSI simulations are presented for the case of model reconstructed from ruptures aneurysms. Several cases of patients who arrived at the Stony Brook university hospital ER with contained ruptured AAA (where the AAA shape was still contained), were studied as described above. Layered masks were automatically segmented in Mimics software with some manual adjustment in cases where ambiguities made it difficult to discern between soft tissue boundary of the ILT and outer wall. Those were carefully reviewed by a physician. Three-dimensional structures were subsequently rendered in Mimics with further iterative smoothing. Masks for the outer wall, intraluminal thrombus, calcifications, lumen, and surrounding bone tissue, allowed for the generation of a cohesive three-dimensional image set which portrayed the pathological details of each particular case in proper anatomical context. After observation of the three-dimensional isosurfaces (Fig. 2), high variability in AAA geometries was apparent

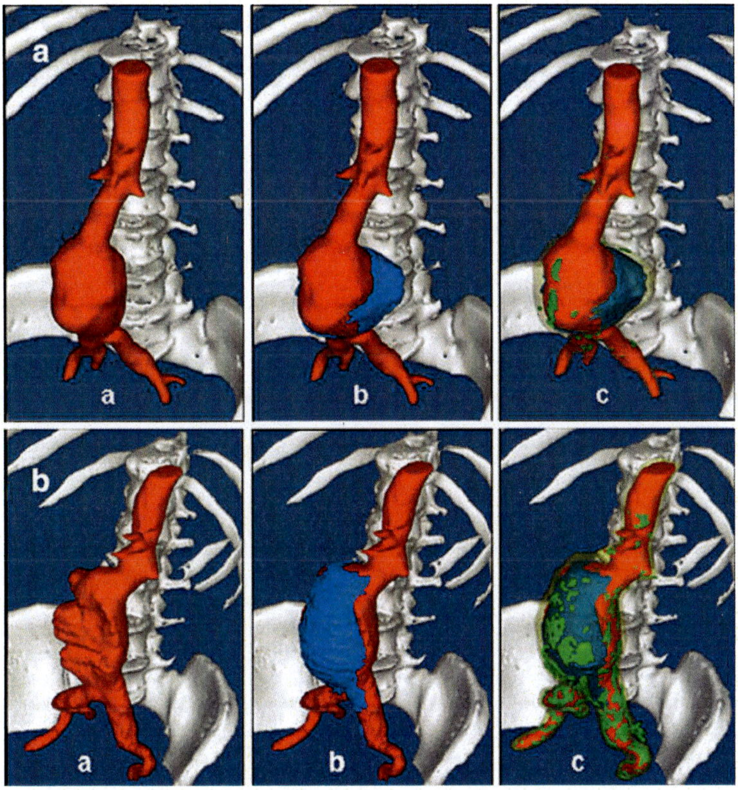

Fig. 3 Structure and components of ruptured AAA (rAAA) used for the numerical simulations, reconstructed from patients CT scans, **a** Rendering of the first rAAA where a lateral rupture occurred at the AAA *bottom*, from left of center above the iliac bifurcation till above the left iliac, **b** Rendering of the second rAAA where a lateral rupture occurred superior to the ILT across the widening inlet of the aneurysmal neck. The composite of the AAA components depicts the (*a*) lumen of the AAA (*red*) relative to the skeletal bone structure, (*b*) intraluminal thrombus added (ILT-*blue*) and (*c*) calcifications (*green*) and the outer wall (*yellow*, transparent)

from these rAAA reconstructed cases, highlighting the need for a patient-specific approach.

Reconstructed geometries of two rAAA cases are depicted in Fig. 3, classified as fusiform aneurysms. The abdominal aorta descends normally slightly to the left of the midline in both cases until approximately *L2* vertebra, after which the aneurysmal bulge can be observed. The aneurysmal rupture line of the first patient (Fig. 3a) was located at the iliac bifurcation superior to the left common iliac artery and inferior to the aneurysmal bulge and ILT. The first patient had several calcifications embedded circumferentially below the ILT at the region of rupture. The second patient (Fig. 3b) exhibited rupture superior to the ILT along the widening neck inlet of the aneurysmal bulge, with a single micro-calcification (Ca) at the region of rupture. Larger calcified plaques were observed in this patient, surrounding the aneurysmal bulge and the iliac arteries. The ILT of the first patient was posterior to the luminal aneurysmal bulge while the ILT of the second patient was anterior to the luminal aneurysmal bulge. Figure 3a and b depict the lumen (red) in relation to the skeletal bone structure. In this figure we also show the ILT (blue) in relation to the lumen, the calcifications (green) with the aneurysmal wall (yellow and transparent) surrounding the various ruptured AAA components.

4 Characterization of the Biomechanical Behavior of the AAA Wall Under Stress

4.1 The Need for Anisotropic Wall Properties in the Prediction of Risk of Rupture in AAAs

In recent years FSI simulations, in which the dynamic interaction between the AAA hemodynamics and wall deformation is modeled, were conducted by our group and others to simulate the biomechanical behavior of the AAA walls [8, 19, 25, 58, 77, 99]. Those were initially based on the isotropic assumption, in which the directional ambiguities associated with abdominal aortic tissue mechanical response to stresses-which may play a major role in the behavior of the tissue under elevated stresses, cannot be resolved. There was a need for modeling these directional ambiguities and the actual stress field on the aneurysmal wall due to the collagen fiber orientation. Several studies to address the role of anisotropy of the aortic wall in pathological conditions were performed in recent years by our group and by others [65, 66, 69, 70, 118]. In this section we describe the isotropic and anisotropic formulations for the wall material models used in the FSI simulations performed in the patient specific AAA geometries described above, as well as the mathematical formulation for solving the blood flow dynamics and the coupling with the aneurysmal wall in patient-based AAAs models reconstructed for the FSI modeling.

4.2 Stress Distribution of the Aortic Wall

The arterial wall is a composite deformable structure that exhibits a non-linear stress–strain relationship with exponential stiffening at high stresses. This stiffening effect results from the embedded mesh of collagen fibers which lead to the characteristic anisotropic mechanical behavior of arteries [8]. Figure 4 depicts typical stiffening effect in the aortic wall for normal aortic wall and pathological aneurysms [93]. Loading beyond the viscoelastic limit (exceeding the normal physiological range), is characteristic of pathologies such as an expanding aneurysm and results in deformation that is associated with inelastic effects.

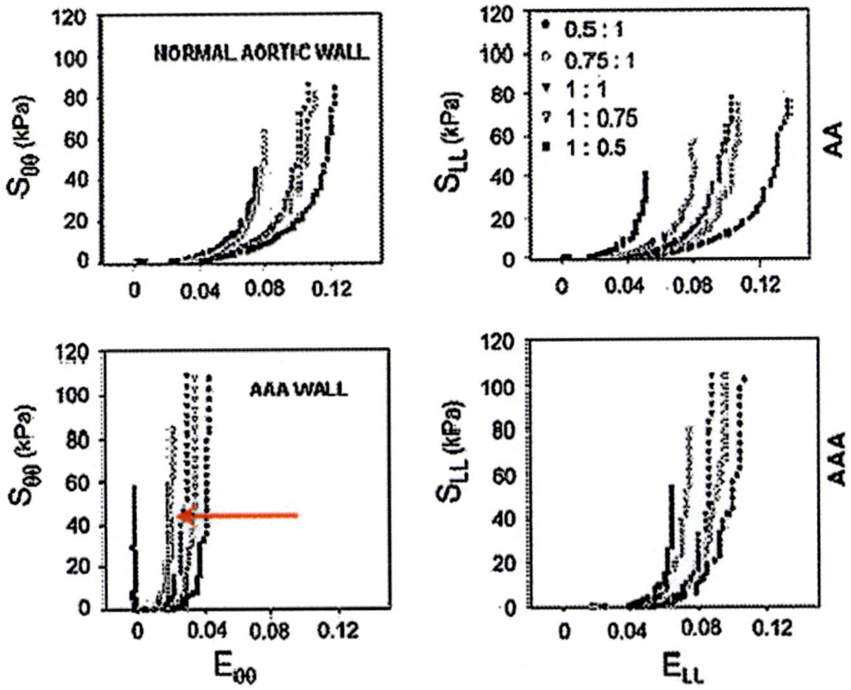

Fig. 4 The stiffening effect of abdominal aortic aneurysm (AAA) wall as compared to healthy abdominal aortic (AA) tissue. Stress–strain curves shift to the left (*bottom*), adapted from [93]

This overstretching leads to dissipation effects where elastoplastic effects may become evident, following a general continuum description of the hyperelastic stress response of the material to the deformation [34]. The strain energy function was used to model the vessel wall as fiber-reinforced composite material with the fibers corresponding to the substructures components of the AAA that influence the vessel remodeling such as the ILT and embedded calcifications (Fig. 3).

4.3 Material Properties of the Aortic Wall

Both isotropic and orthotropic material model formulations were employed in this study in order to compare the predictions of the anisotropic model to the more widely used Mooney Rivlin type of hyperelastic isotropic material model. The orthotropic material formulation models the arterial wall as a composite deformable structure that exhibits non-linear stress–strain relationship with exponential stiffening at high stresses. This stiffening effect results from the embedded mesh of collagen fibers which lead to the characteristic anisotropic mechanical behavior of arteries [27, 34, 68]. A schematic representation of the approximate collagen fiber orientation

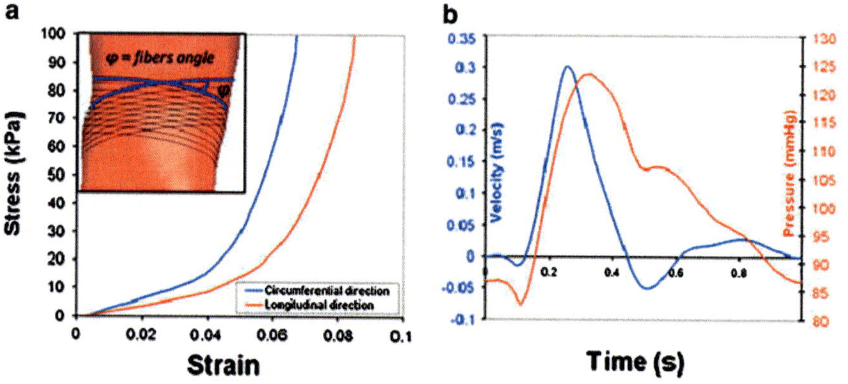

Fig. 5 **a** Anisotropic vessel wall material stress–strain curves, using the orthotropic model formulation (directions of the fibers defined by the angles, a_a and a_b, which are offset from the material axes-inset). A replication of bi-axial testing of AAA arterial wall specimens was conducted numerically in order to validate the model parameters. **b** Boundary conditions for the fluid domain. Velocity and pressure waveforms imposed at the inlet and outlet. Inlet peak systolic flow occurs at 0.25 s and peak systolic pressure at 0.32 s

in the abdominal aortic wall is depicted in Fig. 5. A typical value of the angle between the two families of fibers was extracted from the literature [66, 69, 70].

We employ higher order terms in the isotropic Mooney–Rivlin (M–R) strain energy function [12] to account for the departure from neo-Hookean/Gaussian behavior at large stretches, combined with the Holzapfel orthotropic [34, 70] material formulations. The Holzapfel orthotropic formulation models the vessel wall as a fiber-reinforced composite material, corresponding to the collagenous component of the material. This orthotropic material formulation has been used for modeling the biomechanical behavior of various arterial walls, such as coronary and carotid arteries [34, 38]. Similar models that combine higher order isotropic terms with Fung type exponential terms have been used recently to describe the properties of atherosclerotic coronary arteries [9, 83, 119]. Those are described below.

Higher order terms were used to formulate I_1 and in some cases I_2 to account for the departure from neo-Hookean/Gaussian behavior at large stretches. Such models work well in capturing different deformation states at moderate to large deformations [9]. For the isotropic part, several combinations of higher order terms (including Fung type exponential terms) were used. The family of generalized polynomial strain energy models containing higher order I_1 and I_2 terms offer a better fit to experimental data. These phenomenological models may yield unstable constants, unless Drucker stability is achieved when the tangential stiffness matrix (Hessian) is positive. Drucker stability was achieved for our model (while helping us achieve a much better fit to experimental specimen biaxial stretching data-as shown later on below). The justification for using such formulations follows:

A general Rivlin formulation [67] can be expressed as: $\bar{\Psi}_{iso} = \sum_{i,j=0}^{\infty} C_{ij}(I_1 - 3)^i (I_2 - 3)^j$ where a first term only represents the neo-Hookean behavior. The classic Mooney Rivlin formulation: $\Psi = C_1(I_1 - 3) + C_2(I_1 - 3)^2$ added a second order term. Using the higher order I_1 terms works well for moderate to large deformations. Accordingly, higher order terms were added since in many models, e.g., the *Yeoh model*. Fung type exponential material models either replaced the higher order terms, or added to them. Several investigators have used higher order terms in I_1 and in some cases in I_2 to account for the departure from neo-Hookean/Gaussian behavior at large stretches. It has been proposed to use an increasing order terms- with caution, for capturing better the different deformation states at moderate to large deformations [9]. Few examples follow:

$$\text{TheYeoh model}: \bar{\Psi}_Y = C_1(I_1 - 3) + C_2(I_1 - 3)^2 + C_3(I_1 - 3)^3,$$

or the *Gent model* which is composed of *n + 1* terms of the first invariant:

$$\bar{\Psi}_G = \frac{E}{6}\left[(I_1 - 3) + \frac{1}{2J_M}(I_1 - 3)^2 + \frac{1}{2J_M^2}(I_1 - 3)^3 + \ldots + \frac{1}{(n+1)J_M^n}(I_1 - 3)^{n+1}\right]$$

Other researchers have used an isotropic incompressible Fung type exponential potential energy function [12, 15]: $\bar{\Psi}_F = \frac{a}{b}\left[\exp\left(\frac{b}{2}(I_1 - 3)\right) - 1\right]$

Recently, several researchers employed a mixed formulation (similar to our formulation), e.g., for the isotropic material of carotid arteries:

$$\Psi = C_1(I_1 - 3) + C_2(I_2 - 3) + D_1\left(e^{D_2(I_1 - 3)} - 1\right),$$
$$I_1 = \sum C_{ii}, I_2 = 1/2[I_2^1 - C_{ij}C_{ij}]$$

where I_1 and I_2 are the first and second strain invariants, C is the right Cauchy-Green deformation tensor [85, 119]. One caution regarding the use of phenomenological higher order I_1 models is that the constants chosen should be realistic and create stable constitutive responses in all deformation states. Stability (Drucker stability) is achieved when the tangential stiffness matrix (Hessian):

$$Hi_{jkl} = \frac{\partial^2 W}{\partial \varepsilon_{ij} \partial \varepsilon_{kl}} \text{ is positive.}$$

Such continuum mechanics invariant-based constitutive models were found to be successful in correlating better to experimental results in the specimen. Strain energy expressions which contain a polynomial series including higher order I_1 terms are, in effect, capturing the non-Gaussian nature of the network stretch behavior. Strain energy expressions which contain the second invariant of stretch, I_2, and mixed formulation as above overcome the problem of formulations such as the Mooney–Rivlin model which are found to be overly stiff in certain types of

Table 1 Material properties for the fibrous wall used in the FSI simulations

	Isotropic coefficients				Anisotropic coefficients			
Material coefficients	C_1 (Pa)	C_2 (Pa)	D_1 (Pa)	D_2	k_1 (Pa)	k_2	a_a°	a_b°
Wall	8,888	164,900	48.7	53.46	1,886	94.75	5°	265°
Calcification	92,000	–	36,000	2.0	–	–	–	–

deformation. As described above, caution is advised whenever a phenomenological model is fitted to data because the best fits may be achieved with unstable constants. Precise fits are indeed possible- as demonstrated by our model (below), but such precision comes at the expense of increased model complexity viz the many constants and data sets required.

We indeed expanded the Holzapfel model to include a higher term and exponential term in the isotropic part, in addition to Holzafpel's orthotropic terms:

The isochoric elastic response for our isotropic material model formulation, $\bar{\Psi}_{iso}$, is given in Eq. 1.

$$\bar{\Psi}_{iso} = C_1(I_1 - 3) + C_2(I_1 - 3)^2 + D_1\left(e^{D_2(I_1-3)} - 1\right), \quad I_1 = tr(C_{ij}). \tag{1}$$

Where I_1 is the first invariant of the Cauchy-Green deformation tensor. The isochoric elastic response for the orthotropic material model formulation, $\bar{\Psi}_{aniso}$, is given in Eq. 2.

$$\begin{gathered}\bar{\Psi}_{aniso} = \bar{\Psi}_{iso} + \frac{k_1}{2k_2}\sum_{i=4,6}\left[e^{k_2(J_i-1)^2} - 1\right], \\ J_4 = J^{-1/3}C_{ij}(n_a)_i(n_a)_j, \quad J_6 = J^{-1/3}C_{ij}(n_b)_i(n_b)_j, \quad J = \det(C_{ij}).\end{gathered} \tag{2}$$

Where J, is the third invariant of the Cauchy-Green tensor C_{ij}, n_α and n_b are the directions of the fibers defined by two angles, a_α and a_b, which are offset from the material axes [70]. A displacement-pressure finite element formulation was used where the pressure is not a part of the potential function, $\bar{\Psi}$, and is separately interpolated [83]. The combined strain energy formulation above [66] produced excellent fit to previously published biaxial stretching experimental data with AAA specimen [95]. The parameters that best fit the model are shown in Table 1 ($R^2 = 0.99$). The ILT material was modeled as linear elastic with a Young's modulus of 0.11 MPa and a Poisson ratio of 0.45 [16, 58]. Calcifications (Ca) were assumed to behave as a stiff isotropic material with properties summarized in Table 1 [7, 98].

4.4 Adjustment of the Material Model to Experimental Data

After guaranteeing that the constants chosen are realistic and stable constitutive responses in all deformation states, the Drucker stability was tested by calculating the tangential stiffness matrix (Hessian) and establishing that it is positive. The combined strain energy formulation described above produced excellent fit [66] to previously published AAA arterial wall specimens that were tested by biaxial stretching [95, 100]. The stress–strain relationship in the circumferential and longitudinal directions was matched against the experimental results [66] to produce the stress–strain curves in Fig. 5 that were employed in the FSI simulations.

5 FSI Formulation for AAA Numerical Simulations

5.1 Governing Equations

The fluid domain is governed by the Navier–Stokes and the continuity equations. The Arbitrary Lagrangian–Eulerian (ALE) moving mesh approach was utilized for re-meshing the fluid domain at each time step. The momentum and continuity equations for the fluid domain are shown in Eqs. 3 and 4 for a moving reference frame.

$$\vec{\nabla} \cdot \vec{q} = 0, \quad \text{continuity,} \tag{3}$$

$$\rho\left(\frac{\partial \vec{q}}{\partial t} + \left(\vec{q} - \vec{q}_g\right) \cdot \vec{\nabla}\vec{q}\right) = -\vec{\nabla}p + \mu\vec{\nabla}^2\vec{q} \quad \text{fluid momentum.} \tag{4}$$

Where $\vec{q}$ is the fluid velocity vector and $\vec{q}_g$ is the local coordinate velocity vector, p is the static pressure, ρ the blood density and μ is the dynamic viscosity. The numerical simulations utilized direct coupling between the fluid (blood) and solid (vessel wall) domains. Large strains and large deformations were considered. The dynamics of the flexible wall were calculated using the linear dynamics response of the system shown in Eq. 5.

$$M\ddot{\mathbf{U}} + C\dot{\mathbf{U}} + K\mathbf{U} = \mathbf{R}, \quad \text{solid momentum} \tag{5}$$

M, C and K, represent mass, damping and stiffness matrices respectively. $\mathbf{R}$ is the vector of externally applied loads. $\ddot{\mathbf{U}}$, $\dot{\mathbf{U}}$ and $\mathbf{U}$ are the vectors of acceleration, velocity and displacement of the structural domain. A first order finite-element scheme was used to solve the set of solid motion and fluid flow equations using the commercial software Adina (ADINA, ADINA R&D Inc, Watertown, MA).

5.2 Boundary Conditions

For the fluid domain, the pressure and velocity waveforms were prescribed at the outlet and inlet of the AAA geometry, respectively. At the inlet, the velocity waveform spans the entire cardiac cycle with velocity range from 0.3 m/s to

−0.05 m/s, and outlet pressure boundary condition with range between 123 and 83 mmHg, as shown in Fig. 5b. Blood was modeled as a Newtonian fluid, with a density of 1,050 kg/m3 and a viscosity of 3.5cP [66] and the flow was considered laminar. The flow and pressure waveforms applied were extracted from Olufsen et al. [57]. A no-slip condition was applied at the wall boundaries. Due to the complexity of the geometries, residual stress field in the unloaded configurations was not considered. Instead, all models were assumed to be initially at zero stress state, followed by initial stress loading that was achieved by pressurizing the AAA from 0 to 90 mmHg with zero flow for 1 s, before the FSI waveforms were applied at the inlet and outlet. For the solid domain, all degrees of freedom were fixed at the inlet and outlets.

5.3 FSI Interface Conditions

For the FSI interface the following conditions were applied: (i) displacements of the fluid and solid domain are compatible, (ii) stresses at this boundary are at equilibrium, and (iii) no-slip condition for the fluid. These interface conditions are given in the Eqs. 6–8.

$$\vec{d}_s = \vec{d}_f \tag{6}$$

$$\sigma_s \times \vec{n} = \sigma_f \times \vec{n} \tag{7}$$

$$\vec{q}_s = \vec{q}_f \tag{8}$$

where σ, $\vec{d}$, $\vec{q}$, $\vec{n}$ are the vector of displacement, the stress tensor, the velocity vector and the normal vector on the boundary. The subscripts s, f indicate the solid and the fluid [91].

5.4 Wall Strength and Rupture Potential Index

A statistical model based on non-invasive means to assess the regional in vivo wall strength distribution was employed [96]. It lumps together significant clinical and geometric predictors to yield a local value of the wall strength, Eq. 9:

$$\begin{aligned} Strength =& 71.9 - 37.9\left(ILT^{1/2} - 0.81\right) - 15.6(NORD - 2.46) - 21.3\ HIST \\ &+ 19.3\ SEX. \end{aligned} \tag{9}$$

where *ILT* is the local attached ILT thickness in cm, *NORD* is the local diameter normalized to the diameter of non-aneurysmal aorta (infrarenal) estimated from the patient's age and sex [63]. *HIST* is the family history (½ with history, −½ without history) and *SEX* is patient's gender (½ male, −½ female). More details about the model can be found elsewhere [96].

This model is used to generate 3D mapping of the wall strength that is then compared to the local stress distribution, to generate a map of a rupture potential index (RPI), defined as the ratio between the locally acting wall stress (calculated by FSI) and the local wall strength:

$$RPI = \frac{Local\ wall\ Stress\ obtained\ from\ FSI\ [N/cm^2]}{Local\ wall\ Strength\ [N/cm^2]}. \tag{10}$$

.

Each global predictor variable, e.g., family history and AAA size, was obtained from each patient's hospital chart, while spatially varying predictor variables, local diameter and ILT thickness, were measured from CT images.

5.5 Blood Flow Velocity Measurements within the AAA

Our current studies were retrospective for the pathological cases, meaning that the patient specific boundary conditions at the time of CT scan were not known. The velocity at the inlet of the aneurysm and pressure at the outlet, and the corresponding specific waveforms for AAA were taken from the literature [57].

6 FSI Simulations of AAA

The main goal of the FSI methodology is to employ patient specific AAA geometries, including its pathological components and hemodynamic conditions, and incorporate in the modeling biomechanical considerations in order to augment and refine AAA diagnostics as to be able to predict the AAA risk of rupture. We present herein a series of numerical studies that elucidate the effect of various pathological parameters like the AAA geometry, the material properties of the aortic wall, the ILT, the calcifications (Ca). In addition, we present parametric studies performed in idealized streamline AAA model intended to examine the effects that the iliac arteries bifurcation angle, the neck angulation, and hypertension and may have on aneurysmal disease and its progression.

6.1 Role of ILT

Patients, with different AAA geometries and ILT distribution (Fig. 6), were studied and compared in detail. The complete geometry included both the vessel wall and the intraluminal thrombus. The ILT in the first AAA occupied a significant part of the aneurysm volume. The second AAA ILT mostly occupied the

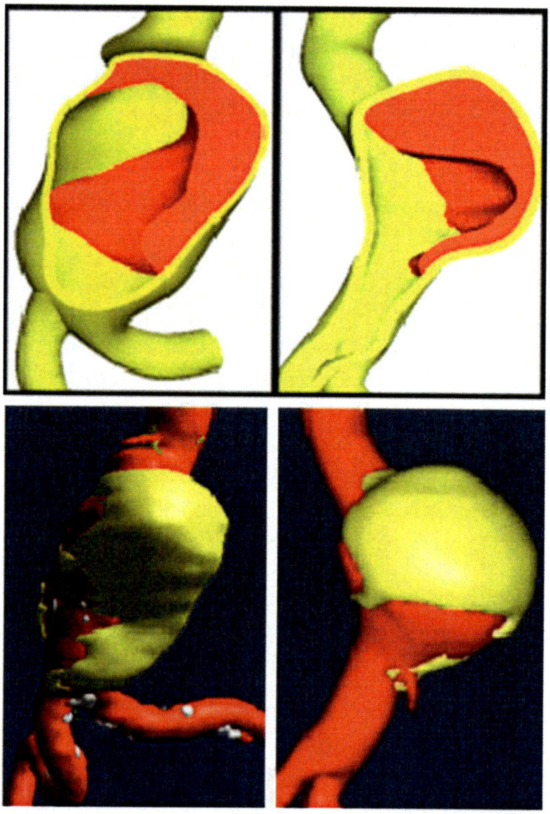

Fig. 6 Two patient-specific 3D reconstructed CT images of abdominal aortic aneurysms (AAA) (*top*) with ILT indicated in red (*bottom*)

upper half of the aneurysm. In order to address the question whether the presence of an ILT serves to protect the aneurysm against rupture, both patients' AAAs were studied with and without ILT. Blood particulates flow pathlines and wall shear stresses (WSS) in one of the AAA are depicted (Fig. 7). In both AAAs a large amount of particles were recirculating along complex flow trajectories with a long residence time within the AAA sac. The deposition patterns resulting from such flow recirculation regions offer a viable mechanism for the formation and growth of the observed ILT in both AAAs [8]. The wall shear stress contours (Fig. 8, right), depicted lower values within the AAA sac, and higher values on the wall opposing it. Von Mises stresses within the vessel walls, with and without ILT, are depicted in Fig. 8 for the first AAA. The peak stress was 87 N/cm^2 without ILT and 58 N/cm^2 with ILT. For the second AAA, peak wall stress reduced from 41 N/cm^2 without ILT, to 31 N/cm^2 with ILT, indicating that the ILT does offer some reduction in peak stress, and should be taken into account when surgical intervention is decided [8]. The proposed methodology clearly indicates that risk of rupture is dictated by the interaction between hemodynamic forces and the wall stresses, and that aneurysm size is too crude of an indicator for an informed decision regarding the need for surgical intervention. Further, it points to the

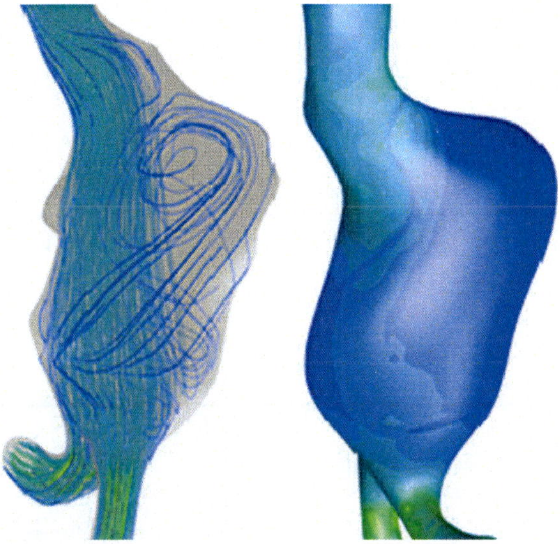

Fig. 7 Complex flow patterns within AAA without ILT (*left*), and the resulting flow induced wall shear stresses (*right*)

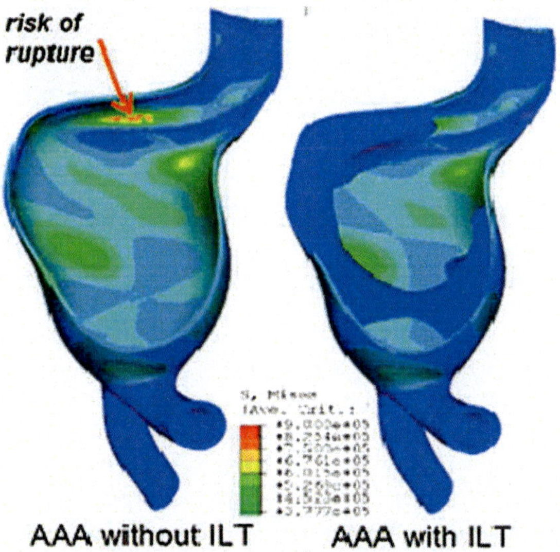

Fig. 8 Higher von Mises stress concentration within AAA wall in the absence of ILT, indicating potential rupture

locations were rupture may be expected to occur as depicted in Fig. 8 for the AAA without ILT, giving clinicians a refined clinical diagnostic tool.

6.2 Anisotropic Versus Isotropic Wall Properties Modeling

In order to accurately model the AAA wall stresses, patient specific AAA geometries were studied employing a fully coupled FSI approach using the anisotropic material model described earlier. This model was matched to experimental measurements of

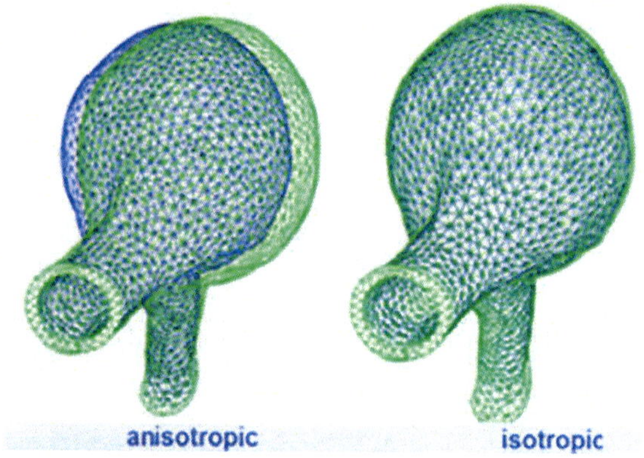

Fig. 9 Marked differences in wall motion between anisotropic and isotropic material models for a saccular aneurysm

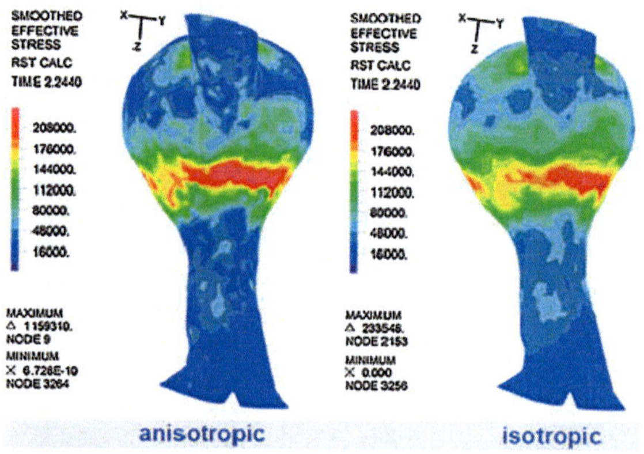

Fig. 10 Stress distributions for isotropic and anisotropic models (with ILT) during peak systole on the aneurysmal wall showing higher stresses for the anisotropic model

AAA wall specimens [95] and incorporates directional stiffness due to fiber orientation within the arterial wall [33, 34]. The anisotropic model was compared to a Mooney–Rivlin type of isotropic hyperelastic model [66]. Comparing the results of the FSI simulations demonstrate a marked difference in the wall motion between the two models, with the anisotropic one exhibiting larger rotation and deformation (Fig. 9). For this specific patient based geometry the maximal wall stresses were observed in the areas where the ILT was not present. It predicted higher wall stresses indicating that the use of an isotropic material model may underestimate the AAA risk of rupture (Fig. 10). These marked differences are also shown in Table 2 for the case of fusiform aneurysms and normal subjects.

6.3 Blood Flow Through Normal and Aneurysmal Aortas

Velocity vector fields for all cases are presented in Fig. 11a, 0.15 s after peak systole of the cardiac cycle. This specific time was chosen because distinct flow patterns that are relevant to aneurysmal disease are prominent at this phase of the cardiac cycle.

Table 2 Summary of the von Mises and principal axis p_1, stresses for the normal and pathological aortas

Subject	Max diameter (cm)	Peak wall stress (isotropic, von Mises, MPa)	Peak wall stress (anisotropic, p_1 stress, MPa)	RPI
N1	1.63	0.35	0.37	–
N2	1.80	0.44	0.54	–
n-rAAA	4.60	0.50 wILT	0.65 wILT	–
		–wILT & 1 Ca	0.70 wILT & 1 Ca	–
rAAA 1	10.6	1.09 wILT	1.37 wILT	0.8
		1.23 wILT & 3 Ca	1.54 wILT & 3 Ca	–
rAAA 2	7.8	2.10 wILT	2.17 wILT	0.95
		2.16 wILT & plaque Ca	2.18 wILT & plaque Ca	–

During peak systole, flow patterns in normal aortas show similar characteristic velocity profiles, with peak values reaching $v_{N1} = 0.36$ m/s and $v_{N2} = 0.37$ m/s, respectively in the abdominal aorta region. At end diastole, peak velocities decrease to $v_{N1} = 0.03$ m/s and $v_{N2} = 0.023$ m/s. The FSI simulation of normal subject 1, a 19 y/o female, does not show any visible recirculation zones during diastole. On the other hand for normal subject 2, a 66 y/o female, a distinct recirculation zone appears in the abdominal aorta during diastole, due to slight widening of the lumen superior to the iliac bifurcation as compared to the vessel inlet [118].

For all aneurysmal cases the flow patterns during peak systole change dramatically as compared to healthy aortas. The abrupt expansion from the aneurysm neck to the dilated AAA lumen, typical of AAA disease induces a rapid decrease in the velocities- coupled with the formation of complex flow patterns. Peak velocities for the rAAA simulations reduce to $v_{rAAA1} = 0.04$ m/s in the aneurysmal area (0.36 m/s at the neck of the aneurysm) and $v_{rAAA2} = 0.07$ m/s in the aneurysmal area (0.34 m/s at the neck of the aneurysm). During diastole, large, diffuse, recirculation zones are formed in both rAAAs models. These large recirculation zones span almost the entire diameter of the aneurysmal bulge. In both pathological cases, smaller recirculation zones can be observed in the flow field, occurring either at the inlet of the aneurysm or close to the stagnation area of the iliac bifurcation during diastole, producing highly complex flow patterns. Maximum velocity during diastole is $v_{rAAA1} = 0.015$ m/s and $v_{rAAA2} = 0.023$ m/s, respectively.

The abrupt changes in the velocity field within the aneurysm and the apparent changes in the flow patterns result also in significant changes in the luminal pressure distribution as compared to the normal aortas at peak systole. At peak systole the typical blood pressure acting on the wall at the normal aortas 1 cm above the iliacs was 120.4 mmHg, where in the pathological cases it was, 122.5 mmHg for rAAA1 (2.1 mmHg higher) and 122.8 mmHg for rAAA2 (2.4 mmHg higher). The blood pressure during diastole was also higher in the pathological cases as compared to the normal cases. For instance at 60% of the cardiac cycle (early diastole) the pressure in the normal aorta 1 cm above the iliacs was 104.5 mmHg whereas the pressure at the same location was 104.9 mmHg (0.4 mmHg higher) for both ruptured AAAs.

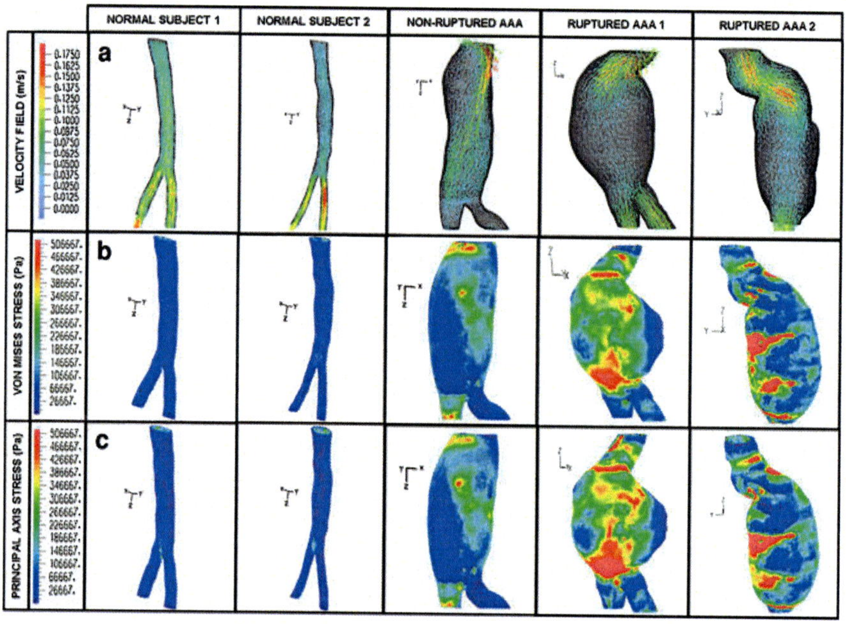

Fig. 11 **a** Coronal cross section view of velocity vector fields 0.15 s after peak systole depicting aneurysmal disease characterization. Marked differences are evident in the flow patterns and the velocities of normal to pathological aortas **b** Von Mises stress in the aortic wall for the isotropic material model formulation and **c** Principal axis stress, p_1 in the aortic wall for the anisotropic material model formulation

The pressure in the ruptured AAAs also remained almost constant throughout the aneurysm and drops abruptly in the iliacs, in contrast to the normal cases where the pressure was dropping linearly downstream. This indicates that in AAAs there is an increase in the pressure acting on the aortic wall that is not compensated by the expected downstream pressure drop.

WSS distributions during peak systole indicate that rAAA have significantly lower overall WSS (peak values $WSS_{rAAA1} = 0.3$ Pa, $WSS_{rAAA2} = 0.27$ Pa) as compared to normal aortas (peak values $WSS_{N1} = 1.87$ Pa, $WSS_{N2} = 2.4$ Pa), at peak systole. However, there are small local regions in rAAAs close to the aneurysmal necks and superior to the ILT locations where the WSS levels remain high throughout the cardiac cycle [118].

6.4 Wall Stress Distributions within the Abdominal Aortic Wall in Ruptured AAA

Results of FSI simulations during peak systole with isotropic and anisotropic wall formulations are presented in Table 2 and Fig. 11b and c, with von-Mises stresses reported for the isotropic simulations and principal axis (p_1) wall stresses reported for the anisotropic simulations, respectively. There is a general trend of rising wall stresses from normal aortas to non-ruptured and ruptured AAAs. Isotropic simulations for normal aortas reach a maximum peak systolic von Mises stress of 0.35 and 0.44 MPa, respectively. Ruptured AAA cases produce much higher wall stresses, with maximum von Mises stresses reaching $\sigma_{vM\ rAAA1} = 1.09$ MPa and $\sigma_{vM\ rAAA2} = 2.1$ MPa during peak systole, respectively. The FSI anisotropic material simulations (Fig. 11c) follow the same trend with principal axis stress, p_1 significantly increasing from normal to pathological cases. Normal subject models predict peak values of $p_{1\ N1} = 0.37$ MPa and $p_{1\ N2} = 0.54$ MPa, respectively. Ruptured AAA cases reach peak values of $p_{1\ rAAA1} = 1.37$ MPa and $p_{1\ rAAA2} = 2.17$ MPa, respectively. It should be noted that the model of the smaller ruptured aneurysm (rAAA2 = 7.8 cm) exhibited much higher peak stress (58% higher) than the larger one (rAAA1 = 10.6 cm).

All anisotropic simulations showed higher peak wall stresses as compared to the isotropic material model formulation. Specifically, in the normal aortas there was a stress increase of 6% for the first case and 22.7% for the second case when applying the anisotropic material model formulation. For the pathological cases this increase was 25.7% for the first rAAA and 3.3% for the second rAAA. On average, there was a 378% increase in peak wall stress when comparing the first normal case to the two ruptured AAA cases (for the anisotropic material formulation). Both von Mises distribution for the isotropic material formulation and principal axis stresses distribution for the anisotropic material formulation became highly non-uniform as the vessel wall progressed from a normal to an aneurysmal pathology.

6.5 Prediction of the Rupture Location-Studies in Patients with Ruptured Aneurysms

The FSI simulations performed in both patients with ruptured AAA indicated highest stresses along the actual rupture line (Fig. 12, first rAAA), with excellent agreement between the two. In these simulations the highest stresses occurred on the anterior side of the AAA (opposite to the ILT). For the first patient with ruptured AAA configuration (first rAAA), the ILT seemed to offer a significant protective effect by reducing the aneurysm (Fig. 12, inset). The calcifications (green) embedded along the rupture line create stress concentration (unlike the bigger calcifications that appear deposited (not embedded) on the wall). While indicating that the neck area could have been a potential location for a secondary rupture, the peak stress values (1.37 MPa for the anisotropic formulation, occurring at peak systole at 305 ms from the start of the cardiac cycle) were prominent in the rupture area. For the second patient with ruptured AAA (second rAAA 2, Fig. 11) the FSI simulations predicted multiple locations of stress concentrations. One of the locations of stress concentration close to the neck of the aneurysm coincided with the rupture line and had the highest values (2.17 MPa for the

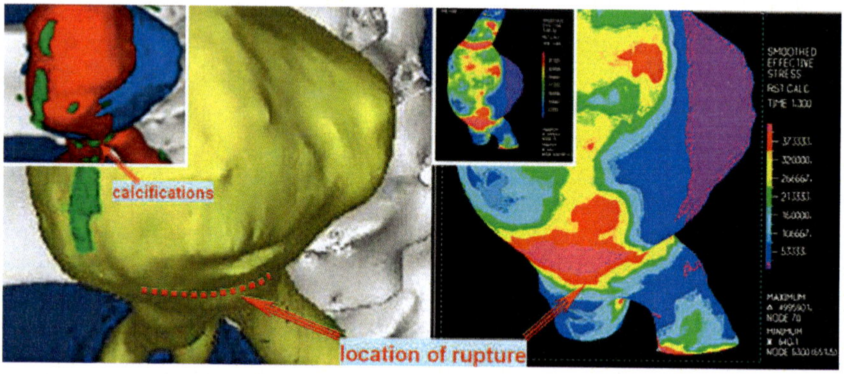

Fig. 12 Location of the maximal wall stresses overlaps the actual rupture region. The inner top panel (*left*) shows the lumen with calcifications above the bifurcation and the left iliac branch. The actual rupture line (*dashed*) is superimposed on the wall surrounding it (*yellow*). The right panel shows the wall stress contours, with stress concentration along and around the rupture line

anisotropic formulation, occurring at peak systole at 305 ms from the start of the cardiac cycle). The other stress concentration locations were predicted on the anterior side of the AAA (opposite to the ILT) indicating that the ILT seemed to offer a protective effect for this patient.

6.6 Role of Calcifications in AAA

Calcification within the aneurysmal wall may have a deleterious effect on the wall integrity and may further increase the risk the AAA risk of rupture. We have incorporated wall calcifications obtained from patient specific scans in our models [65]. To study the effect of calcifications in AAAs we divided the calcified spots into two broad categories. The first type of calcifications is defined as small calcified spots (small-Ca)—such as those appearing near the location of rupture, with a maximum diameter smaller than 4.0 mm. The second type of calcifications is the large plaque-like calcified spots. FSI simulations were performed for the non-ruptured and ruptured AAA with and without the presence of small or plaque-like Ca for comparison purposes. In the non-ruptured AAA, a small-Ca embedded in the wall, distal to the ILT location, increased the local stress within the wall by 36.8% as compared to the FSI simulation without the small-Ca for the anisotropic material wall model formulation. In the case of the first rAAA, a series of three small-Ca in close proximity to the rupture location (based on the CT images) also increased the peak stresses by 12.8% as compared to the FSI simulation without the small-Ca for the isotropic material model wall formulation and by 12.4% as compared to the FSI simulation without the small-Ca for the anisotropic formulation, respectivley. An additional FSI simulation performed for a larger, plaque-like Ca, reconstructed from CT data for the second rAAA, indicated a reduction of

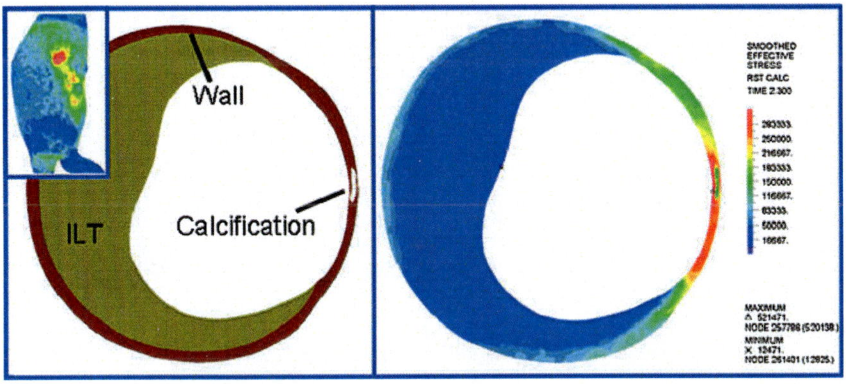

Fig. 13 AAA cross-section showing the wall, ILT, and wall calcification (*left*). Wall stresses extracted from the FSI patient-based simulation shows high stress concentration propagating around the calcification

the local wall stresses near the calcification by 21% as compared to the FSI simulation without the plaque-like Ca (for both material model formulations). Results for the effect of calcified spots in rAAAs at the location of rupture are summarized in Table 2.

Figure 13 depicts a cross-sectional area of the patient-specific AAA (shown in the detail on the top left corner). The AAA modeled is composed of the thin tissue wall, (red) ILT (olive), and the calcification (embedded in the wall—white). The wall stresses extracted from the FSI patient-specific simulation are presented (right). The lowest stress levels are found within the ILT. There is a high stress concentration region propagating around the calcified spot, corresponding to the highest wall stress level found in the AAA wall. Comparing the peak stress in the case with the calcification inclusion (0.65 MPa) to a simulation in which the calcification was removed (0.5 MPa) presented a 30% increase in the peak stress, illustrating the role that calcifications play in weakening the wall and increasing its risk of rupture. This strongly suggests that exclusion of calcifications may underestimate the risk of rupture.

6.7 Wall Strength and RPI

The wall strength estimation for the ruptured AAA and the resulting rupture potential index (RPI—the ratio between the locally acting wall stress from the FSI simulations and the local wall strength) for these cases are shown in Fig. 14. The RPI calculations (Eq. 10) are based on stress distribution derived from FSI simulations without the effect of the calcifications. The range of strength (Eq. 9) for the first rAAA was from 50 to 109.2 N/cm^2, and for the sec rAAA 2 from 68.7 to 131.6 N/cm^2. The mapping of the RPI indicated several distinct regions of high RPI, with at least one coinciding with the actual location of rupture. The RPI for both rAAA cases was close to unity, indicating a very high risk of rupture for both

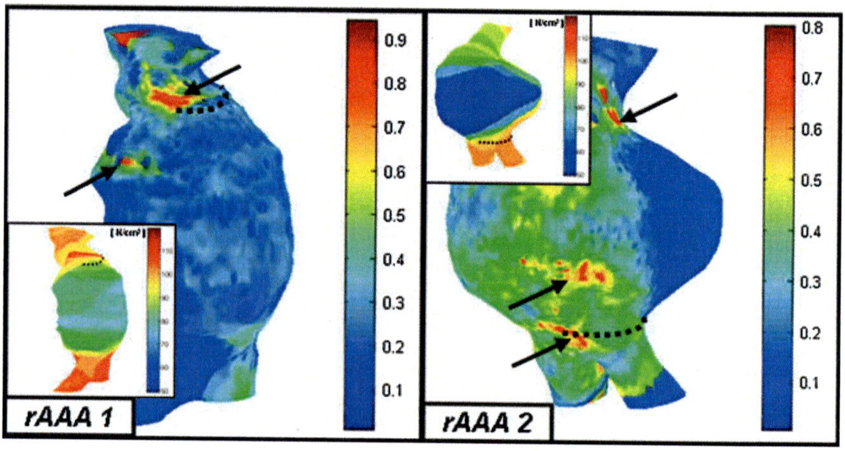

Fig. 14 Wall strength mapping and the corresponding rupture potential index (RPI)—representing the local wall stress to wall strength ratio, for the ruptured AAA cases. Multiple distinct regions of high RPI are observed, with that the highest coinciding with the actual location of rupture. The RPI for both cases is close to unity indicating a very high risk of rupture, with a higher RPI (0.95) for the smaller rAAA

rAAAs (Table 2, $RPI_{rAAA1} = 0.8$, $RPI_{rAAA2} = 0.95$). Similar to the results reported above for the peak wall stresses, it is remarkable that the smaller ruptured aneurysm (second rAAA) reached much higher peak RPI values [118].

6.8 *Parametric Studies in Idealized AAA Geometries-Effects of AAA Neck and Iliac Arteries Angulation*

In these parametric studies, the effects of AAA iliac bifurcation and neck angles on stresses developing within the AAA wall was studied using streamlined ideal AAA geometries with 5 cm diameter aneurysms (Fig. 15). The iliac bifurcation angles ranged between $\varphi = 30°$ and 150° and AAA length, l, was 12.5 cm from the neck to the base of the bifurcation. The vessel wall was composed from a single material with uniform thickness of 2 mm. In addition, the effect of the AAA neck angulation was considered (ranging from $\theta = 0°$, to 40°).

Typical pressure and velocity waveforms were prescribed at the outlet and inlet of the AAA geometry, respectively, covering the complete cardiac cycle (velocity ranging from 0.3 to −0.05 m/s and outlet pressure boundary condition ranging between 123 and 83 mmHg). Blood was modeled as Newtonian fluid, with a density of 1035 kg/m^3 and a viscosity of 3.5 cP [66]. All AAA models were assumed to be initially at zero stress state. FSI simulations were performed for nine iliac bifurcation angles (30°–150°). While the peak stresses at 50°, 100° and 120° bifurcation angles decreased (Fig. 16), the trend of increasing peak stress with

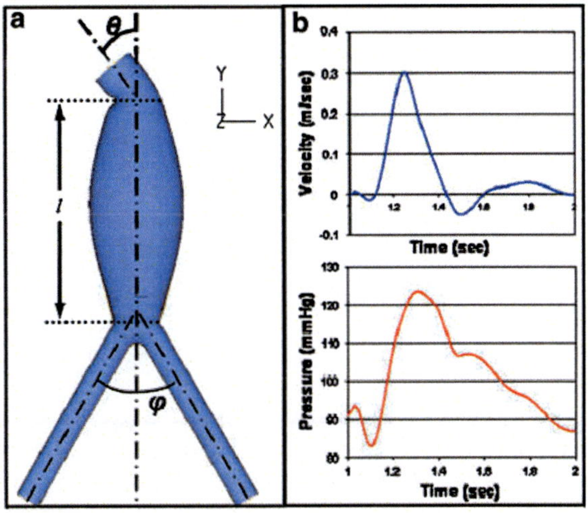

Fig. 15 **a** Ideal AAA geometry and the angles for the iliac bifurcation, φ, and the neck of the aorta, θ. **b** Boundary velocity waveform used at the inlet of the AAA and pressure waveform at the outlet of the iliacs

increasing angulation remained. The mean wall stress in the whole domain showed a different behavior, with an overall decrease of the mean with increasing iliac angle. The complex relationship between iliac angulation, the peak stress (as well as its location-not shown), and the mean stress values, results from the complex fluid structure interaction where different hemodynamic patterns are generated within the AAA lumen because of the varying angulation [116]. It strongly suggests that iliac angulation should be considered when diagnosing an AAA patient condition. This may also offer a glimpse into possible mechanisms of adaptation response of the abdominal aorta with aneurysmal disease progression.

The contribution of AAA neck angulation (at the AAA inlet) on the wall stresses was studied (Fig. 16). The FSI simulations predict a general trend of increasing peak wall stress with increasing neck angulation, indicating an increase in rupture risk. Analysis was performed for the peak von Mises stress and the mean stress for the whole wall domain. At zero inlet angle the peak stress was 0.57 MPa. As the neck angle increased the peak von Mises stress increased, with a maximum of 0.93 MPa at 20°. After that, a slight decrease of the peak stress was observed (0.81 MPa at 40°). Conversely, a substantial decrease of the mean stress was observed as the inlet angle increased (9.5% for the case of 40° inlet angulation in comparison with 0° inlet angle) [116].

6.9 Parametric Studies in Idealized AAA Geometries-Velocity Field

Analysis on the velocity field for two iliac angles (60° and 120°) shows that the maximum velocity magnitude for the 120° angle geometry was 3% lower than the maximum velocity magnitude for the angulation of 60°. As the cardiac cycle

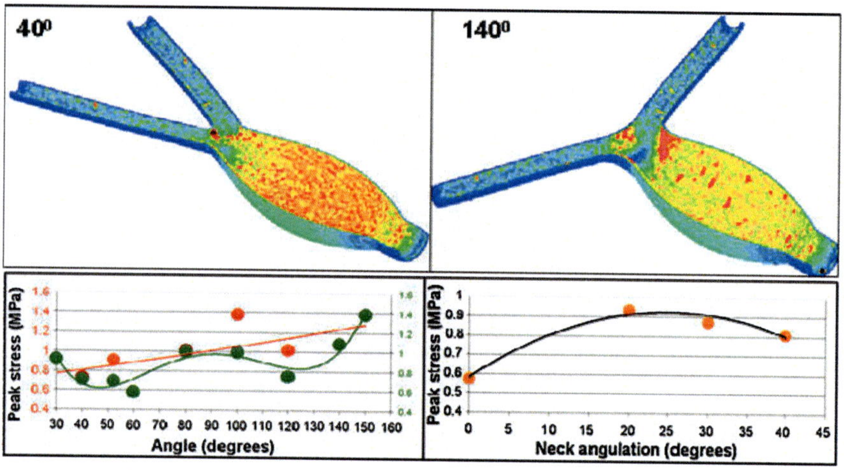

Fig. 16 Wall stress distribution within the AAA wall and iliacs for 2 representative iliac angles (*top*). The peak stress versus the iliac angle for the whole wall domain, red, and for the aneurysmal area only, green, (*left*). Peak von Mises stress versus inlet angle of the AAA. A substantial increase of the peak stress occurs for 20° with a decrease to slightly smaller peak stresses for larger angles (*right*)

progressed, the flow field developed during diastole into several recirculation zones. At early diastole two counter rotating vortices were observed near the inlet influencing the inlet jet of the blood entering the aneurysm. These vortices were retained throughout the diastolic period but lost energy towards the early systole, with the forward flow restored. As the cardiac cycle was progressing, this flow reversal was reduced and became almost zero during late diastole. This phenomenon was more pronounced (by 19.5%) for smaller iliac angles (e.g., 60°), as compared to the larger ones (e.g., 120°). The velocity field is presented in four different time frames of the cardiac cycle for the two iliac angles (60° and 120°, Fig. 17a and b). The four time frames are: (1) after the peak systole, 0.33 s, (2) early diastole, 0.55 s, (3) mid diastole, 0.78 s and (4) late diastole, 0.9 s. The velocity fields were similar for both angles at peak systole.

Analysis on the velocity field for neck angulation studies it appears that a breakdown from symmetric vortices at the AAA inlet into a single larger vortex significantly increases the wall stress. The velocity field is presented at four different time frames of the cardiac cycle for inlet angle of 40° (Fig. 17c). The flow field for 40° inlet angle was substantially different from that of 0° (see Fig. 17a and b). During systole a stronger jet was observed entering the structure. The maximum velocity magnitude difference between 0° and 40° inlet angles was of the order of 7%. A small recirculation zone was formed near the inlet. During diastole this recirculation zone became more pronounced and dominant. During early diastole the vortex near the inlet became broader and influenced the inlet jet of blood entering the aneurysm, as compared with the case of zero inlet angle where two counter rotating vortices were observed. The vortex was retained during

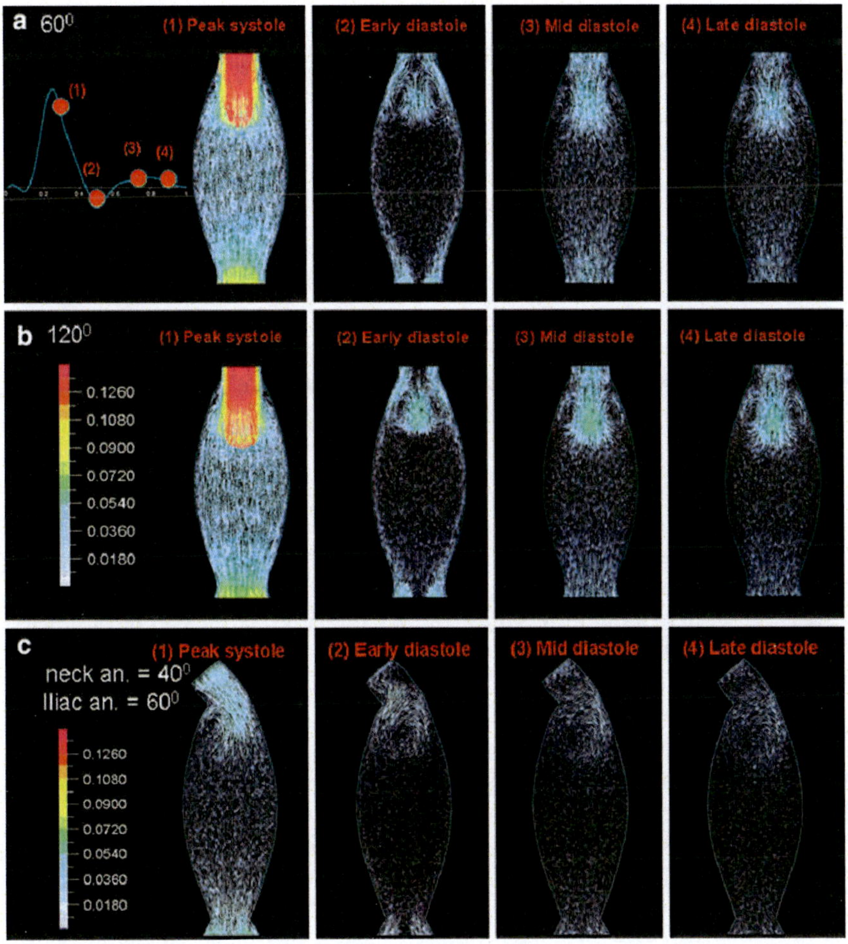

Fig. 17 **a, b** Velocity field in the aneurysmal area for two different angulations and at four different time frames of the cardiac cycle. The four locations for each iliac angle are marked on the inflow velocity during the cardiac cycle. **c** Velocity field in the aneurysmal area for 40° inlet angulation in comparison to the flow field of (**a**) and (**b**) at four different time frames of the cardiac cycle

diastole but lost its energy towards early systole, with forward flow restored thus reducing the influence of the vortex.

6.10 Effects of Blood Pressure on AAA Dynamics

Hypertension plays a very important role in AAA formation, aneurysmal disease progression and increased risk of rupture. Accordingly, physicians try to reduce systolic/diastolic pressure in patients with AAA. Several studies point to the

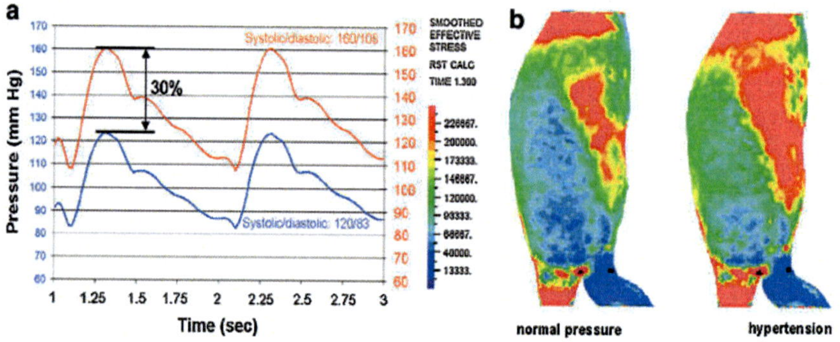

Fig. 18 The role of hypertension in a fusiform aneurysm. **a** 30% increase over normal BP to emulate hypertension. **b** 40% increase of wall stress was observed for the hypertensive case

significant role hypertension has on the rate of growth and the increased risk of rupture [28, 81, 97]. We have studied the effect of hypertension on AAA wall by performing preliminary FSI simulations that demonstrate the sensitivity of our methodology to important clinical parameters that may play a part in aneurysmal disease. The geometry used was a patient based fusiform aneurysm reconstructed from CT data using MMS image reconstruction software (Medical Metrx Solutions, Inc.). The pressure was increased by 30% from normal blood pressure (120 mmHg systolic over 83 mmHg diastolic) to emulate hypertension with a systolic pressure of 160 mmHg and diastolic of 108 mmHg (Fig. 18). FSI simulation was conducted over two cardiac cycles. The simulation for the hypertensive case showed a drastic increase in the wall stresses, predicting a peak stresses on the wall of more than 1.0 MPa, as compared to 0.6 MPa for normal blood pressure (40% increase in peak wall stress). Higher wall stresses regions (in red, Fig. 18, right) appeared larger, especially at the center of the aneurysmal area and along the left iliac branch. These simulations clearly indicate that blood pressure readings are needed for more accurate prediction of a specific patient condition.

7 Summary and Conclusions

The work described in this chapter establishes a patient specific diagnostic methodology that integrates medical imaging with FSI simulations, aimed to describe aneurysmal disease progression from a biomechanical point of view and to refine diagnostic capabilities for predicting AAA risk of rupture. Cases of contained ruptured aneurysms were specifically studied in order to test the ability of the methodology to predict the rupture location. The effect of various pathological parameters like the AAA geometry, the material properties of the aortic wall, the ILT, the Ca, the iliac bifurcation angle, hypertension and other pertaining clinical parameters are discussed accordingly.

The goal of the methodology is to complement the current clinical practice by predicting more accurately the risk of rupture in AAAs, augmenting traditional AAA diagnostics with a refined quantitative and more reliable rupture risk indicators. As a proof of concept, we have demonstrated that our methodology is capable of predicting the location of rupture in several cases of ruptured AAA that were studied. FSI simulations with advanced material models showed an increasing peak wall stress pattern at progressing stages of the disease, demonstrating the capability of the methodology to characterize aneurysmal disease. The anisotropic wall simulations showed higher peak wall stresses as compared to isotropic material models, clearly indicating that the latter may underestimate the AAA risk of rupture. The ILT appeared to provide a cushioning effect reducing the stresses, while small calcifications appeared to weaken the wall and contribute to the rupture risk. FSI simulations with reconstructed rAAA demonstrated that the location of the maximal wall stresses overlaps the actual rupture region. The estimated wall strength was reduced at distinct regions of the AAA, and the maximum RPI also overlapped the location of the actual rupture.

Parametric studies of the Iliac bifurcation angle on the pertinent factors of aneurysmal disease showed that peak von Mises stress within the AAA indicates a general trend of increasing stress as the iliac angle increased. A closer examination, by excluding stresses at the iliac bifurcation stagnation point region where rupture is unlikely to occur, indicates that there are two local minima at 60° and 120° iliac bifurcation angles. Anisotropic material model formulation predicted higher principal axis stresses than the isotropic formulation. The strong correlation between angle and stress is also observed in the anisotropic formulation. Parametric studies of the inlet aneurysmal neck angle showed that peak wall stress increased as the inlet angle increased. However, there was a decrease of the mean stress as the inlet angle increased. The FSI simulations suggest that the adaptation response of the body during aneurysmal disease progression may lead to preferred Iliac bifurcation and inlet neck angles that reduce the peak wall stresses in the interim.

These results clearly indicate that the methodology can predict the potential location of the rupture, and that when specific geometric and biomechanical characteristics of the AAA are incorporated it can indicate the rupture potential of smaller aneurysms that would otherwise not be considered prone to rupture. The methodology will be readily applicable for analyzing and diagnosing other cardiovascular pathologies in which rupture may be imminent, such as intracranial aneurysms and vulnerable plaques in coronary and carotid artery disease. We plan to validate the proposed methodology by establishing its statistical significance with a large enough number of AAA cases. Once validated the clinical endpoint could be improved surgical outcomes, reduced mortality rates, and considerable savings in healthcare costs.

Aneurysmal disease and its progression is a very complex multifactorial process. The models intentionally adopt a reductionist approach to make the computations practical and feasible. However, the FSI methodology is able to offer a pretty faithful representation of the aneurysmal wall behavior. We base this assessment on our success with the retrospective analysis of wall stress distributions in the patient-specific reconstructed ruptured AAAs, where we were able to accurately predict the rupture location. The strength of our approach stems from the fact that we extract all the information that may be needed to predict AAA risk of rupture from data that can be obtained non-invasively, augmenting advanced clinical modalities with numerical modeling.

We envision development of a patient specific diagnostic system that integrates medical imaging with biomechanical simulations, to accurately predict the risk of rupture in AAA and evaluate therapeutic treatments and improve surgical outcomes. Such diagnostic capabilities are essential for procedures in which intervention is directed at those lesions most likely to become symptomatic. It will provide clinicians and surgeons with a refined diagnostic and decision toolkit to determine the need for a surgical intervention. This may have a major impact on the clinical treatment of patients with aneurysmal cardiovascular diseases. By refining the indications for elective surgery and eliminating unnecessary or low benefit surgery, the clinical endpoint will be improved surgical outcomes, reduced mortality rates, and considerable savings in healthcare costs.

References

1. Adolph, R., Vorp, D.A., Steed, D.L., Webster, M.W., Kameneva, M.V., Watkins, S.C.: Cellular content and permeability of intraluminal thrombus in abdominal aortic aneurysm. J. Vasc. Surg. **25**(5), 916–926 (1997)
2. Antiga, L., Ene-Iordache, B., Caverni, L., Cornalba, G.P., Remuzzi, A.: Geometric reconstruction for computational mesh generation of arterial bifurcations from CT angiography. Comput. Med. Imag. Graph **26**(4), 227–235 (2002)
3. Antiga, L., Ene-Iordache, B., Remuzzi, A.: Computational geometry for patient-specific reconstruction and meshing of blood vessels from MR and CT angiography. IEEE. Trans. Med. Imaging **22**(5), 674–684 (2003)
4. Antiga, L., Ene-Iordache, B., Remuzzi, G., Remuzzi, A.: Automatic generation of glomerular capillary topological organization. Microvasc. Res. **62**(3), 346–354 (2001)
5. Bengtsson, H., Bergqvist, D.: Ruptured abdominal aortic aneurysm: a population-based study. J. Vasc. Surg. **18**(1), 74–80 (1993)
6. Bettermann, K., Toole, J.: Diagnostic evaluation and medical management of patients with ischemic cerebrovascular disease. Vascular surgery. In: Rutherford, R. (ed.) pp. 1904–1905. Elsevier, Philadelphia (2005)
7. Bluestein, D., Alemu. Y., Avrahami, I., Gharib, M., Dumont, K., Ricotta, J.J., Einav, S.: Influence of microcalcifications on vulnerable plaque mechanics using FSI modeling. J. Biomech. **41**(5), 1111–1118 (2008)
8. Bluestein, D., Dumont, K., De Beule, M., Ricotta, J., Impellizzeri, P., Verhegghe, B., Verdonck, P.: Intraluminal thrombus and risk of rupture in patient specific abdominal aortic aneurysm-FSI modelling. Comput. Methods Biomech. Biomed. Eng. **12**(1), 73–81 (2008)
9. Boyce, M.C., Arruda, E.M.: Constitutive models of rubber elasticity: A review. Rubber Chem. Technol. **73**(3), 504–523 (2000)
10. Brown, L.C., Powell, J.T.: Risk factors for aneurysm rupture in patients kept under ultrasound surveillance. UK small aneurysm trial participants. Ann. Surg. **230**(3), 289–296 (1999)

11. Calhoun, P.S., Kuszyk, B.S., Heath, D.G., Carley J.C., Fishmanm, E.K.: Three-dimensional volume rendering of spiral CT data: theory and method. Radiographics **19**(3), 745–764 (1999)
12. Chuong, C.J., Fung, Y.C.: On residual stresses in arteries. J. Biomech. Eng. **108**(2), 189–192 (1986)
13. Cronenwett, J.L., Murphy, T.F., Zelenock, G.B., Whitehouse, W.M., Jr., Lindenauer, S.M., Graham, L.M., Quint, L.E., Silver, T.M., Stanley, J.C.: Actuarial analysis of variables associated with rupture of small abdominal aortic aneurysms. Surgery **98**(3), 472–483 (1985)
14. de Putter, S., Wolters, B.J., Rutten, M.C., Breeuwer, M., Gerritsen, F.A., van de Vosse, F.N.: Patient-specific initial wall stress in abdominal aortic aneurysms with a backward incremental method. J. Biomech. **40**, 1081–1090 (2006)
15. Delfino, A., Stergiopulos, N., Moore, J.E., Meister, J.J.: Residual strain effects on the stress field in a thick wall finite element model of the human carotid bifurcation. J. Biomech. **30**(8), 777–786 (1997)
16. Di Martino, E., Mantero, S, Inzoli, F., Melissano, G., Astore, D., Chiesa, R., Fumero, R.: Biomechanics of abdominal aortic aneurysm in the presence of endoluminal thrombus: experimental characterisation and structural static computational analysis. Eur. J. Vasc. Endovasc. Surg. **15**(4), 290–299 (1998)
17. Di Martino, E.S., Bohra, A., Vande Geest, J.P., Gupta, N., Makaroun, M.S., Vorp, D.A.: Biomechanical properties of ruptured versus electively repaired abdominal aortic aneurysm wall tissue. J. Vasc. Surg. **43**(3), 570–576 (2006)
18. Di Martino, E.S., Bohra, A., Vande Geest, J.P., Gupta, N., Makaroun M.S., Vorp, D.A.: Biomechanical properties of ruptured versus electively repaired abdominal aortic aneurysm wall tissue. J. Vasc. Surg. **43**(3), 570–576, discussion 576 (2006)
19. Di Martino, E.S., Guadagni, G., Fumero, A., Ballerini, G., Spirito, R, Biglioli, P., Redaelli, A.: Fluid-structure interaction within realistic three-dimensional models of the aneurysmatic aorta as a guidance to assess the risk of rupture of the aneurysm. Med. Eng. Phys. **23**(9), 647–655 (2001)
20. Di Martino, E.S., Vorp, D.A.: Effect of variation in intraluminal thrombus constitutive properties on abdominal aortic aneurysm wall stress. Ann. Biomed. Eng. **31**(7), 804–809 (2003)
21. Dobrin, P.B.: Pathophysiology and pathogenesis of aortic aneurysms. Current concepts. Surg. Clin. N. Am. **69**(4), 687–703 (1989)
22. Durmowicz, A.G., Badesch, D.B., Parks, W.C., Mecham, R.P., Stenmark, K.R.: Hypoxia-induced inhibition of tropoelastin synthesis by neonatal calf pulmonary artery smooth muscle cells. Am. J. Respir. Cell Mol. Biol. **5**(5), 464–469 (1991)
23. Elger, D.F., Blackketter, D.M., Budwig, R.S., Johansen, K.H.: The influence of shape on the stresses in model abdominal aortic aneurysms. J. Biomech. Eng.-Trans. Asme **118**(3), 326–332 (1996)
24. Fillinger, M.: The long-term relationship of wall stress to the natural history of abdominal aortic aneurysms (finite element analysis and other methods). Ann. N. Y. Acad. Sci. 1085: 22–28 (2006)
25. Fillinger, M.F., Marra, S.P., Raghavan, M.L., Kennedy, F.E.: Prediction of rupture risk in abdominal aortic aneurysm during observation: wall stress versus diameter. J. Vasc. Surg. **37**(4), 724–732 (2003)
26. Fillinger, M.F., Racusin, J., Baker, R.K., Cronenwett, J.L., Teutelink, A., Schermerhorn, M.L., Zwolak, R.M., Powell, R.J., Walsh, D.B., Rzucidlo, E.M.: Anatomic characteristics of ruptured abdominal aortic aneurysm on conventional CT scans: Implications for rupture risk. J. Vasc. Surg. **39**(6), 1243–1252 (2004)
27. Fillinger, M.F., Raghavan, M.L., Marra, S.P., Cronenwett, J.L., Kennedy, F.E.: In vivo analysis of mechanical wall stress and abdominal aortic aneurysm rupture risk. J. Vasc. Surg. **36**(3), 589–597 (2002)

28. Gadowski, G.R., Ricci, M.A., Hendley, E.D., Pilcher, D.B.: Hypertension accelerates the growth of experimental aortic aneurysms. J. Surg. Res. **54**(5), 431–436 (1993)
29. Geest, J.P.V., Sacks, M.S., Vorp, D.A.: The effects of aneurysm on the biaxial mechanical behavior of human abdominal aorta. J. Biomech. **39**(7), 1324–1334 (2006)
30. He, C.M., Roach, M.R.: The composition and mechanical properties of abdominal aortic aneurysms. J. Vasc. Surg. **20**(1), 6–13 (1994)
31. Holzapfel, G., Stadler, M., Gasser, T.C.: Changes in the mechanical environment of stenotic arteries during interaction with stents: Computational assessment of parametric stent designs. J. Biomech. Eng.-Trans. Asme **127**(1), 166–180 (2005)
32. Holzapfel, G.A., Gasser, T.C.: A viscoelastic model for fiber-reinforced composites at finite strains: Continuum basis, computational aspects and applications. Comput. Methods Appl. Mech. Eng. **190**(34), 4379–4403 (2001)
33. Holzapfel, G.A., Gasser, T.C.: Computational stress-deformation analysis of arterial walls including high-pressure response. Int. J. Cardiol. **116**(1), 78–85 (2007)
34. Holzapfel, G.A., Gasser, T.C., Ogden, R.W.: A new constitutive framework for arterial wall mechanics and a comparative study of material models. J. Elast. **61**(1–3), 1–48 (2000)
35. Holzapfel, G.A., Gasser, T.C., Ogden, R.W.: Comparison of a multi-layer structural model for arterial walls with a fung-type model, and issues of material stability. J. Biomech. Eng.-Trans. Asme **126**(2), 264–275 (2004)
36. Holzapfel, G.A., Gasser, T.C., Stadler, M.: A structural model for the viscoelastic behavior of arterial walls: Continuum formulation and finite element analysis. Eur. J. Mech. A-Solids **21**(3), 441–463 (2002)
37. Holzapfel, G.A., Sommer, G, Gasser, C.T., Regitnig, P.: Determination of layer-specific mechanical properties of human coronary arteries with nonatherosclerotic intimal thickening and related constitutive modeling. Am. J. Physiol.-Heart Circ. Physiol. **289**(5), H2048–H2058 (2005)
38. Holzapfel, G.A., Weizsacker H.W.: Biomechanical behavior of the arterial wall and its numerical characterization. Comput. Biol. Med. **28**(4), 377–392 (1998)
39. Hua, J., Mower, W.R.: Simple geometric characteristics fail to reliably predict abdominal aortic aneurysm wall stresses. J. Vasc. Surg. **34**(2), 308–315 (2001)
40. Hunter, G.C., Leong, S.C., Yu, G.S., McIntyre, K.E., Bernhard, V.M.: Aortic blebs: possible site of aneurysm rupture. J. Vasc. Surg. **10**(1), 93–99 (1989)
41. Inzoli, F., Boschetti, F., Zappa, M., Longo, T., Fumero, R: Biomechanical factors in abdominal aortic aneurysm rupture. Eur J. Vasc. Surg. **7**(6), 667–674 (1993)
42. Johnson, P.T., Halpern, E.J., Kuszyk, B.S., Heath, D.G., Wechsler, R.J., Nazarian, L.N., Gardiner, G.A., Levin, D.C., Fishman, E.K.: Renal artery stenosis: CT angiography–comparison of real-time volume-rendering and maximum intensity projection algorithms. Radiology **211**(2), 337–343 (1999)
43. Kazi, M., Thyberg, J., Religa, P., Roy, J., Eriksson, P., Hedin, U., Swedenborg, J.: Influence of intraluminal thrombus on structural and cellular composition of abdominal aortic aneurysm wall. J. Vasc. Surg. **38**(6), 1283–1292 (2003)
44. Kyriacou, S.K., Humphrey, J.D.: Influence of size, shape and properties on the mechanics of axisymmetric saccular aneurysms. J. Biomech. **29**(8), 1015–1022 (1996)
45. Lanne, T., Sonesson, B., Bergqvist, D., Bengtsson, H., Gustafsson, D.: Diameter and compliance in the male human abdominal aorta: influence of age and aortic aneurysm. Eur. J. Vasc. Surg. **6**(2), 178–184 (1992)
46. Lederle, F.A., Johnson, G.R., Wilson, S.E., Ballard, D.J., Jordan, W.D., Blebea, J., Littooy, F.N., Freischlag, J.A., Bandyk, D., Rapp, J.H., Salam, A.A., Invest, V.A.C.S.: Rupture rate of large abdominal aortic aneurysms in patients refusing or unfit for elective repair. Jama-J. Am. Med. Assoc. **287**(22), 2968–2972 (2002)
47. Lederle, F.A., Wilson, S.E., Johnson, G.R., Reinke, D.B., Littooy, F.N., Acher, C.W., Ballard, D.J., Messina, L.M., Gordon, I.L., Chute, E.P., Krupski, W.C., Bandyk, D., Vet, A.D.M.: Immediate repair compared with surveillance of small abdominal aortic aneurysms. N. Engl. J. Med. **346**(19), 1437–1444 (2002)

48. Levene, C.I., Kapoor, R., Heale, G.: The effect of hypoxia on the synthesis of collagen and glycosaminoglycans by cultured pig aortic endothelium. Atherosclerosis **44**(3), 327–337 (1982)
49. Limet, R., Sakalihassan, N., Albert, A.: Determination of the expansion rate and incidence of rupture of abdominal aortic aneurysms. J. Vasc. Surg. **14**(4), 540–548 (1991)
50. Long, Q., Xu, X.Y., Collins, M.W., Bourne, M., Griffith, T.M.: Magnetic resonance image processing and structured grid generation of a human abdominal bifurcation. Comput. Methods Programs Biomed. **56**(3), 249–259 (1998)
51. MacSweeney, S.T., Young, G., Greenhalgh, R.M., Powell, J.T.: Mechanical properties of the aneurysmal aorta. Br. J. Surg. **79**(12), 1281–1284 (1992)
52. McMillan, W.D., Tamarina, N.A., Cipollone, M., Johnson D.A., Parker, M.A., Pearce, W.H.: Size matters: the relationship between MMP-9 expression and aortic diameter. Circulation **96**(7), 2228–2232 (1997)
53. Mooney, M.: A theory of large elastic deformation. J. Appl. Phy. **11**(9), 582–592 (1940)
54. Mower, W.R., Baraff, L.J., Sneyd, J.: Stress distributions in vascular aneurysms: factors affecting risk of aneurysm rupture. J. Surg. Res. **55**(2), 155–161 (1993)
55. Mower, W.R., Quinones, W.J., Gambhir, S.S.: Effect of intraluminal thrombus on abdominal aortic aneurysm wall stress. J. Vasc. Surg. **26**(4), 602–608 (1997)
56. Nicholls, S.C., Gardner, J.B., Meissner, M.H., Johansen, H.K.: Rupture in small abdominal aortic aneurysms. J. Vasc. Surg. **28**(5), 884–888 (1998)
57. Olufsen, M.S., Peskin, C.S., Kim, W.Y., Pedersen, E.M., Nadim, A., Larsen, J.: Numerical simulation and experimental validation of blood flow in arteries with structured-tree outflow conditions. Ann. Biomed. Eng. **28**(11), 1281–1299 (2000)
58. Papaharilaou, Y., Ekaterinaris, J.A., Manousaki, E., Katsamouris, A.N.: A decoupled fluid structure approach for estimating wall stress in abdominal aortic aneurysms. J. Biomech. **40**(2), 367–377 (2007)
59. Papp, Z., Patel, M., Ashtari, M., Takahashi, M., Goldstein, J., Maguire, W., Herman, P.G.: Carotid artery stenosis: optimization of CT angiography with a combination of shaded surface display and source images. AJNR Am. J. Neuroradiol. **18**(4), 759–763 (1997)
60. Pietila, K., Jaakkola, O.: Effect of hypoxia on the synthesis of glycosaminoglycans and collagen by rabbit aortic smooth muscle cells in culture. Atherosclerosis **50**(2), 183–190 (1984)
61. Powell, J.T., Brady, A.R., Brown, L.C., Forbes, J.F., Fowkes, F.G.R., Greenhalgh, R.M., Ruckley, C.V., Thompson, S.G., Participants, U.S.A.T.: Mortality results for randomised controlled trial of early elective surgery or ultrasonographic surveillance for small abdominal aortic aneurysms. Lancet **352**(9141), 1649–1655 (1998)
62. Raghavan, M.L., Kratzberg, J., de Tolosa, E.M.C., Hanaoka, M.M., Walker, P., da Silva, E.S.: Regional distribution of wall thickness and failure properties of human abdominal aortic aneurysm. J. Biomech. **39**(16), 3010–3016 (2006)
63. Raghavan, M.L., Vorp, D.A.: Toward a biomechanical tool to evaluate rupture potential of abdominal aortic aneurysm: identification of a finite strain constitutive model and evaluation of its applicability. J. Biomech. **33**(4), 475–482 (2000)
64. Raghavan, M.L., Webster, M.W., Vorp, D.A.: Ex vivo biomechanical behavior of abdominal aortic aneurysm: assessment using a new mathematical model. Ann. Biomed. Eng. **24**(5), 573–582 (1996)
65. Ricotta, J.J., Pagan, J., Xenos, M., Alemu, Y., Einav, S., Bluestein, D.: Cardiovascular disease management: the need for better diagnostics. Med. Biol. Eng. Comput. **46**(11), 1059–1068 (2008)
66. Rissland, P., Alemu, Y., Einav, S., Ricotta, J., Bluestein, D.: Abdominal aortic aneurysm risk of rupture-patient specific FSI simulations using anisotropic model. J. Biomech. Eng. **13**(3), 031001–031010 (2009)
67. Rivlin, R.S.: Large Elastic Deformations of Isotropic Materials .1. Fundamental Concepts. Philos. Trans. Royal Soc. London Ser. A-Mat. Phy. Sci. **240**(822), 459–508 (1948)

68. Roach, M.R., Burton, A.C.: The reason for the shape of the distensibility curves of arteries. Can. J. Biochem. Physiol. **35**(8), 681–690 (1957)
69. Rodriguez, J.F., Martufi, G., Doblare, M., Finol, E.A.: The effect of material model formulation in the stress analysis of abdominal aortic aneurysms. Ann. Biomed. Eng. **37**(11), 2218–2221 (2009)
70. Rodriguez, J.F., Ruiz, C., Doblare, M., Holzapfel, G.A.: Mechanical stresses in abdominal aortic aneurysms: influence of diameter, asymmetry, and material anisotropy. J. Biomech. Eng. **130**(2), 021023 (2008)
71. Sacks, M.S., Vorp, D.A., Raghavan, M.L., Federle, M.P., Webster, M.W.: In vivo three-dimensional surface geometry of abdominal aortic aneurysms. Ann. Biomed. Eng. **27**(4), 469–479 (1999)
72. Sakalihasan, N., Heyeres, A., Nusgens, B.V., Limet, R., Lapiere, C.M.: Modifications of the extracellular matrix of aneurysmal abdominal aortas as a function of their size. Eur. J. Vasc. Surg. **7**(6), 633–637 (1993)
73. Sakalihasan, N., Limet, R., Defawe, O.D.: Abdominal aortic aneurysm. Lancet **365**(9470), 1577–1589 (2005)
74. Scannel, G., Waxman, K., Kaml, G.J., Ioli, G., Gatanaga, T., Yamamoto, R., Granger, G.A.: Hypoxia induces a human macrophage cell line to release tumor necrosis factor-alpha and its soluble receptors in vitro. J. Surg. Res. **54**(4), 281–285 (1993)
75. Schermerhorn, M., Zwolak, R., Velazquez, O., Makaroun, M., Fairman, R., Cronenwett, J.: Ultrasound screening for abdominal aortic aneurysm in medicare beneficiaries. Ann. Vasc. Surg. **22**(1), 16–24 (2008)
76. Schurink, G.W.H., van Baalen, J.M., Visser, M.J.T., van Bockel, J.H.: Thrombus within an aortic aneurysm does not reduce pressure on the aneurysmal wall. J. Vascul. Surg. **31**(3), 501–506 (2000)
77. Scotti, C., Shkolnik, A., Muluk, S., Finol, E.: Fluid-structure interaction in abdominal aortic aneurysms: effects of asymmetry and wall thickness. Biomed. Eng. Online **4**(1), 64 (2005)
78. Scotti, C.M., Finol, E.A.: Compliant biomechanics of abdominal aortic aneurysms: A fluid-structure interaction study. Comput. Struct. **85**(11–14), 1097–1113 (2007)
79. Scotti, C.M., Jimenez, J., Muluk, S.C., Finol, E.A.: Wall stress and flow dynamics in abdominal aortic aneurysms: finite element analysis vs. fluid-structure interaction. Comp. Methods Biomech. Biomed. Eng. **11**(3), 301–322 (2008)
80. Scoutt, L.M., Zawin, M.L., Taylor, K.J.: Doppler US. Part II. Clinical applications. Radiology **174**(2), 309–319 (1990)
81. Spittell, P.C., Ehrsam, J.E., Anderson, L., Seward, J.B.: Screening for abdominal aortic aneurysm during transthoracic echocardiography in a hypertensive patient population. J. Am. Soc. Echocardiogr. **10**(7), 722–727 (1997)
82. Steinman, D.A., Thomas, J.B., Ladak, H.M., Milner, J.S., Rutt, B.K., Spence, J.D.: Reconstruction of carotid bifurcation hemodynamics and wall thickness using computational fluid dynamics and MRI. Magn. Reson. Med. **47**(1), 149–159 (2002)
83. Sussman, T., Bathe, K.J.: A finite-element formulation for nonlinear incompressible elastic and inelastic analysis. Comput. Struct. **26**(1–2), 357–409 (1987)
84. Szilagyi, D.E., Smith, R.F., Elliott, J.P.: Clinical fate of patient with asymptomatic abdominal aortic aneurysm and unfit for surgical treatment. Arch. Surg. **104**(4), 600–606 (1972)
85. Tang, D., Yang, C., Zheng, J., Woodard, P.K., Sicard, G.A., Saffitz, J.E., Yuan, C.: 3D MRI-based multicomponent FSI models for atherosclerotic plaques. Ann. Biomed. Eng. **32**(7), 947–960 (2004)
86. Thubrikar, M.J.: Vascular Mechanics and Pathology. New York, Springer (2007)
87. Thubrikar, M.J., al-Soudi, J., Robicsek, F.: Wall stress studies of abdominal aortic aneurysm in a clinical model. Ann. Vasc. Surg. **15**(3), 355–366 (2001)
88. Thubrikar, M.J., Labrosse, M., Robicsek, F., Al-Soudi, J., Fowler, B.: Mechanical properties of abdominal aortic aneurysm wall. J. Med. Eng. Technol. **25**(4), 133–142 (2001)

89. Thubrikar, M.J., Robicsek, F., Labrosse, M., Chervenkoff, V., Fowler, B.L.: Effect of thrombus on abdominal aortic aneurysm wall dilation and stress. J. Cardiovasc. Surg. (Torino) **44**(1), 67–77 (2003)
90. Truijers, M., Pol, J.A., Schultzekool, L.J., van Sterkenburg, S.M., Fillinger, M.F., Blankensteijn, J.D.: Wall stress analysis in small asymptomatic, symptomatic and ruptured abdominal aortic aneurysms. Eur. J. Vasc. Endovasc. Surg. **33**(4), 401–407 (2007)
91. Valencia, A., Morales, H., Rivera, R., Bravo, E., Galvez, M.: Blood flow dynamics in patient-specific cerebral aneurysm models: the relationship between wall shear stress and aneurysm area index. Med. Eng. Phys. **30**(3), 329–340 (2008)
92. Valentine, R.J., DeCaprio, J.D., Castillo, J.M., Modrall, J.G., Jackson, M.R., Clagett, G.P.: Watchful waiting in cases of small abdominal aortic aneurysms-appropriate for all patients. J. Vascul. Surg. **32**(3), 441–448 (2000)
93. Vande Geest, J.P., Di Martino, E.S., Bohra, A., Makaroun, M.S., Vorp, D.A.: A biomechanics-based rupture potential index for abdominal aortic aneurysm risk assessment: demonstrative application. Ann. N. Y. Acad. Sci. **1085**, 11–21 (2006)
94. Vande Geest, J.P., Sacks, M.S., Vorp, D.A.: Age dependency of the biaxial biomechanical behavior of human abdominal aorta. J. Biomech. Eng. **126**(6), 815–822 (2004)
95. Vande Geest, J.P., Sacks, M.S., Vorp, D.A.: The effects of aneurysm on the biaxial mechanical behavior of human abdominal aorta. J. Biomech. **39**(7), 1324–1334 (2006)
96. Vande Geest, J.P., Wang, D.H., Wisniewski, S.R., Makaroun, M.S., Vorp, D.A.: Towards a noninvasive method for determination of patient-specific wall strength distribution in abdominal aortic aneurysms. Ann. Biomed. Eng. **34**(7), 1098–1106 (2006)
97. Vardulaki, K.A., Walker, N.M., Day, N.E., Duffy, S.W., Ashton, H.A., Scott, R.A.: Quantifying the risks of hypertension, age, sex and smoking in patients with abdominal aortic aneurysm. Br. J. Surg. **87**(2), 195–200 (2000)
98. Vengrenyuk, Y., Cardoso, L., Weinbaum, S.: Micro-CT based analysis of a new paradigm for vulnerable plaque rupture: cellular microcalcifications in fibrous caps. Mol. Cell Biomech. **5**(1), 37–47 (2008)
99. Venkatasubramaniam, A.K., Fagan, M.J., Mehta, T., Mylankal, K.J., Ray, B., Kuhan, G., Chetter, I.C., McCollum, P.T.: A comparative study of aortic wall stress using finite element analysis for ruptured and non-ruptured abdominal aortic aneurysms. Eur. J. Vasc. Endovasc. Surg. **28**(2), 168–176 (2004)
100. Vito, R.P., Hickey, J.: The Mechanical-properties of soft-tissues .2. The elastic response of arterial segments. J. Biomech. **13**(11), 951–957 (1980)
101. Vorp, D.A., Federspiel, W.J., Webster, M.W.: Does laminated intraluminal thrombus within abdominal aortic aneurysm cause anoxia of the aortic wall? J. Vasc. Surg. **23**(3), 540–541 (1996)
102. Vorp, D.A., Lee, P.C., Wang, D.H., Makaroun, M.S., Nemoto, E.M., Ogawa, S., Webster, M.W.: Association of intraluminal thrombus in abdominal aortic aneurysm with local hypoxia and wall weakening. J. Vasc. Surg. **34**(2), 291–299 (2001)
103. Vorp, D.A., Mandarino, W.A., Webster, M.W., Gorcsan, J., 3rd: Potential influence of intraluminal thrombus on abdominal aortic aneurysm as assessed by a new non-invasive method. Cardiovasc. Surg. **4**(6), 732–739 (1996b)
104. Vorp, D.A., Raghavan, M.L., Muluk, S.C., Makaroun, M.S., Steed, D.L., Shapiro, R., Webster, M.W.: Wall strength and stiffness of aneurysmal and nonaneurysmal abdominal aorta. Ann. N. Y. Acad. Sci. **800**, 274–276 (1996c)
105. Vorp, D.A., Raghavan, M.L., Webster, M.W.: Mechanical wall stress in abdominal aortic aneurysm: influence of diameter and asymmetry. J. Vasc. Surg. **27**(4), 632–639 (1998)
106. Vorp, D.A., Vande Geest, J.P.: Biomechanical determinants of abdominal aortic aneurysm rupture. Arterioscler. Thromb. Vasc. Biol. **25**(8), 1558–1566 (2005)
107. Vorp, D.A., Wang, D.H., Webster, M.W., Federspiel, W.J.: Effect of intraluminal thrombus thickness and bulge diameter on the oxygen diffusion in abdominal aortic aneurysm. J. Biomech. Eng. **120**(5), 579–583 (1998b)

108. Wang, D.H., Makaroun, M., Webster, M.W., Vorp, D.A.: Mechanical properties and microstructure of intraluminal thrombus from abdominal aortic aneurysm. J. Biomech. Eng. **123**(6), 536–539 (2001)
109. Wang, D.H., Makaroun, M.S., Webster, M.W., Vorp, D.A.: Effect of intraluminal thrombus on wall stress in patient-specific models of abdominal aortic aneurysm. J. Vasc. Surg. **36**(3), 598–604 (2002)
110. Wang, K.C., Dutton, R.W., Taylor, C.A.: Improving geometric model construction for blood flow modeling. IEEE Eng. Med. Biol. Mag. **18**(6), 33–39 (1999)
111. Wilson, K., Bradbury, A., Whyman, M., Hoskins, P., Lee, A., Fowkes, G., McCollum, P., Vaughan Ruckley, C.: Relationship between abdominal aortic aneurysm wall compliance and clinical outcome: a preliminary analysis. Eur. J. Vasc. Endovasc. Surg. **15**(6), 472–477 (1998)
112. Wilson, K., Whyman, M., Hoskins, P., Lee, A.J., Bradbury, A.W., Fowkes, F.G., Ruckley, C.V.: The relationship between abdominal aortic aneurysm wall compliance, maximum diameter and growth rate. Cardiovasc. Surg. **7**(2), 208–213 (1999)
113. Wilson, K.A., Lee, A.J., Hoskins, P.R., Fowkes, F.G., Ruckley, C.V., Bradbury, A.W.: The relationship between aortic wall distensibility and rupture of infrarenal abdominal aortic aneurysm. J. Vasc. Surg. **37**(1), 112–117 (2003)
114. Wilson, K.A., Lindholt, J.S., Hoskins, P.R., Heickendorff, L., Vammen, S., Bradbury, A.W.: The relationship between abdominal aortic aneurysm distensibility and serum markers of elastin and collagen metabolism. Eur. J. Vasc. Endovasc. Surg. **21**(2), 175–178 (2001)
115. Wolters, B.J., Rutten, M.C., Schurink G.W., Kose, U., de Hart, J., van de Vosse,F.N.: A patient-specific computational model of fluid-structure interaction in abdominal aortic aneurysms. Med. Eng. Phys. **27**(10), 871–883 (2005)
116. Xenos, M., Alemu, Y., Zamfir, D., Einav, S., Ricotta, J., Labropoulos, N., Tassiopoulos, T., Bluestein, D.: The effect of angulation in abdominal aortic aneurysms: experimental measurements and FSI simulations. Med. Biol. Eng. Comput. **48**:1175–1190 (2010)
117. Xenos, M., Rambhia, S.H., Alemu, Y., Einav, S., Labropoulos, N., Tassiopoulos, A., Ricotta, J.J., Bluestein, D.: Mimics based image reconstruction augments diagnosis and management of vascular pathologies: a study of ruptured abdominal aortic aneurisms. Mimics Innovation Awards 2009
118. Xenos, M., Rambhia, S.H., Alemu, Y., Einav, S., Labropoulos, N., Tassiopoulos, A., Ricotta, J.J., Bluestein, D.: Patient based abdominal aortic aneurysm rupture risk prediction with fluid structure interaction modeling. Ann. Biomed. Eng. **38**(11), 3323–3337 (2010)
119. Yang, C., Bach, R.G., Zheng, J., Naqa, I.E., Woodard, P.K., Teng, Z., Billiar, K., Tang, D.: In vivo IVUS-based 3-D fluid-structure interaction models with cyclic bending and anisotropic vessel properties for human atherosclerotic coronary plaque mechanical analysis. IEEE Trans. Biomed. Eng. **56**(10), 2420–2428 (2009)
120. Yushkevich, P.A., Piven, J., Hazlett, H.C., Smith, R.G., Ho, S., Gee, J.C., Gerig, G.: User-guided 3D active contour segmentation of anatomical structures: significantly improved efficiency and reliability. Neuroimage **31**(3), 1116–1128 (2006)
121. Zankl, A.R., Schumacher, H., Krumsdorf, U., Katus, H.A., Jahn, L., Tiefenbacher, C.P.: Pathology, natural history and treatment of abdominal aortic aneurysms. Clin. Res. Cardiol. **96**(3), 140–151 (2007)

Computational Analysis of Displacement Forces Acting on Endografts Used to Treat Aortic Aneurysms

C. Alberto Figueroa and Christopher K. Zarins

Abstract Endovascular repair has greatly reduced the perioperative morbidity and mortality of abdominal aortic aneurysm repair compared to open surgery (Zarins et al., J. Vasc. Surg 38(6): 1264–1272, 2003). However, endovascular stent-grafts are exposed to a number of clinical complications, such as endograft migration (i.e., loss of positional stability), stent fractures and endoleaks (i.e., persistence of blood flow into the aneurysm sac after device placement). These complications may result in life-threatening and costly events such as aneurysm growth, rupture, need for secondary procedures, and life-long follow-up with imaging studies. Understanding the biomechanical environment experienced by endografts in vivo is a critical factor in improving their performance (Figueroa et al., J. Endovasc. Ther 16(3):350–358, 2009; 284–294, 2009). The loads experienced by aortic endografts are greatly dependent on the tortuosity and size of the endograft, as well as on the hemodynamic state of the patient. The fixation response of the endograft is determined by factors such as the fixation mechanism (radial pressure vs. hooks and barbs), amount of "landing zone" (i.e., area where the endograft can physically attach to the aorta), and the disease state of the vessel wall at the landing zone. The purpose of this study is to review the most common complications associated with endovascular repair of abdominal and thoracic aneurysms, and to provide a summary of the state of the art of the computational tools used to perform patient-specific modeling of endograft dynamics.

C. A. Figueroa
Department of Bioengineering, Stanford University, Stanford, CA, USA
e-mail: cafa@stanford.edu

C. K. Zarins (✉)
Department of Surgery, Stanford University, Stanford, CA, USA
e-mail: zarins@stanford.edu

Stud Mechanobiol Tissue Eng Biomater (2011) 7: 221–246
DOI: 10.1007/8415_2011_73

Published Online: 22 April 2011

1 Introduction

The prevalence of abdominal aortic aneurysms (AAA) has increased 300% over the past 40 years. Currently, aneurysm disease affects 5–7% of Americans over age 60 and the number of aneurysms is expected to increase dramatically in the next years as the population ages [12, 14]. Abdominal aneurysms can be repaired using open surgery or endovascular techniques.

1.1 Open Repair

In this procedure, a large transperitoneal or retroperitoneal incision is made to expose the diseased section of the aorta. Then, the aortic neck and iliac arteries are clamped to temporarily interrupt blood flow into the aneurysm. The aneurysmal aorta is cut open and the intraluminal thrombus is removed. A Dacron or PTFE graft is anastomosed to the infrarenal aorta and common iliac arteries (see Fig. 1) and the thrombus-free aneurysm wall is closed over the graft [49]. This procedure usually requires a 3–5 h long operation, which is physiologically stressful for the patient. Despite the numerous technological advances and improvement in perioperative management advances during the last 55 years of clinical practice, this approach is still associated with significant operative mortality rates (5 and 50% for elective and ruptured repairs, respectively) [7, 18].

1.2 Endovascular Repair

Endovascular repair techniques have revolutionized the treatment of AAA disease, greatly reducing the perioperative mortality and morbidity associated with open surgical repair [5]. These techniques were first introduced in 1990 by Juan Parodi [32] using a home-made device consisting of a modified Palmaz stent sutured to a Dacron vascular graft to exclude an abdominal aneurysm. The stent-graft is deployed in the abdominal aortic region using a catheter guided fluoroscopically. Once deployed, the stent-graft forms a new conduit for the blood, effectively excluding the weakened aneurysm wall from the direct action of blood pressure. The technique requires only small incisions in the groin to expose the femoral artery and thus is a safer procedure for the patient than open repair. Endovascular aneurysm repair has experienced rapid development and wide acceptance during the last decade and has become the primary treatment for AAA disease in the USA [15]. While initially the technique was used primarily in older patients who could not tolerate the risk of invasive open repair, endovascular repair is now used to treat a broad range of patients and aneurysm morphologies [8, 16]. This has been possible due to the evolution in

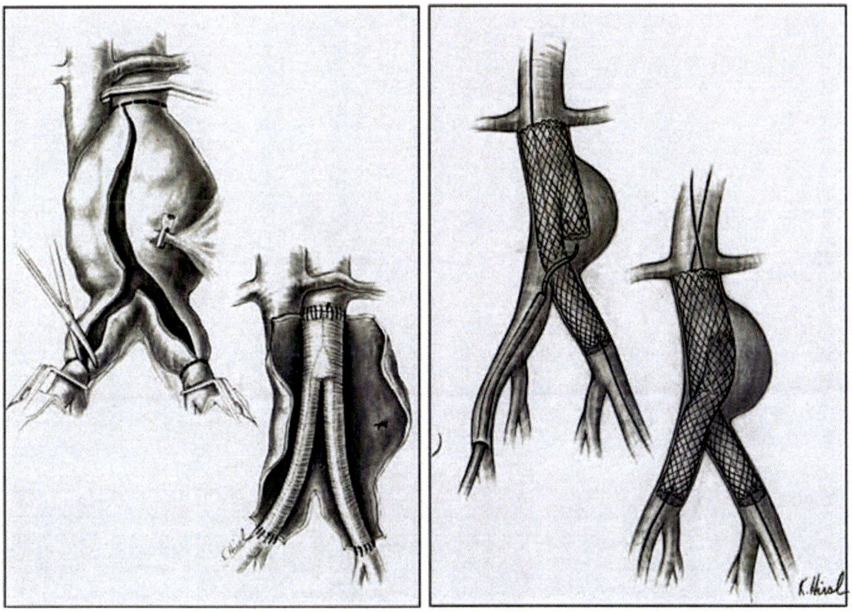

Fig. 1 *Left* In an open surgical AAA repair, a Dacron graft is manually sutured to the aortic neck and common iliac arteries. *Right* In an endovascular repair, a stent-graft is deployed in the aneurysm cavity using a catheter. Adapted from Zarins and Gewertz [49]

design and fixation characteristics of endografts through the years. Most modern devices include supra-renal fixation that enables the use of the device in patients with short aortic necks or with aneurysms that extend close to the renal arteries. There are currently multiple devices in the market that are very different in design (modular vs. unibody), materials (Dacron vs. PTFE; nitinol vs. stainless steel or cobalt-chromium), deployment mechanisms (self-expanded vs. balloon-expanded) and attachment mechanisms (radial force vs. hooks & barbs), (see Fig. 2).

Despite the aforementioned advantages, endovascular aneurysm repair is not free of significant complications, such as late endograft migration (see Fig. 3), endoleak formation, fracture of device components, etc. that may result in continued aneurysm expansion and the need for long-term imaging surveillance and secondary interventions. Furthermore, the EVAR-1 clinical trial showed that hospital costs are higher for patients treated with endovascular repair than for those treated with open repair [16].

Recent data from the DREAM prospective randomized clinical trial comparing long-term outcomes of aneurysm repair in patients treated with open and endovascular techniques indicates that re-intervention rates are significantly smaller in the open repair group with similar survival rates after 7 years [8]. It is therefore fair to say that while endovascular procedures greatly reduce the problems that

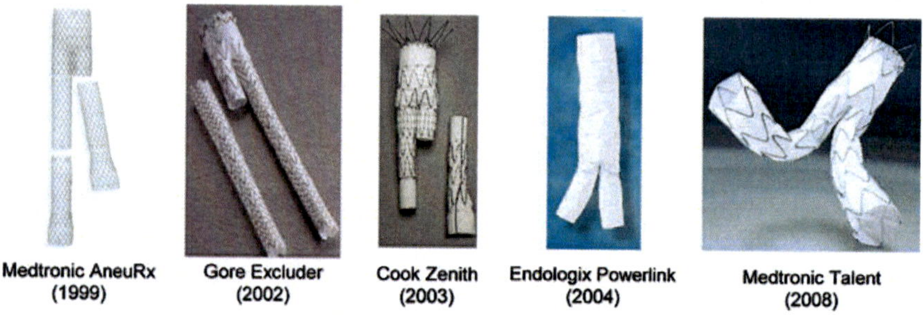

Fig. 2 Current FDA approved endografts for abdominal aortic aneurysm repair. Year of FDA approval is listed in parenthesis

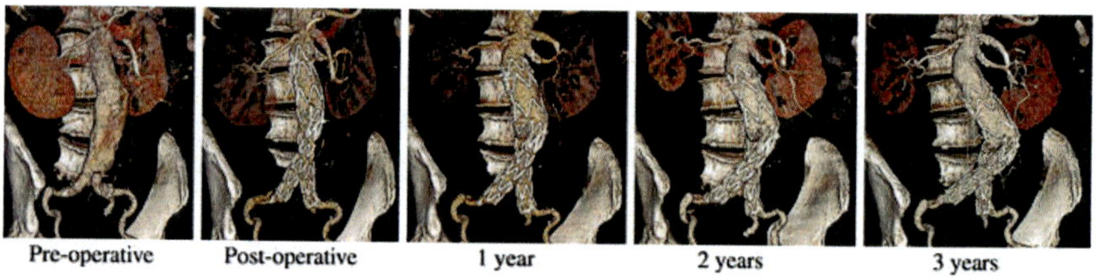

Fig. 3 Medical image follow-up studies of an abdominal aortic aneurysm repaired using a stent-graft. The images show the pre-operative, post-operative and several follow-up configurations of the abdominal aorta and the endograft, clearly demonstrating the migration of the device. The patient eventually required a secondary procedure where a proximal extender cuff was placed in the aortic neck

open repair originates in the short-term; they generate a number of drawbacks that negatively impact the long-term well-being of the patient. It is thus of critical importance to enhance the performance of endografts by improving their design based on a deeper understanding of the hemodynamic conditions that the devices experience in vivo.

2 Endograft Failure Modes

The most common post-operative complications in endovascular repair are endoleaks (i.e., persistence of blood flow into the aneurysm sac after device placement) and endograft migration (i.e., loss of positional stability). Other modes of failure are graft rupture, infection or thrombosis; and endotension (i.e., continuous aneurysm sac expansion in the absence of endoleaks).

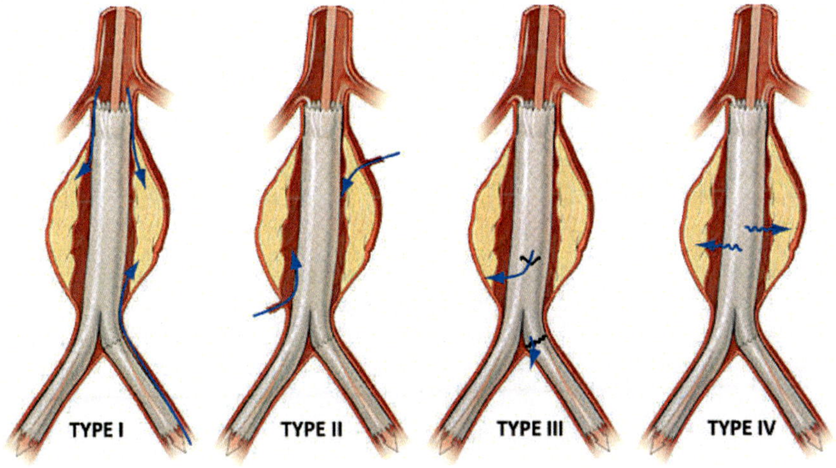

Fig. 4 Schematic representation of the endoleak types following AAA endovascular repair

2.1 Endoleaks

Endoleaks are present in 20–40% of patients following endovascular aneurysm repair. Endoleaks are considered to be dangerous because in their presence, the intra-saccular pressure remains elevated and therefore the risk of rupture of the aneurysm persists. Endoleaks can be classified according to their temporal nature (i.e., short-term or long-term) or to the origin of the endoleak [46] (see Fig. 4):

- Type I endoleaks are the result of poor apposition of the device to the aortic wall. Type Ia refers to blood flow coming into the aneurysm sac from the proximal fixation area, whereas Type Ib refers to flow coming into the aneurysm from the distal fixation zone. Type I endoleaks are considered to be evidence of a poorly performed procedure. They usually account for less than 10% of all endoleak cases [42].
- Type II endoleaks occur when there is reverse flow into the aneurysm sac from aortic branch vessels such as the inferior mesenteric artery or lumbar arteries. This is the most common type of endoleak, accounting for more than half of the cases. Some Type II endoleaks thrombose spontaneously shortly after the procedure. Others persist, and may lead to aneurysm enlargement. If the physician determines that these persistent Type II endoleaks represent a threat, they are treated by coil embolization of the arteries responsible for the retrograde flow [3].
- Type III endoleaks may be due to a defect in the fabric of the graft, such as a tear resulting from a fractured stent, or due to separation of the modular components of the endograft. These endoleaks usually account for less than 5% of the cases.

- Type IV endoleaks are due to filtration of plasma through the fabric of the graft. This is usually the result of a design feature of the graft fabric.

2.2 Endograft Migration

All endografts are subject to migration as a result of the pulsatile action of blood flow and pressure. Endograft migration represents a serious adverse event that usually requires a secondary intervention to restore proper fixation of the device to the abdominal aorta. This secondary procedure may involve the insertion of a proximal or distal extender cuff deployed endovascularly, or open surgical repair procedures such as aortic neck plication and partial or complete endograft removal [31].

To date, there has not been a consistent definition of endograft migration. Some device manufacturers have defined migration as a proximal or distal endograft movement more than 5 mm, whereas others set the threshold at 10 mm. Furthermore, the term migration has often referred to situations where endograft movement results in a clinical event. Under this definition, numerous cases of significant endograft movement that do not result in an adverse clinical event are not regarded as migration [11]. Therefore, the concept of positional stability may be understood differently in clinical and engineering settings.

There are multiple clinical studies that have examined the rates of migration of various devices over time. In general, migration rates increase over time for all devices [27, 40, 48]. These studies have examined the proximal and distal motion of the endograft, but they have neglected the sideways motion. This sideways component has been recently shown to be a predictor of endograft migration and late adverse events [33].

Most current endograft designs rely on two different mechanisms to fixate the device to the arterial wall: radial force and hooks or barbs (see Fig. 5). In the case of radial force, this is achieved by oversizing the device relative to the nominal aortic diameter. The larger the oversizing of the device, the larger the radial force developed against the wall and therefore the larger the fixation response of the device. The degree of oversizing in clinical practice varies significantly, ranging from as little as 5% to as large as 30%. Aortic neck dilation and Type Ia endoleaks have been associated with excessive device oversizing [6, 39, 40, 48]. This is perhaps due to remodeling in the vessel wall in response to the larger circumferential stresses originated by excessive device oversizing. In the case of hooks and barbs, these components penetrate the vessel wall, thus clamping the device to the arterial lumen. A basic design parameter is the length of the hooks and barbs: the longer the hooks, the deeper the penetration through the aortic wall. However, a possible downside of this fixation approach may be the injury that these components cause to the endothelial and intima layers. This may result in vessel wall remodeling that can potentially lead to aortic neck dilation and Type Ia endoleaks.

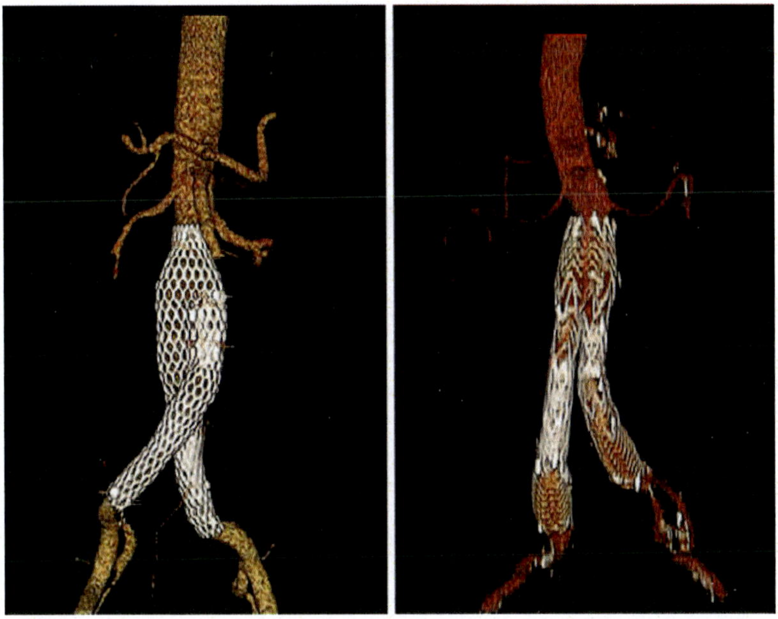

Fig. 5 The attachment of the endograft to the aortic neck wall is achieved via radial force created by device oversizing (*left*) or hooks and barbs that clamp the device to the arterial lumen (*right*)

A number of factors which may affect device migration have been clinically investigated including aortic neck diameter [37], length and angulation [13, 19], neck calcification and thrombus, inadequate proximal and distal fixation length [17, 48] and aneurysm tortuosity [39]. In general, it has been found that longer, disease-free aortic necks are associated with lower incidence of endoleaks and endograft migration since they offer a better surface for the device to stay attached to the arterial wall. Conversely, short, calcified, tortuous necks are associated with a higher incidence of endograft complications.

In addition to the aforementioned clinical studies, experimental in vitro, in vivo, and cadaveric aorta studies have measured the pull-out forces of different devices [2, 25, 30, 35]. In all cases, these experiments considered a rather unrealistic planar configuration of the abdominal aorta and the endograft, in part due to the difficulty of reproducing an anatomically and hemodynamically-realistic experimental environment. In all these studies, the stent-graft was displaced by applying a force in the downwards direction (see Fig. 6), and the amount of force required to dislodge the endograft was recorded. This force is known as displacement force (DF).

The range of measured displacement forces for various devices was (4–30) Newton (N). These investigations are fundamentally limited because their experimental conditions fail to reproduce critical components that define the

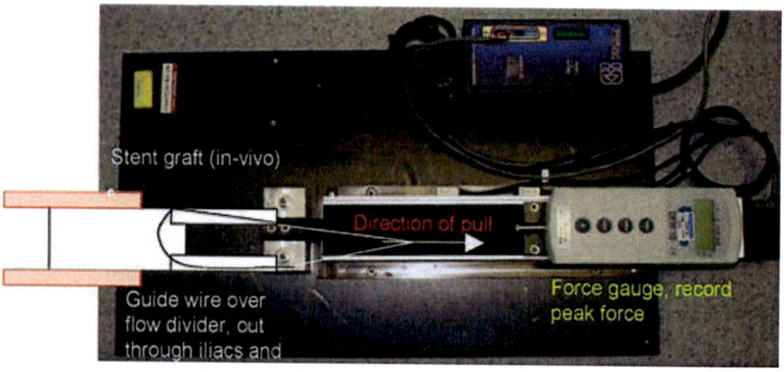

Fig. 6 Experimental setup to measure the pull-out force required to dislodge an endograft inserted in a bovine AAA model. Note the downwards direction of pull, reproduced from Arko et al. [2]

in vivo hemodynamics experienced by the endograft, such as tortuous anatomy, curvature, and pulsatile flow and pressure. These conditions can potentially be better investigated with the help of computational modeling techniques.

3 CFD Modeling Tools for Endograft Dynamics

Computational modeling tools provide an opportunity to evaluate multiple situations where it is extremely difficult to generate an experimental setup that adequately reproduces the environment conditions of the problem at hand. In our case, in order to understand the principles that govern endograft stability, one must be able to reproduce realistic geometries, flow and pressure pulsatile conditions, and material properties of the components in the system. From a mechanical standpoint, the problem of endograft positional stability can be regarded as a competition between de-stabilizing forces or loads acting on the device (i.e., the displacement forces DF) and stabilizing forces that keep the device attached to the wall (i.e., fixation forces FF), see Fig. 7.

The displacement force DF is determined by the endograft geometry (i.e., length, diameter, etc.), the hemodynamic state of the patient (i.e., hypertension, cardiac output, etc.) and the anatomy of the aneurysm (i.e., tortuosity, neck angulation, etc.). The fixation forces FF must balance the displacement forces DF to keep the device attached to the vessel wall. The FF depends on the specific endograft fixation mechanism (i.e., radial force versus hooks and barbs), the structural stiffness of the device, the amount of proximal and distal landing zone (i.e., "fixation real estate"), and the level of disease of the vessel wall in the attachment areas (i.e., healthy wall versus atherosclerotic wall, presence of calcifications, etc.). The larger the loads exerted by the blood on the device, the larger

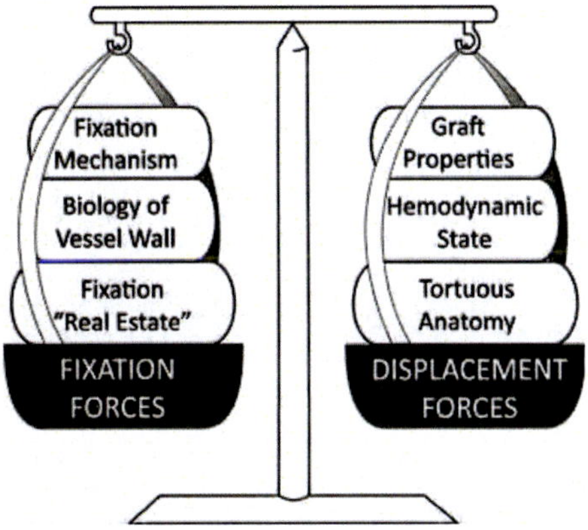

Fig. 7 The problem of endograft long-term durability seen as a competition of de-stabilizing forces or loads acting on the device (displacement forces *DF*) and stabilizing forces (fixation forces *FF*)

the fixation forces required in order to keep the device in place. Computational fluid dynamics (CFD) tools can be used to evaluate the displacement forces that the fluid (blood) exerts on the device. Likewise, computational solid mechanics (CSM) can be used to estimate the fixation forces FF that the device develops to counteract the actions of blood flow and pressure. It is therefore of critical importance to have a good understanding of the magnitude of the loads that the endograft experiences in vivo in order to ensure a design that will guarantee the long-term stability of the device.

Numerous CFD and theoretical analyses have investigated the magnitude of the DF loads exerted by blood flow on endografts using simplified computational models and boundary conditions [20, 24, 28, 29]. In most of these computational models, the geometry utilized was a simple bifurcated graft not including any of the abdominal blood vessels, usually in a planar configuration, and with simplistic outflow boundary conditions such as prescribed pressure waves in the outlet faces. Lastly, almost no studies have investigated the contact mechanics problem between the endograft and vessel wall [1]. In this setting, a 3-body contact problem involving the aortic wall, the stent, and the graft subjected to the CFD DF seeks to investigate the positional stability of the stent-graft. There are 3 possible outcomes in this stability analysis:

- The device and the aortic wall remain attached to each other. This situation corresponds to a stable endograft.
- There is slip between the endograft and the aortic wall surface. This corresponds to a situation of endograft positional instability and possible migration.
- There is separation between the endograft and the aortic wall surfaces. This corresponds to a situation of an endoleak.

In the remainder of this chapter, we review several CFD tools we have used to characterize displacement forces acting on anatomically-realistic models of aortic endografts under physiologically-relevant flow and pressure conditions.

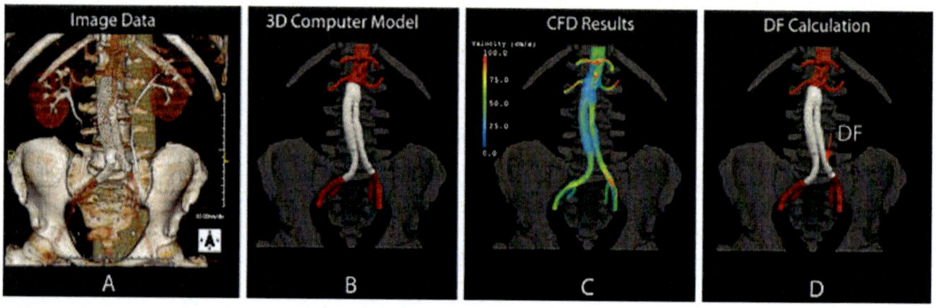

Fig. 8 Patient-specific computer modeling for displacement force calculation: Starting with the patient image data (**A**), 3D computer models of the endograft and the abdominal aorta are built (**B**). Then, CFD analyses calculating blood flow velocity and pressure in the computer model are performed (**C**). Lastly, the displacement force is computed using the results of the CFD analysis (**D**). Reproduced from Figueroa et al. [11]

3.1 Computational Fluid Dynamics Tools

The computation of DF in realistic models of aortic endografts under realistic conditions of flow and pressure relies on the so-called *patient-specific computer modeling* paradigm [41] (see Fig. 8). Here, a computer model of the aorta and endograft is built from medical image data using automatic segmentation techniques. Once the model is adequately discretized into a grid suitable for numerical computation, and adequate boundary conditions are applied to the model, a computer simulation representing blood flow and pressure in the domain of interest is obtained. At this point, the CFD results can be used to investigate specific quantities of interest, such as the overall displacement force exerted by blood flow on the endograft.

3.1.1 Geometric Modeling

Patient-specific geometric models of blood vessels have been constructed using 2 or 3D-based semi-automatic segmentation techniques [45, 47]. These techniques use algorithms such as the level set method to detect the lumen boundary [26]. 2D-based techniques provide higher control in the segmentation process especially in situations of significant noise in the image data, but 3D techniques are better suited to handle complex geometrical features, such as bifurcations and irregular surfaces. Contrast-enhanced medical image data such as Magnetic Resonance Angiography (MRA) or Computed Tomography Angiography (CTA) provide a sharper, clearer boundary of the lumen and therefore facilitate the segmentation task significantly. Both CTA and MRA currently provide sub-millimeter resolutions that allow for accurate reconstructions of the thoracic aorta, endograft and main branches, such as renal arteries, mesenteric and iliac arteries, etc.

3.1.2 CFD Analysis

Once the geometric model of the aortic endograft and the vessels of interest is created, a discretization of the model into a grid suitable for numerical simulation is needed. Here, it is important to generate a grid that represents accurately both the geometry and the solution field (i.e. velocity and pressure). Some desirable features in the grid-generation process are boundary layer meshing, curvature-based refinement and anisotropic, field-based mesh adaptation techniques [36].

A critical step in the CFD analysis is the boundary condition specification. In this process, one must provide adequate information on flow, pressure and potentially vessel wall dynamics on the boundaries of the geometry. This is an active area of research, and currently the most sophisticated approaches rely on coupling reduced-order mathematical models of the proximal and distal circulation to the inlet(s) and outlets of the model [41, 43]. Typically, a supraceliac waveform mapped to a Womersley velocity profile is prescribed at the inlet face of an abdominal aortic endograft model. This waveform can be obtained either directly from Phase-Contrast Magnetic Resonance Imaging (PC–MRI) techniques or from morphometric measurements [23]. In the case of thoracic aortic endograft modeling, a lumped-parameter heart model representing the interactions between the heart and the thoracic aorta [21] may be utilized on the inlet face of the model. For the outflow boundaries, a powerful and versatile solution consists of coupling a three-element Windkessel lumped parameter model representing the resistance and compliance of the vascular beds that are not physically included in the 3D geometrical model [11, 44]. This approach presents important conceptual advantages:

- It does not rely on the specification of any of the primary blood flow variables (flow or pressure), which are generally not known and are part of the desired solution.
- It avoids potentially serious synchronization issues between flow and pressure waveforms in different parts of the domain. These lumped-parameter outlet conditions adapt naturally to the flow "collected" on each of the outlet faces and calculate a pressure based on the specific downstream model. This pressure is then applied as a weak traction on each of the boundary faces.

Once the process of generating a geometrically-accurate aortic endograft geometric model is finished, and the corresponding computational grid and inflow and outflow boundary conditions are determined, a CFD analysis is performed whereby the Navier–Stokes equations of an incompressible fluid are solved in the domain of interest with the ultimate goal of obtaining a characterization of blood flow velocity and pressure. This analysis is usually very time-consuming since it requires obtaining the solution for velocity and pressure in thousands of time steps for grids that are usually well over one million degrees-of-freedom. Hence, a parallel Finite-Element or Finite-Volume implementation of a solver for the Navier–Stokes equations running in a computer cluster is required in order to bring

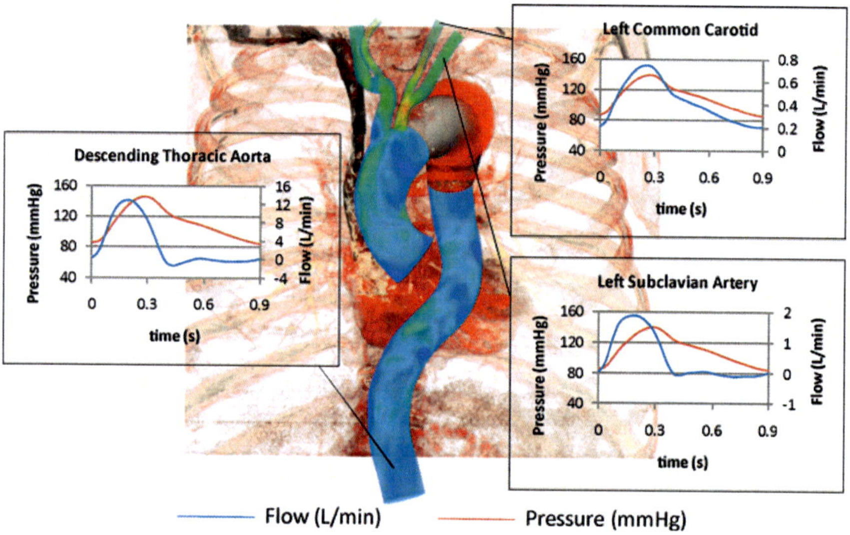

Fig. 9 Flow and pressure waveforms in selected vessels obtained in the CFD analysis of a proximal descending thoracic aortic aneurysm (*TAA*) model. Note the physiologic range of the waveforms, presenting features such as retrograde flow in the descending aorta during early systole, and forward flow through the cycle in the common carotid artery. Reproduced from Figueroa et al. [9]

the computation time down to the point where the CFD results can be generated in a matter of hours [41].

By choosing a suitable distribution of inflow and outflow boundary conditions, it is possible to achieve physiologically-relevant distributions of flow and pressure in the computational model (see Fig. 9). The figure shows the pressure and flow waveforms in the descending thoracic aorta, left subclavian artery, and left common carotid artery obtained in the CFD simulations of a proximal descending thoracic aortic aneurysm (TAA) treated with an endograft. The variables defining the hemodynamic state used in the CFD analysis were typical values for volumetric flow and pressure for the ascending thoracic aorta: The mean flow and heart rate is 4.9 L/min and 67 beats per minute, respectively. The aortic systolic, diastolic, and mean pressures were 145, 85, and 111 mmHg, respectively. The figure clearly shows the differences between the descending thoracic aorta and common carotid artery flow waveforms, with forward flow in the carotid artery throughout the cardiac cycle and reversed flow in the descending aorta during diastole, which represent normal physiological variations.

3.1.3 Displacement Force Calculation

Once the CFD results for velocity and pressure in the computer model are obtained, it is possible to investigate the overall effect of the actions of blood flow

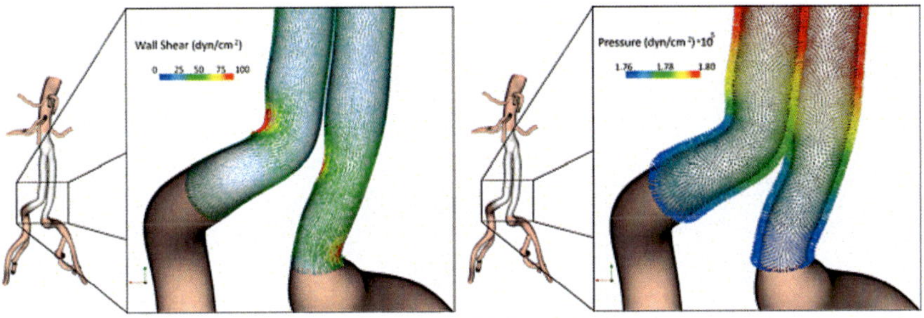

Fig. 10 Wall shear (*left*) and pressure (*right*) stresses representing the actions of the blood on the endograft. These stresses are integrated over the surface of the endograft to calculate the total 3D force exerted by the pulsatile flow. Note that the pressure is several orders of magnitude larger than the shear stress. Reproduced from Figueroa et al. [10]

on the endograft. Specifically, the displacement force DF can be obtained by integrating the total traction (i.e. sum of the wall-shear stress and the total normal stress) over the surface of the endograft. Figure 10 illustrates how the contribution of the pressure to the total load experienced by the device is several orders of magnitude larger than the wall shear stress contribution. This implies that the pressure of the patient has a much larger impact than the cardiac output on the forces exerted by the blood flow on the device. Therefore, chronic alterations in pressure as a result of hypertension should be watched carefully during the follow-up of the patient.

Contrary to what is often assumed, the endograft displacement force does not act primarily in the longitudinal direction of flow: in situations of significant aneurysm tortuosity the displacement force has important antero-posterior and lateral components (see Fig. 11). This example illustrates the power of computational methods towards investigating the loads experienced by endografts in vivo: these techniques make it simple to account for complex, subject-specific geometries of aorta and endograft, and once proper boundary conditions for flow and pressure are obtained to match known patient data, one can easily investigate parameters such as mean and temporal variation of displacement forces, spatial components in the longitudinal, anterior, and lateral directions, etc.

The subject of the example depicted in Fig. 11 has a supraceliac aortic flow of 1 L/min, a heart rate of 60 beats per minute, and systolic and diastolic blood pressures of 168 and 92 mmHg, respectively. These conditions result in a pulsatile endograft displacement force ranging from 6.8 N in peak systole and 3.6 N in diastole, with a temporal mean of 5 N. Of this total force, only 28% (1.4 N) is directed in the axial direction. This may have important consequences on the fixation forces that the device must develop in the aortic neck in order to remain attached to the vessel wall.

Other variables than can be easily investigated are the effect on displacement forces of parameters such as endograft diameter and length, aneurysm angulation, graft bifurcation angle, blood pressure, exercise, etc. These variables been have

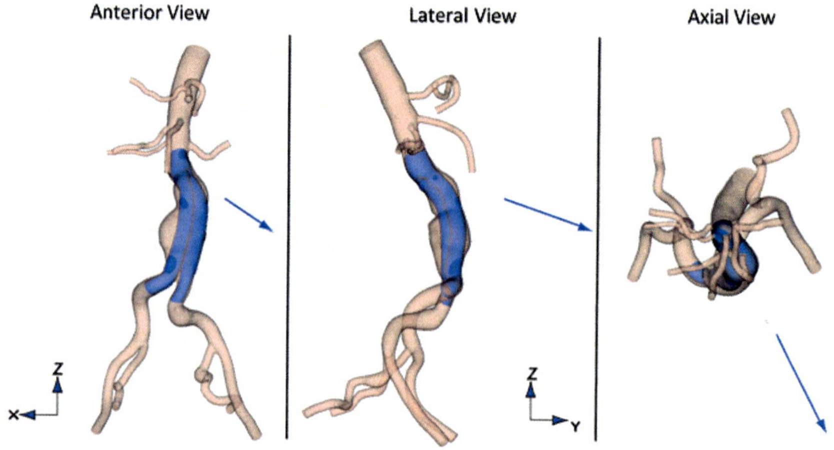

Fig. 11 Anterior, lateral, and axial views of an AAA model with a stent-graft in place. Note the antero-lateral angulation of the endograft, which follows the curved anatomy of the abdominal aorta. The vector of displacement force in drawn to scale in each of the three views

investigated in idealized [22] and patient-specific models [9, 10] of thoracic and abdominal endograft models. Finally, CFD techniques may also provide insight into transport or residence time of thrombogenic agents within the endograft [38].

In the following section, we provide several examples of how computationally determined endograft displacement forces may provide useful insight regarding the hemodynamic conditions experienced by these devices in vivo. We illustrate how these conditions are affected by an array of factors, including curvature, device diameter, blood flow and blood pressure.

3.2 Application Examples

3.2.1 Effect of Curvature on Displacement Forces Acting on Abdominal Endografts

In order to investigate the impact of aortic curvature on the displacement forces experienced by stent-grafts, the computer model depicted in Fig. 11 was modified to accommodate the endograft in an almost flat or planar configuration without changing the dimensions of the device. These changes produced a model with a much smaller curvature in the antero-posterior and lateral directions in the aneurysm region (see Fig. 12).

The aortic and branch flow and pressure boundary conditions remained unchanged between the curved and reduced curvature endograft models. The CFD simulations for both models show that there are virtually no differences in the

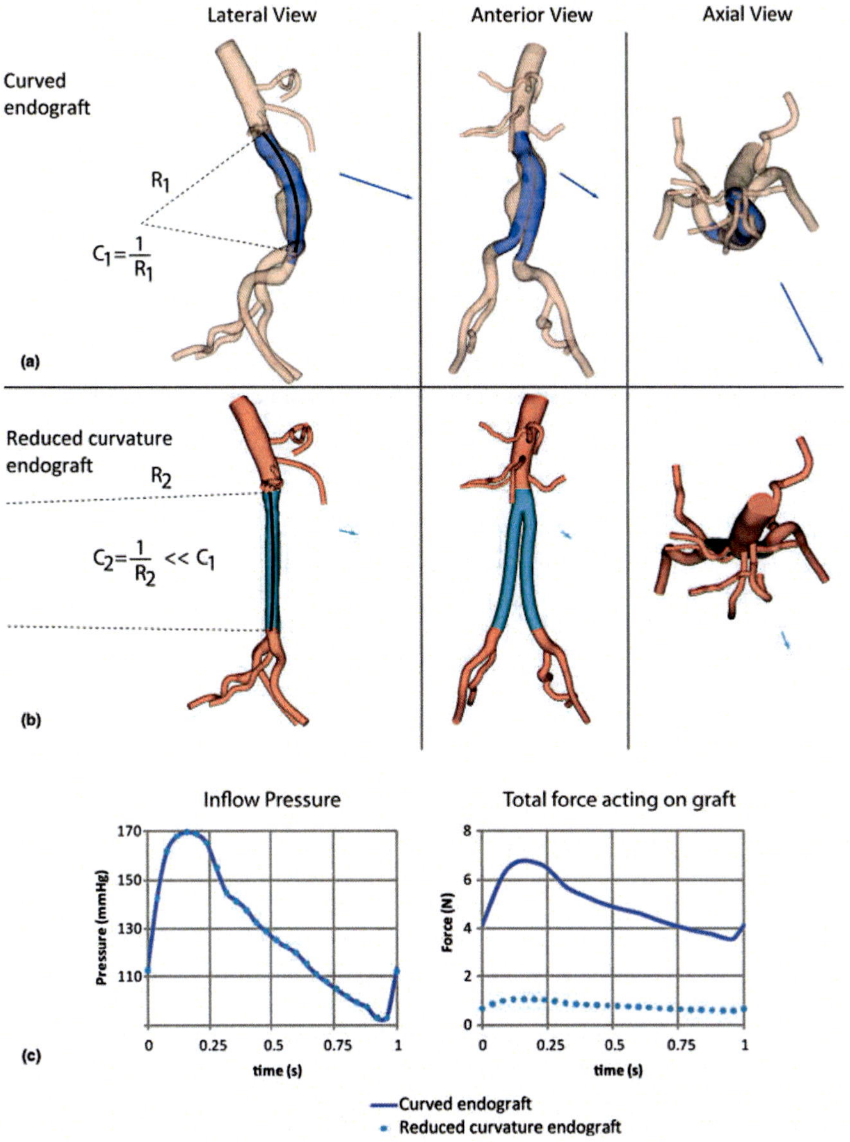

Fig. 12 Lateral, anterior and axial views of the time average of the displacement force acting on the curved endograft model (**a**) and reduced curvature endograft model (**b**). The dimensions of the reduced endograft model are identical to those of the curved endograft. Note the reduction in magnitude of the displacement force acting on the endograft as reflected by the size of the arrows. (**c**) Shows the comparison of the supraceliac and displacement force plots for the two endografts. Note that while the pressures are almost identical, there is a five-fold decrease in pulsatile displacement force (average of 5.01 vs. 0.8 N) in the reduced curvature endograft

Table 1 Total and 3D components of displacement force *DF* for the curved and reduced curvature endografts

	Curved endograft	Reduced curvature endograft
F_x (lateral) (N)	−2.22	−0.29
F_y (anterior) (N)	4.26	0.72
F_z (axial) (N)	−1.42	−0.24
Total force (N)	5.01	0.81
% Downward	28.35	29.54

aortic pressure between the two models: this indicates that local changes in curvature or tortuosity do not significantly affect the pressure field in the aorta unless they are accompanied by geometric changes such as narrowing (stenosis) or enlargement sections that may significantly alter the flow field.

The differences in the total displacement force between the two models are however, significant: the temporal average of the total displacement force in the planar endograft was reduced by more than five-fold from 5 to 0.8 N compared to the normally positioned curved endograft. The total and 3D components of the displacement force for both models are given in Table 1 above.

In both the curved and reduced curvature endografts the largest components of the force was acting in the anterior direction, which can be explained by the angulation of the neck of the abdominal aorta, directing the blood flow to impinge against the anterior face of the endograft and therefore increasing the dynamic component of the pressure on that surface.

3.2.2 Effect of Increased Pressure and Flow on Displacement Forces Acting on Thoracic Endografts

Blood flow and pressure can vary significantly during the course of the day in response to changes in activity level, and chronically due to processes such as hypertension and physical training. It is therefore important to understand the effects of altered flow and pressure on the forces experienced by endografts. We investigate these effects using a patient-specific model of a thoracic aortic endograft. The endograft consists of three modules with an average diameter of 36 mm spanning a length of approximately 30 cm used to repair a mid-descending thoracic aortic aneurysm (TAA) (see Fig. 13).

The variables defining the hemodynamic state used in the CFD analysis represented typical values for volumetric flow and pressure for the thoracic aorta. The mean flow and heart rate were 4.9 L/min and 67 bpm, respectively. The aortic systolic, diastolic, and mean pressures were 145, 85, and 111 mmHg, respectively (pressure pulse of 60 mmHg). Calculation of the endograft displacement force DF produced a vector with significantly larger mean (21.7 N), peak systolic (27.8 N) and end-diastolic (16.7 N) values than those typically found on abdominal endografts. Figure 13 shows the anterior, lateral, and axial views of the thoracic model and the DF vector (blue arrow) resulting from the baseline flow and pressure

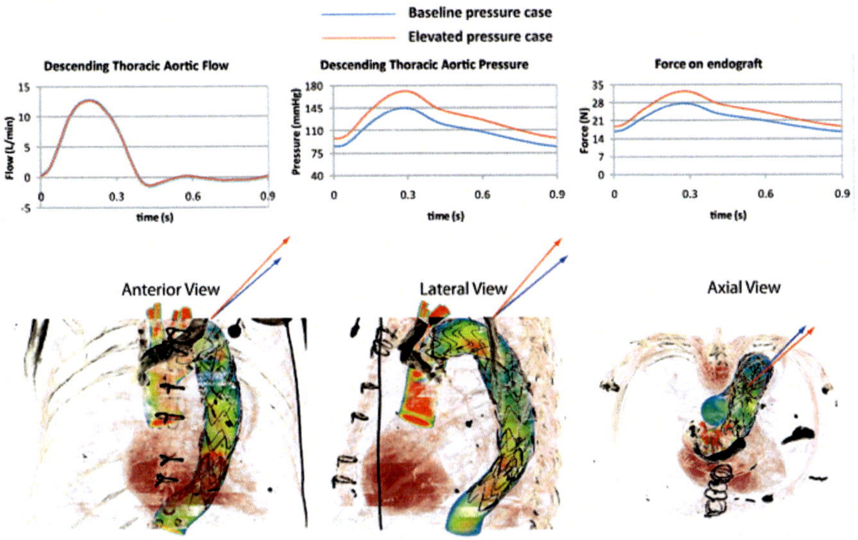

Fig. 13 Flow and pressure waves with pulsatile DF for the baseline pressure (*blue plots*) and the elevated pressure (*red plots*) simulations. The images compare the orientation of the DF vectors for the baseline pressure (*blue arrows*) and the elevated pressure (*red arrows*) scenarios in the anterior, lateral, and axial views. Modified from [9]

conditions given above. Note that the orientation of the DF vector follows the curvature of the endograft and shows a significant component (10.8 N) acting in the craniad direction. The waveforms corresponding to baseline conditions of thoracic aortic flow, descending aortic pressure, and pulsatile DF are given in blue.

We investigated a situation of moderate increase in blood pressure (see red waveforms in Fig. 13) with no changes in cardiac output. The elevated pressure waveform had a mean value of 130 mmHg, with a systolic peak of 171 mmHg, a diastolic minimum of 97 mmHg, and a pulse pressure of 74 mmHg. Thus, this represented an increase of 16.5% in mean pressure with respect to be baseline conditions. The DF vector, shown in red arrows in Fig. 13, had larger mean (25 N), peak systolic (32.5 N), and diastolic (22.6 N) values than in the baseline pressure conditions. This indicates that the increase in mean DF was approximately linearly proportional to the increase in mean pressure over the range of pressure considered. Furthermore, the increase in pulse pressure resulted in a significant change in the orientation of the DF vector. Table 2 compares the axial (craniad) and sideways components of the baseline pressure conditions to the values for the elevated pressure simulation. The greatest increase was in the axial component (37%), whereas the increase in the sideways component was only 7%, which can be explained by the increased acceleration of blood due to the 14 mmHg higher pulse pressure. The blood was therefore pushed against the outer curve of the graft more vigorously than before, which resulted in the larger craniad component of the DF vector.

Table 2 Axial component, transverse component and total mean displacement force DF for the baseline pressure and elevated pressure conditions

	Baseline pressure	Elevated pressure	Increase (%)
Axial component (craniad) (N)	10.8	14.8	37
Transverse component (N)	18.8	20.1	7
Total force (N)	21.7	25	15

We then investigated the impact of increased flow on the DF experienced by the endograft. We considered a situation of light exercise corresponding to a 2.5-fold increase in cardiac output from 4.9 to 12.24 L/min, and 50% increase in heart rate from 67 to 100 bpm. We adjusted the outflow branches boundary conditions to represent realistic changes in thoracic pressure during moderate exercise: 55% increase in pulse pressure from 60 to 93 mmHg; almost no change in diastolic pressure resulting in an increase in mean pressure of 12.5% from 112 to 126 mmHg. These changes of 50% increase in flow rate and the corresponding 12.5% increase in mean blood pressure resulted in a modest 10% increase in the mean DF experienced by the endograft. This indicates that the changes in DF are influenced mostly by changes in blood pressure resulting from the increased level of physical activity. Furthermore, the orientation of the DF vector remained relatively unchanged with respect to the baseline conditions.

3.2.3 Correlation Between Endograft Motion and Post-operative Displacement Forces

To date, there has not been a consistent definition of endograft migration. Migration has been variously defined using an arbitrarily selected distance, such as 5 or 10 mm, or in some cases as an endograft movement that results in the need for a secondary intervention [27, 40, 48]. Most clinical studies quantifying endograft movement have relied on one-dimensional [4, 34, 48] or two-dimensional techniques [33] (see Fig. 14). Measurements have included axial or centerline distances from the renal arteries or superior mesenteric artery to the first appearance of the endograft or to the appearance of the complete fabric-stent ring. However, change in endograft position is actually a complex process in three-dimensional space. Quantification of three-dimensional positional changes of the endograft over time is challenging due to its geometric complexity, the need to co-register two different images in space, and the non-uniform movement of the device, since some parts of the endograft may experience a significant movement while others remain stationary.

We have recently proposed a methodology to quantify the three-dimensional displacement of an endograft based on tracking the position of the centroid of the device over time. The endograft centroid is co-registered in two images, typically a baseline post-operative scan and a follow-up scan, using the center point of the inferior edge of the L3 vertebra as anatomic landmark (yellow dot) (Fig. 15).

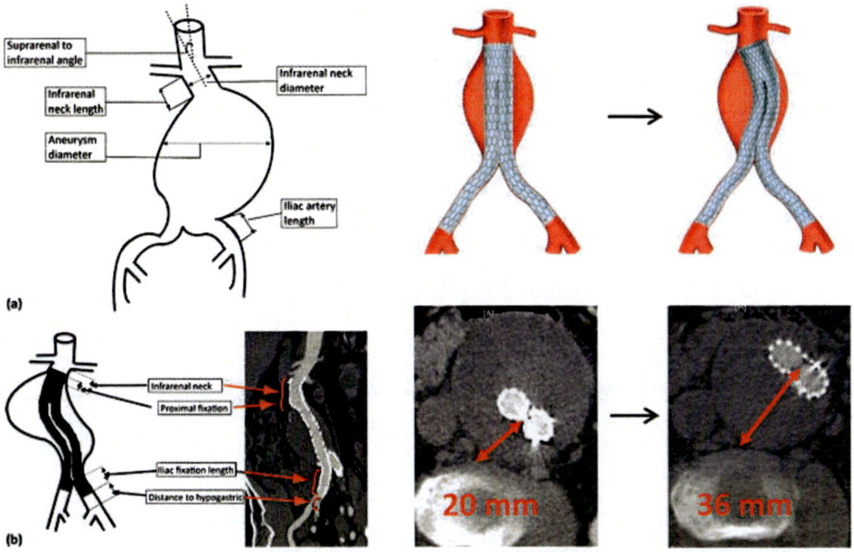

Fig. 14 Studies quantifying endograft movement using 1 or 2D techniques. *Left* Benharash et al. [4] have measured the longitudinal endograft movement with respect to the position of the renal arteries. *Right* Rafii et al. [33] have measured the transverse movement of the endograft with respect to an anatomic landmark in the spine

The distance between the co-registered centroid in the baseline scan (red dot) and follow-up scan (green dot) characterizes the three-dimensional movement of the endograft, which has components in the anterior, lateral, and axial directions.

Once the movement between baseline and follow-up scans is obtained, we investigated the correlation between the orientation of the post-operative endograft displacement force vector and the direction of endograft movement. The metric of the correlation is given by the cosine of the angle between the displacement force and the movement vectors [11]. This correlation can be helpful to predict, using post-operative or even pre-operative imaged data of a patient, the direction in which the endograft is likely to move.

Figure 16 shows the anterior and lateral views of the computed post-operative displacement force vector (red arrows) and measured movement vector (yellow arrows) between baseline and follow-up scans for 5 AAA patients with significant endograft movement that in general resulted in the need for a secondary procedure. Four out the five patients (patient 1, 2, 4 and 5) had late device migration (average time to secondary procedure of 3.3 years). Patient 3 however required a secondary procedure after only 8 months to correct a left iliac type I endoleak.

With the exception of patient 3 who showed a very small correlation between displacement force and displacement vectors, the correlation between the orientation of the post-operative displacement force vector and the movement vector was rather high for the patients with shorter follow-up intervals (average correlation metric of 0.46). This correlation could be further improved if factors such as

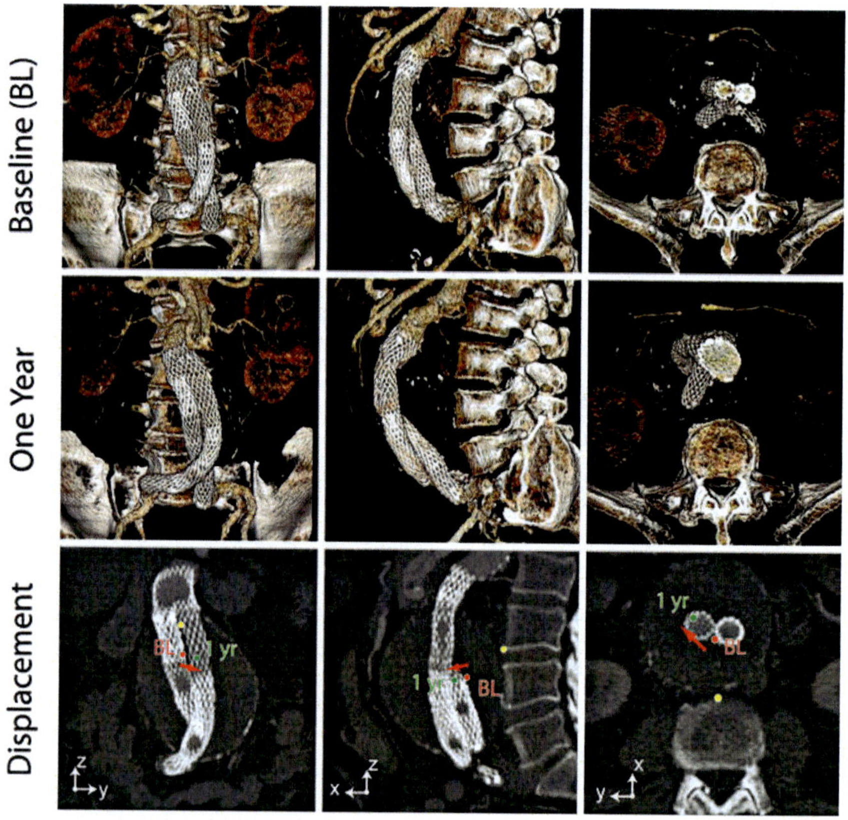

Fig. 15 3D analysis of the endograft centroid movement between a baseline post-operative state (*top row*), and a one-year follow-up state (*middle row*). The endograft centroid moves primarily in the antero-lateral direction (17.6 mm in the *xy* plane), and also in the axial direction (3 mm in *z*). Adapted from Figueroa et al. [11]

the position of the endograft within the aneurysm sac are considered. For instance, patient 3 has the endograft leaning directly against the anterior wall of the aneurysm. Therefore, even though the post-operative displacement force acts primarily in the posterior-anterior direction, the endograft is unlikely to move in that direction due to the constraint provided by the aneurysm wall.

4 Summary and Conclusions

In this article, we have provided an overview of the state of the art of the Computational Fluid Dynamics tools used to characterize the hemodynamics in patients treated with abdominal and thoracic stent-grafts and of the most common complications experienced by these devices. Computational modeling of patient-

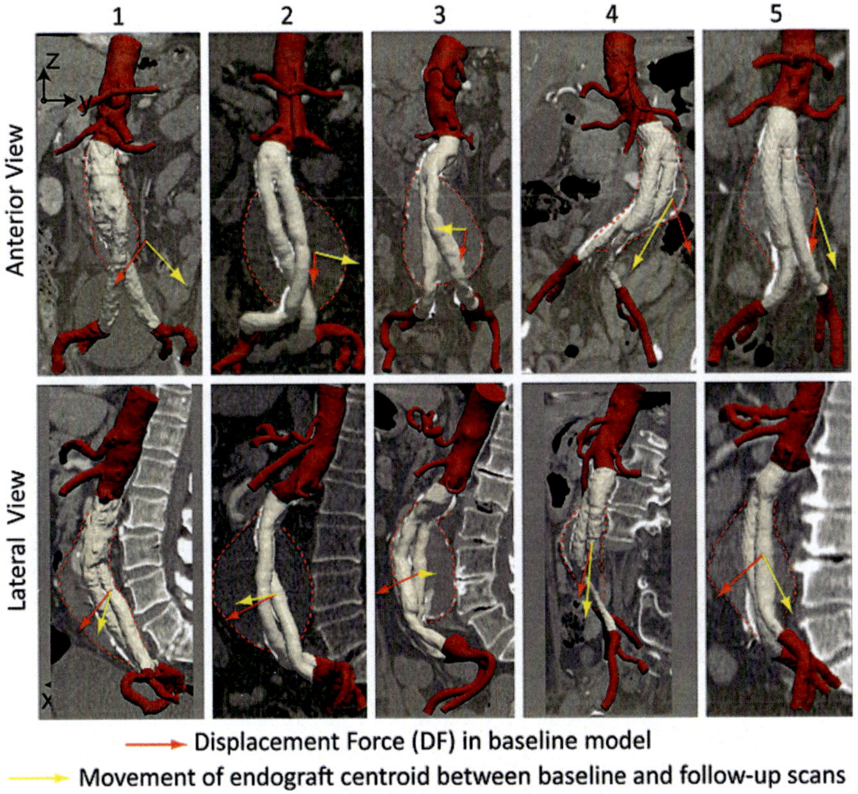

Fig. 16 Displacement force (*red arrows*) and endograft centroid movement (*yellow arrows*) vectors in the anterior and lateral views for the 5 patients who experienced abdominal endograft migration. Relative sizes of arrows reflect magnitude of the vectors. Adapted from Figueroa et al. [11]

specific endograft dynamics provides an unprecedented insight into the in vivo hemodynamic conditions experienced by these devices. These techniques have relied on recent progress in the areas of diagnostic medical imaging, image processing, computational fluid dynamics algorithms for fast, highly scalable parallel iterative solvers, and finally, computer hardware [41]. Computer modeling techniques, although still in need of further improvements, provide a powerful and versatile tool to test and analyze numerous loading conditions and design solutions: for instance, it is easy to investigate alterations in the loading conditions experienced by endografts following changes in blood pressure and blood flow and to virtually modify the size of the various components of the device. Figure 17 provides a schematic representation of how computational, experimental, and clinical tools can be combined to better understand and monitor the performance of aortic stent-grafts. Part (a) represents the CFD analysis performed on the geometry given by the aneurysm and the device and provides the distribution of

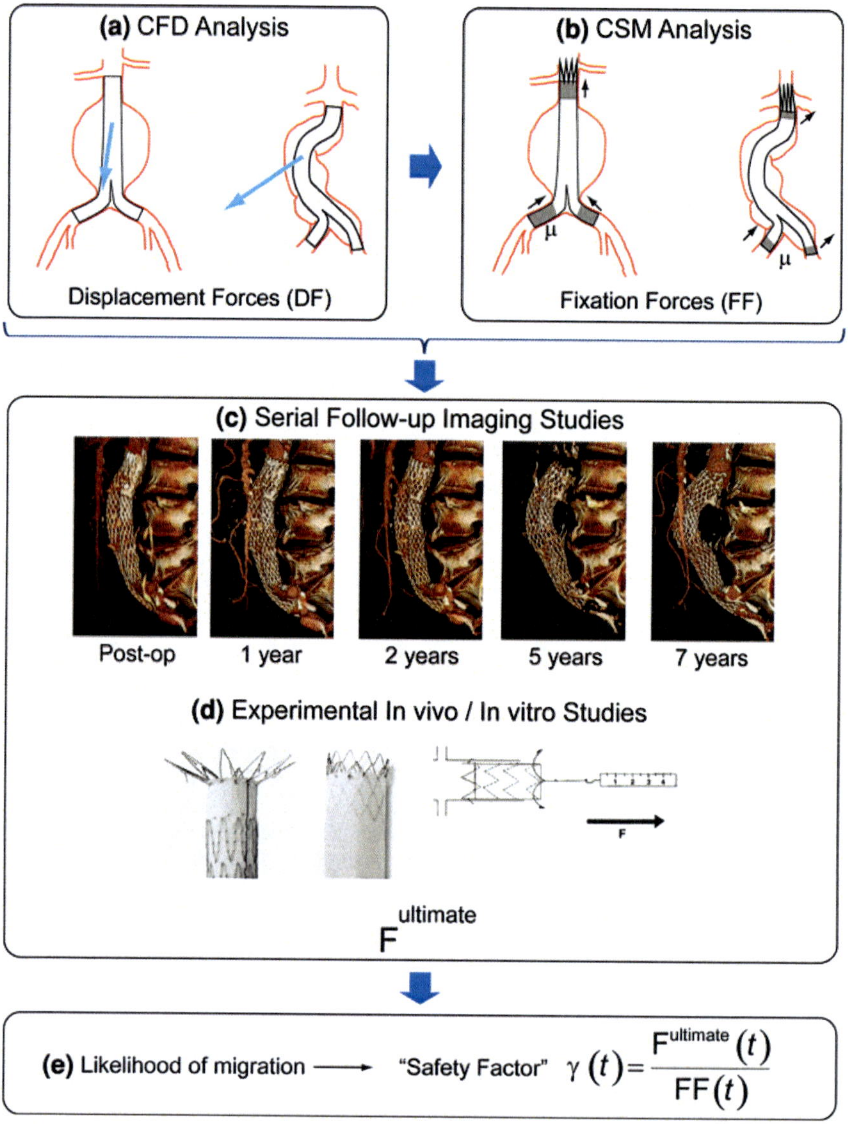

Fig. 17 A computational framework where a combination of computational, clinical, imaging, and experimental tools are used to evaluate the performance of abdominal stent-grafts in vivo

loads or displacement forces DF experienced by the endograft. In part (b) a CSM analysis evaluates the fixation of the device to the vessel wall. The device is subjected to the displacement forces DF obtained in the CFD analysis. The fixation forces FF depend on the fixation characteristics of the device, represented by the coefficient μ in the figure. Part (c) represents the clinical evidence gathered via longitudinal imaging studies of the performance of a given device implanted in a patient: This evidence may be used to define areas of attachment between device and aorta, to quantify the amount of intra-luminal thrombus and its potential constraining effect on the endograft, and to assess the disease state of the vessel wall in terms of atherosclerotic plaque and calcifications. These are all parameters that provide useful information for the CSM analysis depicted in part (b). Finally, in part (d) experimental in vivo and in vitro data provide direct measurement of the fixation response of a given device as a function of oversizing, the characteristics of the luminal surface, etc. Information obtained from (c) and (d) can then be used to estimate the ultimate fixation force that a given device can provide in a specific configuration. Once this ultimate fixation response is evaluated, a likelihood of migration for a given device on a given patient can be estimated by the "safety factor" γ defined as the ratio between the ultimate fixation force the device is able to provide and the actual fixation force.

The paradigm of "virtual prototyping" that has been extensively applied for quite some time in industries such as the automotive and the aeronautical is now beginning to be used in the medical device industry as well. The development and application of new computational and imaging tools and the combination of these tools with clinical and experimental data will further improve the design of devices and will ultimately result in improved patient care and reduced costs.

Acknowledgments This research was supported by the National Institutes of Health (Grants P50 HL083800, and 1 RC1 EB011443-01) and the National Science Foundation (Grant CNS-0619926).

References

1. Amblard, A., Berre, H.W.L., Bou-Said, B., Brunet, M.: Analysis of type I endoleaks in a stented abdominal aortic aneurysm. Med. Eng. Phys. **31**(1), 27–33 (2009)
2. Arko, F.R., Heikkinen, M., Lee, E.S., Bass, A., Alsac, J.M., Zarins, C.K.: Iliac fixation length and resistance to in vivo stent-graft displacement. J. Vasc. Surg. **41**(4), 664–670 (2005)
3. Baum, R.A., Cope, C., Fairman, R.M., Carpenter, J.P.: Translumbar embolization of type 2 endoleaks after endovascular repair of abdominal aortic aneurysms. J. Vasc. Interv. Radiol. **12**(1), 111–116 (2001)
4. Benharash, P., Lee, J.T., Abilez, O.J., Crabtree, T., Bloch, D.A., Zarins, C.K.: Iliac fixation inhibits migration of both suprarenal and infrarenal aortic endografts. J. Vasc. Surg. **45**(2), 250–257 (2007)
5. Brewster, D.C., Cronenwett, J.L., Hallett, J.W., Johnston, K.W., Krupski, W.C., Matsumura, J.S.: Guidelines for the treatment of abdominal aortic aneurysms—Report of a subcommittee of the joint council of the American association for vascular surgery and society for vascular surgery. J. Vasc. Surg. **37**(5), 1106–1117 (2003)
6. Conners, M.S., Sternbergh, W.C., Carter, G., Tonnessen, B.H., Yoselevitz, M., Money, S.R.: Endograft migration one to four years after endovascular abdominal aortic aneurysm repair with the AneuRx device: a cautionary note. J. Vasc. Surg. **36**(3), 476–482 (2002)
7. Dalman, R.L., Tedesco, M.M., Myers, J., Taylor, C.A.: AAA disease—mechanism, stratification, and treatment. In: Tilson, M.D., Kuivaniemi, H., Upchurch, G.R. (eds.) Abdominal Aortic Aneurysm: Genetics, Pathophysiology and Molecular Biology, vol. 1085, pp. 92–109. Blackwell Publishing, Oxford (2006)

8. De Bruin, J.L., Baas, A.F., Buth, J., Prinssen, M., Verhoeven, E.L.G., Cuypers, P.W.M., van Sambeek, M., Balm, R., Grobbee, D.E., Blankensteijn, J.D., Grp, D.S.: Long-Term Outcome of Open or endovascular repair of abdominal aortic aneurysm. N. Engl. J. Med. **362**(20), 1881–1889 (2010)
9. Figueroa, C.A., Taylor, C.A., Chiou, A.J., Yeh, V., Zarins, C.K.: Magnitude and direction of pulsatile displacement forces acting on thoracic aortic endografts. J. Endovasc. Ther. **16**(3), 350–358 (2009)
10. Figueroa, C.A., Taylor, C.A., Yeh, V., Chiou, A.J., Zarins, C.K.: Effect of curvature on displacement forces acting on aortic endografts: a 3-dimensional computational analysis. J. Endovasc. Ther. **16**(3), 284–294 (2009)
11. Figueroa, C.A., Taylor, C.A., Yeh, V., Chiou, A.J., Gorrepati, M.L., Zarins, C.K.: Preliminary 3D computational analysis of the relationship between aortic displacement force and direction of endograft movement. J. Vasc. Surg. **51**(6), 1488–1497 (2010)
12. Fleming, C., Whitlock, E.P., Bell, T.L., Lederle, F.A.: Screening for abdominal aortic aneurysm: a best-evidence systematic review for the US preventive services task force. Ann. Intern. Med. **142**(3), 203–211 (2005)
13. Fulton, J.J., Farber, M.A., Sanchez, L.A., Godshall, C.J., Marston, W.A., Mendes, R., Rubin, B.G., Sicard, G.A., Keagy, B.A.: Effect of challenging neck anatomy on mid-term migration rates in AneuRx endografts. J. Vasc. Surg. **44**(5), 932–937 (2006)
14. Gillum, R.F.: Epidemiology of aortic-aneurysm in the United-States. J. Clin. Epidemiol. **48**(11), 1289–1298 (1995)
15. Greenhalgh, R.M., Brown, L.C., Kwong, G.P.S., Powell, J.T., Thompson, S.G.: Comparison of endovascular aneurysm repair with open repair in patients with abdominal aortic aneurysm (EVAR trial 1), 30-day operative mortality results: randomised controlled trial. Lancet **364**(9437), 843–848 (2004)
16. Greenhalgh, R.M., Brown, L.C., Epstein, D., Kwong, G.P.S., Powell, J.T., Sculpher, M.J., Thompson, S.G., Participants, E.T.: Endovascular aneurysm repair versus open repair in patients with abdominal aortic aneurysm (EVAR trial 1): randomised controlled trial. Lancet **365**(9478), 2179–2186 (2005)
17. Heikkinen, M.A., Alsac, J.M., Arko, F.R., Metsanoja, R., Zvaigzne, A., Zarins, C.K.: The importance of iliac fixation in prevention of stent graft migration. J. Vasc. Surg. **43**(6), 1130–1137 (2006)
18. Heller, J.A., Weinberg, A., Arons, R., Krishnasastry, K.V., Lyon, R.T., Deitch, J.S., Schulick, A.H., Bush, H.L., Kent, K.C.: Two decades of abdominal aortic aneurysm repair: have we made any progress? J. Vasc. Surg. **32**(6), 1091–1098 (2000)
19. Hobo, R., Kievit, J., Leurs, L.J., Buth, J., Collaborators, E.: Influence of severe infrarenal aortic neck angulation on complications at the proximal neck following endovascular AAA repair: a eurostar study. J. Endovasc. Ther. **14**(1), 1–11 (2007)
20. Howell, B.A., Kim, T., Cheer, A., Dwyer, H., Saloner, D., Chuter, T.A.M.: Computational fluid dynamics within bifurcated abdominal aortic stent-Grafts. J. Endovasc. Ther. **14**(2), 138–143 (2007)
21. Kim, H.J., Vignon-Clementel, I.E., Figueroa, C.A., LaDisa, J.F., Jansen, K.E., Feinstein, J.A., Taylor, C.A.: On coupling a lumped parameter heart model and a three-dimensional finite element aorta model. Ann. Biomed. Eng. **37**(11), 2153–2169 (2009)
22. Kleinstreuer, C., Li, Z.: Analysis of biomechanical factors affecting stent-graft migration in an abdominal aortic aneurysm model. J. Biomech. **39**(12), 2264–2273 (2006)
23. Les, A., Yeung, J., Schultz, G., Herfkens, R., Dalman, R., Taylor, C.: Supraceliac and infrarenal aortic flow in patients with abdominal aortic aneurysms: mean flows, waveforms, and allometric scaling relationships. Cardiovasc. Eng. Technol. **1**(1), 39–51 (2010)
24. Li, Z., Kleinstreuer, C., Farber, M.: Computational analysis of biomechanical contributors to endovascular graft failure. Biomech. Model. Mechanobiol. **4**(4), 221–234 (2005)
25. Malina, M., Lindblad, B., Ivancev, K., Lindh, M., Malina, J., Brunkwall, J.: Endovascular AAA exclusion: Will stents with hooks and barbs prevent stent-graft migration? J. Endovasc. Surg. **5**(4), 310–317 (1998)

26. Malladi, R., Sethian, J.A., Vemuri, B.C.: Shape modeling with front propagation—a level set approach. IEEE Trans. Pattern Anal. Mach. Intell. **17**(2), 158–175 (1995)
27. Matsumura, J.S., Brewster, D.C., Makaroun, M.S., Naftel, D.C., Excluder Bifurcated, E.: A multicenter controlled clinical trial of open versus endovascular treatment of abdominal aortic aneurysm. J. Vasc. Surg. **37**(2), 262–271 (2003)
28. Molony, D.S., Callanan, A., Morris, L.G., Doyle, B.J., Walsh, M.T., McGloughlin, T.M.: Geometrical enhancements for abdominal aortic stent-grafts. J. Endovasc. Ther. **15**(5), 518–529 (2008)
29. Morris, L., Delassus, P., Walsh, M., McGloughlin, T.: A mathematical model to predict the in vivo pulsatile drag forces acting on bifurcated stent grafts used in endovascular treatment of abdominal aortic aneurysms (AAA). J. Biomech. **37**(7), 1087–1095 (2004)
30. Murphy, E.H., Johnson, E.D., Arko, F.R.: Device-specific resistance to in vivo displacement of stent-grafts implanted with maximum iliac fixation. J. Endovasc. Ther. **14**(4), 585–592 (2007)
31. Nabi, D., Murphy, E.H., Pak, J., Zarins, C.K.: Open surgical repair after failed endovascular aneurysm repair: is endograft removal necessary? J. Vasc. Surg. **50**(4), 714–721 (2009)
32. Parodi, J.C., Barone, A., Piraino, R., Schonholz, C.: Endovascular treatment of abdominal aortic aneurysms: lessons learned. J. Endovasc. Surg. **4**(2), 102–110 (1997)
33. Rafii, B.Y., Abilez, O.J., Benharash, P., Zarins, C.K.: Lateral movement of endografts within the aneurysm sac is an indicator of stent-graft instability. J. Endovasc. Ther. **15**(3), 335–343 (2008)
34. Resch, T., Ivancev, K., Brunkwall, J., Nyman, U., Malina, M., Lindblad, B.: Distal migration of stent-grafts after endovascular repair of abdominal aortic aneurysms. J. Vasc. Interv. Radiol. **10**(3), 257–264 (1999)
35. Resch, T., Malina, M., Lindblad, B., Malina, J., Brunkwall, J., Ivancev, K.: The impact of stent design on proximal stent-graft fixation in the abdominal aorta: an experimental study. Eur. J. Vasc. Endovasc. Surg. **20**(2), 190–195 (2000)
36. Sahni, O., Muller, J., Jansen, K.E., Shephard, M.S., Taylor, C.A.: Efficient anisotropic adaptive discretization of the cardiovascular system. Comput. Methods Appl. Mech. Eng. **195**(41–43), 5634–5655 (2006)
37. Sampaio, S.M., Panneton, J.M., Mozes, G., Andrews, J.C., Noel, A.A., Kalra, M., Bower, T.C., Cherry, K.J., Sullivan, T.M., Gloviczki, P.: AneuRx device migration: incidence, risk factors, and consequences. Ann. Vasc. Surg. **19**(2), 178–185 (2005)
38. Shadden, S.C., Taylor, C.A.: Characterization of coherent structures in the cardiovascular system. Ann. Biomed. Eng. **36**(7), 1152–1162 (2008)
39. Sternbergh, W.C., Carter, G., York, J.W., Yoselevitz, M., Money, S.R.: Aortic neck angulation predicts adverse outcome with endovascular abdominal aortic aneurysm repair. J. Vasc. Surg. **35**(3), 482–486 (2002)
40. Sternbergh, W.C., Money, S.R., Greenberg, R.K., Chuter, T.A.M., Zenith, I.: Influence of endograft oversizing on device migration, endoleak, aneurysm shrinkage, and aortic neck dilation: results from the Zenith multicenter trial. J. Vasc. Surg. **39**(1), 20–26 (2004)
41. Taylor, C.A., Figueroa, C.A.: Patient-specific modeling of cardiovascular mechanics. Annu. Rev. Biomed. Eng. **11**, 109–134 (2009)
42. Veith, F.J., Baum, R.A., Ohki, T., Amor, M., Adiseshiah, M., Blankensteijn, J.D., Buth, J., Chuter, T.A.M., Fairman, R.M., Gilling-Smith, G., Harris, P.L., Hodgson, K.J., Hopkinson, B.R., Ivancev, K., Katzen, B.T., Lawrence-Brown, M., Meier, G.H., Malina, M., Makaroun, M.S., Parodi, J.C., Richter, G.M., Rubin, G.D., Stelter, W.J., White, G.H., White, R.A., Wisselink, W., Zarins, C.K.: Nature and significance of endoleaks and endotension: summary of opinions expressed at an international conference. J. Vasc. Surg. **35**(5), 1029–1035 (2002)
43. Vignon-Clementel, I.E., Figueroa, C.A., Jansen, K.E., Taylor, C.A.: Outflow boundary conditions for three-dimensional finite element modeling of blood flow and pressure in arteries. Comput. Methods Appl. Mech. Eng. **195**(29–32), 3776–3796 (2006)

44. Vignon-Clementel, I.E., Figueroa, C.A., Jansen, K.E., Taylor, C.A.: Outflow boundary conditions for 3D simulations of non-periodic blood flow and pressure fields in deformable arteries. Comput. Methods Biomech. Biomed. Eng. **13**(5), 625–640 (2010)
45. Wang, K.C., Dutton, R.W., Taylor, C.A.: Improving geometric model construction for blood flow modeling—geometric image segmentation and image-based model construction for computational hemodynamics. IEEE Eng. Med. Biol. Mag. **18**(6), 33–39 (1999)
46. White, G.H., Yu, W.J., May, J., Chaufour, X., Stephen, M.S.: Endoleak as a complication of endoluminal grafting of abdominal aortic aneurysms: classification, incidence, diagnosis, and management. J. Endovasc. Surg. **4**(2), 152–168 (1997)
47. Yushkevich, P.A., Piven, J., Hazlett, H.C., Smith, R.G., Ho, S., Gee, J.C., Gerig, G.: User-guided 3D active contour segmentation of anatomical structures: significantly improved efficiency and reliability. Neuroimage **31**(3), 1116–1128 (2006)
48. Zarins, C.K., Bloch, D.A., Crabtree, T., Matsumoto, A.H., White, R.A., Fogarty, T.J.: Stent graft migration after endovascular aneurysm repair: importance of proximal fixation. J. Vasc. Surg. **38**(6), 1264–1272 (2003)
49. Zarins, C.K., Gewertz, B.L.: Atlas of vascular surgery. Elsevier, Philadelphia (2005)

Experimental Analysis of Endovascular Treatment of AAA and Predictors of Long Term Outcomes

Timothy Corbett, David Molony, Eamon Kavanagh, Pierce Grace, Michael Walsh and Tim McGloughlin

Abstract This chapter describes experimental investigations of parameters which are likely to reduce the ability of an implanted stent-graft for the treatment of Abdominal Aortic Aneurysm (AAA) to resist migration. Idealised AAA analogues were manufactured with realistic wall properties. Both proximal stents and complete stent-graft devices were deployed inside these models and the force required to cause migration during physiological flow was investigated. The effect of stent-graft morphology on the columnar rigidity generated by a stent-graft and on the migration force transmitted to the proximal end of the device was also investigated. Lower wall compliance and pulsatile wall motions due to physiological flow were seen to reduce the fixation of an implanted proximal stent from 8.4 ± 0.32 to 3.7 ± 0.06 N. The results also show that high systolic pressure or low proximal fixation length reduce the force required to migrate a graft from 4.62 ± 0.25 to 2.57 ± 0.11 N in a flexible stent-graft with little longitudinal rigidity. In a fully stented device these correlations were less clear due to the complex compressive behaviour of the device and the increase in iliac fixation when the proximal fixation length was reduced. Longitudinal rigidity was measured in terms of the amount of force to cause 5 mm compression of the graft and was found to provide up to 11.53 N of resistance to migration in a fully stented device which is greater than the resistance afforded by passive proximal stents alone. Even the flexible stent-graft was shown to require up to 5.88 N of

T. Corbett · D. Molony · E. Kavanagh · P. Grace · M. Walsh · T. McGloughlin (✉)
Department of Mechanical and Aeronautical Engineering, MSSi,
Centre for Applied Biomedical Engineering Research (CABER),
University of Limerick, Limerick, Ireland
e-mail: Tim.mcgloughlin@ul.ie

P. Grace
Department of Vascular Surgery, Mid-Western Regional Hospital,
Dooradoyle, Limerick, Ireland

Stud Mechanobiol Tissue Eng Biomater (2011) 7: 247–284

DOI: 10.1007/8415_2011_74

Published Online: 24 April 2011

compressive force to cause 5 mm of device compression due to internal pressure assisting the device in holding its shape. Increasing iliac bifurcation angle or placing the devices in a tortuous configuration was found to reduce the longitudinal rigidity of both devices. The results also showed that the drag force acting on a stent-graft may be somewhat attenuated by compressive forces set up in a non rigid stent-graft model. Both an increase in iliac bifurcation angle and tortuosity was found to increase the migration force on the proximal end of the device from 2.56 to 4.92 N. Tortuosity and higher iliac leg angle were both found to have the double disadvantage of increasing the migration force and decreasing device longitudinal rigidity, while longitudinal rigidity was shown to be crucial to the success of passively fixated stent-grafts. The test methods described in this chapter could be useful in the future preclinical evaluation of stent-grafts and could be useful in the design phase of next generation EVAR devices.

1 Introduction

Clinical management of abdominal aortic aneurysm (AAA) has been evolving for over half a century since DuBost et al. performed the first open surgical repair (OR) in 1952 [15]. During this time OR has demonstrated excellent long term results in patients who were considered healthy enough for this invasive procedure. As with all surgical techniques the desire existed for a less invasive approach which would allow patients who were unfit for OR undergo potentially lifesaving treatment for their aneurismal disease and reduce the cost associated with long ICU stays for postoperative patients of OR. In the late 1980's Volodos et al. performed the first Endovascular AAA repair (EVAR) in Russia [60]. In this technique a collapsed graft is guided to the aneurysm site and allowed to expand creating a new conduit for blood to flow thereby shielding the aneurysm from systemic pressure. These first attempts demonstrated that the surgical technique was feasible but the failure of these first devices also showed that they were operating in a very challenging environment in vivo.

After the genesis of Volodos the first commercially available devices were launched towards the end of the last decade of the twentieth century. EVAR quickly gained a reputation as a procedure with substantially reduced mortality, morbidity, length of ICU stay, operating time and blood loss compared with OR. However these short term benefits were hampered by longer term device failures such as endoleak, device migration, suture and stent breakages, graft tears, endotension and limb kinks/occlusions which in some cases led to aneurysm rupture and death. In less serious cases these failures often necessitated reintervention to remedy their symptoms. Second and third generation devices learned the lessons from previous devices and device performance has improved dramatically. Active fixation in the form of hooks or barbs at the proximal and distal ends gave increased fixation. Columnar rigidity was designed into other devices to aid the passive fixation of the proximal stents. Increased redundancy in terms of the number of sutures or barbs as well as improved metal cutting and finishing techniques has drastically reduced the incidence of graft failures due to breakages of these components. Less porous graft fabric has reduced the prevalance of endotension. Full stenting of device limbs has been responsible for less graft kinks and limb occlusions.

In addition to these device improvements the surgical techniques and limitations have also been perfected. Early devices were oversized in the proximal neck substantially more than is practised today. This served to increase the radial force and therefore fixation of the stent-graft but was shown to be responsible for postoperative neck dilation and migration. Iliac fixation length has also been shown to be important in reducing migration of the proximal end of devices with longitudinal rigidity. Clearer limits of neck length, diameter and angulations that allow for successful placement of a stent-graft have also been defined [6, 16, 55]. With the ability to reconstruct AAA geometries from CT scans surgeons also now have the tools and experience to pick the right device for the anatomy in question.

The result of these improved device designs and surgical experiences has been a major reduction in postoperative failures after EVAR and the endovascular approach is now the preferred treatment in many patients. However, incidents of migration are still reported. Distal migration of the stent-graft is arguably the most serious failure mechanism encountered after EVAR as it can compromise the proximal seal between the stent-graft and the aorta, allowing blood at systemic pressure into the previously excluded aneurysm sack and placing the aneurysm at high risk of rupture.

In the past decade many numerical and experimental studies have been conducted to investigate different aspects of migration. However many of these studies have inherent limitations which are discussed throughout this chapter. In this chapter, new experimental techniques aimed at gaining a better understanding of the parameters influencing migration are presented. Firstly the development of aortic wall analogues that mimic the mechanical behavior of human aortic tissue is presented. These analogues are then used for experimental studies to investigate the force required to dislodge stent-grafts and proximal stents during pulsatile flow and the results from these techniques are discussed in detail. Novel techniques for determining the longitudinal rigidity of stent-grafts and for measuring the migration force that is transmitted to the proximal end of the device during pulsatile flow are also described.

2 Development of Silicone Rubber Aortic Analogues

2.1 Materials Selection and Characterisation

Silicone rubbers have been widely used as aorta analogues in in vitro studies [4, 13, 14, 19, 46]. Wacker Elastosil RT601, (Wacker-Chemie GMBH, Munich, Germany), was used by Doyle et al. in a previous study to manufacture AAA models [14]. This silicone rubber was employed as the base material for this study. Different quantities of Dow Corning 200/5CS silicone fluid (Midland, MI, USA), were added to this material to alter its material properties.

Ten Type 2 dumb bell samples of two different wall materials, 100% RT601 (hereafter "WS") and 90% RT601 (hereafter "WC") by mass, were created by injection moulding and tested in accordance with BS ISO 37 on a uniaxial extensometer with a 1 kN load cell (Tinius Olsen, Surrey, UK). To ensure repeatable results and to minimise the Mullins effect the specimens were preconditioned by deforming the gauge length by 40% ten times before carrying out an actual test. This preconditioning serves to stabilise the stress–strain response of the material [42]. A video extensometer (MESSPHYSIK, Fürstenfeld, Austria) was used to measure the deformation of the specimen gauge length of the material during the test.

Material characterisation was carried out using previously described methods [13]. Briefly, the force extension data from the uniaxial tensile tests were converted to engineering stress and strain. Best fit polynomial trendlines were then fitted to the data for each material. Using these polynomials, twenty representative engineering stress–strain data points were calculated for engineering strains from 0 to 1 for each material and used to define the material in the finite element solver ABAQUS 6.7-1 (Dassault Systems, SIMULA, Providence, RI, USA). The force extension data from the uniaxial tensile tests was converted to true stress (σ) and true strain (ε) using theory for large deformation of incompressible materials (Eqs. 1 and 2).

$$\sigma = \frac{f(1+\varepsilon)}{a_o} \tag{1}$$

$$\varepsilon = \ln\left(\left(\frac{\Delta l}{l_o}\right) + 1\right) \tag{2}$$

Where f is force, a_o is undeformed cross sectional area of the test length of the dumb-bell test pieces and l is length.

The uniaxial test was modelled in FEA with identical geometry to the test specimens. Boundary conditions mimicking the uniaxial test were used. Different strain energy functions were examined, and the most suitable strain energy function selected. To ensure that the optimum strain energy function had been selected, a comparison was made between experimentally determined true stress and strain and the true stress and strain predicted by the finite element analysis of the test specimen. The initial Young's modulus (E_{init}) [48] of the wall materials was calculated by fitting a best fit line to the linear portion of the stress–strain curves. Values of E_{init} for each material are listed in Table 1.

It was determined that for the wall materials a sixth order reduced polynomial strain energy function was most suitable. Due to the incompressible nature of these hyperelastic materials, the reduced polynomial strain energy function takes the form:

$$\text{Reduced polynomial:} \quad W = \sum_{i=1}^{N} C_{i0}(\bar{I}_1 - 3)^i \tag{3}$$

Table 1 Coefficients for strain energy functions and initial Young's modulus for each material

Material	E_{init} (MPa)	SEF type	Coefficients for strain energy functions (MPA)
WS	2.22	R. Poly (6th order)	$C_{10} = 0.2259$, $C_{20} = 0.1821$, $C_{30} = -0.3302$, $C_{40} = 0.3071$, $C_{50} = -0.1178$, $C_{60} = 0.01788$
WC	1.57	R. Poly (6th order)	$C_{10} = 0.2813$, $C_{20} = 0.2263$, $C_{30} = -0.4017$, $C_{40} = 0.3829$, $C_{50} = -0.0148$, $C_{60} = 0.0255$

Where $\bar{I}$ is the left Cauchy strain tensor, and C is a material constant.

The coefficients for the respective strain energy functions for each material are listed in Table 1.

These material models were then used in ABAQUS 6.7-1 to predict the pressure-diameter behaviour of the silicone rubber AAA models (see below).

The wall materials were found to have initial Young's modulus of 2.22 MPa (WS) and 1.57 MPa (WC). These values are comparable to the incremental modulus of aortic tissue at physiological stresses [8, 58]. The compliance and stiffness of the models are also physiological compared to literature at the pressures examined [8, 53, 61].

2.2 Silicone Rubber AAA Model Manufacture

Six idealised wall (AAA_W) models (Fig. 1.) were manufactured for this study (three manufactured from WS and three from WC). The model dimensions were taken from the EUROSTAR database [24] and are an averaged set of dimensions based on a 3413 AAAs. This idealised AAA geometry has been used extensively in previous publications [14, 39, 44–46]. The lumen diameters of the ideal geometry are: neck: 24 mm, aneurysm: 50 mm legs: 12 mm, the ideal AAA geometry has a bifurcation angle of 60° and wall thickness of 2 mm.

The models were manufactured using the previously reported lost wax technique [14, 46]. Briefly, a wax lumen was cast from an aluminium mould. This was placed within a wall mould with a 2 mm cavity into which the silicone rubber wall material was injected. The wall material was allowed to cure at 40°C for 4 h. After curing the wax was melted out of the wall model. The models were then cleaned to remove any remaining wax. The full method is outlined in appendix A.

2.3 Compliance Testing

Each model was placed in the pressurised air system shown in Fig. 2. A precision regulator (Norgren, Warwickshire, UK) was used to adjust the pressure

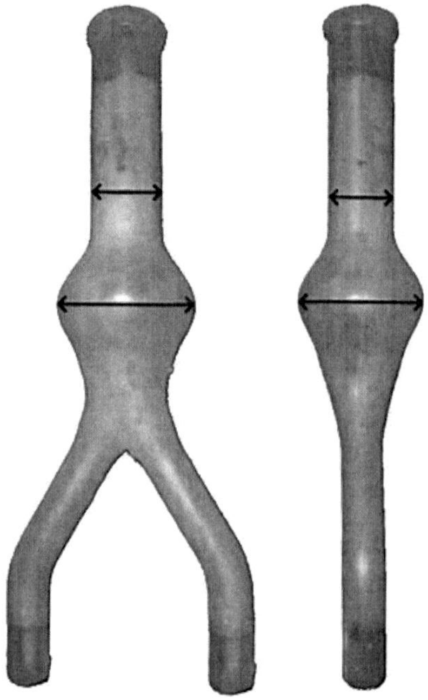

Fig. 1 AAA model viewed from front and side, *arrows* show where measurements were taken during compliance testing

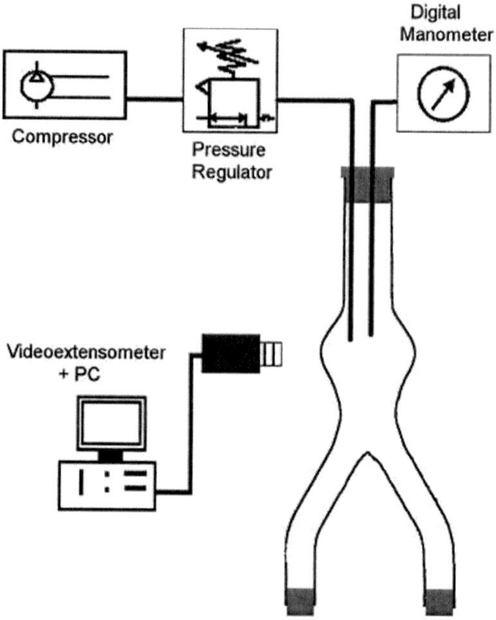

Fig. 2 Pressure-diameter change measurement setup for investigating model compliance

inside the model. A digital manometer (Druck, Leicester, UK) was used to measure the pressure in the system. The models were pressurised from 20 to 160 mmHg in steps of 20 mmHg. A video extensometer (MESSPHYSIK, Fürstenfeld, Austria) was used to measure the change in diameter at each pressure increment.

The models were held at each pressure for 30 s before a diameter measurement was taken. The first location was in the proximal neck of the model, 30 mm above the beginning of the aneurysm section and the second location was the area of maximum diameter of the aneurysm (Fig. 1). The model response to pressurisation was investigated in two planes to allow for the mean compliance to be determined for each model, minimising the effect of any variations in wall thickness on the results. These locations were chosen to determine the compliance of both the aneurysm sac and the proximal neck of the model. The biomechanics in the proximal neck of abdominal aortic aneurysms are of interest as endovascular stent-grafts rely on the proximal neck for fixation within the vessel [6]. Therefore the behaviour in the proximal neck is directly related to the success or failure of EVAR.

When the pressure diameter data was collected for all the models compliance was computed using Eq. 4 [61].

$$\text{Compliance} = \frac{(\Delta A)}{A_{\max}(\Delta P)} \tag{4}$$

Stiffness was then computed using Eq. 5 [53].

$$\text{Stiffness } \beta = \ln\left(\frac{P_{\max}}{P_{\min}}\right)\frac{D_{\min}}{\Delta D} \tag{5}$$

Where D is the outer diameter of the vessel, A is the cross sectional area of the vessel and P is the pressure within the vessel.

Figure 3 shows a comparison of the FEA predicted and mean experimental pressure–diameter behaviour of the models while Table 2 shows the compliance and stiffness in the neck and aneurysm computed for pressures of 120/80 mmHg.

Sonesson et al. [53] reported that the stiffness of the aorta in healthy, non-smoking individuals of age ~69 years was 17.5 ± 5.5. While the stiffness values for the WS models are outside this range, the WC models are within this range. Even though the WS models have stiffness outside the range determined by Sonesson et al. [53] is possible that they may be still physiological. This is because the proximal necks of AAAs are frequently heavily calcified and therefore stiffer than healthy aorta [16].

Vorp et al. [61] reported that the compliance of AAAs of similar size to the models in this study and with ILT was $1.8 - 9.4 \times 10^{-4}$/mmHg (mean 4×10^{-4}/mmHg). All the models in this study had aneurysm compliance within this range.

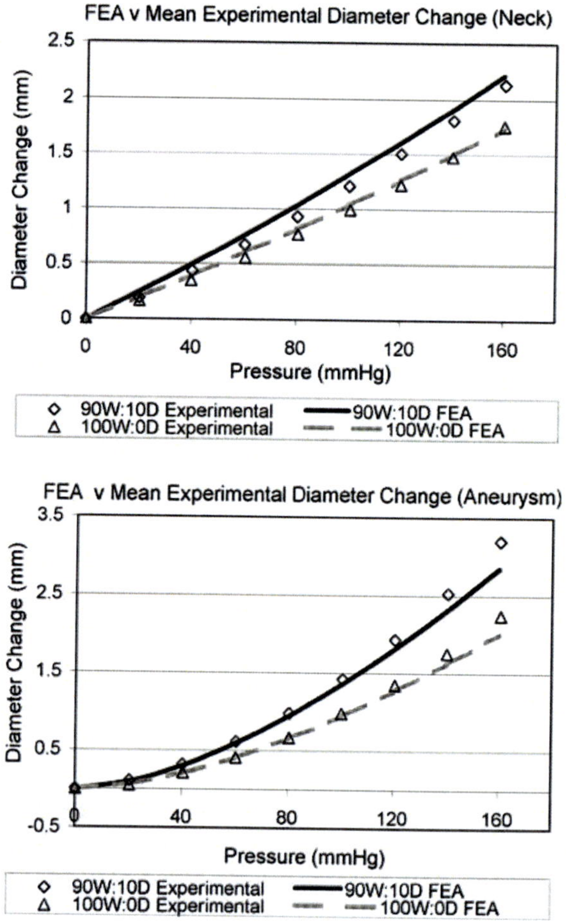

Fig. 3 Comparison of FEA predicted and experimental pressure-diameter behaviour in the neck (*above*) and the aneurysm (*below*) regions of the model

Table 2 Mean experimental stiffness and compliance of the neck and aneurysm section of the models computed for pressures of 120/80 mmHg

Model set	β_{neck}	$\beta_{aneurysm}$	C_{neck} ($\times 10^{-4}$/mmHg)	$C_{aneurysm}$ ($\times 10^{-4}$/mmHg)
WS	25.39 (23.77 – 26.95)	32.38 (31.76 – 33.36)	7.79 (7.31 – 8.29)	6.12 (5.94 – 6.23)
WC	20.44 (19.64 – 21)	23.26 (22.05 – 23.6)	9.61 (9.38 – 9.97)	8.01 (8.01 – 8.9)

3 Use of Silicone Rubber Models for Stent and Stent and Stent-Graft Fixation Tests

A significant amount of research has been conducted into the ability of stents and stent-grafts to resist migration. These have included experiments in sections of cadaver or bovine aorta to in vivo studies in ovine specimens [2, 25, 34, 43, 50, 59].

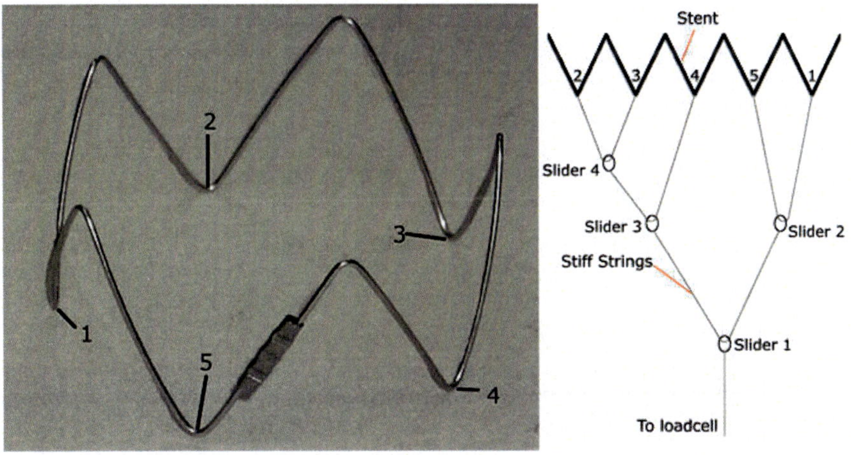

Fig. 4 Proximal talent stent used in the displacement study (*left*) and slider system for distally loading the implanted stent (*right*)

Both of these methods have inherent drawbacks. In the former method the device is inserted in a cylindrical section of cadaver or bovine aorta and the force required to dislodge the device is measured. Here there is no flow in the system and therefore there is no drag force when the full stent-graft is being analysed. Neither is the effect of pulsatile wall motions accounted for. In addition to this the fixation generated by longitudinally stiff stent-grafts due to iliac implantation of the device limbs is not considered [25, 34, 50, 60]. In the latter method the iliac fixation and flow are considered. However the limitations of the ovine model lead to a non-physiological graft-artery anastomosis. In humans endovascular grafts are usually oversized in the region of 10–20% but due to the size of the ovine aorta the oversizing of the graft in these studies was in the region of 30–50% [2, 43]. Such excessive oversizing could greatly increase the radial force generated by a passively fixated stent (no hooks or barbs). This increased radial force could increase the fixation of the stents in contact with the artery wall, thereby increasing the force required to dislodge the stent-graft [2, 43].

The studies described in this section include similar studies to those performed in cylindrical sections of cadaver aorta to validate the use of silicone rubber for displacement force experiments and also studies in pulsatile flow to allow the effect of stent-graft longitudinal stiffness, drag force and pulsatile wall motions to be investigated.

3.1 Stents and Stent-Grafts Investigated

The proximal stent used in this study is shown in Fig. 4. It is a 32 mm diameter (fully expanded) nitinol proximal stent from the Talent (Medtronic Vascular,

Santa Rosa, CA, USA) AAA system. The vertical height of the stent was 15 mm. When this stent was implanted in our 24 mm AAA model the degree of oversizing was in the region of 33% which is higher than the recommended range (10–20%) for EVAR. However these stents are usually uncovered at the proximal end of the Talent device and flare outwards. This allows them to extend beyond the maximum diameter of the graft proper and therefore the oversizing percentage can be higher for these stents.

Stiff polymer strings were attached to the 5 points at the distal end of the stent shown in Fig. 4 using an improved clinch knot. A slider system was used to allow for a load cell to load all 5 points simultaneously using one string. Figure 4 illustrates this setup, for illustration purposes the stent is depicted as if it were cut at one of the points on the proximal end (e.g. between point 1 and 2 in Fig. 4) and uncoiled. Care was taken when assembling this setup so that the strings attached to the stent struts were almost completely longitudinally orientated. This was achieved by placing sliders 2 and 3 in the iliac limbs of the model, as far from the stent as possible. The stents were collapsed in iced water and inserted from the proximal end of the model to a position 27 mm proximal to the beginning of the aneurysm section. When they were in the correct position the model was flushed with water at 37°C to fully expand the stents.

Two stent-grafts were also investigated in this study (Fig. 5). The commercially available AneuRx device (Medtronic Vascular, Santa Rosa, CA, USA) is a fully stented, self-expanding, modular device with a high degree of longitudinal rigidity. A laboratory manufactured device similar to first-generation stent-grafts manufactured by suturing 3 proximal AneuRx nitinol stents to the proximal end of a custom-made open surgery graft (Culzean Medical Devices, Ayrshire, Scotland) was also investigated as a comparison. Laboratory manufactured nitinol Z stents were used at the ends of the iliac legs to ensure that they did not collapse. The stents were manufactured from 0.4 mm nitinol wire and had an outside diameter of 25 mm uncrimped. The vertical height of the stents was 25 mm. The devices had an outside proximal diameter of 28 mm and outside iliac diameters of 14 mm. Therefore, the oversizing of these devices was 16.67% in both the aneurysm neck and the iliac legs, which is similar to the degree of oversizing seen with EVAR.34 It should be noted that although the proximal attachment areas of each stent-graft were similar, they were not identical because the nitinol stent-rings of the AneuRx taper inward near the bifurcation, as can be seen in Fig. 5.

3.2 Stent Displacement Experiments

For the displacement tests the model with the implanted stent in situ was inserted in the system shown in Fig. 6 and described by Corbett et al. [9]. Briefly, the system allows a fluid (at systemic temperature and with physiological pressure and flow waves) to be circulated to provide a realistic aortic flow environment for

Fig. 5 Stent-grafts used in the study: **a** laboratory manufactured device, **b** AneuRx device. Close-up of proximal attachment zone shown below

laboratory testing. Displacement was carried out both during pulsatile flow (PF) and when there was no flow in the system (NF).

For the NF tests the steady flow pump was turned on for 5 min prior to testing to ensure that the stent reached a physiological temperature approaching 37°C. Just before the motorised load cell was switched on the steady pump was stopped and the haemostasis valve for introducing the pressure catheter was opened to allow the pressure in the system to fall to ambient levels. When this had occurred the stent was dislodged from the AAA model without delay to ensure that the temperature remained close to 37°C. The load cell was used to record the maximum force required to completely dislodge the stent from the neck section of the model. This test was performed six times in both the WS and the WC model.

Displacement during PF occurred at pressures of 120/80 and 140/100 mmHg. For these tests both the piston pump and the steady flow pump were switched on and the pressure was adjusted to the correct level. The system was allowed to run for 5 min before the stent was dislodged. Again the load cell was used to measure the maximum force required to dislodge the stent from the neck of the model and each test was performed six times in each model.

3.3 Stent-Graft Displacement Experiments

Displacement force testing was also carried out in a no flow (NF) situation for the stent-grafts to determine the fixation generated by the proximal stents alone [10]. These test were performed similarly to previously described displacement

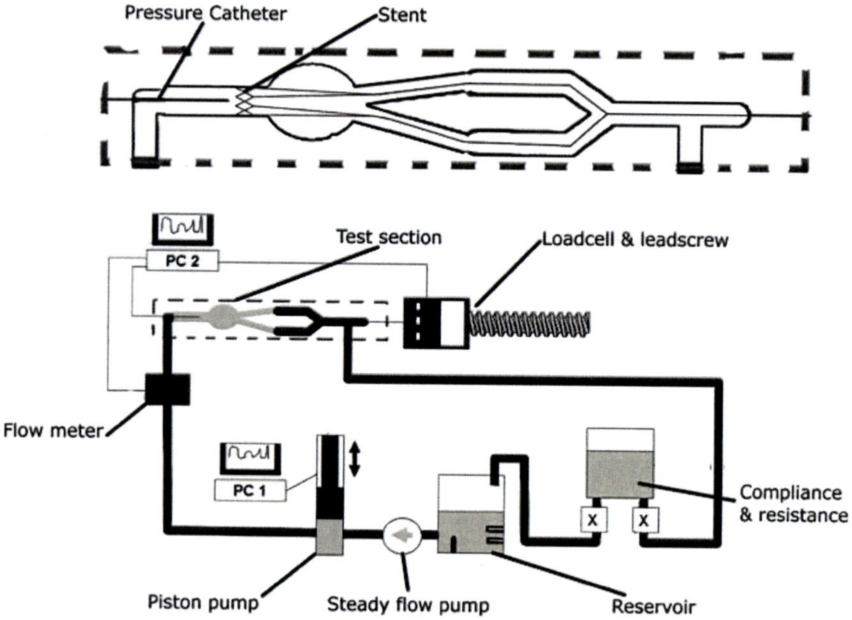

Fig. 6 Flow loop used for displacement tests

experiments undertaken in bovine and cadaver aortic segments [25, 34, 50, 59]. These tests were conducted in the cylindrical neck section of a silicone rubber AAA model. Prior to testing, both stent-grafts were coated with a layer of silicone rubber thrombus analogue to mimic the formation of a thrombus layer in vivo. In vivo a thin layer of thrombus forms on the graft fabric and usually completely seals the graft [21]. This has been mimicked previously by clotting stent-graft devices with materials like gelatine [5]. Gelatine however has been shown to have properties that are not similar to human thrombus [20]. Prior to testing the stent-grafts were pre clotted with an ILT analogue consisting of 45% Wacker RT601 silicone rubber and 55% Dow Corning 200/5CS silicone fluid. This material was found to have similar properties to the medial layer of ILT in a previous study [8]. Each stent-graft was inserted 27 mm into the cylindrical neck section of a silicone rubber model with the aneurysm and iliac legs of the model removed. The silicone rubber neck section with the deployed stent-graft was immersed in water at 37°C to promote full expansion of the nitinol stents. To determine the peak neck fixation force generated by the proximal stents, each stent-graft was dislodged by placing a distal load on the bifurcation as the system was continually flushed with 37°C water. The load mechanism was created using "V" shapes from 2 pieces of 0.5 mm (inner diameter) Harvard syringe pump tubing; a stiff polymer string was passed through these tubes and attached to a load cell on a movable lead screw, and the devices were pulled out at a constant rate of 40 mm/min (Fig. 7). These experiments were repeated 6 times and conducted without any flow in the system.

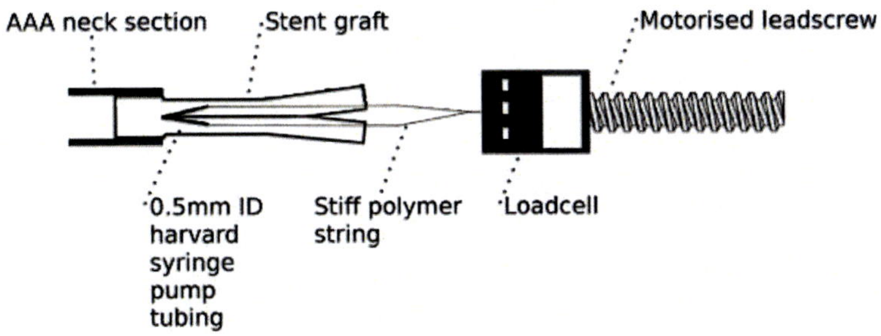

Fig. 7 Test apparatus for in vitro neck fixation displacement experiments; note that the ends of the stent-graft were open to atmospheric pressure

Table 3 List of NF fixation length configurations and PF fixation length and systolic pressure configurations tested

Device	P (mmHg)	Fixation length (mm)
AneuRx	–(NF)	27
	120 (PF)	27, 20, 15
	140 (PF)	27,20
Laboratory manufactured	–(NF)	27
	120 (PF)	27, 20
	140 (PF)	27

N = 6 in all cases except * where N = 3
NF tests conducted in cylindrical neck section with aneurysm and iliacs removed and at atmospheric pressure

The stent-grafts were also tested using a similar setup to Fig. 6 shown with the stent shown in Fig. 6 being replaced with the stent-grafts (for full description of system see [10]. The maximum force required to cause 5 mm displacement was recorded using the load cell 6 times for each device. The 5 mm migration criterion was chosen as it has been frequently used to define migration in clinical trials [7, 54, 55]. A 10 mm migration criterion has also been used [55] but the 5 mm criterion was selected to allow sufficient data to be collected from the 2 stent-grafts without causing severe damage that could affect results.

To investigate the effect of neck fixation length and systolic pressure, the devices were tested in a number of different configurations of pressure and neck fixation length (Table 3).

3.4 Distraction Forces

3.4.1 Stent Distraction Forces

Figure 8 presents a box and whisker plot of the displacement forces for the stent from the WS and WC models in the NF, 120/80 mmHg PF and 140/120 mmHg PF conditions. Table 4 lists the maximum minimum and mean displacement forces from both WS and WC models in the NF and PF conditions.

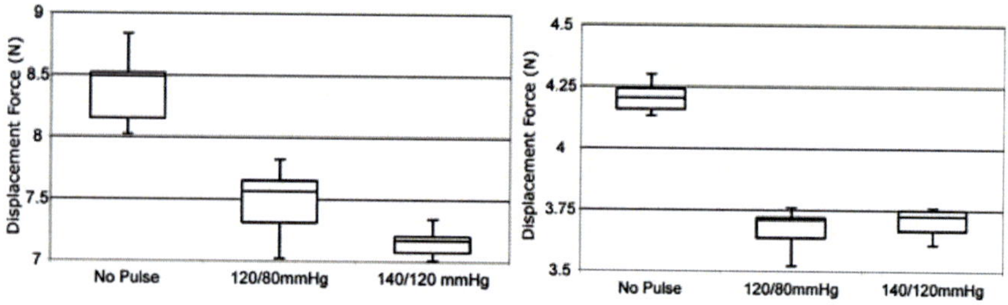

Fig. 8 Box and whisker plot of the displacement forces recorded in the WC model for all configurations

Table 4 Mean and range of displacement forces recorded in all configurations

Model	Pressure (mmHg)	Displacement force ± std dev (N) (range in parenthesis) n = 6
WS	No flow	8.4 ± 0.32 (8.02 – 8.84)
	120/80	7.48 ± 0.3 (7.02 – 7.82)
	140/120	7.16 ± 0.12 (7.01 – 7.35)
WC	No flow	4.21 ± 0.06 (4.13 – 4.3)
	120/80	3.67 ± 0.09 (3.52 – 3.76)
	140/120	3.7 ± 0.06 (3.61 – 3.76)

There was a significant difference between the displacement force in each flow configuration between the WS model and the WC model. On average the displacement force was in the region of 50% lower in the WC model compared to the WS model for each flow configuration.

In the WS models the mean displacement force when there was no flow (no pulsatile wall deformations) in the system was 8.4 ± 0.32 N. There was a significant reduction of 11% (to 7.48 ± 0.3 N) in the mean displacement force from this value when there was pulsatile flow (and pulsatile deformations of the wall) at a pressure of 120/80 mmHg. Further increasing the pressure to 140/100 mmHg was found to slightly decrease the mean displacement force by 4.3% (to 7.16 ± 0.12 N). However this change was found to not be statistically significant.

In the WC models the mean displacement force when there was no flow in the system was 4.21 ± 0.06 N. Similar to the WS models this force dropped significantly to 3.67 ± 0.09 N (12.8%) when there was pulsatile flow at a pressure of 120/80 mmHg in the system. Increasing the pressure to 140/100 mmHg during pulsatile flow was not found to have any significant effect on the displacement force (3.7 ± 0.06 N) when using the WC model.

3.4.2 Stent-Graft Distraction Forces

Figure 9 shows the mean force required to dislodge both the laboratory manufactured device and the AneuRx under neck fixation conditions. In the neck

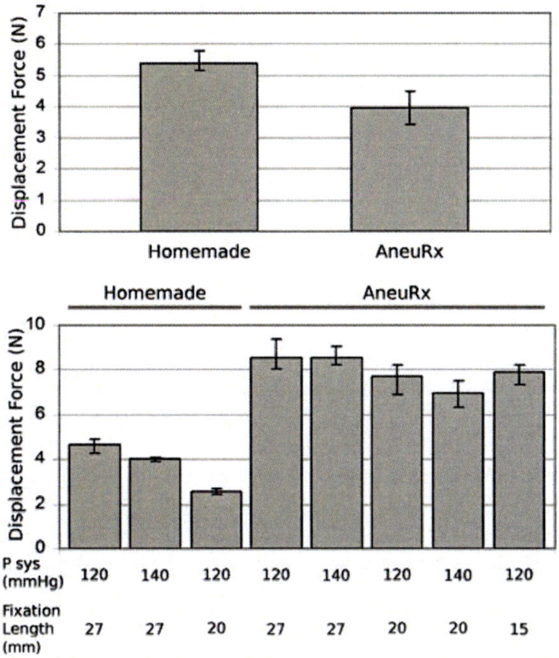

Fig. 9 Stent-graft displacement forces for NF tests (*above*) and stent-graft displacement forces for PF tests (*below*)

fixation tests, once migration of the devices was initiated, there was no recovery of fixation (measured force peaked immediately prior to migration).

Figure 9 also shows the mean force required to dislodge the AneuRx and laboratory manufactured devices under pulsatile flow conditions in the different systolic pressure and fixation length configurations. In all tests undertaken with the laboratory manufactured device, there was again no recovery of fixation once migration was initiated. However, once migration was initiated in the AneuRx device, the force required to migrate the device further continued to increase. Table 5 summarizes the results of all the displacement experiments.

There was a significant difference between the displacement forces for the AneuRx (3.97 ± 0.42 N) and the laboratory manufactured device in the neck fixation tests, with the laboratory manufactured device (5.41 ± 0.23 N) performing better. However, in the pulsatile flow experiments, the worst performing AneuRx configuration performed significantly better than the best performing laboratory manufactured configuration. Subjecting the laboratory manufactured device to the pulsatile flow tests significantly reduced the displacement force from the neck fixation. The inverse was true for the AneuRx device; when subjected to the pulsatile flow tests, the displacement force was significantly increased. In the pulsatile flow tests, an increase in systolic pressure or a decrease in proximal fixation length to 20 mm significantly reduced the displacement force for the laboratory manufactured. Reducing the proximal fixation length to 15 mm further reduced the displacement.

When the proximal fixation of the AneuRx device was 27 mm, there was almost no difference in the mean displacement force when the systolic pressure was 120 and

Table 5 Results of NF and PF displacement tests in all configurations of fixation length and pressure

Device	Proximal fixation mm	Pressure mmHg	Displacement force ± std dev (N) (range in parenthesis)
Laboratory manufactured	27	−(NF)	5.41 ± 0.23 (5.15 – 5.81)
		120 (PF)	4.62 ± 0.25 (4.28 – 4.89)
		140 (PF)	4 ± 0.06 (3.89 – 4.06)
	20	120 (PF)	2.57 ± 0.11 (2.44 – 2.7)
AneuRx	27	−(NF)	3.97 ± 0.42 (3.45 – 4.51)
		120 (PF)	8.52 ± 0.5 (8.03 – 9.38)
		140 (PF)	8.51 ± 0.3 (8.19 – 9.04)
	20	120 (PF)	7.68 ± 0.48 (6.9 – 8.2)
		140 (PF)	6.95 ± 0.49 (6.3 – 7.49)
	15	120 (PF)	7.89 ± 0.39 (7.3 – 8.21)

140 mmHg. However, when the systolic pressure was 120 mmHg and the proximal fixation length was reduced to 20 mm, there was a significant reduction in displacement force. Keeping the systolic pressure at 120 mmHg but reducing the proximal fixation length to 15 mm caused a slight increase in mean displacement force over the previous configuration, but this was not significant. A significant reduction in displacement force was seen when the systolic pressure was increased from 120 to 140 mmHg while keeping the proximal fixation length at 20 mm.

Previous tests by Veerapen et al. [59] and Malina et al. [34] have shown that the displacement force for passively fixated stents and stent-graftsis in the range of 2.5–8 N. In the NF tests the displacement force in the silicone rubber models was within this range, thereby validating the use of these materials for such displacement studies.

4 Investigation of Stent-Graft Longitudinal Rigidity and Influencing Parameters

It is clear that implanting the iliac limbs of the AneuRx stent-graft into the iliac arteries of the model did increase the resistance to migration. This is because the longitudinally stiff device acted like a column and transferred some of the drag force to the distal ends of the device. This in turn acted together with the fixation offered by the proximal stents to provide extra fixation. To quantify this longitudinal rigidity the two stent-grafts already described were subjected to a series of compression tests in this part of the study.

4.1 Compression System

The stent-grafts clotted with the silicone rubber ILT analogue were inserted in the system shown in Fig. 10. A reservoir with a heater element and thermocouple-PID

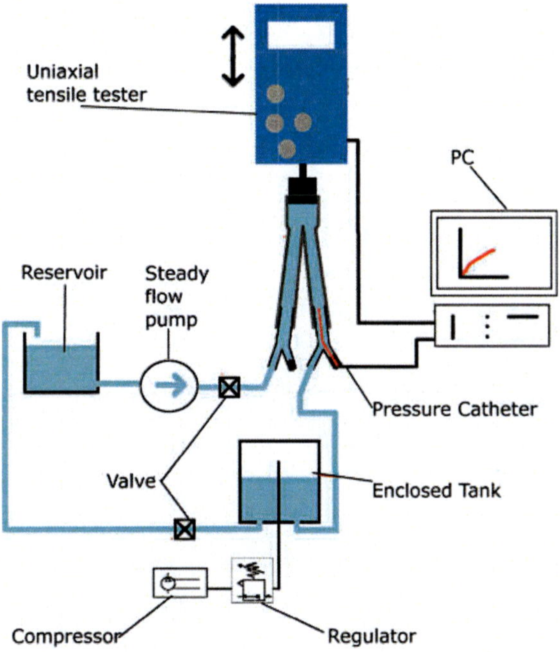

Fig. 10 System used for compressing stent-grafts by 5 mm

temperature control loop contained water at 37°C. Silicone tubing connected the reservoir to a circulating pump (Greylor, Cape Coral, FL, USA). This was connected via silicone tubing via a valve to a "Y" type connector. The "Y" type connectors were held rigidly in place in an adjustable clamping rig (not shown). One leg of the "Y" connectors was inserted into the iliac limbs of the stent-graft. The third leg of each "Y" connector was closed using haemostasis valves. At the proximal end of the stent-graft a 26 mm cylindrical plug was inserted 10 mm (the vertical length of one nitinol stent-ring) into the neck of the stent-graft. The plug was connected to a portable uniaxial tensile tester. This was used to compress the stent-grafts. A pressure catheter was used to monitor pressure in the system and was also connected to the data acquisition PC. Pressure in the system was maintained at a constant level by maintaining a pressurised column of air in the enclosed tank. The air column was connected via a 0.5 bar precision regulator to a compressor. When pressurising the system, clamps were closed to prevent fluid flowing from the stent-graft and enclosed tank back to the reservoir.

4.2 Compression Testing

To determine the effect of morphology and angulations on longitudinal stiffness each stent-graft was compressed by 5 mm in different configurations at physiological temperature and pressure. The pressure catheter was zeroed at ambient

Table 6 Stent-graft configurations tested for longitudinal stiffness

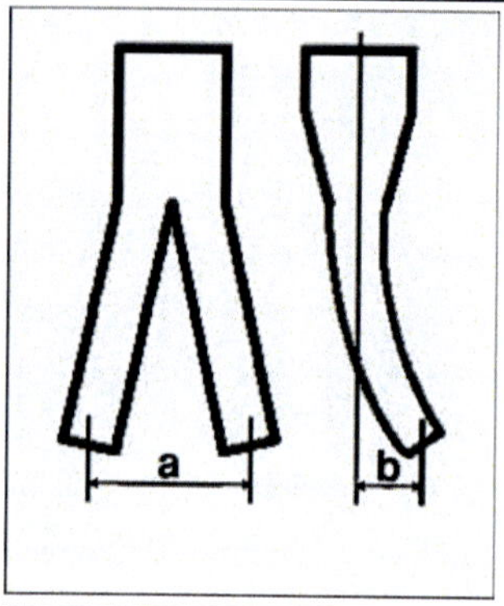

a (mm)	b (mm)	Pressure (mmHg)
25	0	120
75	0	100, 120, 140, 180
75	0	140
75	75	120
75 (limbs crossed)	0	120
100	0	120

pressure and inserted into one of the iliac limbs of the stent-graft via a haemostasis valve. The valves on the silicone tubing between the pump and the stent-graft and between the enclosed tank and the reservoir were opened and water at 37°C was allowed to flow through the system for 5 min. After 5 min the pump was turned off and the tubing valves were closed. The compressor was switched on and the pressure in the stent-graft was adjusted to 120 mmHg. The load cell and height scale were zeroed and then the motorised test stand was used to compress the stent-graft by 5 mm at a constant rate. Once 5 mm compression had been achieved the direction of movement of the load cell and height scale was reversed and the stent-graft was returned to its original length. The test speed was 40 mm/min. During testing the pressure in the device was monitored to ensure that it did not change. During the test the PC recorded the force-deflection behaviour of the devices.

Both stent-grafts were compressed three times in each configuration shown in Table 6. The standard configuration was considered to be the situation where there was no curvature of the body and the L_{iliac} was 75 mm. This configuration was based on both stent-graft geometry and AAA geometry from population means [24].

One configuration involved crossing the limbs of the stent-graft as this has been previously reported as a technique that may be useful in aneurysms with difficult geometry [49]. It has also been reported that it may reduce drag force in some circumstances [22]. The stent-grafts were also tested in one configuration to determine if higher pressure (140 mmHg) had an effect on their longitudinal stiffness.

In most cases when the graft compressed, the iliac bifurcation was also moved downwards and therefore the bifurcation angle had to increase signifying some bending in the iliac limbs. As the devices were compressed they stored energy, somewhat similar to a spring. This energy was released as the devices returned to their original length. Provided that the force-deflection behaviour is linear and that there is minimal buckling of the devices, the spring stiffness (k) which is the gradient of the force deflection curve may be used to characterise the resistance to migration of the devices in different configurations.

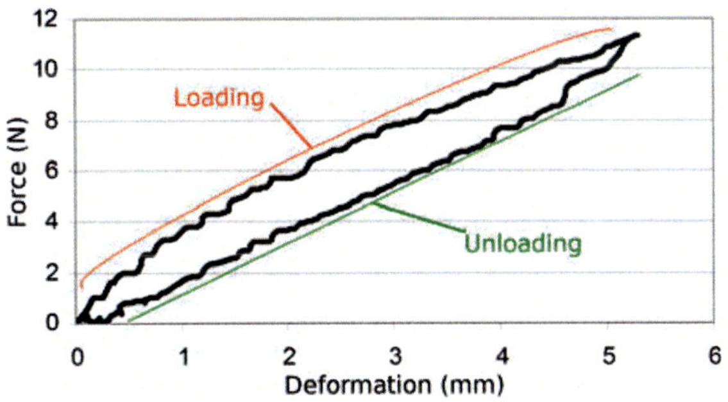

Fig. 11 Typical force-deformation curve for 5 mm stent-graft compression

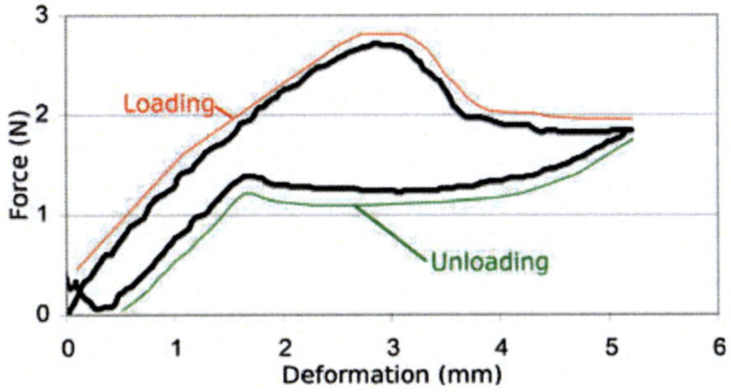

Fig. 12 Force-deformation curve for laboratory manufactured device when L_{iliac} was 100 mm, note the drop in compression force due to kink after ~2.9 mm of compression

The force-deformation behaviour of both devices was broadly similar in all but one configuration. A typical force-deformation plot is shown in Fig. 11. The devices had different load-deformation profiles for the compression and relaxation phases due to hysteresis (Fig. 11). Also it was evident that the compression loading phase had an initial stiff state and after ~2.5 mm of compression the stiffness of the device reduced somewhat. Therefore the initial stiffness (k_i) of the devices is expressed as the slope of the best fit line to the loading curve of force-deflection behaviour for deflections of 0–2.5 mm.

In the case of the laboratory manufactured device when L_{iliac} was 100 mm there was a marked change in the load deformation profile. During initial compression in this configuration the device behaved as before, with the force required to compress the device increasing with deformation. However after ~2.9 mm of compression a kink developed in the iliac limbs of the device causing a buckling effect close to the bifurcation which dramatically reduced the resistance of the device to further compression (Fig. 12). Because of this behaviour, the maximum compressive force (CF_{max}) was also investigated as another measure of longitudinal stiffness as it was envisioned that different device configurations could result in similar initial stiffness but markedly different values of CF_{max}.

The AneuRx device performed significantly better ($p < 0.05$) in all configurations compared to the laboratory manufactured device. The AneuRx device was hardest to compress when the body was straight and L_{iliac} was 25 mm. In the standard configuration (L_{iliac}: 75 mm) there was a slight but significant drop in the

k_i and CF_{max}. Decreasing the pressure to 100 mmHg caused a significant reduction in CF_{max} however there was no significant difference in k_i. An increase in pressure from 120 to 140 mmHg in the standard configuration did appear to slightly increase the resistance of the device to compression but the difference was not statistically significant. Increasing the internal pressure to 180 mmHg caused a significant reduction in CF_{max} but had no effect on k_i. When the iliac limbs were crossed and L_{iliac} was 75 mm with an internal pressure of 120 mmHg there was a slight but significant increase in k_i and CF_{max} from the standard configuration. Increasing L_{iliac} to 100 mm caused a significant decrease in k_i and CF_{max}. The biggest reduction in longitudinal stiffness was seen when the stent-graft was bent out of plane with the legs 75 mm apart and 75 mm from the central axis of the proximal plug.

The laboratory manufactured device was also hardest to compress when the body was straight and L_{iliac} was 25 mm. Similar to the AneuRx there was a significant reduction in k_i and CF_{max} when the device was in the standard configuration. Similar to the AneuRx there was a significant reduction in CF_{max} but no significant change in k_i when the pressure was reduced to 100 mmHg. There was a slight but not significant change in k_i and CF_{max} when the pressure in the device was increased from 120 to 140 mmHg. Similar to the AneuRx an increase in internal pressure to 180 mmHg caused a significant decrease in CF_{max} but this had no effect on k_i. It should be noted that although there was no significant variation in k_i between each pressure step that there was a significant difference in this measure at 100 and 180 mmHg for both the AneuRx and the laboratory manufactured device.

Unlike with the AneuRx device, crossing the iliac limbs of the laboratory manufactured device caused a significant drop in k_i and CF_{max} from the standard configuration. As with the AneuRx, increasing L_{iliac} to 100 mm significantly reduced the stiffness of the device compared with the standard configuration. This stiffness value is 7% lower than that of the laboratory manufactured device in the standard configuration. However, as previously stated, CF_{max} occurred after ~2.9 mm of compression. After 2.9 mm of compression a kink formed at the bifurcation reducing the force required to compress the device further and the 5 mm compression force was much lower than CF_{max}. In this configuration CF_{max} was 44% less than in the standard configuration. Again the poorest performing configuration was when the stent-graft was bent out of plane with L_{iliac} of 75 and 75 mm from the central axis of the proximal plug.

Table 7 presents the mean initial stiffness (k_i) and maximum force (CF_{max}) generated during 5 mm of compression of both devices in all configurations.

5 Investigation of the Axial Migration Force on the Proximal End of a Stent-Graft Model

It is widely hypothesised that the continual downward haemodynamic forces acting on an implanted stent-graft device could be large enough to overcome the fixation force generated by the device and cause migration [27, 32, 35, 37, 38, 40, 51, 56, 63].

Table 7 Force required to compress laboratory manufactured and AneuRx devices 5 mm in all configurations

n = 3 in all cases				Mean CF_{MAX} (N)		Mean k_i (N/mm)	
	a (mm)	b (mm)	P (mmHg)	AneuRx (range)	Laboratory manufactured (range)	AneuRx (range)	Laboratory manufactured (range)
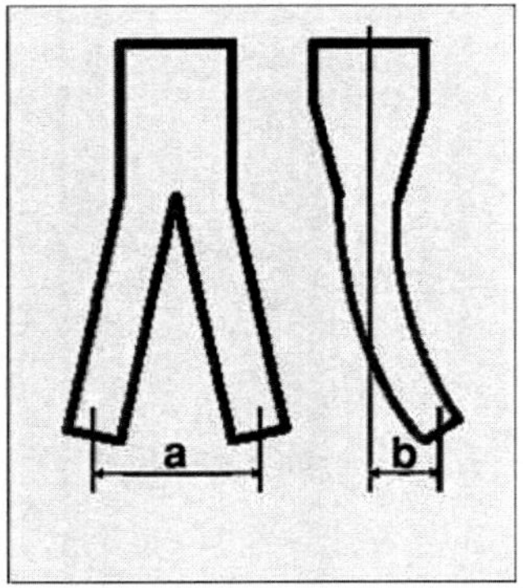	25	0	120	**11.53** (11.28 – 11.96)	**5.88** (5.8 – 5.95)	**3.1** (2.92 – 3.3)	**1.41** (1.35 – 1.44)
	75	0	100	**9.82** (9.67 – 9.97)	**4.7** (4.69 – 4.71)	**2.74** (2.66 – 2.8)	**1.21** (1.19 – 1.22)
	75	0	120	**10.19** (10.08 – 10.4)	**4.83** (4.81 – 4.87)	**2.72** (2.65 – 2.78)	**1.22** (1.21 – 1.23)
	75	0	140	**10.69** (10.38 – 10.63)	**4.9** (4.88 – 4.98)	**2.91** (2.78 – 3.1)	**1.21** (1.16 – 1.24)
	75	0	180	**10.03** (9.95 – 10.08)	**4.74** (4.7 – 4.77)	**2.91** (2.9 – 2.92)	**1.24** (1.24 – 1.25)
	75	75	120	**5.21** (5.1 – 5.32)	**0.92** (0.9 – 0.95)	**1.43** (1.39 – 1.46)	**0.24** (0.21 – 0.25)
	75 (limbs crossed)	0	120	**10.49** (10.42 – 10.64)	**3.63** (3.47 – 3.79)	**2.9** (2.86 – 2.93)	**1.07** (1.05 – 1.09)
	100	0	120	**8.69** (8.21 – 8.99)	**2.72** (2.67 – 2.77)	**2.27** (2.17 – 2.35)	**1.13** (1.12 – 1.13)

Fig. 13 Stent-graft model used in study, note the polymer rib wrapped around the graft to mimic the influence of stents

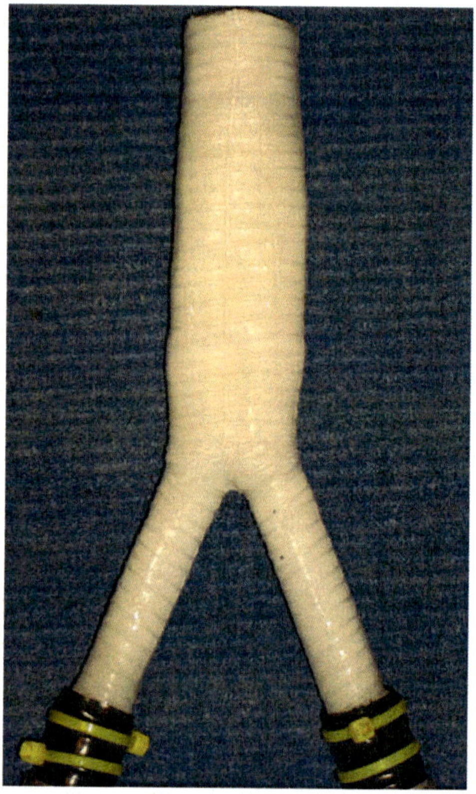

Many studies have detailed numerical analysis to calculate the drag force acting on a device in vivo. It has been shown that pressure, stent-graft diameter and bifurcation angle can influence the drag force experienced by the stent-graft [27, 32, 35, 37, 38, 40, 51, 56, 63].

Experimental studies have also been carried out in rigid stent-graft models and it has been shown that a relatively simple one dimensional steady inviscid analytical model can predict the drag force experienced by the device [35, 56, 63].

While many studies have experimentally and numerically investigated the drag force acting on rigid stent-graft models, the proportion of force that is transferred to the proximal end of a non rigid model has not been fully explored.

Recently we have designed and developed a customised rig for measuring the axial component of proximal migration (ACPM) force acting on a non rigid stent-graft model under physiological conditions. The effect of pressure, bifurcation angle and out of plane geometry on the ACPM force acting on the device is also investigated Corbett et al. [11].

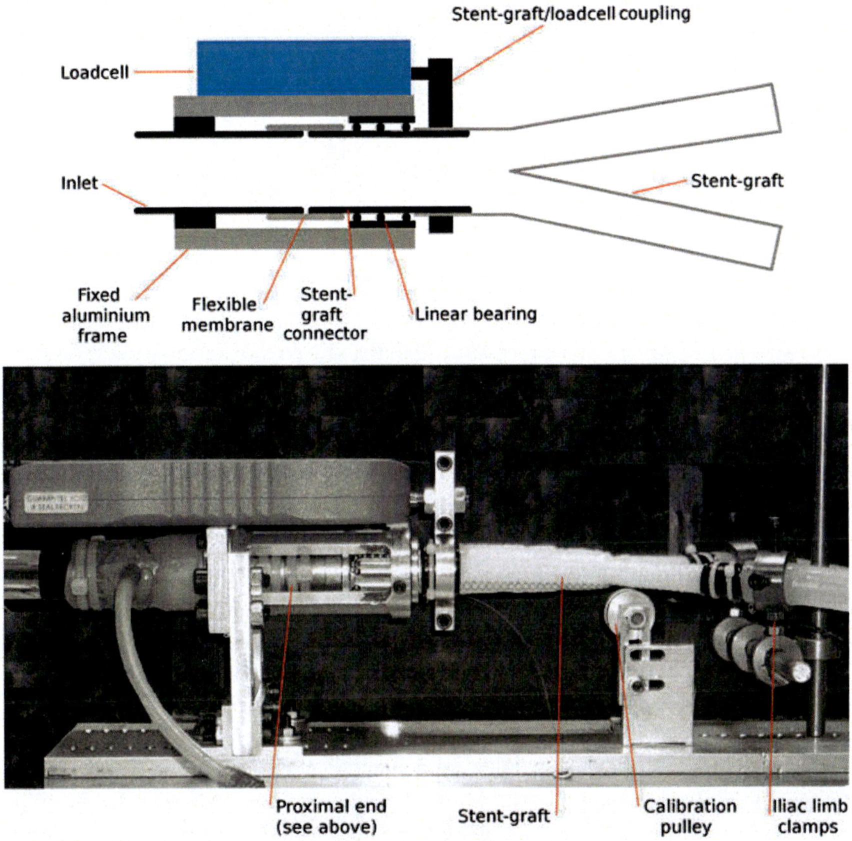

Fig. 14 Migration force rig (proximal end shown above in more detail)

5.1 Stent-Graft Model and Migration Force Rig

The stent-graft model used (Fig. 13) consisted of a custom manufactured Dacron open surgery graft supported by a polymeric rib (Culzean Medical Devices Ayrshire, Scotland). The polymeric rib imparted longitudinal and radial support to the device in a similar fashion to the stents of a stent-graft. The proximal and iliac limb diameters were 24 and 12 mm, respectively. The main body had a length of 100 mm and the iliac limbs were 70 mm long. This geometry was considered as the "average" stent-graft because in the study group of Laheij et al. [24] the mean proximal neck internal diameter was 24 mm and the mean iliac artery internal diameter was 12 mm in a study group of 3,413 patients receiving EVAR. The stent-graft model was again coated with the ILT analogue as previously outlined.

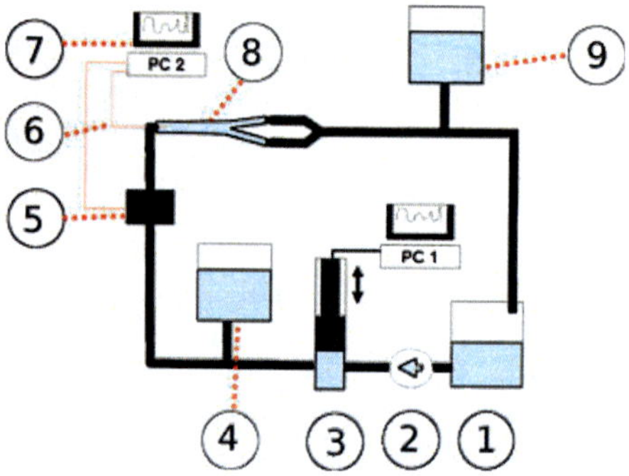

Fig. 15 Flow circuit for ACPM force measurement

The stent-graft model was mounted in the migration force rig shown in Fig. 14 and described by Corbett et al. [11]. At the proximal end an aluminium bracket was bolted to the base. The bracket held a fixed inlet and a linear bearing. The bearing allows for axial motion of the stent-graft outlet so that only the axial component of force was measured by the load cell (BFG 50 N, Mecmesin, West Sussex, UK). To allow the axial force at the proximal end to be measured, the inlet and stent-graft connector were joined by a custom made flexible silicone rubber membrane. The load cell was also mounted on the aluminium bracket and the spur was attached to the proximal end of the stent-graft via a coupling which clamped the stent-graft tightly around the stent-graft connector. A stiff string was also clamped in place by the coupling to allow for the response of the load cell to be calibrated. This was achieved by passing the string over the pulley and out through holes in the base. Calibration weights were attached to the end of the string and the voltage output of the load cell was measured. The iliac limb ends were held rigidly in position by a clamping system that allowed both the bifurcation angle and the body tortuousity to be altered.

5.2 *Flow Loop*

The migration force rig was placed into the flow loop shown in Fig. 15. The flow loop consisted of a reservoir (1) which contained the working fluid. The fluid used was a 42% Glycerine and 58% water mixture. This was found to have a viscosity of 3.16 mPas using a viscometer (Brookfield DV II, Brookfield Engineering Laboratories Inc. Middleboro, MA, USA) and a density of 1100 kg/m^3. A steady flow gear pump (2) (Greylor, Cape Coral, FL, USA) was used to provide the steady component of flow while a computer controlled piston pump (3) provided pulsatile flow. A compliance chamber (4) and an ultrasonic flowmeter probe (ME 19 PXN, Transonic Systems Inc. Ithaca, NY, USA) (5) were placed proximal to a 0.7 m inlet length to ensure that the flow entering the stent-graft was fully developed. This entrance length was calculated based on the mean flow [20]. For confidence the calculated entrance length was doubled.

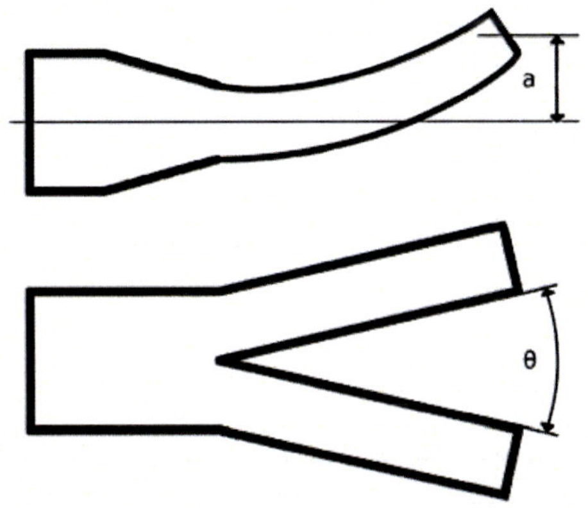

Tortousity, a (mm)	Bifurcation angle, θ (°)
0 (straight)	60
	90
30 (tortuous)	60
	90
60 (very tortuous)	60
	90

Fig. 16 Geometric stent-graft configurations investigated

The flow probe was connected to a Transonic TS410 ultrasonic flowmeter. A pressure catheter (CTC 4F Gaeltec, Isle of Skye, Scotland) (6) was used to monitor pressure in the system. The pressure catheter and the flowmeter were connected to a PC running National Instruments Labview for data acquisition. The test section (8) consisted of the stent-graft model placed in the migration force rig. The analogue output of the load cell was also connected to the data acquisition PC. Distal compliance (9) was placed downstream of the test section. Pressure in the system was tuned by adjusting the volume of air in the compliance chambers and tightening or loosening tubing clamps placed downstream to the distal compliance and at the inlets of both compliances.

To determine the effect of stent-graft morphology on the ACPM force the stent-graft model was analysed in the six configurations shown in Fig. 16. In these configurations the ACPM force was measured at systolic pressures (P_{sys}) of 120 and 140 mmHg.

5.3 System Calibration and ACPM Forces

The output from the load cell on the migration force rig was calibrated in all the configurations of Fig. 16 and at internal pressures of 0, 40, 80, 120 and 160 mmHg.

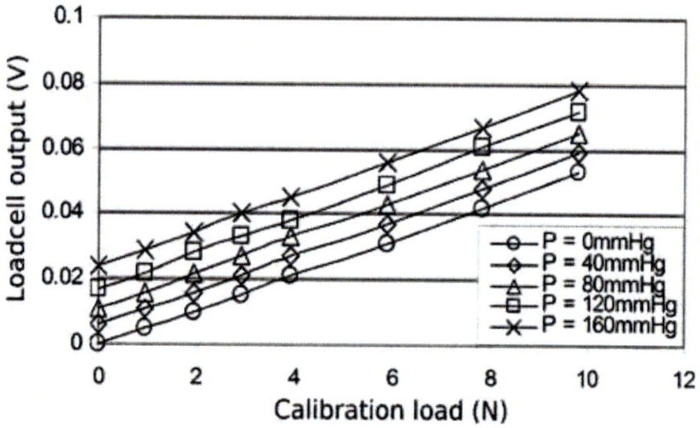

Fig. 17 Calibration of migration force rig. Note linearity at all pressures

Figure 17 shows a sample calibration curve. This curve was generated when the stent-graft was in the tortuous configuration with a bifurcation angle of 90° (see Fig. 16). The voltage output was found to be linear in all configurations and its relationship to applied calibration weight was not affected by pressure except for an initial offset due to the force caused by graft pressurisation (Fig. 17). This initial offset is expected because internal pressurization without any fluid flow actually causes a distal force on the graft. Because the linearity or slope of the calibration curves was not affected by pressure, the load cell voltage could be converted to a force with confidence at all pressures.

A flow wave similar to that described by Fraser et al. [18] (Fig. 18) was generated using the computer controlled piston pump and the steady flow pump shown in Fig. 18. The mean flowrate was 0.8 l/min and the heart rate was ~65BPM [18]. The pressure, flow and ACPM force were sampled at 100 Hz for at least 100 cycles. To reduce the effect of noise on the signal the average flow, pressure and ACPM force waveforms over the 100 cycles were calculated. The pressure in the system was adjusted to values of 120/60 and 140/80 mmHg by adjusting the quantities of water in the compliance chambers and by varying the distal resistance of the system by means of a tubing clamp. The pressure and flow pulses used in the system are shown in Fig. 18.

A typical ACPM force waveform (straight configuration, bifurcation angle 60° and systolic pressure of 120 mmHg) is plotted against the recorded flow and pressure waveforms in Fig. 19. Also shown is a comparison of a raw ACPM force waveform and the averaged ACPM force waveform.

Table 8 lists the maximum, minimum and average ACPM force experienced by the proximal end of the stent-graft model in all configurations during the cycle. Increasing the systolic pressure, the bifurcation angle or the tortousity of the graft caused an increase in measured ACPM force.

To investigate the relationship between drag force and ACPM force, the drag force was calculated for the straight geometries using Eq. 1 [35]. This equation was validated in vitro by Zhou et al. [63] who found that it gave a reasonably accurate estimate of the axial drag force acting on a rigid model of a bifurcated endovascular stent-graft.

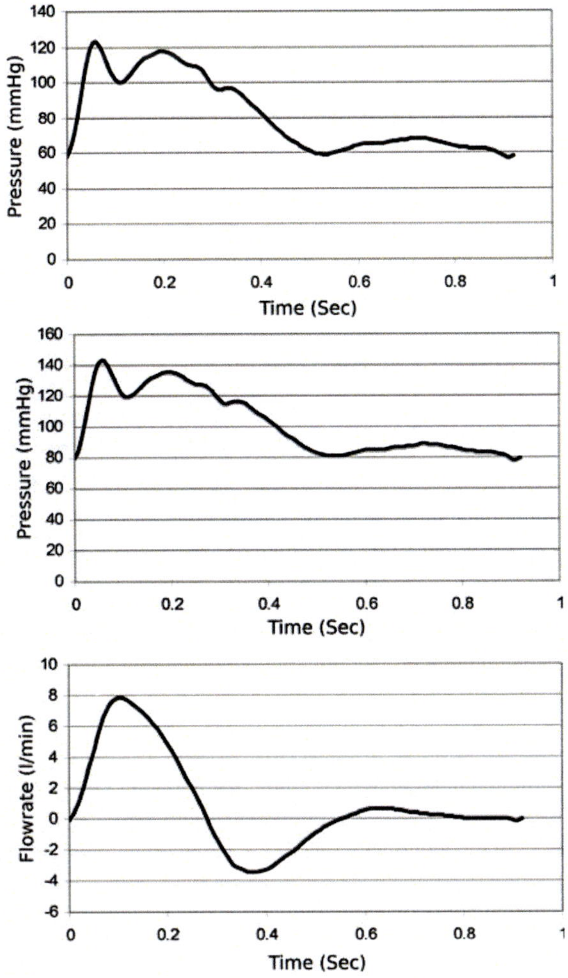

Fig. 18 Pressure pulses (systolic pressures 120 mmHg, *top* and 140 mmHg, *middle*) and flow pulse (*below*) measured in flow circuit during experiment

$$
\begin{aligned}
f_x = {} & P_1A_1 + \rho A_1 U_1^2 - \frac{\rho A_1^2 U_1^2 \cos(\theta/2)}{2A_2} \\
& -2A_2\cos(\theta/2)\left[P_1 + \frac{\rho U_1^2}{2}\left(1 - \frac{A_1^2}{4A_2^2}\right)\right]
\end{aligned} \qquad (6)
$$

Where P is pressure, A is area, U is mean velocity and ρ is density, subscript 1 refers to these parameters at the proximal inlet while subscript 2 refers to these parameters at the distal outlets.

The drag force was computed for the straight geometries using the flow and pressure pulse data shown in Fig. 18 in Eq. 6. Figure 20 shows a comparison of measured ACPM force and calculated drag force for the straight configuration with bifurcation angle of 60^0 and systolic pressure of 120 mmHg. For this configuration

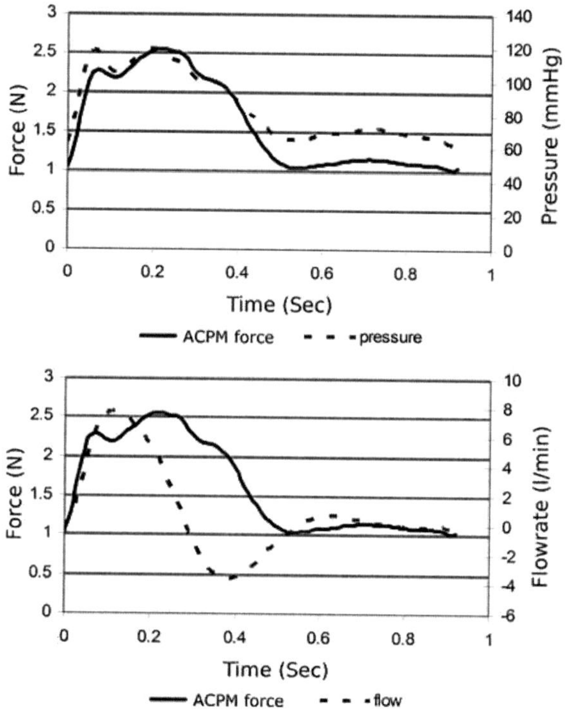

Fig. 19 The measured ACPM force plotted against the pressure pulse (*top*) and against the flow pulse (*below*) for the straight configuration with bifurcation angle of 60° and systolic pressure of 120 mmHg

Table 8 ACPM forces for each configuration of tortousity, bifurcation angle and systolic pressure

Tortousity, a (mm)	Leg angle, θ (°)	ACPM force (N)	
		$P_{sys} = 120$ mmHg	$P_{sys} = 140$ mmHg
0 (straight)	60	Max: 2.56	Max: 3
		Min: 1.02	Min: 1.5
		Mean: 1.61	Mean: 2.09
	90	Max: 3.51	Max: 4.08
		Min: 1.09	Min: 1.82
		Mean: 2.01	Mean: 2.69
30 (tortous)	60	Max: 3.21	Max: 3.81
		Min: 1.36	Min: 2.03
		Mean: 2.07	Mean: 2.72
	90	Max: 3.73	Max: 4.27
		Min: 1.43	Min: 2.11
		Mean: 2.3	Mean: 2.92
60 (very tortous)	60	Max: 3.57	Max: 4.05
		Min: 1.48	Min: 2.07
		Mean: 2.27	Mean: 2.83
	90	Max: 4.29	Max: 4.92
		Min: 2.72	Min: 2.56
		Mean: 1.78	Mean: 3.43

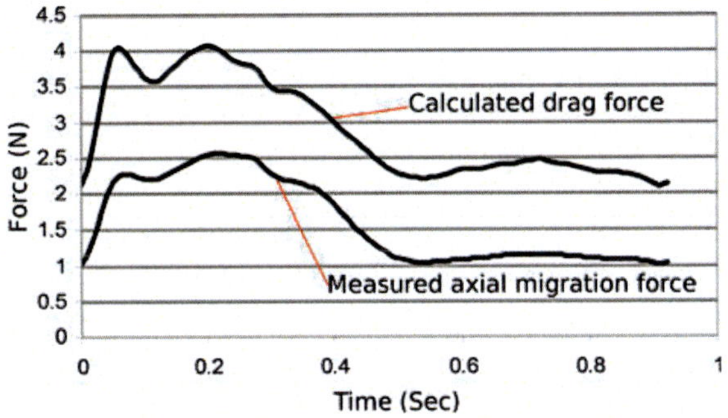

Fig. 20 Comparison of measured ACPM force and calculated drag force for straight configuration with bifurcation angle of 60° and systolic pressure of 120 mmHg

Table 9 Comparison of calculated drag and measured ACPM forces for straight configurations

Leg angle, θ (°)	P_{sys}	Calculated drag force	Measured ACPM force	Drag force attenuated (%)
60	120	4.06	2.56	36.95
	140	4.81	3	37.63
90	120	4.23	3.51	17.02
	140	4.89	4.08	16.56

and pressure the maximum calculated drag force was 4.06 N and the maximum ACPM force was 2.56 N. At these values the ACPM force is 36.95% less than the drag force. Table 9 presents a comparison of the maximum calculated drag force and the maximum measured ACPM force for the two straight configurations.

Using the calibration curves (example curve shown in Fig. 17) it was possible to determine how much of the ACMP force was due to viscous forces and pulsatile flow. This was achieved by comparing the voltage output from the load cell when it was subjected to static pressures of 120 and 140 mmHg (interpolated) with no calibration weights to the voltage when the stent-graft model was subjected to pulsatile flow. The mean difference between static pressure force and flow force was 10.8%. This finding is logical since Zhou et al. [63] found previously that the viscous forces and pulsatility accounted for ~10% of the total maximum drag force acting on their stent-graft model.

6 Discussion and Conclusions

Type IA endoleak is a long term (>one year after EVAR) failure of the seal between the proximal neck of an abdominal aortic aneurysm and an implanted stent-graft. When type I endoleak occurs there is substantial risk of rupture to the aneurysm [6]. The primary cause of this complication is downward movement (migration) of the proximal end of the implanted stent-graft. It has been reported that little or no ingrowth or healing occurs between the implanted stent-graft and the aortic wall [33]. Therefore, migration can be considered as a completely mechanical event. The factors and forces causing migration are not completely understood. However, it can be stated that migration is initiated when the summation of all the downward haemodynamic forces exceeds the summation of all the fixation forces at the proximal end of the device. An extensive amount of research has been undertaken in the past decade to quantitatively investigate the nature of both the downward haemodynamic forces and the fixation force of different devices [2, 25–32, 34, 36, 41, 43, 50, 56, 59, 63]. This research has also extended to an investigation of how parameters such as proximal stent design or fixation lengths increase or decrease these forces. However there were some limitations to previous studies reported in the literature. Chief amongst these were limitations of previously reported experimental fixation experiments. Previous in vitro studies comprised of an investigation of proximal fixation only and previous in vivo studies were limited by unrealistic device oversizing. Because previously reported studies were undertaken with animal tissue with inherent variations in material properties, there was no investigation of how vessel wall properties affected fixation. While it has been known for some time that longitudinal stiffness can assist the fixation of an implanted stent-graft, this parameter has not previously been quantified. The drag force acting on stent-graft models has been previously investigated by numerical, computational and experimental means [26–32, 36, 41, 56, 63]. In previous experimental studies the models were rigid and the effect of this assumption of rigidity on the force transferred to the proximal end was not clear. While some computational studies did include deformable stent-graft models they only reported drag on the total graft and not the force acting on the fixation at the proximal end.

To address the problems with availability and variability of cadaver and animal tissue, a novel adjunct to previously described injection moulding methods was developed. This allowed for the manufacture of models with different wall properties without changing or compromising the geometry of the models. This was facilitated by adding a second material (Dow Corning 200/5CS silicone fluid) to the base material (Wacker RT601). It has been previously demonstrated that a model manufactured from 100% Wacker had physiological properties [41]. The study confirmed this finding and it was also shown that a mixture with a ratio of 90:10 RT601:200/5CS resulted in a physiological model with significantly different properties than the 100% RT601 model. These repeatable models allowed for the model properties to be tightly controlled so that the effect of these properties on fixation could be investigated. In addition to this, the effect of other parameters affecting fixation could be investigated without the inclusion of undesirable variations inherent in cadaver or animal tissue. Uniaxial tensile testing was performed on these materials and strain energy functions representing their behaviour were determined. The pressure-diameter change relationship of the models was investigated using the finite element method and a good correlation was found between the FEA and the experimental results. This validated finite element model and the strain energy functions may be of use in designing future realistic models or numerically modelling stent-graft fixation behaviour.

The use of silicone rubbers for the manufacture of AAA models is not without its limitations. Firstly silicone rubber is known as a material with a high coefficient of static friction. It is possible that the friction behaviour between a stent-graft and the mock artery is different than that between an implanted stent-graft and the tunica intima. However, recent results from collaborative work between the University of Limerick and the University of Arizona suggest that the coefficient of friction between the AneuRx and a silicone rubber tube is somewhat similar to that between an artery and a stent [57]. Vad et al. [57] determined that the coefficient of friction between the AneuRx and a 24 mm ID silicone rubber tube to be 0.065 while previous studies [47, 62] have reported a coefficient of friction of 0.05 between a stent and the artery wall. While the AAA models behaved physiologically at physiological pressures it should be noted that there are differences between the stress–strain behaviour of silicone rubber and aneurysmal tissue. Raghavan et al. [48] found the initial modulus (E_{init}) of aneurysmal tissue to be $\sim$0.56 MPa. It was shown that E_{init} was much higher for the silicone rubber wall materials (1.57 and 2.22 MPa). Therefore care should be taken if using these models at different pressures to ensure that their behaviour remains physiological. While there was no viscoelastic behaviour observed while prestressing the materials during uniaxial tensile testing it is possible that over a long period that these effects might change the behaviour of the models. Therefore it would be prudent to determine how time affects the behaviour of the materials. Until the nature of the time varying behaviour (if any) of the material can be established all models should be used within three months as outlined in BSISO37.

A flow loop was used to investigate what parameters affected the force required to dislodge both stents and stent-grafts from their initial position within silicone rubber models. The in vitro displacement force setup of the flow loop and the silicone models had a number of potential advantages over previous stent and stent-graft fixation tests as discussed previously.

While the potential advantages of using the flow loop and the silicone rubber models were clear, it was not known whether the fixation generated between a the stents or stent-grafts and aortic tissue would be mirrored using the silicone rubber models. To investigate this, previously reported in vitro experiments (similar to those described by Malina et al. [34] and others) were recreated using a cylindrical section of the silicone rubber model rather than aortic tissue. The mean displacement force for the stents and stent-grafts ranged from 3.97 $\pm$ 0.42 to 8.4 $\pm$ 0.32 N which is within the range for passive stents implanted in cadaveric or bovine aorta [34, 59].

Two different stent-grafts were used in the study, a commercially available longitudinally stiff device (AneuRx) and a laboratory manufactured device with little longitudinal stiffness, both with passive fixation. While the latter device was somewhat unrealistic in modern terms it was similar to the first generation of devices and interesting for the purposes of comparison [3]. The oversizing percentage for the proximal end of the devices was within the recommended range for EVAR.

In the silicone rubber AAA models the oversizing percentage was 16.6% in both the proximal neck and the iliac legs when the model was unpressurised. This oversizing percentage was based on the internal diameter of the model. During pulsatile flow the pressure in the models would have reduced the oversized percentage of the stent-graft to 11% at 120 mmHg and to 9.88% at 140 mmHg, thereby reducing the radial force of the proximal stents and therefore reducing the fixation of the proximal stents of the devices. There was a dramatic reduction in the force required to dislodge the laboratory manufactured device due to this reduction and also due to the presence of drag force acting on the device. There was actually an increase in the displacement force for the AneuRx device due to the presence of longitudinal rigidity. Indeed it was shown that the fixation due to longitudinal rigidity of the AneuRx device could be more important than the fixation due to the proximal stents. The experiments during physiological flow also showed that either a reduction in proximal fixation length or an increase in systolic pressure could dramatically reduce the force required to dislodge the laboratory manufactured device. A reduction in proximal fixation length caused the largest drop in displacement force. This is in agreement with previously reported clinical findings that have linked hypertension and short proximal neck length to migration. There was a greater degree of variability in the displacement forces measured for the AneuRx than measured for the laboratory manufactured device. This may have been due in part to slight differences in the compressive behaviour of the AneuRx from test to test. Because of this variability it was somewhat harder to draw conclusions regarding the effect of proximal fixation length and increased systolic pressure on the displacement forces for this device. Shortening the proximal fixation length in our study had the knock on effect of increasing the iliac fixation length. It has been shown that increased iliac fixation length can reduce the risk of migration in the AneuRx device [2]. An initial reduction in proximal fixation length was found to reduce the force required to displace the AneuRx, but the displacement force rose again when the proximal fixation length was further reduced. This was possibly due to increased iliac fixation. The displacement forces measured in this study for the AneuRx were much lower than those reported in the unrealistic case of the ovine model [2, 43]. The results of this work show the benefits of dislodgement analysis in a carefully controlled in vitro environment where large amounts of data can be accumulated quickly and efficiently.

It should be noted that the displacement force measured in these pulsatile flow experiments is the force required to cause migration of the graft in addition to all the haemodynamic forces acting on it. It is the "safety envelope" that must be exceeded before the graft begins to migrate. Alone it is useful for comparing different stent-grafts based on their resistance to migration and could be a useful tool in the preclinical testing of endovascular grafts. However coupled with other information presented in this study it may be possible to determine how changing different parameters could diminish this safety envelope and increase the risk of migration of an implanted device.

To isolate the effect of wall properties and wall pulsatility on fixation, a proximal stent alone was analysed in similar pullout tests but in models with different properties. It was determined that the force required to displace the stent was greater in a less compliant model. It has previously been shown that there is a high amount of variability in the properties of aortic tissue in patients with AAA [52]. Paradoxically this finding could mean that patients with more healthy and compliant fixation sites could possibly be more prone to migration. Obviously care must be taken in analysing the results of this study and while it is an interesting finding it should be further investigated in retrospective clinical trials to determine if the simplifying assumptions of the experiment (geometry, use of stent only) and the use of silicone rubber impinged on the results. This study also showed that the fixation of the stent was reduced due to wall pulsatility and the diameter change due to pressurisation to normal physiological levels. Further increasing the pressure to the hypertensive range was not found to further decrease the fixation of the stent.

Longitudinal stiffness is a secondary fixation mechanism used on a number of commercially available devices and was found to be as important as proximal fixation in the AneuRx device. Previously no methodology for the measurement of longitudinal stiffness existed. A novel test was developed to investigate the increase in fixation due to longitudinal stiffness and to compare the longitudinal stiffness of different devices. A key objective of this test was also to determine how geometrical parameters affect the longitudinal stiffness of endovascular grafts. Two measures of longitudinal stiffness were investigated. The first being the initial stiffness (k_i) of the stent-graft and the second being the maximum force recorded over 5 mm of compression of the device (CF_{max}). The maximum compression force was found to be a more valuable measurement because it could reflect the existence of behaviour (e.g. kinking) that cannot be described by considering the device as a spring. While the better performance of the AneuRx in these tests was expected, an interesting finding was that in an idealised configuration the laboratory manufactured device could actually resist a reasonable amount of compressive force before being compressed to clinically significant levels. Placing the ends of the iliac leg outside the centre plane of the body of the graft significantly reduced the longitudinal stiffness of both devices by almost 50% in the AneuRx and by over 80% in the laboratory manufactured device compared to the ideal configurations. The force required to cause 5 mm compression of the AneuRx device in this configuration was still higher than the fixation due to the proximal stent for this device. Increasing the bifurcation angle was also shown to reduce the amount of force required to cause 5 mm compression of the devices. Another interesting finding from these tests was that a stent-graft kink may significantly reduce the longitudinal stiffness of a stent-graft. Previous studies have linked stent-graft kinking to migration, however it is not known which parameter is the cause and which is the effect [17]. From this finding it appears that if kink occurs in certain areas that it may increase the risk of migration by reducing longitudinal stiffness. It was hypothesised that increasing the pressure might help the devices to hold their shape and resist buckling and compression. However, this proved not to be the case, meaning that increased pressure has no known benefits and many reported disadvantages from the viewpoint of successful aneurysm exclusion.

The drag force acting on bifurcated stent-graft models has previously been calculated both experimentally and numerically [26–32, 36, 40, 56, 63]. However while the drag force acting on the device is an important consideration, it was not clear whether all of the drag force is transferred to the proximal end of the device. The longitudinal stiffness measurements showed that small compressive deformations of the graft could set up relatively large reaction forces which could attenuate the drag force acting on the proximal end of the device. The results of the ACPM force analysis corroborated this hypothesis with the measured axial migration force being much less than the calculated drag force. It was also shown that factors that have previously been found to increase drag (increased pressure, increased bifurcation angle and out of plane tortuosity) also increased the axial migration force The fact that a portion of the drag force is attenuated by compressive forces in the graft, may be an interesting area of investigation when designing new devices, and methods of maximising this attenuation could be investigated. It was also found that on average 10% of the total axial migration force was due to viscous forces and the pulsatile nature of the flow. This finding mirrored the finding of Zhou et al. [63] who experimentally determined that the component of drag force due to these parameters was also 10%. It is important to note that while stiffness is useful in preventing migration that some degree of compressibility may also be important in attenuating the drag force acting on the graft. It is unlikely that any conceivable stent-graft device would be so rigid that it would not compress under normal drag forces and attenuate the drag force somewhat. Indeed it is likely that the AneuRx device is close to the maximum practical stiffness in terms of stent-graft design. This is because stiffer devices do not conform to tortuous vessels easily. As there is always some degree of tortuosity present in the aorta, it follows that stent-graft devices also need some degree of flexibility.

In terms of the safety envelope previously discussed, it is clear that an increase of pressure would not increase the longitudinal stiffness of the device, but would increase the proximal migration force and should decrease the displacement force. This was seen in the laboratory manufactured device but not always in the AneuRx, again possibly due to variability in the results for this stent-graft. Increasing the bifurcation angle or placing the graft in a tortuous configuration would have the double effect of lowering longitudinal stiffness and increasing the proximal migration force, and again this should reduce the size of this safety envelope. It is possible that stent-grafts placed in more compliant arteries could also have a smaller safety envelope than stent-grafts in stiffer arteries.

In this work, symmetrical AAA models were used to determine the force required to displace implanted stent-grafts. In reality AAAs are not symmetrical and there are angulations of the neck, the aneurysm sac and the iliac arteries in both the coronal and the sagittal planes [12]. While these factors could increase the axial migration force and decrease the longitudinal stiffness of an implanted device, they could also have implications for proximal stent attachment. It has been previously shown that high neck angulations can have the effect of “lifting” the proximal stent from the wall in some locations [1]. Methodologies for generating patient specific silicone rubber AAA models have been previously documented [12, 41]. For the purposes of migration investigations, an improved idealised (quasi realistic) geometry would be more useful. This geometry could be based on population means of measures such as neck angle (coronal and sagittal), neck length, neck diameter, length, aneurysm diameter, aneurysm asymmetry [12], iliac angle (coronal and sagittal), and iliac diameter. A model set with variations in critical dimensions (Neck angle, neck diameter, bifurcation angle) from these mean values towards the limit of aneurysm geometry that can be treated by EVAR could be created and the effect of varying these geometrical parameters could be investigated in depth.

Only passively fixated (devices without hooks or barbs on the proximal stent) infrarenal devices were analysed in this study. Silicone rubber has been shown to have relatively low tear strength and if completely penetrated by a number of barbs the model could rupture making testing impossible.

Devices with suprarenal fixation have allowed EVAR to be practised in patients with very short proximal necks. These devices usually have a bare proximal stent that crosses the renal arteries or fenestrations that allow blood to flow into the renal arteries. It has been shown that fenestrated devices with a stent placed in a branching artery resist higher migration forces than non fenestrated devices [64]. Again it would be interesting to determine the amount of force these types of devices could withstand while subjected to physiological flow.

If these proposed tests show that the silicone models cannot withstand rupture due to tearing because of barb penetration and internal pressurisation, a composite model could be developed consisting of a silicone rubber aneurysm sac and iliac arteries and a bovine proximal neck. Models with bovine proximal necks and rigid aneurysms have already been described [23] and an extension of this method would allow for a compliant aneurysm with a biological proximal neck.

References

1. Albertini, J.N., Macierewicz, J.A., Yusuf, S.W., Wenham, P.W., Hopkinson, B.R.: Pathophysiology of proximal perigraft endoleak following endovascular repair of abdominal aortic aneurysms: a study using a flow model. Eur. J. Vasc. Endovasc. Surg. **22**, 53–56 (2001)
2. Arko, F.R., Heikkinen, M., Lee, E.S., Bass, A., Alsac, J.M., Zarins, C.K.: Iliac fixation length and resistance to in vivo stent-graft displacement. J. Vasc. Surg. **41**, 644–671 (2005)
3. Brunkwall, J.: How to design the optimal stent graft—what have we learnt? Scand. J. Surg. **97**, 191–194 (2008)
4. Chong, C.K., How, T.V., Harris, P.L.: Flow visualization in a model of a bifurcated stent-graft. J. Endovasc. Ther. **12**(4), 435–445 (2005)
5. Chong, C.K., How, T.V., Gilling-Smith, G.L., Harris, P.L.: Modelling endoleaks and collateral reperfusion following endovascular AAA exclusion. J. Endovasc. Ther. **10**(3), 424–432 (2003)
6. Chuter, T.A.M.: Stent graft design: the good, the bad and the ugly. Cardiovasc. Surg. **10**, 7–13 (2002)
7. Conners, M.S., Sternbergh, W.C., Carter, G., Tonnessen, B.H., Yoselevitz, M., Money, S.R.: Endograft migration one to four years after endovascular abdominal aortic aneurysm repair with the AneuRx device: a cautionary note. J. Vasc. Surg. **36**, 476–484 (2002)
8. Corbett, T.J., Doyle, B.J., Callanan, A., Walsh, M.T., McGloughlin, T.M.: Engineering silicone rubbers for in vitro studies: creating AAA models and ILT analogues with physiological properties. J. Biomech. Eng. **132**(1), 011008 (2010). doi:10.1115/1.4000156
9. Corbett, T.J., Molony, D.S., Callanan, A., McGloughlin, T.M.: The effect of vessel material properties and pulsatile wall motion on the fixation of a proximal stent of an endovascular graft. Med. Eng. Phys. (2010 in press)

10. Corbett, T.J., Callanan, A., O'Donnell, M.R., McGloughlin, T.M.: An improved methodology for investigating the parameters influencing migration resistance or abdominal aortic stent-grafts. J. Endovasc. Ther. **17**(1), 95–107 (2010)
11. Corbett, T.J., Callanan, A., McGloughlin, T.M.: In vitro measurement of the axial migration force on the proximal end of a bifurcated abdominal aortic aneurysm stent-graft model. Proc. Inst. Mech. Eng. Part H: J. Eng. Med. (2010, in press)
12. Doyle, B.J.: Rupture behaviour of abdominal aortic aneurysms: a computational and experimental investigation. Ph.D. Thesis, University of Limerick (2009)
13. Doyle, B.J., Corbett, T.J., Callanan, A., Walsh, M.T., Vorp, D.A., McGloughlin, T.M.: An experimental and numerical comparison of the rupture locations of an abdominal aortic aneurysm. J. Endovasc. Ther. **16**(3), 322–335 (2009)
14. Doyle, B.J., Morris, L.G., Callanan, A., Kelly, P., Vorp, D.A., McGloughlin, T.M.: 3D reconstruction and manufacture of real abdominal aortic aneurysms: from CT scan to silicone model. J. Biomech. Eng. **130**(3), 034501-1–034501-5 (2008)
15. DuBost, C., Allary, M., Oeconomos, N.: Resection of an aneurysm of the abdominal aorta: reestablishment of the continuity by a preserved human arterial graft, with result after five months. Arch. Surg. **64**, 405–408 (1952)
16. Fairman, R.M., Velazquez, O.C., Carpenter, J.P., Woo, E., Baum, R.A., Golden, M.A., Kritpracha, B., Criado, F.: Midterm pivotal trial results of the talent low profile system for repair of abdominal aortic aneurysm: analysis of complicated versus uncomplicated aortic necks. J. Vasc. Surg. **40**, 1074–1082 (2004)
17. Fransen, G.A.J., Desgranges, P., Laheij, R.J.F., Harris, P.L., Becquemin, J.P.: Frequency, predictive factors, and consequences of stent graft kink following endovascular AAA repair. J. Endovasc. Ther. **10**, 913–918 (2003)
18. Fraser, K., Meagher, S., Blake, J.R., Easson, W.J., Hoskins, P.R.: Characterization of an abdominal aortic velocity waveform in patients with abdominal aortic aneurysm. Ultrasound Med. Biol. **34**, 73–80 (2008)
19. Gawenda M, Knez P, Winter S, Jaschke G, Wassmer G, Schmitz-Rixen T, Brunkwall J.: Endotension is influenced by wall compliance in a latex aneurysm model. Eur. J. Vasc. Endovasc. Surg. **27**(1), 45–50 (2004)
20. Hinnen, J.W., Rixen, D.J., Koning, O.H., van Bockel, J.H., Hamming, J.F.: Development of fibrinous thrombus analogue for in vitro abdominal aortic aneurysm studies. J. Biomech. **40**(2), 289–295 (2007)
21. How, T.V., Guidoin, R., Young, S.K.: Engineering design of vascular prostheses. Proc. Inst. Mech. Eng. Part H. J. Eng. Med. **206**(2), 61–71 (1992)
22. Hyun, S., Hyun, Y.E., Klyachkin, M.: Effects of Endovascular Graft Morphology on the Migration Force. Paper # 192887 Proc. 10th American Society of Mechanical Engineers (ASME) Summer Bioengineering Conference Marco Island, Florida, USA; June 25–29, (2008)
23. Kratzberg, J.A., Golzarian, J., Raghavan, M.L.: Role of graft oversizing in the fixation strength of barbed endovascular grafts. J. Vasc. Surg. 49, 1543–1553 (2009)
24. Laheij, R., Van Marrewijk, C., Buth, J.: Progress report including the data of the overall patient cohort. EUROSTAR Data Registry Centre, p. 8, January (2001)
25. Lambert, A.W., Williams, D.J., Budd, J.S., Horroks, M.: Experimental assessment of proximal stent-graft (InterVascularTM) fixation in human cadaveric infrarenal aortas. Eur. J. Vasc. Endovasc. Surg. **17**, 60–65 (1999)
26. Li, Z., Kleinstreur, C.: Effects of major endoleaks on a stented abdominal aortic aneurysm. J. Biomech. Eng. **128**, 59–68 (2006)
27. Li, Z., Kleinstreur, C.: Analysis of biomechanical factors affecting stent-graft migration in an abdominal aortic aneurysm model. J. Biomech. **39**, 2264–2273 (2006)
28. Li Z, Kleinstreur C (2006) Computational analysis of type II endoleaks in a stented abdominal aortic aneurysm model. J. Biomech. **39**(14), 2573–2582.
29. Li, Z., Kleinstreur C.: Fluid-structure interaction effects on sac-blood pressure and wall stress in a stented aneurysm. J. Biomech. Eng. **127**, 662–671(2005)

30. Li, Z., Kleinstreur, C.: Blood flow and structure interactions in a stented abdominal aortic aneurysm model. Med. Eng. Phys. **27**, 369–382 (2005)
31. Li, Z., Kleinstreur, C.: Computational analysis of biomechanical contributors to possible endovascular graft failure. Biomech. Mod. Mechanobiol. **4**, 221–234 (2005)
32. Liffman, K., Lawerence Brown, M.M.D., Semmens, J.B., Bui, A., Rudman, M., Hartley, D.E.: Analytical modelling and numerical simulation of forces in an endoluminal graft. J. Endovasc. Ther. **8**, 358–371 (2001)
33. Major, A., Guidoin, R., Soulez, G., Gaboury, L.A., Cloutier, G., Saproval, M., Douville, Y., Dionne, G., Geelkerken, R.H., Petrasek, P., Lerouge, S.: Implant degradation and poor healing after endovascular repair of abdominal aortic aneurysms: an analysis of explanted stent-grafts. J. Endovasc. Ther. **13**, 457–467 (2006)
34. Malina, M., Lindblad, B., Ivancev, K., Lindh, M., Malina, J., Brunkwall, J.: Endovascular AAA exclusion: will stents with hooks and barbs prevent stent-graft migration? J. Endovasc. Surg. **5**, 310–317 (1998)
35. Mohan, I.V., Harris, P.L., van Marrewijk, C.J., Laheij, R.J., How, T.V.: Factors and forces influencing stent-graft migration after endovascular aortic aneurysm repair. J. Endovasc. Ther. **9**, 748–755 (2002)
36. Molony, D.S., Callanan, A., Walsh, M.T., Kavanagh, E.K., McGloughlin, T.M.: Fluid-structure interaction of a patient-specific abdominal aortic aneurysm treated with an endovascular stent-graft. BioMed. Eng. OnLine **8**, 24 (2009). doi:10.1186/1475-925X-8-24
37. Molony, D.S., Callanan, A., Morris, L.G., Doyle, B.J., Walsh, M.T., McGloughlin, T.M.: Geometrical enhancements for abdominal aortic stent-grafts. J. Endovasc. Ther. 15, 518–529.
38. Morris, L., Delassus, P., Grace, P., Wallis, F., Walsh, M., McGloughlin, T.: Effects of flat, parabolic and realistic steady flow inlet profiles on idealised and realistic stent graft fits through abdominal aortic aneurysms (AAA). Med. Eng. Phys. **28**, 19–26 (2006)
39. Morris, L., O'Donnell, P., Delassus, P., McGloughlin, T.: Experimental assessment of stress patterns in abdominal aortic aneurysms using the photoelastic method. Strain **40**(4), 165–172 (2004)
40. Morris, L., Delassus, P., Walsh, M., McGloughlin, T.: A mathematical model to predict the in vivo pulsatile drag forces acting on bifurcated stent grafts used in endovascular treatment of abdominal aortic aneurysm (AAA). J. Biomech. **37**, 1087–1095 (2004)
41. Morris, L.G.: Numerical and experimental investigation of mechanical factors in the treatment of abdominal aortic aneurysms. Ph.D. Thesis, University of Limerick (2004)
42. Mullins, L.: Softening of rubber by deformation. Rubber Chem. Technol. **42**(1), 339–362 (1969)
43. Murphy, E.H., Johnson, E.D., Arko, F.R.: Device-specific resistance to in vivo displacement of stent-grafts implanted with maximum iliac fixation. J. Endovasc. Ther. **14**, 585–592 (2007)
44. O' Brien, T.P., Walsh, M.T., Morris, L.G., Grace, P.A., Kavanagh, E.G., McGloughlin, T.M.: Numerical and Experimental Techniques for the Study of Biomechanics in the Arterial System. in Biomechanical Systems Technology. World Scientific Publishing Company, pp. 233–270, Singapore, Chap. 7 (2008)
45. O' Brien, T., Morris, L., McGloughlin, T.: Evidence suggests rigid aortic grafts increase systolic blood pressure: results of a preliminary study. Med. Eng. Phys. **30**(1), 109–115 (2007)
46. O' Brien, T., Morris, L., O' Donnell, M., Walsh, M., McGloughlin, T.: Injection-moulded models of major and minor arteries: the variability of model wall thickness owing to casting technique. Proc. Inst. Mech. Eng. Part H: J. Eng. Med. **219**(5), 381–386 (2005)
47. Petrini, L., Migliavacca, F., Massarotti, P., Schievano, S., Dubini, G., Auricchio, F.: Computational studies of shape memory alloy behaviour in biomedical applications. J. Biomech. Eng. **127**, 716–725 (2005)
48. Raghavan, M.L., Webster, M.W., Vorp, D.A.: Ex vivo biomechanical behaviour of abdominal aortic aneurysm: assessment using a new mathematical model. Ann. Biomed. Eng. **24**, 573–582 (1996)

49. Ramaiah, V.G., Thompson, C.S., Shafique, S., Rodriguez, J.A., Ravi, R., DiMungo, L., Diethrich, E.B.: (2002) Crossing the Limbs: A Useful Adjunct for Successful Deployment of the AneuRx Stent-Graft. J. Endovasc. Ther. **9**(5):583–586
50. Resch, T., Malina, M., Lindblad, B., Brunkwall, J., Ivancev, K.: The impact of stent design on proximal stent-graft fixation in the abdominal aorta: an experimental study. Eur. J. Vasc. Endovasc. Surg. **20**, 190–195 (2000)
51. Roy, S.A., West, K., Rontala, R.S., Greenberg, R.K., Banerjee, R.K.: In vitro measurement and calculation of drag force on iliac limb stent-graft in a compliant arterial wall model. Mol. Cell Biomech. **4**, 211–226 (2007)
52. Sonesson, B., Hansen, F., Lanne, T.: Abdominal aortic aneurysm: a general defect in the vasculature with focal manifestations in the abdominal aorta? J. Vasc. Surg. **26**, 247–254 (1997)
53. Sonesson, B., Lanne, T., Vernersson, E., Hansen, F.: Sex difference in the mechanical properties of the abdominal aorta in human beings. J. Vasc. Surg. **20**(6):959–969 (1994)
54. Sampaio, S.M., Panneton, J.M., Mozes, G., Andrews, J.C., Noel, A.A., Kalra, M., Bower, T.C., Cherry, K.J., Sullivan, T.M., Gloviczki, P.: AneuRx device migration: incidence, risk factors, and consequences. Ann. Vasc. Surg. **19**, 178–185 (2005)
55. Sternbergh, W.C., Money, S.R., Greenberg, R.K., Chuter, T.A.: Influence of endograft oversizing on device migration, endoleak, aneurysm shrinkage, and aortic neck dilation: results from the Zenith Multicenter Trial. J. Vasc. Surg. **39**, 20–26 (2004)
56. Sutalo, I.D., Liffman, K., Lawerence Brown, M.M.D., Semmens, J.B.: Experimental force measurement on a bifurcated endoluminal stent-graft model: comparison with theory. Vascular **13**, 98–106 (2005)
57. Vad, S., Eskinazi, A., Corbett, T., McGloughlin, T., Vande Geest, J.: Determination of coefficient of friction for self expanding stent-grafts. J. Biomech. Eng. **132**, 121007 (2010) doi:10.1115/1.4002798
58. Vallabhaneni, S.R., Gilling-Smith, G.L., How, T.V., Carter, S.D., Brennan, J.A., Harris, P.L.: Heterogeneity of tensile strength and matrix metalloproteinase activity in the wall of abdominal aortic aneurysms. J. Endovasc. Ther. **11**(4), 494–502 (2004)
59. Veerapen, R., Dorandeu, A., Serre, I., Berthet, J.P., Marty-Ane, C.H., Mary, H., Alric, P.: Improvement in proximal aortic endograft fixation: an experimental study using different stent-grafts in human cadaveric aortas. J. Endovasc. Ther. **10**, 1101–1109 (2003)
60. Volodos, N.L., Karpovich, I.P., Troyan, V.I., Kalashnikova, Yu.V., Shekhanin, V.E., Ternyuk, N.E., Neoneta, A.S., Ustinov, N.I., Yakovenko, L.F.: Clinical experience of the use of self-fixing synthetic prostheses for remote endoprosthetics of the thoracic and the abdominal aorta and iliac arteries through the femoral artery and as intraoperative endoprosthesis for aorta reconstruction. VASA Suppl. **33**, 93–95 (1991)
61. Vorp, D.A., Mandarino, M., Webster, M.W., Gorcsan, J.: Potential influence of intraluminal thrombus on abdominal aortic aneurysm as assessed by a new non-invasive method. Cardiovasc. Surg. **4**(6), 732–739 (1996)
62. Wu, W., Qi, M., Liu, X.P., Yang, D.Z., Wang, W.Q.: Delivery and release of nitinol stent in carotid artery and their interactions: a finite element analysis. J. Biomech. **40**, 3034–3040 (2007)
63. Zhou, S.N., How, T.V., Black, R.A., Vallabhaneni, S.R., Mcwilliams, R., Brennan, J.A.: Measurement of pulsatile haemodynamic forces in a model of a bifurcated stent-graft for abdominal aortic aneurysm repair. Proc. Inst. Mech. Eng. Part H: J. Eng. Med. **222**, 543–549 (2008)
64. Zhou, S.S., How, T.V., Rao Vallabhaneni, S., Gilling-Smith, G.L., Brennan, J.A., Harris, P.L., McWilliams, R.: Comparison of the fixation strength of standard and fenestrated stent-grafts for endovascular abdominal aortic aneurysm repair. J. Endovasc. Ther. **14**, 168–175 (2007)

Thoracic Aortic Aneurysms—Clinical Assessment and Treatment

Moqueet A. Qureshi, Brain D. Conway and Roy K. Greenberg

Abstract Aneurysmal dilatation of the aorta can occur along all the 3 divisions of the thoracic aorta and most commonly affect males in the sixth and seventh decade of life. These aneurysms, if remained un-noticed, grow and eventually dissect or rupture (or both). They are mostly degenerative in nature and are often discovered incidentally. A number of diagnostic modalities are used to diagnose them and Computed Tomography (CT) of the Aorta, has revolutionized the way we diagnose, follow and treat these aneurysms using different semi-automated algorithms. Proximal thoracic aneurysms (ascending and arch) are predominantly treated using an open approach, whereas distal aneurysms (descending) can be treated utilizing either an open or an endovascular approach, with the endovascular approach being favored for degenerative descending aortic aneurysms. More extensive aneurysms can be treated using hybrid procedures. The knowledge of different engineering principles, such as biomechanics of the thoracic aorta and computational flow dynamics has allowed us to understand the forces acting on these thoracic endografts and has and will provide further insights on how to modify the currently available devices and eventually patient outcomes.

M. A. Qureshi · R. K. Greenberg
Departments of Vascular and Endovascular Surgery, Cleveland Clinic Foundation, Cleveland, OH, USA

R. K. Greenberg
Thoracic and Cardiovascular Surgery, Cleveland Clinic Foundation, Cleveland, OH, USA

B. D. Conway · R. K. Greenberg (✉)
Biomedical Engineering, Cleveland Clinic Foundation, Desk H32, 9500 Euclid Avenue, Cleveland, OH, USA
e-mail: greenbr@ccf.org

Stud Mechanobiol Tissue Eng Biomater (2011) 7: 285–329
DOI: 10.1007/8415_2011_75

Published Online: 30 April 2011

An aneurysm is defined as a localized dilatation of the aorta in excess of 50%, including all three layers of the vessel (intima, media, and adventitia) [1]. An aneurysm can be localized to a single aortic segment, or it can involve multiple segments. Thoracoabdominal aortic aneurysms, for example, involve both the descending thoracic aorta and the abdominal aorta. In the most extreme cases, the entire aorta is aneurysmal; this condition is often called mega-aorta.

1 Classification

Aortic aneurysms can be either "true" or "false." True aneurysms can take two forms: fusiform and saccular. Fusiform aneurysms are more common and are symmetrical dilatations of the aorta. Saccular aneurysms are localized outpouchings of the aorta. False aneurysms, also called pseudoaneurysms, are leaks in the aortic wall that cause blood to collect in pouches of scar tissue on the exterior of the aorta.

Aneurysms of the thoracic aorta can be classified into four general anatomic categories, although some aneurysms involve more than one segment [2]:

Ascending aortic aneurysms arise anywhere from the aortic valve to the innominate artery—60%
Aortic arch aneurysms include any thoracic aneurysm that involves the brachiocephalic vessels—10%
Descending aortic aneurysms distal to the left subclavian artery—40%
Thoracoabdominal aneurysms—10%

Beyond distinguishing thoracic aneurysms by anatomic position, these broad categories further provide guidance for the operative approach to repair the aneurysm, intraoperative circulatory management, and end-organ injury risk stratification. We will be focusing our details on the assessment of thoracic aortic aneurysms only, for the purpose of this review (Fig. 1).

2 Epidemiology

The incidence of thoracic aortic aneurysm is estimated to be 6–10 cases per 100,000 patient years [3, 4]. They occur most commonly in the sixth and seventh decade of life with males being affected two to four times more commonly than females. Hypertension is an important risk factor and up to 13% of patients diagnosed with an aortic aneurysm are found to have multiple aneurysms; approximately 20–25% of patients with a large thoracic aortic aneurysm also have an abdominal aortic aneurysm [5, 6].

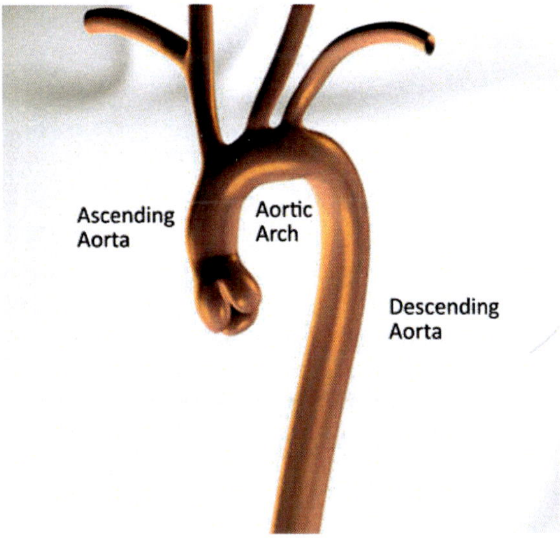

Fig. 1 Schema demonstrating the 3 anatomical divisions of the thoracic aorta

3 Natural History

The natural history of thoracic aortic aneurysms is characterized by progressive aortic dilatation and eventual dissection, rupture, or both. An analysis of 1600 patients with thoracic aortic disease by Elefteriades has helped quantify some well-recognized risks [7]. Although imaging methods were suboptimal, the average expansion rates were 0.07 and 0.19 cm/y in ascending and descending thoracic aortic aneurysms respectively.

The most important determinant of rupture is the aneurysm size [8–10]. One study of 133 patients (with ascending aortic or aortic arch disease alone in 40%, descending thoracic aortic disease alone in 31%, and both in 29%).found that the five year risk of rupture was 0% for aneurysms less than 40 mm, 16% for aneurysms 40–59 and 31% for aneurysms $\geq$60 mm [8]. In a prospective study of 721 patients with ascending disease, the risk of dissection or rupture was 2% per year for aneurysms less than 50 mm, 3% for aneurysms between 50 and 59 mm, and 7% for aneurysms $\geq$60 mm [11].

The rate of growth is related to the initial diameter with larger aneurysms growing more quickly as illustrated by a study of 67 patients with thoracic aortic aneurysm who underwent serial CT imaging [12]. The rate of expansion was much greater for aneurysms more than 50 mm in diameter (7.9 vs. 1.7 mm per year in smaller aneurysms) independent of the history of being hypertensive.

Certain types of aneurysms have an increased propensity for expansion and rupture. For example, aneurysms in patients with Marfan syndrome tend to dilate at an accelerated rate and rupture or dissect at smaller diameters than non-

Marfan-related aneurysms, with aortic root complications causing the majority of deaths [13]. Saccular aneurysms tend to grow more rapidly than fusiform aneurysms.

4 Etiology and Pathogenesis

The normal aorta derives its elasticity and tensile strength from the medial layer, which includes lamellae of elastin, collagen, and smooth muscle cells. During systole, kinetic energy imparted by ventricular ejection is absorbed in the compliance of the aorta, resulting in transient expansion and recoil. The amount of energy absorption is proportional to the proximity to the left ventricle. As such, the ascending, descending, and abdominal aorta have different cellular features to accommodate their unique fluid-mechanical environments. Elastin fiber content is typically higher in the ascending aorta. These fibers are synthesized and degraded continuously by the smooth muscle cells, and a progressive fragmentation of these fibers is associated with aging, which is the reason for gradual dilation of the ascending aorta in the elderly. However, certain acquired conditions can accelerate the process, producing the pathologically enlarged aorta resulting in aneurysms [14].

Hemodynamic factors clearly contribute to the process of aortic dilatation, yet are complex in nature. The vicious cycle of increasing diameter and increasing wall tension, as described in Laplace's law (tension $=$ pressure $\times$ radius), is well established. Turbulent blood flow may also impart a risk for developing significant aortic pathology. This is clearly demonstrated in patients where there is a stenosis (within the aortic valve or at the aortic isthmus), and the region immediately distal to the stenosis appears aneurysmal. Hemodynamic derangements, however, are only one piece of a complex puzzle.

The vast majority of descending thoracic aneurysms are associated with atherosclerosis [2] and the risk factors for aneurysm formation are the same as those for atherosclerosis (e.g., hypertension, hypercholesterolemia, smoking) [15]. However, it remains unclear whether atherosclerosis is actually responsible for aneurysm formation [15]. It seems likely that there is a multifactorial, systemic, nonatherosclerotic causal process, such as a defect in vascular structural proteins, with atherosclerosis occurring secondarily [16].

Some of the pathologies resulting in thoracic aneurysms (Table 1) are discussed as follows:

4.1 Nonspecific Medial Degeneration

Histologic findings of mild medial degeneration, including fragmentation of elastic fibers and loss of smooth muscle cells, are expected in the aging aorta. However, an advanced, accelerated form of medial degeneration leads to progressive

Table 1 Causes of thoracic aortic aneurysms

Non-specific medial degeneration
Aortic dissection
Genetic
Marfan syndrome
Ehlers–Danlos syndrome
Loeys–Dietz syndrome
Familial thoracic aortic aneurysm and dissection syndrome (FTAAD)
Congenital (bicuspid aortic valve)
Mycotic
Autoimmune

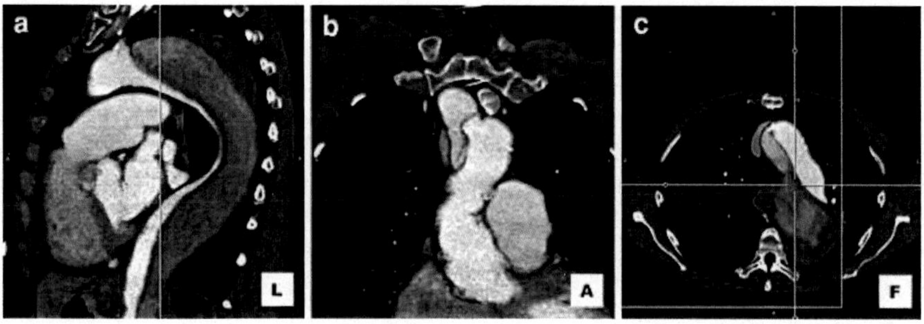

Fig. 2 Dissection of the thoracic aorta extending from the arch to the descending thoracic aorta. **a** Sagittal, **b** coronal and **c** axial views

weakening of the aortic wall, aneurysm formation, and eventual dissection, rupture, or both. The underlying cause of medial degeneration disease, however, remains unknown.

4.2 Aortic Dissection

An aortic dissection usually begins as a tear in the inner aortic wall, which initiates a progressive separation of the medial layers and creates two channels within the aorta. This event profoundly weakens the aortic wall, which often leads to aneurysm formation (Fig. 2).

4.3 Genetic Disorders

4.3.1 Marfan Syndrome

Marfan syndrome is a disorder characterized by a mutation in the fibrillin gene 15. Patients usually present with a tall stature, joint hypermobility, eye lens dislocation, mitral valve prolapse and aortic aneurysms. The aortic wall is weakened by fragmentation of elastic fibers and deposition of extensive amounts of mucopolysaccharides (a process previously called cystic medial degeneration). The traditionally held view is that abnormal fibrillin (caused by a mutation in the FBN-1 gene) in the extracellular matrix decreases connective tissue strength and results in abnormal elasticity, which predisposes the aorta to dilatation from wall tension caused by left ventricular ejection impulses [17]. More recent evidence, however, shows that the abnormal fibrillin causes degeneration of the aortic wall matrix by increasing the activity of transforming growth factor beta (TGF-β) [18]. Aortic root disease, which leads to the formation of aneurysmal dilatation, aortic regurgitation, and dissection, is the main cause of morbidity and mortality in the Marfan syndrome [19].

4.3.2 Vascular Ehlers–Danlos Syndrome (Type IV)

Vascular type Ehlers–Danlos (type IV) syndrome is characterized by a defect in type III collagen synthesis. Thoracic aortic aneurysms and dissections are less common, but when they do occur, they pose a particularly challenging surgical problem because of the reduced integrity of the aortic tissue. Spontaneous arterial rupture, usually involving the mesenteric vessels, is the most common cause of death in these patients.

4.3.3 Loeys–Dietz Syndrome

Loeys–Dietz syndrome is a condition that is distinguished by the triad of arterial tortuosity and aneurysms, hypertelorism (widely spaced eyes), and bifid uvula or cleft palate. It is caused by mutations in the genes encoding TGF-$\beta\beta$ receptors, rather than fibrillin 1 [20, 21].

4.3.4 Familial Aortic Aneurysms

About 20% of patients with thoracic aortic aneurysms and dissections have a genetic predisposition (a first degree relative with aortic aneurysms or dissections) without having any of the aforementioned connective tissue disorders. Thus far, mutations involving the genes for TGF-$\beta\beta$ receptor 2 (TGFβR2), β-myosin heavy chain (MYH11), and α-smooth muscle cell actin (ACTA2) have been identified and other genes responsible for such pathology are under investigation.

4.4 Congenital Bicuspid Aortic Valve

Congenital Bicuspid Aortic Valve is the most common congenital malformation of the heart or great vessels [22]. Patients with such a pathology have an increased incidence of ascending aortic aneurysm formation and usually a more rapid rate of aortic enlargement [23]. Fifty percent to seventy percent of adults with bicuspid aortic valve, but without significant valve dysfunction, have echocardiographically detectable aortic dilatation [24, 25]. This dilatation usually is limited to the ascending aorta and root [26].

Recent findings suggest that aneurysms associated with bicuspid aortic valves have a fundamentally different pathobiologic cause than aneurysms that occur in patients with trileaflet valves [27]. The exact mechanism responsible for aneurysm formation in patients with bicuspid aortic valves remains controversial. The two most popular theories posit that the dilatation is caused by (a) a congenital defect involving the aortic wall matrix that results in progressive degeneration, or (b) ongoing hemodynamic stress caused by turbulent flow through the diseased valve. However, the second theory is not well-supported [24, 28].

4.5 Infection

Bacterial invasion of the aortic wall may result from bacterial endocarditis or extension from an infected laminar clot within a pre-existing aneurysm. The most common causative organisms are Staphylococcus aureus, Staphylococcus epidermidis, Salmonella, and Streptococcus [29, 30]. Infection often produces saccular aneurysms located in areas of aortic tissue destroyed by the infectious process. Syphilitic aortitis used to be a common cause of ascending thoracic aneurysms which is rarely seen these days because of effective antibiotic regimens.

4.6 Aortitis

Aortic Takayasu's arteritis and Giant Cell Arteritis are a few that have been reported to be associated with thoracic aneurysms. Thoracic aneurysm formation is a particular problem in patients with giant cell arteritis that occurs in as many as 11% of patients and may be associated with aortic dissection [31].

5 Clinical Presentation

In many patients with thoracic aortic aneurysms, the aneurysm is discovered incidentally when imaging studies are performed for unrelated reasons and patients often are asymptomatic [6]. However, thoracic aortic aneurysms that initially go undetected eventually create symptoms and signs that correspond with the segment of aorta that is involved and have a wide variety of manifestations.

5.1 Local Compression

Aneurysms of the ascending aorta and transverse aortic arch can cause symptoms related to compression of the superior vena cava, the pulmonary artery, the airway, or the sternum. Rarely, these aneurysms erode into the superior vena cava or right atrium, causing acute high-output failure. Expansion of the distal aortic arch can stretch the recurrent laryngeal nerve, which results in left vocal cord paralysis and hoarseness.

Descending thoracic aneurysms frequently cause back pain localized between the scapulae. When the aneurysm is largest in the region of the aortic hiatus, it may cause middle back and epigastric pain. Thoracic or lumbar vertebral body erosion typically causes severe, chronic back pain. Mycotic aneurysms have a peculiar propensity to destroy vertebral bodies. Descending thoracic aortic aneurysms may cause varying degrees of airway obstruction, manifesting as cough, wheezing, stridor, or pneumonitis. Pulmonary or airway erosion presents as hemoptysis. Compression of the esophagus causes dysphagia.

5.2 Aortic Valve Regurgitation and Increased Heart Strain

Ascending aortic aneurysms can cause displacement of the aortic valve commissures and annular dilatation. The resulting deformation of the aortic valve leads to progressively worsening aortic valve regurgitation. In response to the volume overload, the heart remodels and becomes increasingly dilated. Patients with this condition may present with progressive heart failure.

5.3 Thromboembolic Phenomenon

Aneurysmal compression of branch vessels or the occurrence of embolism to various peripheral arteries due to thrombus within the aneurysm can cause cerebral, renal, mesenteric, lower extremity and spinal cord ischemia, with the patient presentation varying according to the vascular bed involved.

5.4 Dissection or Rupture

The most serious complications of thoracic aortic aneurysm are dissection [31] or rupture, which can cause pain, or extravasation of blood into the left intrapleural space or intrapericardial space. Patients with ruptured thoracic aortic aneurysms often experience sudden, severe pain in the anterior chest (ascending aorta) and upper back or left chest (descending thoracic aorta).

A descending thoracic aortic aneurysm can rupture into the adjacent esophagus, producing an aortoesophageal fistula and presenting with hematemesis. Rupture is often catastrophic, being associated with severe pain and hypotension or shock.

6 Diagnosis

Although certain constellations of symptoms and signs are highly suggestive of thoracic aortic aneurysm, diagnosing and characterizing these aneurysms requires imaging studies. In addition to establishing the diagnosis, imaging studies provide critical information that guides the selection of treatment options.

6.1 Plain Radiography

A Chest x-ray is neither sensitive nor specific for diagnosing aneurysmal disease as many aneurysms are not apparent on the chest x-ray, with radiographic abnormalities being detectable in about 61% of patients [32]. The aneurysm may produce a widening of the mediastinal silhouette, enlargement of the aortic knob, or displacement of the trachea from midline. Other reported features include displaced calcification, aortic kinking, and opacification of the aorticopulmonary window [32].

6.2 Echocardiography

According to the 2003 ACC/AHA practice guidelines for echocardiography, echocardiography has been recommended for the diagnosis of aortic aneurysm [33]. Transthoracic echocardiography (TTE) is the preferred procedure, with trans-esophageal echocardiography (TEE) being performed, especially for examination of the entire aorta, as in emergency situations; and for imaging when coexistent dissection is suspected.

6.3 Computed Tomographic (CT) Scan

CT is the most common and the most useful imaging modality for evaluating thoracic aortic aneurysms [34]. The major disadvantage of contrast-enhanced CT scanning is the possibility of contrast-induced acute renal failure in patients who are at risk (e.g., patients with pre-existing renal disease or diabetes) [35].

Interventional technologies have blossomed over the past two decades. However, advances would not have been possible without similar improvements in noninvasive imaging techniques [36, 37]. The development of multidetector CT systems that allow the acquisition of spiral data, digital storage mediums, and the application of novel reconstruction algorithms in conjunction with user-friendly workstations has revolutionized preoperative patient evaluation and allows interventionalists to properly match patient anatomic factors with device features.

6.3.1 Prophylactic Surveillance

Based on natural history studies and clinical evidence, until the operative intervention threshold is reached (discussed with treatment indications), aneurysms of the thoracic aorta are usually monitored every 6 months and features such as maximum diameter, morphology and rate of increase in aneurysm size are assessed.

6.3.2 Diagnosis and Assessment for Endovascular Suitability

Semi-automated algorithms are commonly used to calculate an aortic center line of flow (CLF). The aortic anatomy can then be visualized in a curved planar reconstruction format or "straightened" to provide a more precise assessment of the extent of aneurysmal disease in addition to the involvement of aortic branches. It also gives us information about the aorta adjacent to the aneurysm, a segment that is critical for open surgical or endovascular reconstructions by defining proximal and distal "landing zones" (Fig. 3).

Additional assessment of the iliac vasculature (such as severity of calcification and diameter) will help to determine the ability to introduce devices using a transfemoral approach. CT is particularly accurate for the determination vessel diameters. Contrast-enhanced CT provides information about the aortic lumen and can detect mural thrombus, aortic dissection, inflammatory periaortic fibrosis, and mediastinal or retroperitoneal hematoma due to contained aortic rupture. A CT scan can also reveal detailed information about aortic calcification.

The essential role of CT scan for post-operative evaluation of endovascular repair is discussed in the section detailing endovascular management of thoracic aneurysms.

6.4 Magnetic Resonance Angiography

Magnetic resonance angiography (MRA) produces aortic images comparable to those produced by contrast-enhanced CT but does not necessitate

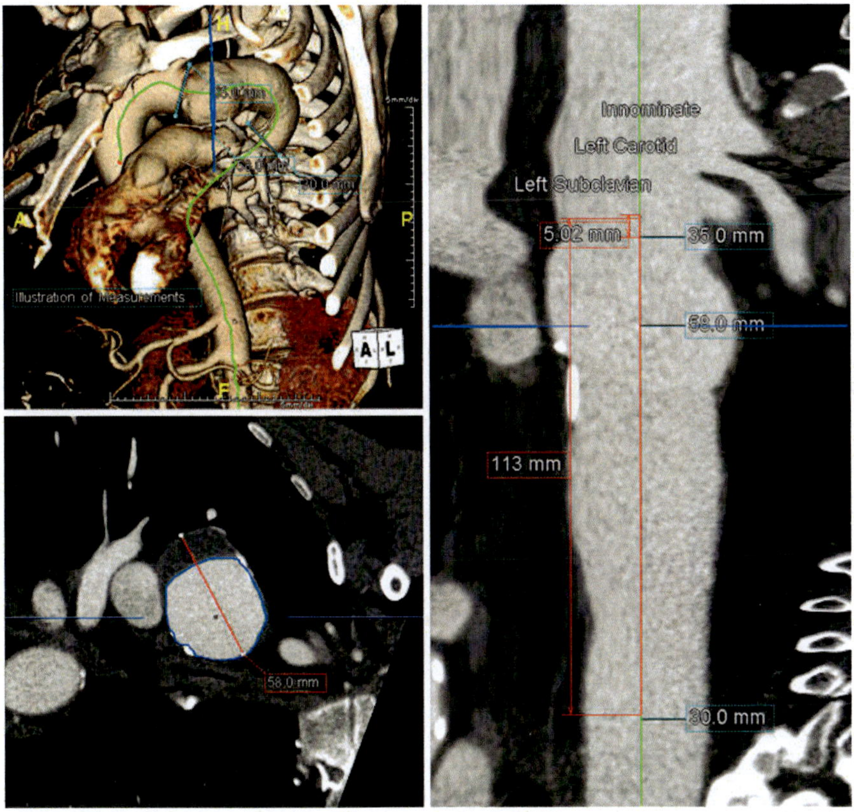

Fig. 3 An analysis of a proximal descending thoracic aneurysm based on the centerline of flow (CLF) analysis. The three supra-aortic vessels are also seen

exposure to ionizing radiation. In addition, MRA offers excellent visualization of branch vessel details, and it is useful in detecting branch vessel stenosis [38].

Current limitations of MRA include high expense, a susceptibility to artifacts created by ferromagnetic materials, gadolinium-linked nephrogenic systemic fibrosis and acute renal failure in patients with advanced renal insufficiency [39]. MRA also fails to depict vessel wall calcification, which has implications for vascular access.

6.5 Aortography and Cardiac Catheterization

In selected cases, aortography is used to gain important information when other types of studies are contraindicated or have not provided satisfactory results. For example, information about obstructive lesions of the brachiocephalic, visceral, renal, or iliac arteries is useful when surgical treatment is being planned; if other imaging studies have not provided adequate detail, aortograms can be obtained in patients with suspected branch vessel occlusive disease.

Unlike standard aortography, cardiac catheterization continues to play a major role in diagnosis and preoperative planning, especially in patients with ascending aortic involvement. Proximal aortography can reveal not only the status of the coronary arteries and left ventricular function but also the degree of aortic valve regurgitation, the extent of aortic root involvement, coronary ostial displacement and the relationship of the aneurysm to the arch vessels.

A key limitation of aortography is that it images only the lumen and may therefore underestimate the size of large aneurysms that contain laminated thrombus. Manipulation of intraluminal catheters can result in embolization of laminated thrombus or atheromatous debris with proximal aortography carrying a 0.6–1.2% risk of stroke. Other risks include allergic reaction to contrast agent, iatrogenic aortic dissection and bleeding at the arterial access site.

7 Basics of Treatment

7.1 Determination of the Extent and Severity of Disease

Patients with thoracic aortic aneurysms will frequently have concomitant remote aneurysms [3]. In such cases, the more threatening lesion usually is addressed first. In many patients, staged operative procedures are necessary for complete repair of extensive aneurysms involving the ascending aorta, transverse arch, and descending thoracic or thoracoabdominal aorta [40].

7.2 Continued Surveillance Versus Intervention

Once a thoracic aortic aneurysm is detected, in the absence of symptoms, management begins with patient education. Medical treatment involves blood pressure control, heart rate control and anti-impulse therapy most commonly with beta-blocking agents. The effect of this is to attempt to reduce the force of cardiac contraction against the weakened aortic wall. Additionally, many clinicians restrict patient activities, advising patients to avoid straining, or any exercise that results in isometric muscle contraction. Cessation of cigarette smoking is also critical, as persistent smokers have been shown to have considerably higher rates of rupture and aortic growth. Patients being followed with ascending or descending aortic aneurysms should receive repeat imaging for signs of interval growth prompting surgical intervention. The timing of such studies is somewhat controversial, with larger aneurysms typically imaged at 6 month intervals, and smaller aneurysms yearly.

Patients with genetic diseases are best managed by dedicated aortic clinics as criteria for intervention differ when contrasted with older patients believed to have non-specific aneurysms.

7.3 Indications for Intervention

Surgical or Endovascular therapy is often recommended prophylactically to prevent the morbidity and mortality associated with aneurysm rupture. However, the optimal timing of surgery for a thoracic aortic aneurysm is uncertain since the natural history is variable, particularly for aneurysms less than 50 mm in size [11], and the majority of patients have concomitant cardiovascular disease that increases the risks associated with surgery. Thus, many patients die of other cardiovascular causes before the aneurysm ruptures.

7.3.1 Open Surgery

On the basis of the natural history studies discussed earlier, elective operation is recommended when the diameter of an ascending aortic aneurysm is >5.5 cm, when the diameter of a descending thoracic aortic aneurysm is >6.5 cm, or when the rate of dilatation is >1 cm/y [7, 41]. In patients with connective tissue disorders, such as Marfan and Loeys–Dietz syndromes, the threshold for operation is lower with regard to both absolute size (5 cm for the ascending aorta and 6 cm for the descending thoracic aorta) and rate of growth. Smaller ascending aortic aneurysms (4–5.5 cm) also are considered for repair when they are associated with significant aortic valve regurgitation, bicuspid valves or other connective tissue disorders.

Most patients are asymptomatic at the time of presentation, so there is time for thorough preoperative evaluation and improvement of their current health status. In contrast, patients who present with symptoms often require urgent operation since they are at an increased risk of rupture and warrant expeditious evaluation. The onset of new pain in patients with known aneurysms is especially concerning, because it may herald significant expansion, leakage, or impending rupture. Because emergent interventions produce worse outcomes than elective procedures do, emergent intervention is reserved for patients who present with rupture or superimposed acute dissection [42].

7.3.2 Endovascular Surgery

The indications for Thoracic Endovascular Aneurysm Repair (TEVAR) of descending aortic aneurysms are similar to those for surgical repair regarding size parameters. A diagnosis of aortic rupture or dissection, leading to malperfusion or rupture, is also indicative of an endovascular rescue. Attempts have been made to extend the use of endovascular therapy to aortic arch and proximal aortic aneurysms and devices are undergoing clinical trials presently.

8 Preoperative Assessment and Preparation

8.1 Assessment of Physiologic Reserve

Given the impact of comorbid conditions on perioperative complications, a careful preoperative assessment of physiologic reserve is critical in assessing operative risk. Therefore, most patients undergo a thorough evaluation—with emphasis on cardiovascular, pulmonary, and renal function—before undergoing elective surgery [43, 44].

8.1.1 Cardiovascular Evaluation

Coronary artery disease is common in patients with thoracic aortic aneurysms and is responsible for a substantial proportion of early and late postoperative deaths in such patients. Similarly, valvular disease and myocardial dysfunction have important implications when one is planning anesthetic management and surgical approaches for aortic repair. Transthoracic echocardiography is a satisfactory noninvasive method for evaluating both valvular and biventricular function. Dipyridamole-thallium myocardial scanning identifies regions of myocardium that have reversible ischemia. Cardiac catheterization and coronary arteriography are performed in patients who have evidence of coronary disease. Patients with ascending aortic and particularly arch disease should undergo carotid artery duplex ultrasound examination.

8.1.2 Pulmonary Evaluation

Patients with descending thoracic aortic disease require left thoracotomy and should undergo pulmonary function testing if clinical symptoms of pulmonary disease are present.

8.1.3 Renal Evaluation

Renal function is assessed preoperatively by measuring serum electrolyte, blood urea nitrogen, and creatinine levels. Information about kidney size and perfusion can be obtained from the CT scan or aortogram used to evaluate the aorta.

8.2 Assessment for Endovascular Suitability

In order to assess the suitability for an endovascular repair, CT Angiography (CTA) of the chest, abdomen and pelvis is performed preoperatively and details about the aneurysm and access vessels is obtained, which has been discussed previously.

9 Open Repair of Thoracic Aortic Aneurysms

9.1 Ascending Aorta and Arch

Operations to repair proximal aortic aneurysms—which involve the ascending aorta, transverse aortic arch, or both—are performed through a midsternal incision and require cardiopulmonary bypass.

9.1.1 Aortic Valve Disease and Root Aneurysms

The spectrum of operations used to treat aortic valve pathologies and root aneurysms vary and are fairly broad. In summary, based on the diseased aortic portion, the following operations have been performed:

- Separate aortic valve and ascending aorta replacement, in cases when the valve and ascending aorta are pathologic but the aortic root morphology is preserved.
- Composite valve-graft conduit, when the aortic root is also ecstatic/aneurysmal.
- Ross procedure, which involves auto-implantation of the pulmonary root homograft to the aortic root position, is seldom done to address aortic root pathology, largely because of its technical demands and because of concerns about the potential for autograft dilatation in patients with connective tissue disorders [45].
- Aortic root replacement can also be done by either sparing the aortic valve (in which case the aortic valve is "re-suspended") or including the aortic valve (as done in the Bentall technique and its other modifications).

9.1.2 Aortic Arch Aneurysms

Several options are also available for handling aneurysms that extend into the transverse aortic arch. The surgical approach depends on the extent of involvement and the need for cardiac and cerebral protection. Exposure of the arch is particularly challenging, as the ascending aorta and proximal arch are best dissected via an anterior approach, while the distal arch and thoracic aorta are typically treated using a left thoracotomy.

- Saccular aneurysms that arise from the lesser curvature of the distal transverse arch and that encompass <50% of the aortic circumference may be treated by patch graft aortoplasty.
- For fusiform aneurysms, when the distal portion of the arch is a reasonable size, a hemiarch replacement can be done through a conventional median sternotomy.
- More extensive arch aneurysms require total arch replacement. The brachiocephalic vessels are reattached to one or more openings made in the graft, or if these vessels are aneurysmal, they are replaced with separate, smaller grafts.
- In the most extreme cases, the aneurysm involves the entire arch and extends into the descending thoracic aorta. Such aneurysms are approached by using Borst's elephant trunk technique of staged total arch replacement [46]. The distal anastomosis is constructed so that a portion of the graft is left suspended within the proximal descending thoracic aorta. During a subsequent operation, this "trunk" is used to facilitate repair of the descending thoracic aorta through a thoracotomy incision.

Vascular Control and Adjuncts used in open repair of Proximal Thoracic Aneurysms. One important difference between thoracic and abdominal aneurysm repair is the means for obtaining proximal vascular control during thoracic aneurysm repair. Thoracic aneurysm repair most commonly utilizes cardiopulmonary bypass to off-load the heart when the descending thoracic aorta is treated, and under circulatory arrest when the arch requires replacement [47].

- In ascending aneurysm repair, protection against end-organ damage is accomplished with cardiopulmonary bypass and antegrade aortic perfusion distal to the aneurysm.
- In arch aneurysm repair, the need for open vascular access to the brachiocephalic orifices mandates interruption of cerebral blood flow and the need for cerebral protection [48, 49] using circulatory arrest.
- In descending thoracic aortic repair, the afterload on the heart induced by a proximal cross-clamp coupled with ischemia of end-organs distal to the aortic cross-clamp can be mitigated using perfusion adjuncts.

Cerebral protection. There are three common methods of providing cerebral protection:

- With selective cerebral perfusion (SCP), antegrade cerebral blood flow is reestablished via a graft anastomosed to the brachiocephalic vessels [50]. A variation of SCP with perfusion via the right axillary artery has also been described and may reduce embolic complications [51, 52].
- Deep Hypothermic circulatory arrest (DHCA) involves the use of cardiopulmonary bypass with the establishment of profound systemic hypothermia, followed by controlled exsanguination and total body circulatory standstill

[53, 54]. DHCA provides the best operative field visualization; however, stroke rates increase significantly after 45 min of arrest [55–57].

- Retrograde cerebral perfusion (RCP) is the one in which hypothermic circulatory arrest is used in conjunction with retrograde jugular venous perfusion with cold oxygenated blood [58–61] with a significantly reduced risk of stroke with RCP compared to DHCA alone. Controversy exists over whether RCP functions by providing neuronal metabolic support, by preserving cerebral autoregulation, or by enhancing cerebral cooling and permitting retrograde evacuation of air and debris from the intracranial circulation [62–68].

9.2 Descending Thoracic Aorta

In patients with descending thoracic or thoracoabdominal aortic aneurysms, several aspects of treatment—including preoperative risk assessment, anesthetic management, choice of incision, and use of protective adjuncts—are dictated by the overall extent of aortic involvement.

Descending thoracic aortic aneurysms are repaired through a left thoracotomy. Use of a double-lumen endobronchial tube allows selective ventilation of the right lung and deflation of the left lung. During a period of aortic clamping, the diseased segment is replaced with a polyester tube graft. Important branch arteries—including intercostal arteries are reattached to openings made in the side of the graft.

Vascular Control and Adjuncts used in open repair of Distal Thoracic Aneurysms. In descending aortic aneurysm repair, the utility of adjunctive end-organ protective measures is unclear [69]. The degree of risk for this patient was illustrated in a series of over 1500 patients: paraparesis or paraplegia developed in 16%, while acute renal failure severe enough to require dialysis occurred in 7% [70, 71].

Initial reports found that a quick operation without adjunctive measures was sufficient for thoracoabdominal aneurysm repair [71–75]. However, newer studies suggest that measures such as distal aortic perfusion via bypass circuitry [76]; selective perfusion of renal, segmental, and visceral arteries [47, 77]; cerebrospinal fluid drainage combined with intrathecal papaverine in high-risk patients [78]; intercostal reimplantation [79]; and profound hypothermia [80, 81] markedly decrease the likelihood of renal, mesenteric and spinal ischemia, which can result in paraplegia [82, 83]. As an example, one retrospective study of 132 patients found that reimplantation of the vessels between the eighth thoracic intercostal and the second lumbar arteries reduced operative mortality (4.9 vs. 13.2% for no reimplantation) and the incidence of postoperative paraplegia (0 vs. 8.8%) or the overall rate of spinal cord dysfunction (2.4 vs. 9.9%) [84].

Perioperative monitoring of spinal pressure and active management of spinal perfusion (mean arterial pressure—spinal pressure of greater than 80 mm Hg) has been shown to dramatically reduce and even reverse spinal cord injury in both open surgical as well as closed (endovascular) surgery, especially in higher risk patients including those with renal dysfunction, PAD, long segments surgery and prior abdominal aortic aneurysm [85].

10 Endovascular Repair of Thoracic Aortic Aneurysms

10.1 Ascending Aorta and Arch

Endovascular grafts to exclude focal ascending aneurysms have been placed as a part of clinical trials in selected high risk patients and patient outcomes are being assessed. Covering the coronary or supra-aortic vessels can prove fatal and the high hemodynamic pulsatile force of the heart requires a brief period of cardiac stand-still for accurate device deployment. Placing endografts at this location, therefore, requires a highly skilled and experienced physician due to the complexity of the procedure.

For distal arch aneurysms, devices with scallops for the left common carotid or the brachiocephalic artery have been deployed successfully, with an extra-anatomic left carotid-subclavian or carotid-carotid bypass respectively, and are being used as investigational devices.

For distal ascending aortic aneurysms involving the aortic arch, branched arch devices have been used in the context of clinical studies. Kawaguchi et al. [86] have a large experience in dealing with aortic arch aneurysms and have used about 36 versions of home-made devices, including devices based on Z-stents and devices with fenestrations. Inoue et al. [87] and Chuter et al. [88] have also described case reports about endovascular exclusion of the aortic arch. Inoue described a 3 vessel branched unibody stent, whereas Chuter has described a technique involving a modular approach. The graft, in simple terms, is an inverted infrarenal device, inserted via the right common carotid or brachiocepalic artery after a carotid-carotid bypass. Arch exclusion is completed with a thoracic stent graft placed distal to the brachiocephalic take off.

10.2 Descending Thoracic Aorta

Stent graft repair of descending thoracic aortic aneurysms has become an accepted treatment option for selected patients [89]. In 1991, Parodi and associates [90] reported using endovascular stent grafting to repair abdominal aortic aneurysms. Only 3 years after this seminal report was published, Dake and colleagues [91] reported performing endovascular descending thoracic aortic repair with "home-made" stent grafts in 13 patients. Details are discussed in the next section and the devices used for the treatment of descending thoracic aneurysms [92] are summarized in Table 2.

Table 2 Endovascular devices for exclusion of descending thoracic aneurysms being used in the United States

Company	Endograft
Cook	TX2
Gore	TAG
Medtronic	Talent
Medtronic	Valiant
Bolton	Relay

11 Engineering Principles and Technical Considerations for Endovascular Management of Thoracic Aortic Aneurysms

11.1 Endograft Characteristics

Endovascular grafts, also known as endografts, consist of components that provide flexibility and strength integral for device placement and longevity throughout the course of treatment. Such a device must be flexible enough to conform to a plethora of patients anatomies, be compressed to fit in a minimal diameter catheter-style delivery system, and expand upon placement to the original dimensions without deformation. For thoracic endovascular aneurysm repair (TEVAR), stent-graft devices constructed from woven polyester materials like Dacron are combined with strong, yet flexible materials like stainless steel Gianturco Z-stents [93]. Stents placed in gaps along the length of the graft provide rigid scaffolding for the graft material while maintaining flexibility essential for device loading and placement. The Dacron graft, similar to that used in open procedures, provides a conduit for blood flow through the existing vasculature while excluding aneurysmal tissue. In this manner, the pressurized aneurysm sac is cut off from the rest of the vasculature and corresponding sac shrinkage often occurs. Relieving the pressure on the diseased aortic wall mitigates the risk of aneurysm rupture which is the ultimate goal of any TEVAR procedure.

11.2 Overlap Zones with Respect to the Native aorta and Among Modular Components

Some overlap of the graft with native tissue (15 mm) is necessary in order to obtain an adequate endograft—vessel seal. Stent placement along the proximal end of the graft is always internal to the fabric resulting in maximized graft to vessel apposition. When multiple endograft components are used, overlapping of stents is required to assemble a continuous structure to exclude a longer length of diseased aorta. For these modular devices where a distal component is placed within a proximal component, stents are positioned along the endograft to allow for regions of fabric to fabric overlap for maximal seal [94].

11.2.1 Mechanics of Device Migration and its Prevention

Oversizing the endograft diameter a certain extent will help maintain adequate seal and lower the risk of device migration (>10 mm movement of the endograft relative to the native aorta). However, too high an oversizing may cause the device to collapse or damage the accessed vessel [95]. Typically, oversizing 10–20% is called for to maintain adequate sealing. The radial force from the oversized stent-graft acting on the lumen of the arterial wall must provide enough force to hold the endograft in place with blood flowing along the inside of the device. Blood flow through these devices imparts a shear stress on the inner wall of the endograft that translates down the entire lumen of the device. Without proper endograft fixation along the native vessel, the shear stress can cause the entire device to move from its placement. Devices combating this effect often utilize an anchoring mechanism such as metal barb fixation.

11.2.2 Mechanics of Component Separation and its Prevention

As morphologic changes occur following aneurysm exclusion, the potential for component separation exists [96]. Component separation (defined as inadequate overlap length according to "Instructions for use", clinical symptoms related to movement or the need for a secondary intervention to address the issue) can result from inadequate stent overlap, stent fractures or deformation, aneurysm sac shrinkage, or any combination of these. This can lead to complications, including sac repressurization, which may eventually lead to rupture. Careful pre-operative planning is necessary to minimize this risk and a stent overlap of 3 stents has been recommended [97].

11.3 Navigating Tortuous Vessels

Tortuous vessels may contain regions that are extremely difficult for a physician to feed a catheter or delivery system through and severely hinder device placement if not rendering it impossible. If a device could be placed in a region of high tortuosity, it may become susceptible to kinking, stent fracture, inadequate sealing, component separation, or any combination of these which may lead to eventual device failure. Additionally, vessel straightening algorithms may not accurately predict vessel dimensions in severely tortuous regions. This can severely impact pre-procedural planning and potentially exclude patients from endovascular repair.

Innovations in the design and construction of both the introducer sheath and endografts have made for lower profile systems that have a higher degree of

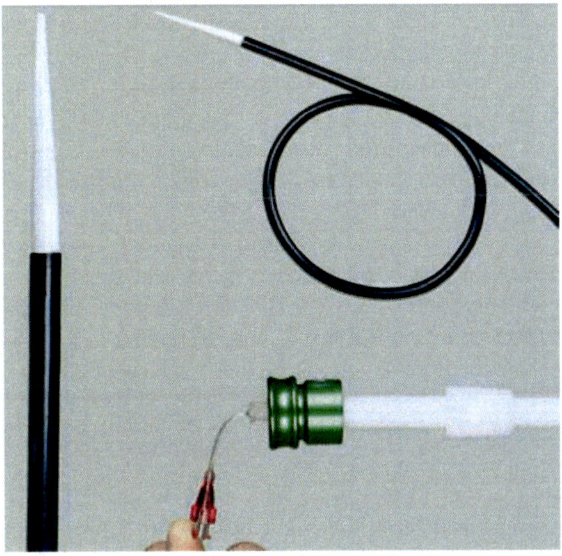

Fig. 4 The braided flexor sheath helps to navigate tortuous vessels and the acute angulations often encountered in the thoracic aorta

maneuverability. For example, the delivery system available for the Zenith TX2 TAA device consists of a Flexor braided sheath with hydrophilic coating. The braided sheath is designed to resist kinking in highly tortuous regions. It also allows for a greater torque to be placed on the sheath by the physician and for that torque to be resolved at the other end of the sheath. When treating vessels closer to the heart (i.e., the ascending aorta) a much longer sheath must be employed than for treating more distal vessels. Introducer sheaths that allow applied torsional forces to be fully resolved at the opposite end of the graft without causing plastic material deformation are essential for device delivery. A hydrophilic coating makes the system easier to feed through vessels due to interactions between the outer surface of the polymeric sheath and water in the blood. The coefficient of friction is lowered due to these interactions making the force required to slide the device over the surface of the vessel lumen minimal. This is especially important for device feeding through smaller diameter vessels where the vessel wall interacts with the introducer system. Sheath adherence to the lumen walls would be highly unfavorable in this situation (Fig. 4).

11.4 Computational Analysis and its Role in Device Design

The use of analytical models can shed light on the behavior of fluids within a control volume where boundary conditions can be applied and computational fluid dynamics (CFD) used. These techniques have already been utilized from the aortic arch [98, 99] to the aortic bifurcation [100] as well as for analyzing endovascular devices [101, 102]. Determining realistic inlet and outlet flow conditions respective to specific aneurysm locations along with flow pulsatility has been calculated and confirmed experimentally [103–108]. Currently CFD can be applied for simulation of blood flow in any number of patient specific geometries [109, 110]. Importing clinical images and applying physiological flow rates provides accurate boundary conditions for these simulations [109]. Wall stress, compliance, velocity stream lines, drag forces, and the influence of the stent-grafts Young's modulus have been measured for both preoperative and postoperative patients. Wall shear stress on the aneurysm wall have been shown to decrease by 92% after endovascular aneurysm repair, compliance at the repair region significantly decreased, and a potential increase in stent drag force due to aneurysmal remodeling [110]. The highly pressurized zone inside the aneurysm sac has also been shown to markedly decrease with the incorporation of stent-grafts in CFD, which is expected in vivo [99]. CFD has been very helpful in providing quantitative evidence for in vivo situations that may eventually help predict the outcome of a given treatment.

Virtual aneurysm models for the AAA have been created using complex linear hexahedral elements which have been imported into finite element analysis software and analyzed under transient flow conditions [111]. Finite element analyses attempt at determining stress and strain along aneurysm models in an effort to predict rupture risk related to aneurysm biomechanics [112]. Limitations from utilizing linear elastic models for truly non-homogenous aneurysmal regions with variable wall thicknesses result in data that can lead to inaccurate interpretation [113]. This limits the utility of finite element analysis at this time, however, future work in this field will strive to produce accurate, reproducible results from realistic models employing physiological boundary conditions.

Figueroa et al. have recently published computational evidence for the hemodynamic forces acting on endografts in both the abdominal and the thoracic aorta [114, 115]. Such analyses have led to dramatic breakthroughs in our understanding of the forces acting on endografts in vivo which may help with the development of endovascular devices in the future. For example, abdominal aortic devices have been shown to have a greater outward force imparted on them from hemodynamics than a caudal directed force [114]. Thoracic devices have been shown to have a higher cranial force acting to displace these endografts [115]. This knowledge will further aid engineers when designing devices to fit these locations.

11.5 Thoracic Hemodynamic Forces and the Impact on Future Device Design

Computational modeling has been used widely to predict aneurysm formation at the ascending aortic level including the arch [116, 117]. A steady state flow analysis has found that the aortic arch curvature increases the force on the ascending aorta by a factor over ten-fold and is the ultimate cause of vascular pathology at this region [117]. Another potential cause of ascending aortic pathology is from aortic valve asymmetry. Bicuspid aortic valve (BAV) is the most common form of congenital defect which results in different morphological characteristics for the aortic valve. Computationally determined wall shear stress for models incorporating BAV show localization in the region of the mid-ascending aorta due to the resulting asymmetrical higher velocity profile [117].

The geometry of the right aortic arch has been shown through Doppler calculations to impact the flow of the ascending aorta while not altering flow in the descending aorta [118]. Attempting to model fluid flow for the aortic arch has given different results in the form of helical fluid flow developing in both the ascending and descending portions [98]. While this data differs slightly, both agree that wall shear stress is focused at the top of the arch. Careful comparison of computational results with experimental evidence is required to corroborate both sets. Improvements for computational modeling discussed earlier may provide accurate and reproducible results that will become more helpful for clinicians and engineers alike when designing devices and administering treatment to patients with thoracic pathology.

11.6 Thoracic Device Evolution

11.6.1 Cook Zenith—TX2

Recent advances in endograft design from procedural feedback and improved methods in graft manufacturing have had a great impact on these devices. For example, the newly available TX2 thoracic stent-graft with Proform consists of the same materials as the TX2 endograft but with a subtle yet very important improvement for device deployment [93]. Patients with a proximal landing zone falling in the more tortuous region of the aortic arch still have problems with endograft conformation to the vessel wall and poor sealing [119–122]. Inadequate proximal sealing of the descending device to the lesser curvature of the arch is common for acute aortic angles. This poor seal may lead to late complications, device failure, or even death [123]. To address this issue, the proform device has diameter reducing ties around the most proximal stent during deployment. When proper orientation is determined with the proximal device wall parallel to the aortic inner curvature, the constraining mechanism is released and the proximal device is deployed. Limited experience with this technology has shown to be safe for treating a variety of pathologies in regions of the thoracic aorta with even the highest degrees of aortic angulation [93] (Figs. 5 and 6).

The newest Zenith TAA devices are called TX2-LP devices (LP: Low Profile). The graft material is the same Dacron used in the regular TX2 devices, but tighter weaving patterns have allowed for fabrication of a graft with smaller thickness and comparable strength. This improves device availability by making larger diameter grafts able to fit inside smaller delivery systems. This advance in graft construction has allowed physicians to treat patients with larger vessel diameters endovascularly that would have otherwise required an open procedure. Continuing advances in synthesizing lower profile materials that hold up to the stresses present in the

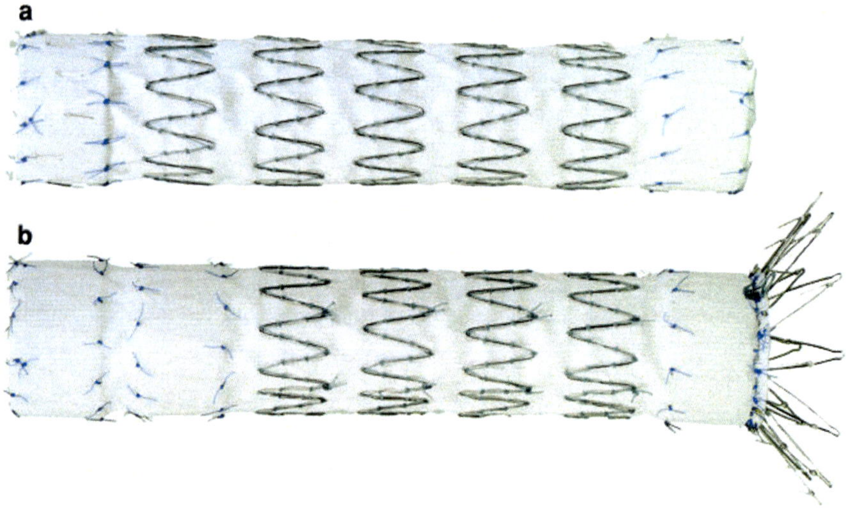

Fig. 5 A regular TX2 device with proximal (**a**) and distal (**b**) pieces. The distal piece has an uncovered distal stent with cranially-oriented barbs which act as a fixation mechanism to prevent migration

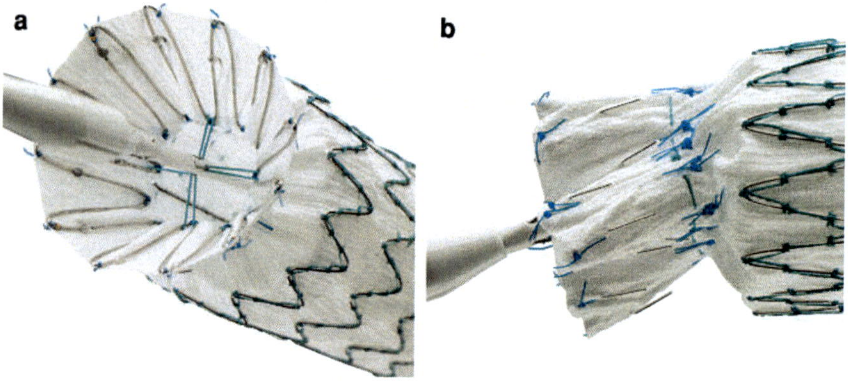

Fig. 6 An illustration of the proform device which helps the graft to conform well to the lesser curvature of the arch. **a** Luminal view **b** side view

body will continue in the future allowing for improvements in device availability for patients with abnormal anatomy (Fig. 7).

The TX2-LP not only has a thinner fabric but also contains nitinol stents in place of the stainless steel stents in the TX2. Nitinol stents make the device easier to pack into a smaller system and able to deploy without deformation. Stainless steel has the potential to deform when placed under the stress used to pack nitinol stents into the smaller delivery systems. This improved packing ability does not

Fig. 7 The TX2 low-profile device with nitinol stents. Proximal (**a**) and distal (**b**) pieces

come without a consequence. While stainless steel stents may deform under considerable stress, their ability to withstand cyclic loading, which is present in the body throughout the cardiac cycle, is significantly better then nitinol. The amount of force that nitinol can withstand before fracture is not well understood for cyclic loading at this point. Continued observation of outcomes will be necessary to determine the safety and efficacy of the nitinol stents.

11.6.2 Gore—TAG

In 1998, the Gore Excluder TAG thoracic endovascular device (W.L. Gore and associates, Flagstaff, AZ) was the first endovascular graft to undergo United States feasibility trials with pivotal trials following in 1999 [124]. Upon completion of study enrollment, the device was voluntarily withdrawn from distribution in May 2001 due to longitudinal stent fractures [124]. After structural modification, the redesigned TAG endovascular device was enrolled in a confirmatory trial at the end of 2003 and gained United States Food and Drug Administration (FDA) approval in March 2005 [95, 124]. FDA approval was only defined for treatment of aneurysms of the descending thoracic aorta which remains as the only on-label indication [125].

The TAG endograft is a tube of expanded polytetrafluoroethylene (ePTFE) reinforced with external nitinol self self-expanding stents and a film of ePTFE and fluorinated ethylene propylene (FEP). The nitinol stents are attached to the graft surface with ePTFE/FEP bonding tape. To enhance device conformation to tortuous anatomy, scalloped flares were incorporated at each end with a PTFE cuff covering the base of the flares. The original TAG device was constructed using two ePTFE layers with two longitudinally placed wires providing columnar support. Fractures in these wires led to the previously mentioned device modification. The current TAG endoprosthesis consists of three ePTFE layers with the third layer sandwiched between the other two replacing the longitudinal wires. This third ePTFE layer provides the longitudinal stiffness previously maintained by the longitudinal wires.

The unique deployment mechanism of the TAG endoprosthesis begins with the ePTFE/FEP constraining sleeve. Turning and pulling a deployment knob detaches a deployment line connected to the sleeve. The device then begins rapid expansion from the center along the length of the device in a bidirectional progression. However, this rapid deployment significantly limits de novo adjustment. A trilobed balloon designed to allow for continuous blood flow upon inflation is used to assist with device fixation.

11.6.3 Medtronic—Valiant

The Valiant endoprosthesis (Medtronic, Inc, Santa Rosa, CA) was introduced into clinical practice in 2005 as an evolution of the Talent thoracic endograft

(Medtronic, Inc, Santa Rosa, CA). Improved design features including modifications to the design of stents, endograft configuration, delivery sheath, deployment method and markers incorporated into the Valiant device. The most significant improvements between the two endoprostheses were in the removal of the spring connecting bar, increased choice in device lengths, a redesign of the proximal and supporting stent configuration, and redesign of the delivery system [126].

The Valiant device is constructed from woven, mono-filament polyester graft with a self-expanding nitinol stent sewn onto the outside of the graft. This nitinol support system comprises of 5 peaked "springs" arranged longitudinally along the body of the graft. The proximal and distal stents are unique from the rest of the assembly due to their 8-peak configuration which may be 15 or 17 mm in height [126]. The spacing of the springs allows for contact between adjacent spring peaks when loaded into the Delivery System. Spring arrangement also provides the necessary columnar strength for delivery. Removal of the longitudinal connecting bar present on the Talent endograft provides increased flexibility during tracking and device conformability to the aortic wall [127]. Platinum-iridium radiopaque markers are sewn to the proximal and distal ends as well as the midpoint of the device for surveillance. Currently, the Valiant endoprosthesis is available in diameters ranging from 24 to 46 mm with a maximum length of 227 mm [126, 128].

The delivery system utilized for deployment is called the xcelerant delivery system. This consists of a single use catheter with an integrated handle intended to provide a mechanical advantage while lowering user deployment force [127]. The system ranges in outer diameter from 22 to 25 Fr and tracks over a 0.035 inch glide wire [126]. For delivery, there is an inner membrane with a tapered tip, a middle membrane with a stent-stop, and an outer cover. The middle membrane is flexible while allowing tracking thorough tortuous anatomy. The outer cover is strengthened with a stainless steel braid. The cover restrains the device during tracking and releases the endograft upon retraction. Rotation of the delivery handle allows for slow deployment useful for confirming the proximal position. Once correctly positioned, a ratchet mechanism rapidly releases the endograft via a trigger and slider assembly on the delivery catheter.

11.6.4 Bolton Medical—Relay

The Relay and Relay NBS (Non Bare Stent) Thoracic Stent Graft with the Plus Delivery System (Bolton Medical, Sunrise, FL) were designed for deployment in the thoracic aorta near the arch [129]. The bare stent graft was released in 2005 while the Relay NBS was released later in March 2008. The straight stent grafts are available in diameters ranging from 26 to 46 mm and lengths in 50 mm increments from 100 to 250 mm. The tapered endoprosthesis currently ranges in diameter from 34 to 46 mm with the same lengths.

The Relay NBS Thoracic Stent Graft is composed of a polyester vascular graft fabric with Duralloy self expanding nitinol stents sutured along the length of the device. Longitudinal support and aorta conformability is ensured with a curved

Table 3 Endoleak types and descriptions

Endoleak	Description
Type I	Aneurysm sac repressurization due to inadequate sealing at the proximal or distal zone resulting in inflow around the endograft
Type II	Aneurysm sac repressurization due to inflow from collateral vessels
Type III	Aneurysm sac repressurization due to graft material failure or inadequate seal of overlapping endograft segments
Type IV	Aneurysm sac repressurization due to blood passing through graft pores

nitinol Spiral Support Strut. Placement and surveillance of the graft is ensured through radiopaque platinum/iridium markers sutured onto the stent graft. All materials used in device construction allow the Relay NBS Thoracic stent graft to be MRI compatible (at 3 Tesla or below).

The Relay Plus System employs a delivery system designed for easy advancement and accurate device deployment. A series of sheaths and catheters arranged in a coaxial fashion compose the introducer system. The braided outer sheath and tip are coated with a hydrophilic material which provides easier advancement and enhanced radiopacity to facilitate visibility in highly tortuous anatomical regions. A nitinol inner catheter has been recently introduced to the system to aid in aligning the system.

Recent modifications present in the new NBS were made primarily to facilitate deployment around a smaller radius of curvature. During deployment, the previous system clasped the entire proximal end until the entire device was released. The new system leaves the two apices on the inferior side of the graft unclasped to allow this section to expand at the beginning of deployment and proved greater control of the segment facing the inner curvature of the arch [129]. Two nitinol wires were added to the delivery system to control expansion of the inferior portion of the device for proper apposition to the anatomic inner curvature. The inferior proximal portion of the stent graft contains suture loops tethered to the distal ends of the support wires. The distal end of the constraining sleeve present in the introducer system was increased to allow the proximal end of the device to expand within the sleeve while permitting longitudinal adjustment of the device in its partially deployed state [130].

11.7 Concluding Steps of Endovascular Repair

In the end, after device deployment, a final angiogram is done to assess for proper aneurysm exclusion and absence of endoleaks (leakage of blood into the sac after endovascular exclusion). There are four distinct types of endoleak that may occur and are summarized in Table 3 [96, 131–133]. Type I endoleak is the most common endoleak for thoracic endovascular procedures requiring re-intervention [134].

12 Central Role of Computed Tomography in Post-Operative Assessment of Endovascular Thoracic Aortic Aneurysm Repair and Evaluation of Device Integrity

12.1 Assessment of Aneurysm Sac and Endoleaks

The prevention of wall tension following endovascular repair is incumbent upon the depressurization of the aneurysm. This, in turn, is accomplished by complete exclusion of the aneurysm sac by the endovascular device. The failure to accomplish this may be associated with a continued risk of rupture. The presence of continued blood flow into the aneurysm sac following an endovascular repair is termed an endoleak. This is assessed by evaluating the arterial and delayed phase studies from a CT scan. These are reconstructed with smoothing algorithms, providing information regarding the arterial wall, stentgraft-arterial wall apposition, and opacification of the aneurysm sac with contrast in either the arterial or delayed phase of the study. When the endoleak is in continuity with the proximal or distal sealing zone of the endovascular device, it is termed a type I endoleak (type Ia for the proximal portion, or type Ib for the distal portion). If the contrast in the aneurysm sac arises from a collateral vessel remote from the stentgraft it is termed a type II endoleak. When contrast in the sac results from poor apposition of modular components or defects in the fabric, the leak is termed a type III endoleak. One must remember that endoleaks do not imply rupture, as the contrast remains in the aneurysm sac, however, the risk for rupture is likely increased when an endoleak is noted (particularly if the endoleak is type I or III in nature) (Figs. 8 and 9).

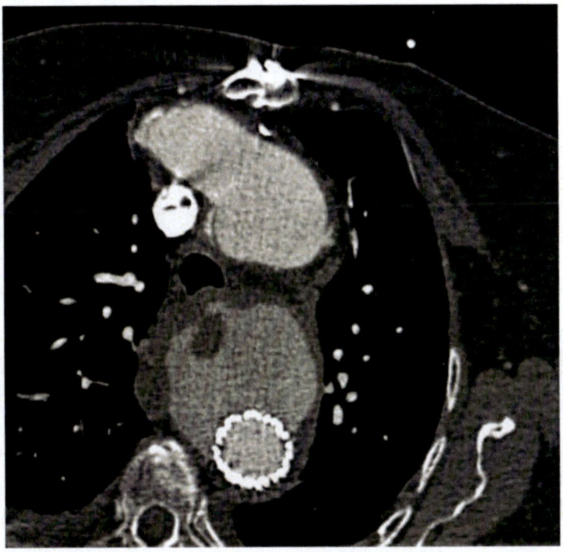

Fig. 8 Axial view a type I endoleak seen on a CT scan

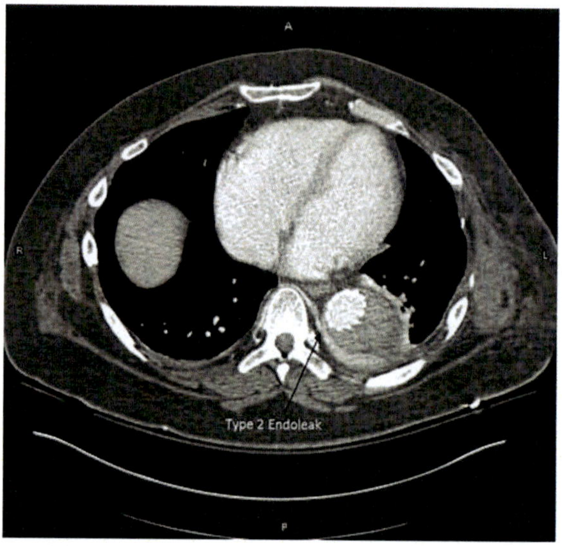

Fig. 9 Axial view a Type II Endoleak seen on a CT Scan resulting from retrograde flow from an intercostal vessel

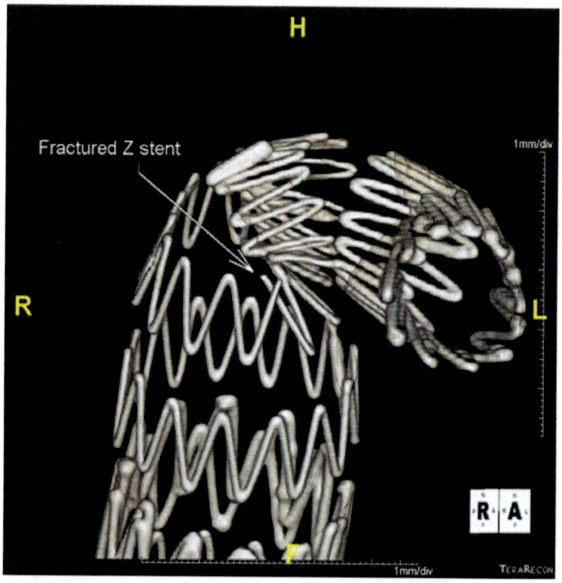

Fig. 10 Three dimensional demonstration of a fractured stainless steel Z-stent

12.2 Device Integrity

After image acquisition, the reconstruction of the native (precontrast) study can be performed using a high-resolution edge-detection algorithm [135]. This provides a means to assess for stent fracture and intercomponent relationships (Figs. 10 and 11).

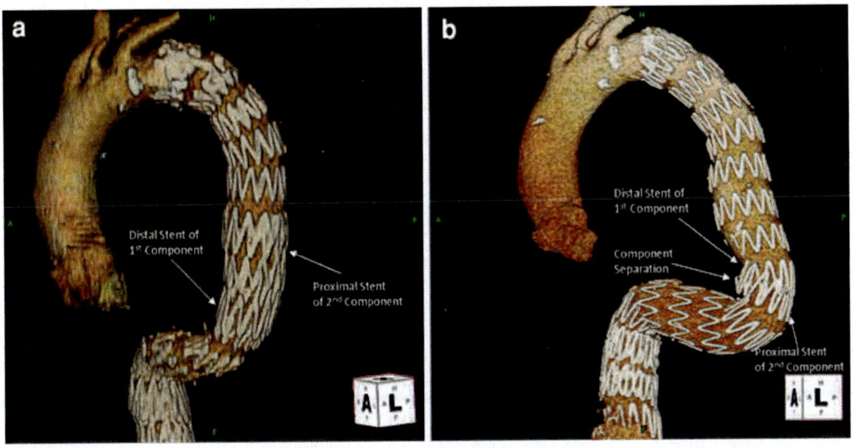

Fig. 11 **a** 3 stent overlap between the 1st and 2nd components of thoracic endografts. **b** Aortic remodeling overtime leads to movement of the distal component leading to component separation

12.3 Assessment of Device Migration

Movement of a device relatively to the arterial system at a fixation site represents device migration. Thin-slice axial computed tomography (CT) and plain abdominal radiography may be a reasonable method of analyzing migration in a non-tortuous abdominal aorta given that the abdominal aorta is a relatively uniplanar structure with a favorable orientation perpendicular to the axis of the CT [136, 137]. However, the inherent tortuosity of the thoracic aorta causes most image analysis without the use of three dimensional technology, including centerline of flow (CLF) analyses, multiplanar reconstructions, and other tools to be inaccurate. Thus, false assurances may be obtained with respect to device stability, aneurysm size, or the relational positions of components when axial images are used in isolation.

The use of CLF algorithms provides a reproducible means of determining distance within tortuous aortic segments. However, it is apparent that the aortic length between fixed landmarks is not always static. It is for this reason that pure CLF analysis cannot be used in isolation to determine device migration. The most obvious examples of this come from our experiences with elephant trunk grafts [138]. The stretching of the graft proximal to the endovascular device results in tortuosity distally, that was not present when the device was placed. This is an exaggeration of what happens with the native vasculature, as diseased arteries grow in length and diameter. Therefore, if the aorta is lengthening on either side of an implant, the measurement of a CLF length from vessel to another will not be static (left common carotid to celiac artery length). In such cases, one must observe surface landmarks neighboring the repair to differentiate device motion with respect to the implanted vasculature from lengthening of the entire aortic segment housing the device. However, this is a labor intensive process, thus patients are rapidly screened to assess for aortic length consistencies between measurements. In the absence of a lengthening aorta, the CLF distances are used to calculate migration. When the length measurements are inconsistent from scan to scan, the surface landmarks are used instead.

13 Hybrid Repair of Thoracic Aortic Aneurysms

Hybrid arch repairs involve some form of "debranching" of the brachiocephalic vessels, followed by endovascular exclusion of some or all of the aortic arch. Although this technique has many variants, they often involve sewing a branched graft to the proximal ascending aorta with the use of a partial aortic clamp. The branches of the graft are then sewn to the arch vessels. Once the arch is "debranched", the arch aneurysm can be excluded with endografts. The arguments for using a hybrid approach to treat aortic arch aneurysms include the elimination of cardiopulmonary bypass, circulatory arrest, and cardiac ischemia. In another technique, the arch can be replaced by using the open elephant trunk technique, after which the descending thoracic aorta can be excluded endovascularly by using the elephant trunk as the proximal landing zone (Fig. 12).

14 Outcomes

14.1 Open Repair

14.1.1 Proximal Aortic Aneurysms

All varieties of aortic root replacement have shown acceptable early mortality rates and few complications. Two groups with 20 and 27 years experience with composite valve graft replacement reported early mortality rates of 5.6 and 1.9%, respectively, with the more recent repairs having better outcomes [139, 140]. Repairs incorporating the ascending aorta and aortic arch have acceptable outcomes, with risk increasing as larger sections of the aortic arch are incorporated into the repair [141, 142]. Reported early mortality rates after stage 1 elephant trunk repairs range from 2.3 to 13.9% [143–147].

The risk of operative death and stroke in these repairs is additionally increased by severe atherosclerosis of the ascending aorta, and a revised surgical strategy is often needed to avoid clamping this section of the aorta. In a review of literature accompanying Zingone and colleagues for 36 such patients, pooled early death

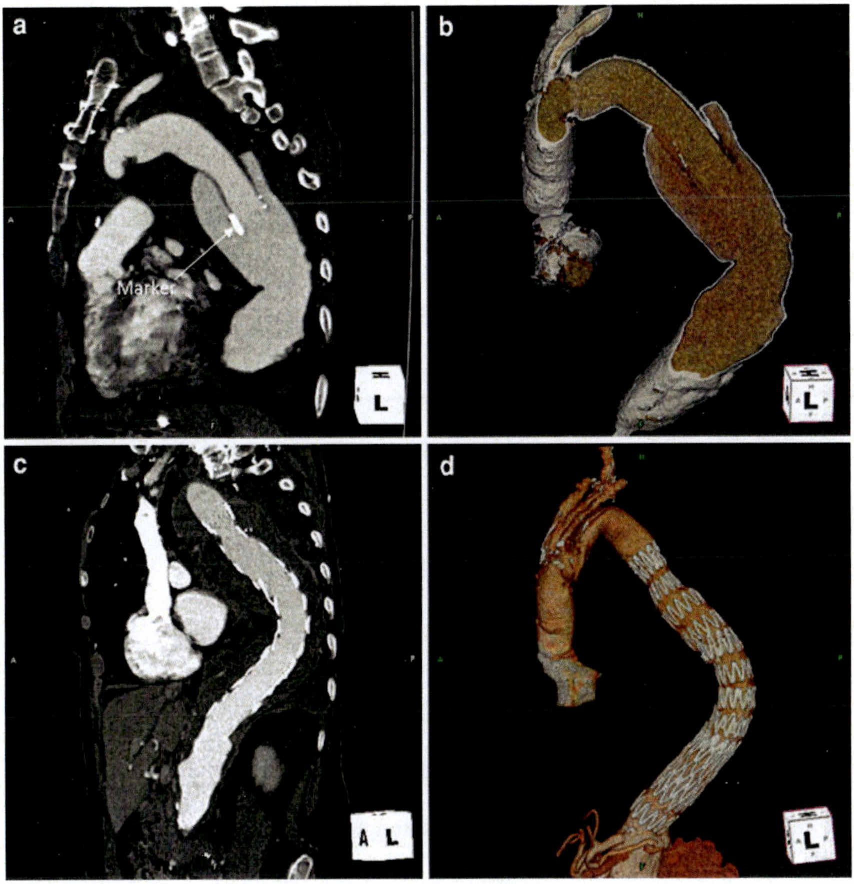

Fig. 12 Elephant trunk procedure—hybrid approach, the arch is replaced during open surgery (**a**, **b**) and the "trunk" is accessed endovascularly to complete the repair (**c**, **d**)

rates averaged 9% (ranged from 3.7 to 25%) and pooled stroke rates averaged 5.3% (ranged from 0 to 17.6%) [148].

In a report by Kazui and Bashar [149] covering 20 years of experience and 472 consecutive patients who underwent aortic arch repair with selective antegrade cerebral perfusion, early mortality was 9.3% for all repairs and 4.1% for more recent repairs. The stroke rate was 3.2% for permanent deficits and 4.7% for temporary deficits. Recent innovations in aortic arch replacement, such as the use of extra-anatomic grafts and moderate hypothermia, have also produced good results, including early mortality rates ranging from 0 to 4.7%, permanent stroke rates ranging from 0 to 4%, and rates of transient neurologic dysfunction ranging from 0 to 4.9% [150–152].

14.1.2 Distal Aortic Aneurysms

Contemporary results of open repairs of descending thoracic aortic aneurysms indicate that early mortality rates range from 4.4 to 8% and paraplegia rates range from 2.3 to 5.7%; stroke rates are generally lower, ranging from 1.8 to 2.1% [153–155]. The risk of mortality is increased for those undergoing repair of the proximal two-thirds of the descending aorta [153]. As expected, stroke rates after distal aortic repairs were highest when the clamp site was near the left subclavian artery.

Several studies have compared endovascular and surgical approaches to descending thoracic aortic repair. Some studies found no significant differences in rates of early death, stroke, and paraplegia [156–158] whereas others found that surgical patients had higher rates of early mortality (27%) [159] and paraplegia (14%) [160].

14.2 Endovascular Repair

14.2.1 Proximal Aortic Aneurysms—Ascending Aorta and Arch

Endovascular exclusion of isolated ascending aneurysms is being done as part of investigational trials and no reports have been published yet.

Case reports about successful exclusion of arch aneurysms by Chuter [88] and Inoue [87], using modular and unibody configurations respectively, have been published. Kawaguchi et al. [86] has reported a 92.5% success rate, with a 3.8% stroke and 2.6% SCI incidence reported over 60 months, in excluding arch aneurysms.

14.2.2 Distal Aortic Aneurysms—Descending Thoracic Aneurysms

While little comparative data exists between TEVAR and medical management, it is reasonable to assume that outcomes will also be better with TEVAR in those patients with indications for OS, since TEVAR compares favorably with OS. The outcomes presented below suggest that TEVAR is a promising alternative to open surgical (OS) repair [156, 157, 161].

Stroke. Because the proximal seal zone is in proximity to the carotid arteries, embolic strokes can occur following TEVAR. Risk factors for embolic stroke include the need for proximal deployment of the graft, presence of mobile atheromata in the arch, and prior stroke [162]. The vertebral arteries arising from the subclavian may be the source for posterior circulation strokes [163]. Perioperative stroke has ranged from 4 to 8% [164–166], comparable to OS [89].

Spinal cord ischemia. Extensive coverage of the thoracic aorta as well as prior history of AAA repair places the patient at increased risk for spinal cord ischemia (SCI) with the potential for paraplegia [167–169]. The risk of spinal cord ischemia has been reported to be between 3 and 11% [157, 160, 166, 170, 171], comparable to the rate of OS [89].

Some studies have demonstrated a lower rate of SCI with TEVAR than with OS. In a well-performed, retrospective review of 724 patients at a single institution who were treated with either TEVAR ($n = 352$) or OS ($n = 372$) for thoracic or thoracoabdominal aneurysms, no statistically significant difference in the rate of SCI was found between the two approaches (4.3 vs. 7.5% respectively) [170]. The extent of aortic disease was the strongest predictor of SCI.

Access complications. Because of the obligatory large sheath size for delivery of the device, passage of the sheath through a small diameter, tortuous, or excessively calcified external iliac artery can lead to iliac artery disruption. Anticipation of the need for adjunctive access measures is thus important and is required in a significant percentage of patients, ranging from 9.4 to 23.8% in published reports [156, 157]. A pseudoaneurysm or hematoma may also form.

Survival Analysis. The largest of the early series, for which there is now medium-term follow up, was a prospective, uncontrolled study of a first-generation, custom-fabricated self-expanding stent graft, including 103 patients with descending thoracic aortic aneurysms, 60% of whom were not candidates for conventional surgery [166]. After a 1.8 year follow up, late stent graft complications occurred in 38% of patients and included stent graft misdeployment or removal, endoleak, aortic dissection, distal embolization, gut ischemia, and infection. Fatal complications occurred in 4%, including rupture of the treated aneurysm, stent graft erosion into the esophagus (aortoesophageal fistula), arterial injury, and excessive bleeding.

In a subsequent report at 4.5 years of follow-up, actuarial survival at one, five and eight years were 82, 49, and 27%, respectively [164]. Patients who had been identified as suitable surgical candidates at the time of stent graft placement had significantly better survival at one year (93 vs. 74%) and five years (78 vs. 31%).

In a series of 100 patients (81 aneurysms) by Greenberg et al. [172], with a mean follow up duration of 14 months, the overall mortality was 17%, and aneurysm-related mortality was 14% at 1 year. Sac regression (>5 mm maximum diameter decrease) was observed in 52 and 56% of patients at 12 and 24 months, respectively.

Device migration and endoleak. Migration of the graft (>10 mm) caudally can occur, with a published incidence of 1–2.8% over a 6–12 month period [157, 161]. Factors predisposing to migration include excessive oversizing and tortuous seal zone anatomy. The incidence of endoleak following thoracic aortic stent placement is less common than that for endovascular repair of the abdominal aorta and is estimated at 3.9–15.3% [157, 161, 173]. The incidence of endoleak at five-year follow up with the Gore TAG device was 4.3% with Type I attachment site leaks being the most common type. The rate of secondary intervention required following stent grafting due to either endoleak or device migration is 3.6–4.4% [157].

15 Open Repair Versus Endovascular Repair

Endovascular repair of thoracic aortic aneurysms has become an accepted treatment option in selected patients, particularly patients with isolated degenerative descending thoracic aortic aneurysms [157, 172, 174]. According to the Gore TAG 5 year results [174], comparing endovascular to open repair, even though the all-cause mortality was same but aneurysm-related mortality, major adverse events and secondary procedures not directly related to aneurysm repair, were all low in TAG vs the open repair group at 5 years. This demonstrated the superiority of endovascular vs open repair for descending thoracic aneurysmal disease. Some of these recommendations may differ from published guidelines regarding the use of thoracic endografts for aneurysmal disease [175], however, the prospective studies leave little question that the morbidity and aneurysm related mortality for properly selected patients treated with TEVAR are superior to open surgical approaches. One must recognize that study limitations do exist, and the absence of a randomized trail precludes definitive conclusions.

According to a meta-analysis of 17 observational studies (1,109 patients) which compared TEVAR to OS [176], there was a significant reduction in perioperative mortality (OR 0.36; 95% CI 0.23–0.58) and in major neurological injury (OR 0.39; 95% CI 0.25–0.62). There was no difference in major reintervention rates, but a significant reduction in hospital and critical care stay. These benefits were seen predominantly in stable patients. Non-randomized comparisons published after the meta-analysis has suggested equivalent or better outcomes with the TEVAR. In a single center, retrospective study of over 700 patients who received either TEVAR or OS, mortality was not significantly different at 30 days (5.7 vs. 8.3% respectively) and 12 months (15.6 vs. 15.9% respectively) [170].

The evidence cited above argues for the use of TEVAR as opposed to OS in those patients who are candidates for both approaches. Even if patients' important outcomes such as mortality, and the risks of stroke and paraplegia are equivalent, patients undergoing TEVAR have a shorter length of hospital stay, quicker rehabilitation, lower aneurysm-related mortality and a longer average number of months lived (due to a reduction in perioperative mortality).

16 Conclusion

Both TEVAR and OS are extremely challenging procedures and patients undergoing either procedure should be referred only to health care teams that are both experienced in the preoperative and intraoperative management of these patients.

In the years to come, these endografts will certainly be explored in great depth. The opportunity to one day be able to treat aneurismal disease from the ascending aorta to the iliac arteries using a fully endovascular approach is near. Advances in imaging, device construction, and improved materials will play a major role in designing the next endovascular grafts to achieve this goal.

References

1. Johnston, K.W., Rutherford, R.B., Tilson, M.D., et al.: Suggested standards for reporting on arterial aneurysms. Subcommittee on reporting standards for arterial aneurysms, ad hoc committee on reporting standards, society for vascular surgery and north american chapter, international society for cardiovascular surgery. J. Vasc. Surg. **13**, 452 (1991)
2. Isselbacher, E.G.: Thoracic and abdominal aortic aneurysms. Circulation **111**, 816 (2005)
3. Bickerstaff, L.K., Pairolero, P.C., Hollier, L.H., et al.: Thoracic aortic aneurysms: a population-based study. Surgery **92**, 1103 (1982)
4. Conrad, M.F., Cambria, R.P.: Contemporary management of descending thoracic and thoracoabdominal aortic aneurysms: endovascular versus open. Circulation **117**, 841 (2008)
5. Crawford, E.S., Cohen, E.S.: Aortic aneurysm: a multifocal disease. Arch. Surg. **117**, 1393 (1982)
6. Pressler, V., McNamara, J.J.: Aneurysms of the thoracic aorta: review of 260 cases. J. Thorac. Cardiovasc. Surg. **89**, 50 (1985)
7. Elefteriades, J.A.: Natural history of thoracic aortic aneurysms: indications for surgery, and surgical versus nonsurgical risks. Ann. Thorac. Surg. **74**, S1877 (2002)
8. Clouse, W.D., Hallett, J.W., Schaff, H.V., et al.: Improved prognosis of thoracic aortic aneurysms: a population-based study. JAMA **280**, 1926 (1998)
9. Juvonen, T., Ergin, M.A., Galla, J.D., et al.: Prospective study of the natural history of thoracic aortic aneurysms. Ann. Thorac. Surg. **63**, 1533 (1997)
10. Coady, M.A., Rizzo, J.A., Hammond, G.L., et al.: Surgical intervention criteria for thoracic aortic aneurysms: a study of growth rates and complications. Ann. Thorac. Surg. **67**, 1922 (1999)
11. Davies, R.R., Goldstein, L.J., Coady, M.A., et al.: Yearly rupture or dissection rates for thoracic aortic aneurysms: simple prediction based on size. Ann. Thorac. Surg. **73**, 17 (2002)
12. Dapunt, O.E., Galla, J.D., Sageghi, A.M., et al.: The natural history of thoracic aortic aneurysms. J. Thorac. Cardiovasc. Surg. **107**, 1323 (1994)
13. Murdoch, J.L., Walker, B.A., Halpern, B.L., et al.: Life expectancy and causes of death in the Marfan syndrome. N. Engl. J. Med. **286**, 804 (1972)
14. Haft Jonathan, W: The heart: I. surgical treatment of acquired cardiac disease. In: Doherty Gerard, M. (ed.) Current Diagnosis & Treatment: Surgery, 13th edn. http://www.accesssurgery.com/content.aspx?aID=5214744 (2010)
15. Reed, D., Reed, C., Stemmermann, G., et al.: Are aortic aneurysms caused by atherosclerosis? Circulation **85**, 205 (1992)
16. Tilson, M.D.: Aortic aneurysms and atherosclerosis (editorial). Circulation **85**, 378 (1992)
17. Segura, A.M., Luna, R.E., Horiba, K., et al.: Immunohistochemistry of matrix metalloproteinases and their inhibitors in thoracic aortic aneurysms and aortic valves of patients with Marfan's syndrome. Circulation **98**(19 Suppl), II331 (1998)
18. Neptune, E.R., Frischmeyer, P.A., Arking, D.E. et al.: Dysregulation of TGF-β activation contributes to pathogenesis in Marfan syndrome. Nat. Genet. **33**, 407 [PubMed: 12598898] [Full Text] (2003)
19. Adams, J.N., Trent, R.J.: Aortic complications of Marfan's syndrome. Lancet **352**, 1722 (1998)
20. Loeys, B.L., Schwarze, U., Holm, T., et al.: Aneurysm syndromes caused by mutations in the TGF-β receptor. N. Engl. J. Med. **355**, 788 (2006)
21. LeMaire, S.A., Pannu, H., Tran-Fadulu, V., et al.: Severe aortic and arterial aneurysms associated with a TGFBR2 mutation. Nat. Clin. Pract. Cardiovasc. Med. **4**, 167 (2007)
22. Hoffman, J.I., Kaplan, S.: The incidence of congenital heart disease. J. Am. Coll. Cardiol. **39**, 1890 (2002)
23. Keane, M.G., Wiegers, S.E., Plappert, T., et al.: Bicuspid aortic valves are associated with aortic dilatation out of proportion to coexistent valvular lesions. Circulation **102**(19 Suppl 3), III35 (2000)

24. Nistri, S., Sorbo, M.D., Marin, M., et al.: Aortic root dilatation in young men with normally functioning bicuspid aortic valves. Heart **82**, 19 (1999)
25. Cecconi, M., Manfrin, M., Moraca, A., et al.: Aortic dimensions in patients with bicuspid aortic valve without significant valve dysfunction. Am. J. Cardiol. **95**, 292 (2005)
26. Sabet, H.Y., Edwards, W.D., Tazelaar, H.D., et al.: Congenitally bicuspid aortic valves: a surgical pathology study of 542 cases (1991 through 1996) and a literature review of 2,715 additional cases. Mayo. Clin. Proc. **74**, 14 (1999)
27. LeMaire, S.A., Wang, X., Wilks, J.A., et al.: Matrix metalloproteinases in ascending aortic aneurysms: bicuspid versus trileaflet aortic valves. J. Surg. Res. **123**, 40 (2005)
28. Yasuda, H., Nakatani, S., Stugaard, M., et al.: Failure to prevent progressive dilation of ascending aorta by aortic valve replacement in patients with bicuspid aortic valve: comparison with tricuspid aortic valve. Circulation **108**(Suppl 1), II291 (2003)
29. Johnson, J.R., Ledgerwood, A.M., Lucas, C.E.: Mycotic aneurysm: new concepts in therapy. Arch. Surg. **118**, 577 (1983)
30. Brown, S.L., Busuttil, R.W., Baker, J.D., et al.: Bacteriologic and surgical determinants of survival in patients with mycotic aneurysms. J. Vasc. Surg. **1**, 541 (1984)
31. Nuenninghoff, D.M., Hunder, G.G., Christianson, T.J., et al.: Incidence and predictors of large-artery complication (aortic aneurysm, aortic dissection, and/or large-artery stenosis) in patients with giant cell arteritis: a population-based study over 50 years. Arthritis Rheum. **48**, 3522 (2003)
32. von Kodolitsch, Y., Nienaber, C.A., Dieckmann, C., et al.: Chest radiography for the diagnosis of acute aortic syndrome. Am. J. Med. **116**, 73 (2004)
33. Cheitlin, M.D., Armstrong, W.F., Aurigemma, G.P., et al.: ACC/AHA/ASE 2003 guideline update for the clinical application of echocardiography: summary article: a report of the American college of cardiology/American heart association task force on practice guidelines (ACC/AHA/ASE committee to update the 1997 guidelines for the clinical application of echocardiography). Circulation **108**, 1146 (2003)
34. Fillinger, M.F.: Imaging of the thoracic and thoracoabdominal aorta. Semin. Vasc. Surg. **13**, 247 (2000)
35. Weisbord, S.D., Palevsky, P.M.: Radiocontrast-induced acute renal failure. J. Intensive Care Med. **20**, 63 (2005)
36. Rubin, G.D.: CT angiography of the thoracic aorta. Semin Roentgenol. **38**, 115–134 (2003)
37. Rubin, G.D.: Data explosion: the challenge of multidetector-row CT. Eur. J. Radiol. **36**, 74–80 (2000)
38. Danias, P., Eldeman, R., Manning, W.: Magnetic resonance angiography of the great vessels and the coronary arteries. In: Pohost, G.M. (ed.) Imaging in Cardiovascular Disease, p. 449. Lippincott Williams & Wilkins, Philadelphia (2000)
39. Ergun, I., Keven, K., Uruc, I., et al.: The safety of gadolinium in patients with stage 3 and 4 renal failure. Nephrol. Dial. Transplant. **21**, 697 (2006)
40. Coselli, J.S., LeMaire, S.A., Buket, S.: Marfan syndrome: the variability and outcome of operative management. J. Vasc. Surg. **21**, 432 (1995)
41. Coady, M.A., Rizzo, J.A., Elefteriades, J.A.: Developing surgical intervention criteria for thoracic aortic aneurysms. Cardiol. Clin. **17**, 827 (1999)
42. LeMaire, S.A., Rice, D.C., Schmittling, Z.C., et al.: Emergency surgery for thoracoabdominal aortic aneurysms with acute presentation. J. Vasc. Surg. **35**, 1171 (2002)
43. LeMaire, S.A., Miller III, C.C., Conklin, L.D., et al.: A new predictive model for adverse outcomes after elective thoracoabdominal aortic aneurysm repair. Ann. Thorac. Surg. **71**, 1233 (2001)
44. LeMaire, S.A., Miller III, C.C., Conklin, L.D., et al.: Estimating group mortality and paraplegia rates after thoracoabdominal aortic aneurysm repair. Ann. Thorac. Surg. **75**, 508 (2003)
45. Oury, J.H.: Clinical aspects of the Ross procedure: indications and contraindications. Semin. Thorac. Cardiovasc. Surg. **8**, 328 (1996)

46. Borst, H.G., Frank, G., Schaps, D.: Treatment of extensive aortic aneurysms by a new multiple-stage approach. J. Thorac. Cardiovasc. Surg. **95**, 11 (1988)
47. Jacobs, M.J., Eijsman, L., Meylaerts, S.A., et al.: Reduced renal failure following thoracoabdominal aortic aneurysm repair by selective perfusion. Eur. J. Cardiothorac. Surg. **14**, 201 (1998)
48. Crawford, E.S., Coselli, J.S.: Replacement of the aortic arch. Semin. Thorac. Cardiovasc. Surg. **3**, 194 (1991)
49. Coselli, J.S., Buket, S., Djukanovic, B.: Aortic arch operation: current treatment and results. Ann. Thorac. Surg. **59**, 19 (1995)
50. Kazui, T., Inoue, N., Yamada, O., Komatsu, S.: Selective cerebral perfusion during operation for aneurysms of the aortic arch: a reassessment. Ann. Thorac. Surg. **53**, 109 (1992)
51. Spielvogel, D., Strauch, J.T., Minanov, O.P., et al.: Aortic arch replacement using a trifurcated graft and selective cerebral antegrade perfusion. Ann. Thorac. Surg. **74**, S1810 (2002)
52. Strauch, J.T., Spielvogel, D., Lauten, A., et al.: Technical advances in total aortic arch replacement. Ann. Thorac. Surg. **77**, 581 (2004)
53. Griepp, R.B., Stinson, E.B., Hollingsworth, J.F., Buehler, D.: Prosthetic replacement of the aortic arch. J. Thorac. Cardiovasc. Surg. **70**, 1051 (1975)
54. Livesay, J.J., Cooley, D.A., Duncan, J.M., et al.: Open aortic anastomosis: improved results in the treatment of aneurysms of the aortic arch. Circulation **66**, I122 (1982)
55. Coselli, J.S., Crawford, E.S., Beall Jr., A.C., et al.: Determination of brain temperature for safe circulatory arrest during cardiovascular operation. Ann. Thorac. Surg. **45**, 638 (1988)
56. Svensson, L.G., Crawford, E.S., Hess, K.R. et al.: Deep hypothermia with circulatory arrest. Determinants of stroke and early mortality in 656 patients. J. Thorac. Cardiovasc. Surg. **106**, 19 (1993)
57. Ergin, M.A., Galla, J.D., Lansman, S.L., et al.: Hypothermic circulatory arrest in operations on the thoracic aorta. Determinants of operative mortality and neurologic outcome. J. Thorac. Cardiovasc. Surg. **107**, 788 (1994)
58. Safi, H.J., Brien, H.W., Winter, J.N., et al.: Brain protection via cerebral retrograde perfusion during aortic arch aneurysm repair. Ann. Thorac. Surg. **56**, 270 (1993)
59. Lin, P.J., Chang, C.H., Tan, P.P.C., et al.: Protection of the brain by retrograde cerebral perfusion during circulatory arrest. J. Thorac. Cardiovasc. Surg. **108**, 969 (1994)
60. Deeb, G.M., Jenkins, E., Bolling, S.F., et al.: Retrograde cerebral perfusion during hypothermic circulatory arrest reduces neurologic morbidity. J. Thorac. Cardiovasc. Surg. **109**, 259 (1995)
61. Bavaria, J.E., Woo, Y.J., Hall, R.A., et al.: Retrograde cerebral and distal aortic perfusion during ascending and thoracoabdominal aortic operations. Ann. Thorac. Surg. **60**, 345 (1995)
62. Okita, Y., Takamoto, S., Ando, M., et al.: Mortality and cerebral outcome in patients who underwent aortic arch operations using deep hypothermic circulatory arrest with retrograde cerebral perfusion: no relation of early death, stroke, and delirium to the duration of circulatory arrest. J. Thorac. Cardiovasc. Surg. **115**, 129 (1998)
63. Levy, W.L., Levin, S.K., Bavaria, J.E.: Cerebral oxygenation during retrograde perfusion. Ann. Thorac. Surg. **60**, 184 (1995)
64. Nojima, T., Mori, A., Watarida, S., Onoe, M.: Cerebral metabolism and effects of pulsatile flow during retrograde cerebral perfusion. J. Cardiovasc. Surg. (Torino) **34**, 483 (1993)
65. Filgueiras, C.L., Winsborrow, B., Ye, J., et al.: A 31P-magnetic resonance study of antegrade and retrograde cerebral perfusion during aortic arch surgery in pigs. J. Thorac. Cardiovasc. Surg. **110**, 55 (1995)
66. Oohara, K., Usui, A., Murase, M., et al.: Regional cerebral tissue blood flow measured by the colored microsphere method during retrograde cerebral perfusion. J. Thorac. Cardiovasc. Surg. **109**, 772 (1995)
67. Boeckxstaens, C.J., Flameng, W.J.: Retrograde cerebral perfusion does not perfuse the brain in nonhuman primates. Ann. Thorac. Surg. **60**, 319 (1995)

68. Neri, E., Sassi, C., Barabesi, L., et al.: Cerebral autoregulation after hypothermic circulatory arrest in operations on the aortic arch. Ann. Thorac. Surg. **77**, 72 (2004)
69. Fehrenbacher, J.W., McCready, R.A., Hormuth, D.A., et al.: One-stage segmental resection of extensive thoracoabdominal aneurysms with left-sided heart bypass. J. Vasc. Surg. **18**, 366 (1993)
70. Svensson, L.G., Crawford, E.S., Hess, K.R., et al.: Experience with 1509 patients undergoing thoracoabdominal aortic operations. J. Vasc. Surg. **17**, 357 (1993)
71. Svensson, L.G., Coselli, J.S., Safi, H.J., et al.: Appraisal of adjuncts to prevent acute renal failure after surgery on the thoracic or thoracoabdominal aorta. J. Vasc. Surg. **10**, 230 (1989)
72. Livesay, J.J., Cooley, D.A., Ventemiglia, R.A., et al.: Surgical experience in descending thoracic aneurysmectomy with and without adjuncts to avoid ischemia. Ann. Thorac. Surg. **39**, 37 (1985)
73. Mauney, M.C., Blackbourne, L.H., Langenburg, S.E., et al.: Prevention of spinal cord injury after repair of the thoracic or thoracoabdominal aorta. Ann. Thorac. Surg. **59**, 245 (1995)
74. Crawford, E.S., Mizrahi, E.M., Hess, K.R., et al.: The impact of distal aortic perfusion and somatosensory evoked potential monitoring on prevention of paraplegia after aortic aneurysm operation. J. Thorac. Cardiovasc. Surg. **95**, 357 (1988)
75. Crawford, E.S., Svensson, L.G., Hess, K.R., et al.: A prospective randomized study of cerebrospinal fluid drainage to prevent paraplegia after high-risk surgery on the thoracoabdominal aorta. J. Vasc. Surg. **13**, 36 (1991)
76. Coselli, J.S., LeMaire, S.A.: Left heart bypass reduces paraplegia rates after thoracoabdominal aortic aneurysm repair. Ann. Thorac. Surg. **67**, 1931 (1999)
77. Ueda, T., Shimizu, H., Mori, A., et al.: Selective perfusion of segmental arteries in patients undergoing thoracoabdominal aortic surgery. Ann. Thorac. Surg. **70**, 38 (2000)
78. Svensson, L.G., Hess, K.R., D'Agostino, R.S., et al.: Reduction of neurologic injury after high-risk thoracoabdominal aortic operation. Ann. Thorac. Surg. **66**, 132 (1998)
79. Hollier, L.H., Money, S.R., Naslund, T.C., et al.: Risk of spinal cord dysfunction in patients undergoing thoracoabdominal aortic replacement. Am. J. Surg. **164**, 210 (1992)
80. Kouchoukos, N.T., Rokkas, C.K.: Hypothermic cardiopulmonary bypass for spinal cord protection: Rationale and clinical results. Ann. Thorac. Surg. **67**, 1940 (1999)
81. Kouchoukos, N.T., Masetti, P., Rokkas, C.K., et al.: Safety and efficacy of hypothermic cardiopulmonary bypass and circulatory arrest for operations on the descending thoracic and thoracoabdominal aorta. Ann. Thorac. Surg. **72**, 699 (2001)
82. Connolly, J.E.: Hume memorial lecture. Prevention of spinal cord complications in aortic surgery. Am. J. Surg. **176**, 92 (1998)
83. Robertazzi, R.R., Cunningham, J.N.: Intraoperative adjuncts of spinal cord protection. Semin. Thorac. Cardiovasc. Surg. **10**, 29 (1998)
84. Ross, S.D., Kron, I.L., Parrino, P.E., et al.: Preservation of intercostal arteries during thoracoabdominal aortic aneurysm surgery: a retrospective study. J. Thorac. Cardiovasc. Surg. **118**, 17 (1999)
85. McGarvey, M.L., Cheung, A.T., Szeto, W., Messe, S.R.: Management of neurologic complications of thoracic aortic surgery. J. Clin. Neurophysiol. **24**, 336 (2007)
86. Kawaguchi, S., et al.: Thoracic endovascular aneurysm repair in Japan: experience with fenestrated stent grafts in the treatment of distal arch aneurysms. J. Vasc. Surg. **48**(6 Suppl), 24S–29S (2008)
87. Inoue, K., et al.: Aortic arch reconstruction by transluminally placed endovascular branched stent graft. Circulation **100**(19 Suppl), II316–II321 (1999)
88. Chuter, T.A., Schneider, D.B.: Endovascular repair of the aortic arch. Perspect. Vasc. Surg. Endovasc. Ther. **19**(2), 188–192 (2007)
89. Svensson, L.G., Kouchoukos, N.T., Miller, D.C., et al.: Expert consensus document on the treatment of descending thoracic aortic disease using endovascular stent-grafts. Ann. Thorac. Surg. **85**(1 Suppl), S1 (2008)

90. Parodi, J.C., Palmaz, J.C., Barone, H.D.: Transfemoral intraluminal graft implantation for abdominal aortic aneurysms. Ann. Vasc. Surg. **5**, 491 (1991)
91. Dake, M.D., Miller, D.C., Semba, C.P., et al.: Transluminal placement of endovascular stent-grafts for the treatment of descending thoracic aortic aneurysms. N. Engl. J. Med. **331**, 1729 (1994)
92. Lin, P.H., El Sayed, H.F., Kougias, P., et al.: Endovascular repair of thoracic aortic disease: overview of current devices and clinical results. Vascular **15**, 179 (2007)
93. Melissano, G., Civilini, E., Bertoglio, L., Logaldo, D., Chiesa, R.: Initial clinical experience with the modified zenith "Pro-Form" TX2 thoracic endograft. J. Endovasc. Ther. **17**(4), 463–470 (2010)
94. Qureshi, M.A., Greenberg, R.K.: New results with the Zenith graft in the treatment of aortic aneurysms. J. Cardiovasc. Surg. (Torino) **51**(4), 503–514 (2010)
95. Knepper, J., Upchurch Jr., G.R.: A review of clinical trials and registries in descending thoracic aortic aneurysms. Semin. Vasc. Surg. **23**(3), 170–175 (2010)
96. Pua, U., Tay, K.H., Tan, B.S., Htoo, M.M., Sebastian, M., Sin, K., Chua, Y.L.: CT appearance of complications related to thoracic endovascular aortic repair (TEVAR): a pictorial essay. Eur. Radiol. **19**(5), 1062–1068 (2009)
97. Dowdall, J.F., Greenberg, R.K., West, K., Moon, M., Lu, Q., Francis, C. et al.: Separation of components in fenestrated and branched endovascular grafting-branch protection or a potentially new mode of failure? Eur. J. Vasc. Endovasc. Surg. **36**(1), 2–9. Epub 2008 Apr 25
98. Mori, D., Yamaguchi, T.: Computational fluid dynamics modeling and analysis of the effect of 3-D distortion of the human aortic arch. Comput. Method. Biomech. Biomed. Eng. **5**(3), 249–260 (2002)
99. Qiao, A., Liu, Y.: Medical application oriented blood flow simulation. Clin. Biomech. (Bristol, Avon) **23** Suppl 1, S130-S136 (2008). Epub 2007 Nov 19
100. Wu, X., Xu, K., Chen, R., Xiao, L., Zhang, X., Su, H., Feng, B.: Hemodynamics analysis of a stented aortic bifurcation. Sheng Wu Yi Xue Gong Cheng Xue Za Zhi **26**(6), 1250–1254 (2009)
101. Liffman, K., Lawrence-Brown, M.M., Semmens, J.B., Bui, A., Rudman, M., Hartley, D.E.: Analytical modeling and numerical simulation of forces in an endoluminal graft. J. Endovasc. Ther. **8**(4), 358–371 (2001)
102. Pless, D., Boll, D.T., Goerich, J., Fleiter, T.R., Scharrer-Pamler, R., Brambs, H.J.: Multislice CT and Computational fluid dynamics: a new technique to visualize, analyze and simulate hemodynamics in aortic aneurysms and stentgrafts. Stud. Health. Technol. Inform. **81**, 393–395 (2001)
103. Moore Jr., J.E., Ku, D.N., Zarins, C.K., Glagov, S.: Pulsatile flow visualization in the abdominal aorta under differing physiologic conditions: implications for increased susceptibility to atherosclerosis. J. Biomech. Eng. **114**(3), 391–397 (1992)
104. Heldt, T., Shim, E.B., Kamm, R.D., Mark, R.G.: Computational modeling of cardiovascular response to orthostatic stress. J. Appl. Physiol. **92**(3), 1239–1254 (2002)
105. Taylor, C.A., Hughes, T.J., Zarins, C.K.: Effect of exercise on hemodynamic conditions in the abdominal aorta. J. Vasc. Surg. **29**(6), 1077–1089 (1999)
106. Gin, R., Straatman, A.G., Steinman, D.A.: A dual-pressure boundary condition for use in simulations of bifurcating conduits. J. Biomech. Eng. **124**(5), 617–619 (2002)
107. Ferrandez, A., David, T., Brown, M.D.: Numerical models of auto-regulation and blood flow in the cerebral circulation. Comput. Method. Biomech. Biomed. Eng. **5**(1), 7–19 (2002)
108. Westerhof, N., Elzinga, G., Sipkema, P.: An artificial arterial system for pumping hearts. J. Appl. Physiol. **31**(5), 776–781 (1971)
109. Karmonik, C., Bismuth, J., Davies, M.G., Lumsden, A.B.: Computational fluid dynamics as a tool for visualizing hemodynamic flow patterns. Methodist. Debakey Cardiovasc. J. **5**(3), 26–33 (2009)
110. Molony, D.S., Callanan, A., Kavanagh, E.G., Walsh, M.T., McGloughlin, T.M.: Fluid-structure interaction of a patient-specific abdominal aortic aneurysm treated with an endovascular stent-graft. Biomed. Eng. Online **6**(8), 24 (2009)

111. Scotti, C.M., Shkolnik, A.D., Muluk, S.C., Finol, E.A.: Fluid-structure interaction in abdominal aortic aneurysms: effects of asymmetry and wall thickness. Biomed. Eng. Online **4**(4), 64 (2005)
112. Scotti, C.M., Jimenez, J., Muluk, S.C., Finol, E.A.: Wall stress and flow dynamics in abdominal aortic aneurysms: finite element analysis vs. fluid-structure interaction. Comput. Method. Biomech. Biomed. Eng. **3**, 301–322 (2008)
113. Beller, C.J., Gebhard, M.M., Karck, M., Labrosse, M.R.: Usefulness and limitations of computational models in aortic disease risk stratification. J. Vasc. Surg. **52**(6), 1572–1579 (2010)
114. Figueroa, C.A., Taylor, C.A., Yeh, V., Chiou, A.J., Zarins, C.K.: Effect of curvature on displacement forces acting on aortic endografts: a 3-dimensional computational analysis. J. Endovasc. Ther. **16**(3), 284–294 (2009)
115. Figueroa, C.A., Taylor, C.A., Chiou, A.J., Yeh, V., Zarins, C.K.: Magnitude and direction of pulsatile displacement forces acting on thoracic aortic endografts. J. Endovasc. Ther. **16**(3), 350–358 (2009)
116. Viscardi, F., Vergara, C., Antiga, L., Merelli, S., Veneziani, A., Puppini, G., Faggian, G., Mazzucco, A., Luciani, G.B.: Comparative finite element model analysis of ascending aortic flow in bicuspid and tricuspid aortic valve. Artif. Organs. **34**(12), 1114–1120 (2010)
117. Poullis, M.P., Warwick, R., Oo, A., Poole, R.J.: Ascending aortic curvature as an independent risk factor for type A dissection, and ascending aortic aneurysm formation: a mathematical model. Eur. J. Cardiothorac. Surg. **33**(6), 995–1001 (2008) Epub 2008 Apr 22
118. Fogel, M.A., Weinberg, P.M., Haselgrove, J.: The impact of aortic arch geometry on flow dynamics using a simplified approach with magnetic resonance velocity mapping. Congenit. Heart Dis. **1**(6), 300–308 (2006)
119. Canaud, L., Alric, P., Desgranges, P., Marzelle, J., Marty-Ané, C., Becquemin, J.P.: Factors favoring stent-graft collapse after thoracic endovascular aortic repair. J. Thorac. Cardiovasc. Surg. **139**(5), 1153–1157 (2010) Epub 2009 Jul 29
120. Lazar, H.L., Varma, P.K., Shapira, O.M., Soto, J., Shaw, P.: Endograft collapse after thoracic stent-graft repair for traumatic rupture. Ann. Thorac. Surg. **87**(5), 1582–1583 (2009)
121. Sze, D.Y., Mitchell, R.S., Miller, D.C., Fleischmann, D., Frisoli, J.K., Kee, S.T., Verma, A., Sheehan, M.P., Dake, M.D.: Infolding and collapse of thoracic endoprostheses: manifestations and treatment options. J. Thorac. Cardiovasc. Surg. **138**(2), 324–333 (2009)
122. Lee, W.A.: Failure modes of thoracic endografts: prevention and management. J. Vasc. Surg. **49**(3), 792–799 (2009)
123. Kölbel, T., Resch, T.A., Dias, N., Björses, K., Sonesson, B., Malina, M.: Staged proximal deployment of the Zenith TX2 thoracic stent-graft: a novel technique to improve conformance to the aortic arch. J. Endovasc. Ther. **16**(5), 598–602 (2009)
124. Cho, J.S., Haider, S.E., Makaroun, M.S.: Endovascular therapy of thoracic aneurysms: gore TAG trial results. Semin. Vasc. Surg. **19**(1), 18–24 (2006)
125. Hughes, G.C., Daneshmand, M.A., Swaminathan, M., Nienaber, J.J., Bush, E.L., Husain, A.H., Wolfe, W.G., McCann, R.L.: "Real world" thoracic endografting: results with the gore TAG device 2 years after U.S. FDA approval. Ann. Thorac. Surg. **86**(5), 1530–1537 (2008) (discussion)
126. Brooks, M., Loftus, I., Morgan, R., Thompson, M.: The valiant thoracic endograft. J. Cardiovasc. Surg. (Torino) **47**(3), 269–278 (2006)
127. Dagenais, F.: Commentary: the valiant thoracic stent graft for endovascular repair of thoracic aortic aneurysms and dissections: looking good so far. J. Endovasc. Ther. **17**(2), 151–152 (2010)
128. Torsello, G.B., Torsello, G.F., Osada, N., Teebken, O.E., Ratusinski, C.M., Nienaber, C.A.: Midterm results from the TRAVIATA registry: treatment of thoracic aortic disease with the valiant stent graft. J. Endovasc. Ther. **17**(2), 137–150 (2010)
129. Funovics, M., Blum, M., Langenberger, H., Plank, C., Schoder, M., Edelhauser, G., et al.: Endovascular repair of the descending aorta and the aortic arch with the relay stent graft. Ann. Thorac. Surg. **88**(2), 637–640 (2009)

130. Ferro, C., Rossi, U.G., Seitun, S., Guastavino. A., Scarano, F., Passerone, G.C.: Relay NBS graft with the plus delivery system to improve deployment in aortic arch with small radius curve. Cardiovasc. Intervent. Radiol. **34**(2), 401–405 (2010)
131. Ueda, T., Fleischmann, D., Dake, M.D., Rubin, G.D., Sze, D.Y.: Incomplete endograft apposition to the aortic arch: bird-beak configuration increases risk of endoleak formation after thoracic endovascular aortic repair. Radiology **255**(2), 645–652 (2010)
132. Melissano, G., Civilini, E., Bertoglio, L., Calliari, F., Setacci, F., Calori, G., Chiesa, R.: Results of endografting of the aortic arch in different landing zones. Eur. J. Vasc. Endovasc. Surg. **33**(5), 561–566 (2007) Epub 2007 Jan 3
133. Melissano, G., Bertoglio, L., Civilini, E., Marone, E.M., Calori, G., Setacci, F., Chiesa, R.: Results of thoracic endovascular grafting in different aortic segments. J. Endovasc. Ther. **14**(2), 150–157 (2007)
134. Beregi, J.P., Marco, M.: Stent-grafting of the thoracic aorta. In: Morgan, R.A., Walser, E. (eds.) Handbook of Angioplasty and Stenting Procedures. Techniques in Interventional Radiology., 1st edn edn. Springer, London (2010)
135. Goel, V.R., Greenberg, R.K., Greenberg, D.P.: Automated vascular geometric analysis of aortic aneurysms. IEEE Comput. Graph. Appl. **28**(3), 76–86 (2008)
136. Greenberg, R.K., Turc, A., Haulon, S., Srivastava, S.D., Sarac, T.P., O'Hara, P.J., et al.: Stent-graft migration: a reappraisal of analysis methods and proposed revised definition. J. Endovasc. Ther. **11**, 353–363 (2004)
137. Murphy, M., Hodgson, R., Harris, P.L., McWilliams, R.G., Hartley, D.E., Lawrence-Brown, M.M.: Plain radiographic surveillance of abdominal aortic stent-grafts: the Liverpool/Perth protocol. J. Endovasc. Ther. **10**, 911–912 (2003)
138. Greenberg, R., Haddad, F., Svensson, L., O'Neill, S., Walker, E., Lyden, S., et al.: Hybrid approaches to thoracic aortic aneurysms: the role of endovascular elephant trunk completion. Circulation **112**, 2619–2626 (2005)
139. Aomi, S., Nakajima, M., Nonoyama, M., et al.: Aortic root replacement using composite valve graft in patients with aortic valve disease and aneurysm of the ascending aorta: twenty years experience of late results. Artif. Organs **26**, 467 (2002)
140. Kindo, M., Billaud, P., Gerelli, S., et al.: Twenty-seven-year experience with composite valve graft replacement of the aortic root. J. Heart Valve Dis. **16**, 370 (2007)
141. Achneck, H.E., Rizzo, J.A., Tranquilli, M., et al.: Safety of thoracic aortic surgery in the present era. Ann. Thorac. Surg. **84**, 1180 (2007)
142. Estrera, A.L., Miller III, C.C., Madisetty, J., et al.: Ascending and transverse aortic arch repair: the impact of glomerular filtration rate on mortality. Ann. Surg. **247**, 524 (2008)
143. LeMaire, S.A., Carter, S.A., Coselli, J.S.: The elephant trunk technique for staged repair of complex aneurysms of the entire thoracic aorta. Ann. Thorac. Surg. **81**, 1561 (2006)
144. Svensson, L.G., Kim, K.H., Blackstone, E.H., et al.: Elephant trunk procedure: newer indications and uses. Ann. Thorac. Surg. **78**, 109 (2004)
145. Heinemann, M.K., Buehner, B., Jurmann, M.J., et al.: Use of the "elephant trunk technique" in aortic surgery. Ann. Thorac. Surg. **60**, 2 (1995)
146. Safi, H.J., Miller III, C.C., Estrera, A.L., et al.: Staged repair of extensive aortic aneurysms: long-term experience with the elephant trunk technique. Ann. Surg. **240**, 677 (2004)
147. Sundt, T.M., Moon, M.R., DeOliviera, N., et al.: Contemporary results of total aortic arch replacement. J. Card. Surg. **19**, 235 (2004)
148. Zingone, B., Rauber, E., Gatti, G., et al.: Diagnosis and management of severe atherosclerosis of the ascending aorta and aortic arch during cardiac surgery: focus on aortic replacement. Eur. J. Cardiothorac. Surg. **31**, 990 (2007)
149. Kazui, T., Bashar, A.H.: Aortic arch replacement using a trifurcated graft. Ann. Thorac. Surg. **81**, 1552 (2006)
150. Suzuki, K., Kazui, T., Bashar, A.H., et al.: Total aortic arch replacement in patients with arch vessel anomalies. Ann. Thorac. Surg. **81**, 2079 (2006)
151. Spielvogel, D., Etz, C.D., Silovitz, D., et al.: Aortic arch replacement with a trifurcated graft. Ann. Thorac. Surg. **83**, S791 (2007)

152. Kamiya, H., Hagl, C., Kropivnitskaya, I., et al.: Quick proximal arch replacement with moderate hypothermic circulatory arrest. Ann. Thorac. Surg. **83**, 1055 (2007)
153. Coselli, J.S., LeMaire, S.A., Conklin, L.D., et al.: Left heart bypass during descending thoracic aortic aneurysm repair does not reduce the incidence of paraplegia. Ann. Thorac. Surg. **77**, 1298 (2004)
154. Chiesa, R., Melissano, G., Civilini, E., et al.: Ten years experience of thoracic and thoracoabdominal aortic aneurysm surgical repair: lessons learned. Ann. Vasc. Surg. **18**, 514 (2004)
155. Estrera, A.L., Miller III, C.C., Chen, E.P., et al.: Descending thoracic aortic aneurysm repair: 12-year experience using distal aortic perfusion and cerebrospinal fluid drainage. Ann. Thorac. Surg. **80**, 1290 (2005)
156. Stone, D.H., Brewster, D.C., Kwolek, C.J., et al.: Stent-graft versus open-surgical repair of the thoracic aorta: Mid-term results. J. Vasc. Surg. **44**, 1188 (2006)
157. Matsumura, J.S., Cambria, R.P., Dake, M.D., et al.: International controlled clinical trial of thoracic endovascular aneurysm repair with the Zenith TX2 endovascular graft: 1-year results. J. Vasc. Surg. **47**, 247 (2008)
158. Dick, F., Hinder, D., Immer, F.F., et al.: Outcome and quality of life after surgical and endovascular treatment of descending aortic lesions. Ann. Thorac. Surg. **85**, 1605 (2008)
159. Brandt, M., Hussel, K., Walluscheck, K.P., et al.: Stent-graft repair versus open surgery for the descending aorta: a case-control study. J. Endovasc. Ther. **11**, 535 (2004)
160. Bavaria, J.E., Appoo, J.J., Makaroun, M.S., et al.: Endovascular stent grafting versus open surgical repair of descending thoracic aortic aneurysms in low-risk patients: a multicenter comparative trial. J. Thorac. Cardiovasc. Surg. **133**, 369 (2007)
161. Najibi, S., Terramani, T.T., Weiss, V.J., et al.: Endoluminal versus open treatment of descending thoracic aortic aneurysms. J. Vasc. Surg. **36**, 732 (2002)
162. Gutsche, J.T., Cheung, A.T., McGarvey, M.L., et al.: Risk factors for perioperative stroke after thoracic endovascular aortic repair. Ann. Thorac. Surg. **84**, 1195 (2007)
163. Feezor, R.J., Martin, T.D., Hess, P.J., et al.: Risk factors for perioperative stroke during thoracic endovascular aortic repairs (TEVAR). J. Endovasc. Ther. **14**, 568 (2007)
164. Demers, P., Miller, D.C., Mitchell, R.S., et al.: Midterm results of endovascular repair of descending thoracic aortic aneurysms with first-generation stent grafts. J. Thorac. Cardiovasc. Surg. **127**, 664 (2004)
165. Svensson, L.G.: Device discordancy: lost cords, quick-fix seekers, quality, and ethics. J. Thorac. Cardiovasc. Surg. **131**, 261 (2006)
166. Dake, M.D., Miller, D.C., Mitchell, R.S., et al.: The "first generation" of endovascular stent-grafts for patients with aneurysms of the descending thoracic aorta. J. Thorac. Cardiovasc. Surg. **116**, 689 (1998)
167. Amabile, P., Grisoli, D., Giorgi, R., et al.: Incidence and determinants of spinal cord ischaemia in stent-graft repair of the thoracic aorta. Eur. J. Vasc. Endovasc. Surg. **35**, 455 (2008)
168. Buth, J., Harris, P.L., Hobo, R., et al.: Neurologic complications associated with endovascular repair of thoracic aortic pathology: Incidence and risk factors. a study from the European collaborators on stent/graft techniques for aortic aneurysm repair (EUROSTAR) registry. J. Vasc. Surg. **46**, 1103 (2007)
169. Cheung, A.T., Pochettino, A., McGarvey, M.L., et al.: Strategies to manage paraplegia risk after endovascular stent repair of descending thoracic aortic aneurysms. Ann. Thorac. Surg. **80**, 1280 (2005)
170. Greenberg, R.K., Lu, Q., Roselli, E.E., et al.: Contemporary analysis of descending thoracic and thoracoabdominal aneurysm repair: a comparison of endovascular and open techniques. Circulation **118**, 808 (2008)
171. Leurs, L.J., Bell, R., Degrieck, Y., et al.: Endovascular treatment of thoracic aortic diseases: combined experience from the EUROSTAR and United Kingdom thoracic endograft registries. J. Vasc. Surg. **40**, 670 (2004)

172. Morales, J.P., Greenberg, R.K., Morales, C.A., Cury, M., Hernandez, A.V., Lyden, S.P., et al.: Thoracic aortic lesions treated with the Zenith TX1 and TX2 thoracic devices: intermediate- and long-term outcomes. J. Vasc. Surg. **48**(1), 54–63 (2008)
173. Preventza, O., Wheatley, G.H. III, Ramaiah, V.G., et al.: Management of endoleaks associated with endovascular treatment of descending thoracic aortic diseases. J. Vasc. Surg. **48**(1), 69–73 (2008)
174. Makaroun, M.S., Dillavou, E.D., Wheatley, G.H., Cambria, R.P., Gore TAG (2008) investigators.:Five-year results of endovascular treatment with the Gore TAG device compared with open repair of thoracic aortic aneurysms. J. Vasc. Surg. **47**(5), 912–918 (2008) Epub 2008 Mar 19
175. Svensson, L.G., Kouchoukos, N.T., Miller, D.C., Bavaria, J.E., Coselli, J.S., Curi, M.A., et al.: Society of thoracic surgeons endovascular surgery task force. expert consensus document on the treatment of descending thoracic aortic disease using endovascular stent-grafts. Ann. Thorac. Surg. **85**(1), S1–S41 (2008)
176. Walsh, S.R., Tang, T.Y., Sadat, U., et al.: Endovascular stenting versus open surgery for thoracic aortic disease: systematic review and meta-analysis of perioperative results. J. Vasc. Surg. **47**, 1094 (2008)

Intracranial Aneurysms: Clinical Assessment and Treatment Options

Manik Mehra, Gabriela Spilberg, Matthew J. Gounis and Ajay K. Wakhloo

Abstract Intracranial aneurysms (IA) represent a challenging pathological entity ubiquitous in 2–6% percent of the population which may present occasionally with a devastating and often fatal intracranial bleed. While the understanding of the pathophysiological mechanisms governing their genesis and rupture are still being elucidated, advancements in the imaging techniques have enabled radiologists to accurately characterize their morphology and relationship with the neighboring vasculature. The past few decades have seen a refinement in the microneurosurgical techniques and made possible permanent exclusion of the aneurysm from the circulation. The development of detachable platinum microcoils to embolize the aneurysmal sac led to the dawn of an era of minimally invasive endovascular surgery. This branch has seen phenomenal progress with the development of an assortment of medical devices. These include guidewires and catheters for access, improvements in the embolic coil design, endovascular balloons, neurovascular stents and flow diverting devices that facilitate obliteration of the aneurysm. Recently concluded clinical trials attest to the remarkable safety profile of endovascular occlusion. However, there is a concern over the durability of treatment necessitating mandatory follow up imaging and retreatments for aneurysm recanalization. Also, controversy exists over the indications for offering treatment for the incidentally discovered IA. Future research into the identification of factors influencing the aneurysms predisposition to rupture and development of markers to assess that risk, will hopefully address this issue.

M. Mehra · G. Spilberg · M. J. Gounis (✉) · A. K. Wakhloo
New England Center for Stroke Research, Neuroimaging and Intervention,
Departments of Radiology, Neurological Surgery and Neurology,
University of Massachusetts, Worcester, USA
e-mail: Matthew.Gounis@umassmed.edu

Stud Mechanobiol Tissue Eng Biomater (2011) 7: 331–372

DOI: 10.1007/8415_2011_78

Published Online: 6 May 2011

Abbreviations

ADPKD	Autosomal dominant polycystic kidney disease
AICA	Anterior inferior cerebellar artery
CAMEO	Cerebral aneurysm multicenter European Onyx trial
CARAT	Cerebral aneurysm rerupture after treatment
CCT	Cerecyte coil trial
CE-MRA	Contrast-enhanced magnetic resonance angiography
CSF	Cerebrospinal fluid
CT	Computed tomography
CTA	Computed tomography angiography
DMSO	Dimethyl sulfoxide
DSA	Digital subtraction angiography
EVOH	Ethylene vinyl alcohol
EVT	Endovascular treatment
FDA	Food and drug administration
FLAIR	Fluid-attenuated inversion recovery
GDC	Guglielmi detachable coils
HELPS	HydroCoil endovascular aneurysm occlusion and packing study
IA	Intracranial aneurysm
ICA	Internal cerebral artery
ICP	Intracranial pressure
ISAT	International subarachanoid aneurysm trial
ISUIA	International study of unruptured intracranial aneurysms
MCA	Middle cerebral artery
MIP	Maximum intensity projection
MMP	Matrix metalloproteinase
MR	Magnetic resonance
MRA	Magnetic resonance angiography
PICA	Posterior inferior cerebellar artery
SAH	Subarachnoid hemorrhage
SNIS	Society of neurointerventional surgery
TCD	Transcranial doppler
TOF	Time of flight
VR	Volume rendered

1 Epidemiology

Intracranial aneurysms (IA) are pathological localized dilatations in the cerebrovascular wall in direct communication with the lumen and therefore the blood flow, resulting from an acquired or less frequently a congenital weakness in the arterial wall. The prevalence of IA remains uncertain and controversial due to its insidious nature. Evidence suggests that 2–6% of the general population may harbor an intracranial aneurysm [122, 152, 156, 164], afflicting approximately 1.5 and 6 million people in Germany and the United States [64], respectively. The mean annual incidence of subarachnoid hemorrhage, its most devastating presentation, in Western Europe and the United States is 10 per 100,000 [53, 54], accounting for 30,000 new patients each year [75] in the United States. The rupture risk in the Finnish population is known to be higher- about 22.5 per 100,000 per year [54]. The clinical outcome of a ruptured aneurysm is abysmal, with thirty day mortality close to 50%. Of the survivors, 25% of the patients suffer severe disability and the rest are at an increased risk for stroke, recurrent bleeding and other complications [100]. However, prophylactic aneurysmal obliteration is not feasible for all unruptured IA and is associated with concurrent risks. This justifies the motivation to aggressively investigate their elusive natural history so as to ascertain and treat those at a higher risk for rupture.

2 Morphology and Pathogenesis of Aneurysms

2.1 Aneurysm Classification

IA can be classified according to their etiology, morphology, size and their location. Based on their etiology, IA can be classified as developmental, degenerative, flow related, atherosclerotic, mycotic, oncotic and of traumatic origin. Morphologically, IA can be classified as saccular, fusiform and dissecting aneurysms [139] (Fig. 1). Saccular aneurysms, defined as an outpouching of the vascular wall with a definable neck are by far the most common, responsible for 80–90% IA. Fusiform cerebral aneurysms are nonsaccular, spindle shaped aneurysms with focal circumferential dilatation, whereas dolichoectatic aneurysms are elongated and tortuous with a uniform circumference. They are rare in comparison to saccular aneurysms having an annual incidence of less than 0.1% [4, 51]. Based on their pathophysiology, they can arise acutely as dissecting aneurysms, from a disruption of the internal elastic lamina resulting in an intramural hemorrhage, or as chronic fusiform/dolichoectatic aneurysms believed to arise from atherosclerotic degeneration of the aneurysmal wall [97].

The saccular aneurysms can be classified based on the largest diameter of the aneurysm dome as small (<10 mm in diameter), large (10–25 mm), and giant (>25 mm) aneurysms [153]. Further, it is important to measure the neck of the aneurysm. Wide-neck aneurysms are defined as having a neck diameter greater than 4 mm, or a dome-to-neck ratio ≤ 2.

Saccular aneurysms are most often located at the arterial bifurcations of the Circle of Willis, with a majority in the anterior circulation at the anterior communicating artery (30%), at the junction of the internal carotid artery (ICA) and the posterior communicating artery (25%), the middle cerebral artery (MCA) bifurcation (20%) and at the ICA bifurcation (7.5%) [15]. Of the aneurysms from the

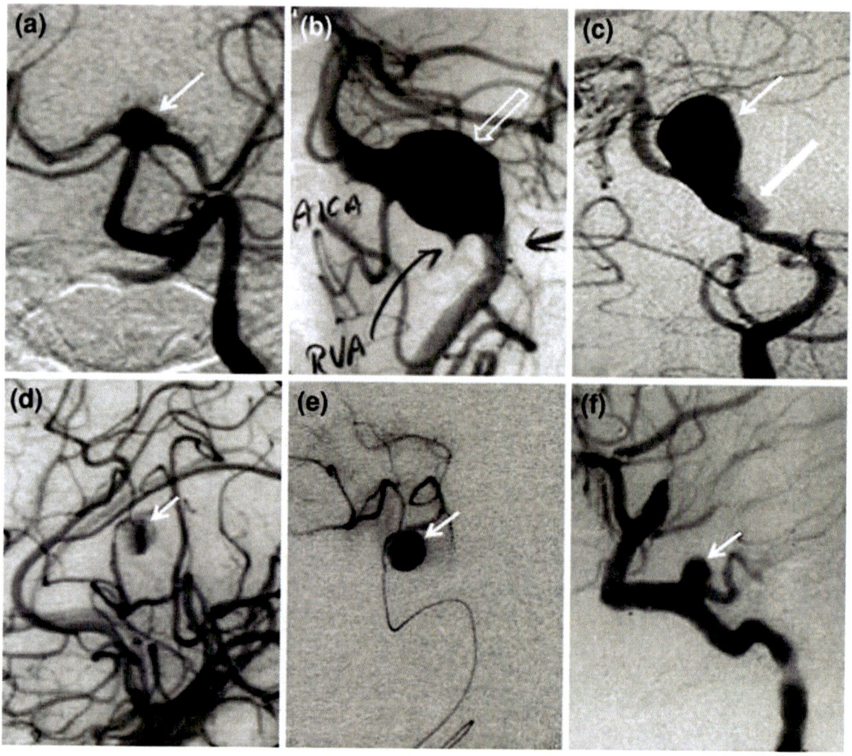

Fig. 1 Digital subtraction angiography images of different aneurysmal morphologies and etiology. **a** Left vertebral artery angiogram depicts a basilar bifurcation aneurysm (*arrow*), involving both origins of the posterior cerebral arteries. **b** Left vertebral artery angiogram, oblique lateral view, demonstrates a fusiform atherosclerotic aneurysm (*arrow*) of the basilar trunk, involving the origin of the right anterior inferior cerebellar arteries(AICA). **c** Right vertebral artery angiogram, oblique lateral view, shows a dissection of the basilar trunk (*thick arrow*), associated with a large pseudo-aneurysm (*arrow*) **d** and **e** Internal carotid artery (ICA) angiogram (**d**) and supraselective microcatheter injection (**e**), lateral view, demonstrates a distal MCA mycotic aneurysm. Internal carotid artery angiogram, lateral view depicts a saccular aneurysm associated with a persistent trigeminal artery (developmental anomaly with the persistence of the short wide fetal connection between the cavernous carotid and upper third of the basilar artery) (**f**)

posterior circulation, 7% arise from the basilar artery bifurcation and 3% from the origin of the posterior inferior cerebellar artery (PICA) [15] (Fig. 2).

2.2 Pathophysiology

A healthy cerebral artery comprises of three circumferential layers: *tunica intima* with endothelial cells and a separating internal elastic lamina; a *tunica media* with elastin fibers and circumferentially organized smooth muscle cells and *tunica*

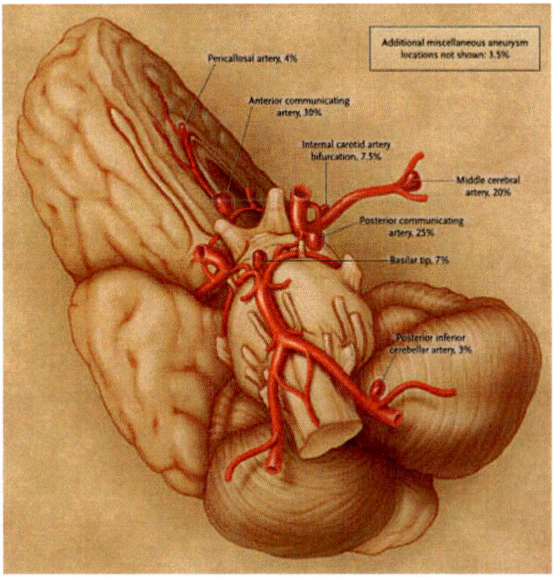

Fig. 2 Intracranial vasculature shows the most frequent locations of intracranial aneurysms. Percentages indicate the incidence of intracranial aneurysms [15]

externa with helical pitches of type I collagen fibers. The collagen provides for the high tensile strength and the elastin for the compliant response of the cerebral arteries [157]. In comparison to systemic arteries, cerebral arteries possess significantly less elastin in the media, lack an external elastic lamina and have medial discontinuities (raphes). These features are believed to make them more vulnerable to aneurysmal dilatation [139].

The emerging pathophysiological hypothesis regarding the initiation of the saccular IA formation, attributes their asymmetric or focalized outward vascular remodeling as an attempt to re-establish homeostasis following hemodynamic and biological insults [46]. Endothelial cell activation results from the sustained and localized impact of the high wall sheer stress experienced by the vascular wall at regions of susceptible arterial geometries, like arterial bifurcations and in vasculature made vulnerable secondary to vasculopathies like connective tissue disorders. This, in combination with myointimal hyperplasia—the proliferation and the luminal migration of the smooth muscle cells, initiates inflammatory cell recruitment, secreting matrix metalloproteinases (MMPs) like elastases [41, 46]. This results in the pathologic degradation of the internal elastic lamina, which accounts for most of the mechanical strength and integrity of the cerebral vessel, resulting in a preaneurysmal change. Histologically, unruptured cerebral aneurysms lack an endothelial layer, have a disrupted internal elastic lamina, thinned media with apoptosis of smooth muscle cells and inflammatory mediators [41]. The early aneurysm undergoes modeling and growth which may lead to its eventual stabilization [161]. It may however become unstable with continual exposure to the inflammatory mediators and hemodynamic impulses. When the intraluminal pressure overcomes the tensile strength of the aneurysmal wall it

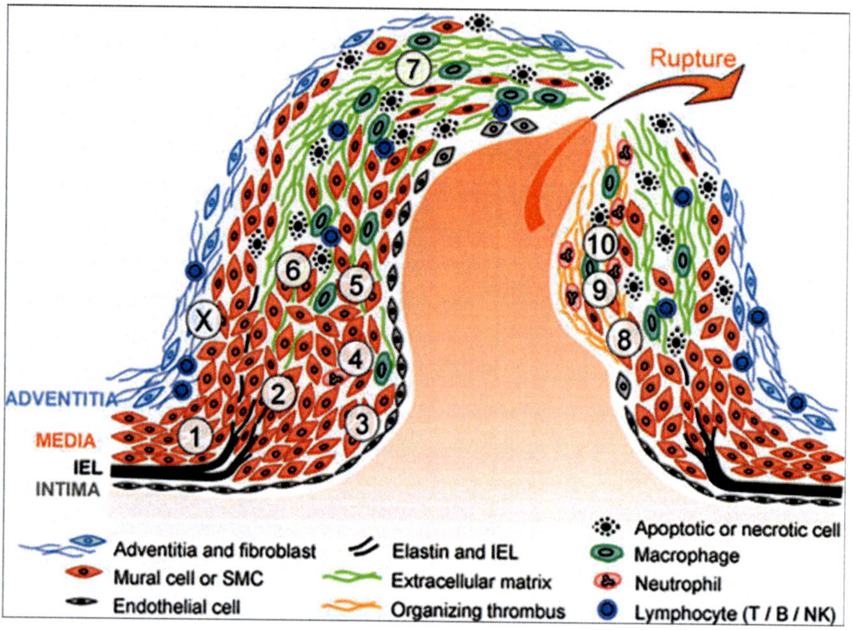

Fig. 3 Pictorial representation of the mechanism of intracranial aneurysm wall degeneration and rupture. Aneurysmal dilatation of the cerebral vessel at the site of the disrupted internal elastic lamina (*1*) leads to the luminal migration and the proliferation of the smooth muscle cells (*2*). These structural modifications in the luminal wall, in combination with the hemodynamic stress lead to endothelial cell dysfunction (*3*) and the chemotactic recruitment of the inflammatory cells (*4*). Growth factors secreted by the macrophages (*5*) promote aneurysm remodeling and stabilization of its mechanical strength and integrity by secreting matrix proteins. However, the cytotoxicity caused by the inflammatory cell infiltration and decreased oxygenation results in apoptotic mural cell death (mostly smooth muscle cells) (*6*). Increased cell lysis from the metalloproteinases (*7*) and decreased production secondary to the decellularization slows the matrix turnover. Endothelial dysfunction and hemodynamic shear stress promote thrombus formation at the aneurysmal wall (*8*). Thrombus attracts neutrophils and other inflammatory cells (*9*) which are potent producers of proteolytic enzymes. Mural cells lead to the organization of the thrombus (*10*). The net effect of these degenerative and reparative mechanisms may result in the critical thinning of the aneurysm wall, which when it overcomes the wall tensile strength results in aneurysm rupture. [145]

succumbs, resulting in rupture. A graphic representation of the pathophysiology of aneurysm rupture is provided in Fig. 3.

2.3 Aneurysm Growth and Rupture

The mechanical behavior of the aneurysm wall relies largely on the strength and organization of the adventitial collagen fibers which have an increased turnover due to the enhanced fibroblastic activity, secondary to elastin and the smooth muscle cell degradation [65, 66, 81]. Although aneurysm size has been implicated as a quantitative predictor of rupture risk [55], this factor alone cannot explain for all the ruptured aneurysms that are typically small, known as the aneurysm size paradox. International Study of Unruptured Intracranial Aneurysms (ISUIA) is the largest study conducted to date to elucidate the natural history and assess the rupture risk for IA. They concluded that the rupture risk was a function of their size, being small for aneurysms smaller than 7 mm and 40% for aneurysms greater than 25 mm in diameter [165]. However, as previously mentioned, most ruptured aneurysms are small. This suggests that although a critical size for a rupture may exist, other factors may also influence its integrity. Irregularities in the aneurysm morphology like lobulations and daughter sacs, best described as undulations in the aneurysmal wall; biomechanically represent regions of stress concentration and mechanical anisotropy [81, 113]. Determination of stress and deformation using non linear finite element analysis on simulated aneurysm models has demonstrated that spatial variations in the wall thickness and material stiffness may lead to wall stress concentration in the 'bleb' region of the aneurysmal wall, which may be responsible for the susceptibility of irregular and multilobular aneurysms to rupture [19, 45]. It is likely that these bleb regions are undergoing an active remodeling process that lacks the mechanical integrity to withstand physiological stresses.

3 Risk Factors

Mechanisms by which various risk factors mediate the formation of the aneurysmal pouch remain to be elucidated. Both genetic constitution and environmental influences have a role in determining the risk for aneurysm occurrence and their rupture.

The genetic risk of an individual developing an intracranial aneurysm can be classified into two categories: hereditary syndromes and familial IA. Hereditary connective tissue disorders like autosomal dominant polycystic kidney disease (ADPKD), Ehlers-Danlos syndrome and Marfans syndrome have a predisposition to IA due to their inherent arterial wall weakness. Other hereditary disorders like Neurofibromatosis Type 1 and Moyamoya disease are also associated with IA [1, 60]. Another rare distinct entity is the familial aneurysm. Familial IA are defined as families in which two or more first or second degree relatives present with either an asymptomatic IA or SAH without any known heritable disease known responsible for IA [156]. The screening of the first degree relatives of patients with IA and sporadic subarachnoid bleed has revealed 4–9% prevalence [110, 123], a risk upto four times greater than the general population, by one estimate. Familial aneurysms exhibit a distinct pattern, having a propensity to rupture at a smaller size and at an earlier age from the acquired aneurysms. Siblings often present with aneurysms at an identical or a mirrored location [77]. Multifactorial, autosomal dominant and autosomal recessive transmission patterns have been described in literature [16, 168].

Several lines of evidence implicate multiple environmental risk factors like hypertension, cigarette smoking, family history of cerebrovascular disease and estrogen deficiency as observed in the postmenopausal female in the pathogenesis and the rupture of IA. The strongest association for a modifiable environmental risk factor identified for both ruptured and unruptured IA is cigarette smoking, with current smokers having a five times greater risk of a subarachnoid bleed. Case control studies have demonstrated a relative risk of 3.8 for men and women combined, with a positive dose-response relationship [11] and a 20% attributable risk of SAH among smokers [125]. The role of cigarette smoking may be attributed to a decrease in the effectiveness of α_1-antitrypsin [42, 57], the main inhibitor of proteolytic enzymes like elastase. The female preponderance of IA at a rate of 3:1 [76, 99, 102, 126] may be attributed to collagen wasting secondary to the postmenopausal estrogen deficiency.

Once the IA is acquired the risk for its rupture is determined by a number of factors. Previous history of SAH, especially in the small aneurysms, and posterior circulation aneurysms are independent risk factors for the IA rupture [55]. Vessel instability, familial associations, and large aneurysm size may also predispose the aneurysm to rupture.

4 Clinical Presentation

4.1 Ruptured IA

Most IAs remain quiescent until they rupture, when it presents as a neurological emergency. Breach in the aneurysmal wall causes extravasation of blood under arterial pressure into the subarachnoid space, rapidly increasing the intracranial pressure and presenting acutely with an excruciating headache ("worst headache of my life"), which may be associated with nausea and vomiting (77%), depressed global consciousness (53%) and nuchal rigidity (35%) [37]. The physical examination may reveal meningismus, ocular hemorrhages and focal and/or generalized neurological signs. Premonitory symptoms referred to as a sentinel headache, occurring 2–8 weeks prior to the subarachnoid bleed may be experienced in 10–43% of the patients [106]. An acute expansion, thrombosis, or dissection of the cerebral aneurysm or a small focal rupture may be responsible for its occurrence [27]. Since headache is a nonspecific and an exceedingly common presentation to the emergency room, a high index of suspicion is warranted to avoid a misdiagnosis and a delay in the treatment [28].

Clinical evaluation of a SAH includes an examination of the patient's level of consciousness, meningismus, focal neurological signs and a fundoscopic examination. Meningismus presenting as nuchal rigidity, photophobia and a positive Kernig's and Brudzinski's sign may be seen within 6–12 h following SAH due to an inflammatory response to the blood breakdown products and is present in half the patients [27]. A funduscopic exam may reveal papilledema due to an increased intracranial pressure and retinal hemorrhages in 40% of the patients [40]. Approximately 80% of the patients with a spontaneous SAH have a ruptured

Table 1 Hunt and Hess scale adapted from surgical risk as related to time of intervention in the repair of intracranial aneurysms, William E. Hunt and Robert M. Hess Journal of Neurosurgery

Category	Criteria
Grade I	Asymptomatic or minimal headache and slight nuchal rigidity
Grade II	Moderate to severe headache, nuchal rigidity, no neurological deficit other than cranial nerve palsy
Grade III	Drowsiness, confusion or mild focal deficit
Grade IV	Stupor, moderate to severe hemiparesis, possibly early decerebrate rigidity and vegetative disturbances
Grade V	Deep coma, decerebrate rigidity, moribund appearance

Table 2 *WFNS = World Federation of Neurological Surgeons; SAH = Subarachnoid Hemorrhage; GCS = Glasgow Coma Scale Surgical risk as related to time of intervention in the repair of intracranial aneurysms. Journal of neurosurgery [0022-3085] Hunt yr:1968 vol:28 iss:1 pg:14–20

WFNS grade	GCS score	Motor deficit
I	15	Absent
II	14–13	Absent
III	14–13	Present
IV	12–7	Present or absent
V	6–3	Present or absent

aneurysm. A milder clinical spectrum and a normal neurological exam may confound the evaluation. This may result in a misdiagnosis of SAH, which has a fourfold higher risk of death and disability at one year [63].

Various grading systems have been devised to guide clinical assessment, provide prognostic information and to quantify changes in the disease severity. The most widely used among them are the Hunt–Hess scale [52] and the World Federation of Neurological Societies grading scale [119] which is an adaptation of the Glasgow Coma Scale (Tables 1 and 2). The Fisher scale was devised as an indicator of the vasospasm risk, based on the distribution of the blood in the subarachnoid, parenchymal and the interventricular space as observed on the initial non-contrast head CT [36].

4.2 Unruptured IA

Asymptomatic unruptured IA are most often discovered incidentally during a neuroimaging workup or during screening of individuals believed to be at risk for IA. When symptomatic they present most often with a headache. They also may present with a mass effect, which is the compression of adjacent neural structures by the aneurysm. The associated neurological deficit may give clues as to the location of the aneurysm. This mass effect manifests as a loss in visual acuity or cranial neuropathies, most commonly as third nerve palsy when the aneurysm is located at predictable locations (posterior communicating artery, the basilar terminus, anterior choroidal artery, and the intracavernous internal carotid artery). On examination, this presents as a drooping eyelid, with a large unreactive pupil and the eye resting in abduction, slight depression and intortion. Finally, some unruptured aneurysms may shed emboli from intraaneurysmal thrombosis and present with ischemic symptoms [39].

5 Diagnosis

5.1 *Acute Presentation*

The imaging workup of the patient is determined by their presentation. A high index of suspicion for SAH should exist when a patient presents with an acute onset of severe headache. Any headache which presents with the red flag neurological signs such as paralysis, drowsiness, confusion or memory impairment warrants a thorough neuroimaging workup [80]. The non-contrast head computed tomography (CT) scan is the initial radiological investigation of choice to establish the diagnosis of SAH. The sensitivity of the CT for detecting SAH within 12 h of symptom onset varies from 93 to 98% for older generation CT scanners [147]. The newer generation 64-slice multidetector CT scanners have demonstrated a sensitivity approaching 100%, within the first five days of an intracranial bleed [10, 21]. If clinical suspicion remains high, but the CT findings are read negative for a subarachnoid bleed, a lumbar puncture is warranted to investigate for the presence of xanthochromia, the yellowish hue due the presence of hemoglobin breakdown products in the cerebrospinal fluid. A normal CT scan and the absence of xanthochromia excludes the SAH, provided if the xanthochromia is investigated by spectrophotometry and lumbar puncture is carried out between 12 h and 2 weeks after the ictus [151]. A subsequent imaging workup is necessary to investigate the presence of a cerebral aneurysm.

5.1.1 CT Angiography

With rapid strides made in the CT technology, CT angiography (CTA) is a fast evolving diagnostic imaging modality in the detection and treatment planning of IA [2, 143]. The procedure involves a rapid intravenous injection of iodinated contrast with image acquisition during the arterial phase in the region of interest. The 64-slice multidetector CT scanners have shown significant improvement in detection of smaller aneurysms due to less venous contamination and faster scanning afforded by the increased detector number. The sensitivity and the specificity of CTA for the diagnosis of the aneurysms depend on the aneurysm size, location and the interpreting radiologist's experience. The sensitivity of CTA in detection of aneurysms smaller than 4 mm is 92.3%, whereas for aneurysms greater than 4 mm it is close to 100% [84]. Addition of the CTA protocol to the initial non-contrast head CT imaging evaluation at the expense of very little extra time, yields invaluable information for the planning of surgical and endovascular procedures by characterizing the cranial and the cervical vasculature, and delineation of aneurysm morphology in relation to the adjacent structures. CTA can also complement the information gathered by catheter angiography since it can better define aneurysmal wall calcification, intraluminal thrombus, orientation of the aneurysm with respect to intraparenchymal hemorrhage; localize the rupture site and the relationship to the bony landmarks [6]. However the CTA at present suffers from some limitations: Failure to appreciate the course of the contrast-enhanced cerebrovasculature in the close proximity to the bony structures. Also the beam hardening artifact due to the attenuation of the neuroimplants (i.e., neurosurgical clips or platinum coils in previously treated aneurysms) may obscure the visualization of the aneurysm, parent vessel and the brain parenchyma thus limiting its diagnostic value.

Catheter digital subtraction angiography (DSA) adds the temporal information regarding the flow of contrast material, which is not appreciated on CTA. However, the 2D images acquired limit the appreciation of complex vascular relationships due to large areas of opacification secondary to vessel overlap. CTA allows the data to be collected in a 3D array. Three dimensional image workstations allow real time manipulation the rotational data, allowing numerous projections and postprocessing options, generating image reconstructions that can be evaluated in any plane [112]. Two most commonly employed reconstruction algorithms are the maximum intensity projection (MIP) and the volume rendered (VR) images. MIP images are more reliable for the measurement of size, stenosis and the aneurysm neck, since they are less susceptible to the windowing/leveling artifact. VR images provide the best overall 3D perspective of the aneurysm morphology and its relation to the adjacent structures [143].

5.1.2 Catheter Angiography

Catheter angiography remains the "gold standard" imaging modality for cerebral aneurysm evaluation, allowing for their detection, spatial and morphometric analysis, assessment of the parent vessel vasospasm and also permitting for adjunctive intraluminal treatment in the same session. Catheter angiography is typically performed using digital subtraction angiography using biplane imaging which reduces the contrast dose administered. Angiography of both the internal carotids and vertebral arteries is performed to adequately evaluate the anterior and the posterior circulation.

Although the neurological morbidity of the invasive catheter technique in evaluating patients with SAH and cerebral aneurysms is quite low [20], risks associated with the groin puncture (for femoral artery access), navigation of the catheters, and ionizing radiation are present. Though rare, two major risks associated with iodinated contrast use are allergic contrast media reaction and contrast mediated nephrotoxicity, which can be life threatening [59]. These risks are associated with both catheter angiography and CTA. Use of low or isoosmolar contrast media, pre-treatment with corticosteroids and antihistamines in patients with a documented history alleviates the allergic reaction [70]. Pre-procedural hydration and cytoprotective drug acetylcysteine in patients with high serum creatinine offers protection from contrast mediated nephropathy.

5.1.3 MR Imaging

Fluid attenuation inverse recovery (FLAIR) and proton density weighted imaging are sensitive MR imaging sequences for the accurate detection of acute SAH [167]. However due to the constraints of acute patient management, the role of MR is better realized in the sub acute setting, as a screening modality and for follow-up of patients after endovascular treatment of aneurysms and in the evaluation of pregnant patients. MR angiography (MRA) technique employs the time-of-flight (TOF) and the contrast-enhanced MRA (CE-MRA) for aneurysm detection. TOF is a gradient echo sequence which suppresses the background tissue by saturating the volume of tissue including blood with radiofrequency (RF) pulses. In contrast, the blood flowing into the slice after the RF pulse will retain its signal intensity and stand out in the suppressed background. The sensitivity and accuracy of VR 3D-TOF MR in detecting IA on a 3T system has been found to be more than 95% when compared with the VR DSA [71]. CE-MRA is almost directly analogous to IV-DSA and IV-CTA. An intra-venous injection of a short dense bolus of a gadolinium based contrast agent is imaged on the first pass through the arterial system. Sequential scans are added to acquire early and late venous images. Because of its speed, lack of sensitivity to flow related artifacts (spin saturation and phase dispersion), and capability of covering a large field of view, contrast-enhanced MR angiography has inherent advantages over TOF MR angiography, and is a practical alternative to CTA. The MRA sensitivity for image interpretation is augmented when source images are used in concert with three dimensional reconstructed images generated using image post processing algorithms [94]. Multimodality cerebral angiography including CT angiography, MR angiography (TOF), catheter DSA and 3D rotational angiogram is represented in Fig. 4.

5.2 Aneurysm Screening

Screening is warranted in high risk individuals with the genetic syndromes associated with IA, familial aneurysm clustering and to proactively seek de novo aneurysms in individuals with previous SAH. It is well conceded that an approach that is balanced between cost effectiveness and with an acceptable risk profile needs to be adopted. Since the risk for the development of an IA remains lifelong, a repeated screening model may be justified [159]. Cost effectiveness simulation

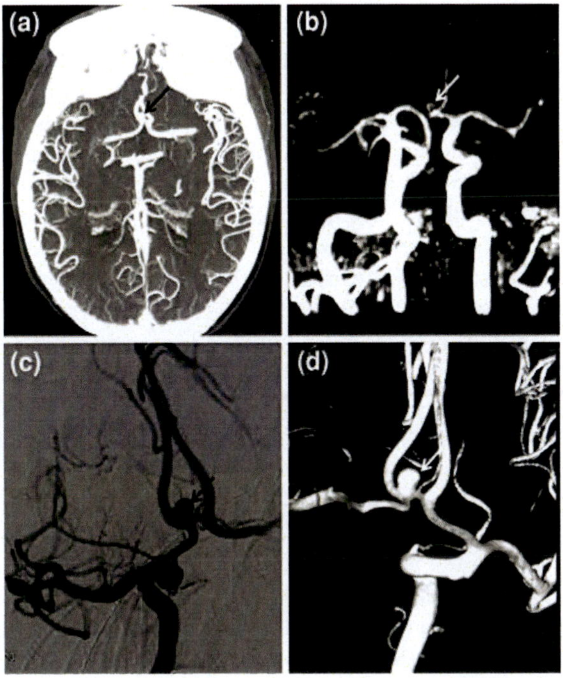

Fig. 4 A 68 year-old male received brain imaging following onset of unsteady gait. **a** CT angiography (axial MIP) and **b** Time-of-flight MR angiography shows a small anterior communicating artery aneurysm (*arrow*) **c** Catheter DSA and **d** 3D angiography depict the aneurysm and adjacent vasculature with high spatial resolution (frontal projection)

analysis in the Dutch health care system provides evidence based recommendations for screening individuals with 2 or more first-degree relatives with SAH. An optimum strategy for screening individuals based on their model is every 7 years from age 20 to 70 years [12]. The 3T 3D-TOF MRA and 64 channel multi detector CTA are both excellent modalities with a high diagnostic accuracy for evaluating unruptured IA. Since both these techniques yield similar diagnostic accuracy and spatiomorphological characterization, it may be more prudent to screen individuals with 3D-TOF-MRA since it spares the exposure of individuals to ionizing radiation and contrast media [50]. The psychosocial consequences of positive screening need to be assessed and appropriate support and professional counseling provided to the individual, making the available evidence known, so that an informed decision regarding treatment and conservative management can be made [160].

6 Management of Unruptured Aneurysms

In this age of diagnostic imaging, more and more unruptured aneurysms are being diagnosed both incidentally and proactively by screening. Prophylactic treatment in certain high risk situations like polycystic kidney diseases, collagen vascular diseases, symptomatic unruptured aneurysms and in patients with history of previous SAH is usually more straightforward. However, the management of a sporadically occurring aneurysm remains controversial, since more definitive aspects of the natural history of these lesions still remain obscure. It is estimated that for most unruptured aneurysms, the annual rupture risk is 0.1% annually. Certain factors are known to increase this rate, and are explained hereafter.

The International Study of Unruptured Intracranial aneurysms (ISUIA) aimed to provide information about the magnitude and determinants of the risks associated with the natural history and repair of unruptured aneurysms [165]. The prospective arm of the study enrolled 4,060 patients; 1,692 patients with 2,686 aneurysms (1,077 patients with no previous history of IA related SAH and 615 patients with previous SAH) received conservative treatment (medical management of patients comorbidities, no treatment of the aneurysm), whereas 1,917 and 451 patients underwent surgical and endovascular intervention, respectively. 3% of the unoperated cohort (51 patients) experienced aneurysmal rupture during the study period. A further characterization of the rupture risk was made based on the site and the size of the aneurysm. In the anterior circulation, the five year cumulative risk of aneurysm rupture was 0, 2.6, 14.5 and 40% for aneurysms less than 7, 7–12, 13–24 and 25 mm or greater, respectively. In contrast, aneurysms found in the posterior circulation or the posterior communicating artery demonstrated a higher cumulative five year rupture risk of 2.5, 14.5, 18.4 and 50%, for the same aneurysmal dimensions, respectively. The study also sought to compare the natural history of the disease with an aneurysm treatment, either endovascular or neurosurgical approaches. The overall one year morbidity and mortality in the surgical cohort was 13.7% for individuals with no SAH and 11% of individuals with SAH. In the endovascular group combined treatment morbidity and mortality at 1 year was 7.1% with prior SAH and 9.8% without SAH. Predictors of poor surgical outcome were age above 50 years, large aneurysm size, and posterior circulation (particularly basilar tip). Endovascular treatment of patients above 50 years-old was safer than craniotomy but statistically conclusive statements could not be made due to the small size of the endovascular cohort. Although a landmark study, with to date the best data surrounding natural history of unruptured brain aneurysms, ISUIA was deeply criticized by both the neurosurgical and endovascular communities alike for the methodological flaws in the study design and its conclusions [116]. Most of the criticism is centered around the lack of randomization that led to bias in the aneurysm population. To settle the controversy regarding the benefits and risks of intervention versus observation in unruptured aneurysms, a randomized trial on Endovascular Aneurysm Management (TEAM) was launched with an objective to recruit 2,002 patients in a three year period and monitor their progress over a 10 year period [117]. However due to insufficient recruitment the trial was suspended. There is still a need for more randomized trials to evaluate the natural history of unruptured IA.

The two available treatment approaches for IA treatment involve minimally invasive endovascular treatment and neurosurgical clipping. To evaluate the effectiveness and to assess the outcomes of endovascular coiling versus neurosurgical treatment in unruptured aneurysms, a large retrospective cohort of 2535 unruptured aneurysms was analyzed, 74% (1881) of whom were surgically treated.

The data base was collected from 429 hospitals, in 18 states representing 58% of the US population during a one year time interval. Endovascular treatment in comparison to neurosurgical clipping ($P < 0.05$) was associated with fewer adverse outcomes (6.6 versus 13.2%), decreased mortality (0.9 versus 2.5%), shorter lengths of stay (4.5 versus 7.4 days) and lower hospital charges ($42,044 versus $47,567) [48].

Due to the lack of definitive guidelines, each case needs to be examined individually, assessing the factors like patient's age, comorbid medical conditions, aneurysm size, location and its relationship to the parent and other adjacent vessels, surgical or endovascular access and patient preference. If a decision for expectant management is made, it is suggested to monitor the patient with CTA or MRA annually for two to three years and there after every 2–5 years if the aneurysms are clinically and radiologically stable. Smoking cessation, avoidance of alcohol overconsumption and stimulant medications should be advised [166].

7 Management of Ruptured Aneurysms

7.1 Emergency Evaluation

The initial medical evaluation of two thirds of the patients presenting with a ruptured aneurysm is made by the emergency medical services. After a quick neurological assessment and an initial stabilization by ensuring an adequate airway, breathing and circulation, a mechanism of swift transport and advanced notification to the emergency department needs to be institutionalized to help minimize the unnecessary on-scene delays [6].

On confirmation of SAH by the initial non-contrast head CT, the focus shifts to definitive diagnosis and prevention of complications. Although most often, the cause for a nontraumatic subarachanoid bleed is a ruptured aneurysm, other rare causes like an intracranial or spinal vascular malformation and perimesencephalic nonaneurysmal SAH need to be ruled out. Consultation with a skilled interdisciplinary team, including neurosurgery, interventional neuroradiology and a neurointensivist should act in concert to determine the future course of action.

7.2 Medical Stabilization

7.2.1 Blood Pressure and Intracranial Pressure Control

The risk of aneurysm rebleeding as a result of rerupture is highest during the first 24 h, particularly during the first 6 h after the SAH [5, 95]. After admission to a neurointensive setting; bed rest, stool softeners and analgesics are administered to minimize the risk of raising the intraarterial pressure. Simultaneous measures to medically stabilize the patient should continue while preparations are made for definitive treatment.

Precise hemodynamic control is critical to sustain the precarious balance between cerebral perfusion and preventing rerupture. The immediate aftermath of SAH causes an intracranial volume loading and a simultaneous increase in the CSF outflow resistance resulting in a raised intracranial pressure (ICP) [13]. As the elevating ICP equalizes the arterial pressure, it arrests the bleeding from the ruptured aneurysm due to tamponade. Although an increased arterial pressure is the only means to maintain cerebral perfusion and prevent infarction, the gradient may unsecure the temporary thrombus plug on the ruptured aneurysm and worsen the SAH.

To control an elevated blood pressure, short acting continuous infusion of intravenous formulations like nicardipine, labetalol and esmolol are preferred as they have a predictable dose-response relationship. Sodium nitroprusside should be avoided since it has a tendency to raise the intracranial pressure and cause toxicity with prolonged infusion [6]. Treatment for pain and anxiety may also help to reduce the elevated blood pressure. A ventriculostomy should be considered in some situations to monitor the ICP, to help guide the antihypertensive measures and treat hydrocephalus.

7.2.2 Prophylactic Oral Nimodipine

Several randomized clinical trials have demonstrated the efficacy of Nimodipine administration, a dihydropyridine calcium antagonist with a favorable reduction of morbidity and mortality [25, 101]. This response of Nimodipine was believed to be due to the prevention of vasospasm, i.e., the narrowing of large capacitance vessels in response to the spasmogenic substances produced by lysed blood. However, on angiography it demonstrates a limited large vessel vascular response to vasospasm. Also, the failure of translation of these favorable outcomes to other potent vasodilators like nicardipine, have led to a hypothesis of alternative cerebroprotective mechanisms due to Nimodipine [101]. Nimodipine is administered orally or through a nasogastric tube every 4 h for 3 weeks [103]. Monitoring of blood pressure and adjustment of antihypertensive therapy is necessary due to the hypotensive response to the drug.

7.2.3 Antifibrinolytic Therapy

The role of antifibrinolytic therapy in reducing the risk of rebleeding has been investigated for decades and numerous randomized clinical studies have concluded that the initial reduction in the rebleeding risk is offset by worse outcomes secondary to cerebral infarction [124]. However, recent evidence from a prospective, randomized trial suggests that the administration of a short course of antifibrinolytic Tranexamic Acid immediately at the time of SAH diagnosis in concert with early intervention to obliterate the aneurysm, followed by the discontinuation of the drug results in a significant reduction of early rebleeding rates without increasing the risk for cerebral infarction [49, 138].

7.3 Therapeutic Intervention for Ruptured Intracranial Aneurysms

Untreated ruptured aneurysms have a 3–4% risk of rebleeding in the first 24 h, a 1–2% risk per day in the first month, and a long term risk of 3% per year after 3 months [6]. Hence early aneurysm obliteration and its exclusion from the parent vessel circulation is the accepted treatment strategy to increase the likelihood of positive outcomes. There are two treatment approaches available. Neurosurgical intervention, involving aneurysmal clipping has been the cornerstone of management of ruptured IA in the past century. The technique was further refined in the 1970 s with the advent of the stereoscopic operating microscope and development of microsurgical techniques. The development of Guglielmi detachable endovascular coils (GDC; Stryker Neurovascular, Fremont, CA) led to the dawn of an era of minimally invasive endovascular coil occlusion of IA [44]. This neurointerventional branch has made phenomenal advancements in the past 20 years with the strides made in diagnostic imaging, medical devices, and highly specialized interventionalists.

The evidence comparing endovascular therapy and surgical clipping came from two randomized, prospective studies. In 1999, a single center study from Finland presented the data from 109 patients with acute SAH, randomized to the endovascular and neurosurgical arm showing an equal benefit in survival and neuropsychological outcomes at three and twelve months after treatment [61, 150]. The International Subarachanoid Aneurysm Trial (ISAT) is a landmark multicenter prospective, randomized study that defined and established the safety and efficacy of endovascular treatment (EVT) of ruptured aneurysms. The trial commenced in 1994 and recruitment was halted early in 2002 due to clear superiority of EVT. 2143 SAH patients with ruptured IA enrolled from 1994 to 2002 were randomly assigned to the neurosurgical clipping ($n = 1070$) and endovascular treatment with detachable platinum coils ($n = 1073$). Outcomes in terms of freedom from death and disability were significantly lower with endovascular coiling (22.6% relative risk reduction, 6.9% absolute risk reduction) [86]. One year after treatment the absolute risk reduction for the endovascular arm was 7.4% and this independent survival benefit continued for seven years [88].

These early results suggest the superiority of endovascular coiling in comparison to neurosurgical clipping. However, concerns regarding treatment durability remained due to the threat of rebleeding secondary to aneurysm reoccurrence. Aneurysm recurrence is defined as an increase in the size of the remnant opacification of the sac on subsequent follow-up angiograms and is attributable to low initial volumetric filling, and aneurysm recanalization due to coil compaction, migration of the coils into thrombus, or aneurysm regrowth. This risk makes it imperative to practice long term imaging follow up and if deemed necessary offer retreatment [17]. The prospective comparative rebleeding risk of coil embolization was 1.9% within the first 30 days of treatment and 0.6% per patient year between 30 days and 1 year of treatment. The corresponding rates for clip ligation were 0.7% within the first 30 days and 0.3% per patient year from 30 days to 1 year. Long term rebleeding risk (>1 year) post treatment for coil embolization and clipping is 0.21% and 0.063% per patient year, respectively. The ISAT investigators concluded that although the risk of rebleeding from the treated aneurysm was greater with coiling than with clipping, the risk remained small and there was no difference between the groups in the number of deaths due to rebleeding [89].

8 Neurosurgical Aneurysmal Clipping

Microvascular neurosurgical clipping involves craniotomy and placing of a titanium clip across the aneurysm neck using an extravascular approach. This aligns and reconstitues the structural components of the parent artery, resulting in the exclusion of the aneurysmal segment from the circulation (Fig. 5). Although neurosurgery is associated with higher procedural morbidity and mortality, it has a low reoccurrence and rebleed rate. In fact surveillance imaging is not standard of care following neurosurgical clipping. This modality is more frequently considered in younger patients with IA, depending on their location and angioarchitecture. MCA aneurysms due to their easier surgical exposure through the sylvian fissure are more amenable for surgical repair [118]. Posterior circulation aneurysms by contrast, represent a major challenge for the neurosurgical approach due to the narrowness of the field and presence of vital perforators. For example, the combined morbidity and mortality in the surgical treatment of giant unruptured basilar apex aneurysms is found to be in the range of 50% [136]. Aneurysm rupture with a massive intracranial bleed constitutes a specific indication for craniotomy to permit the expeditious evacuation of the clot to relieve the mass effect and treat with aneurysm clipping [118].

In preparation of the patient for a neurosurgical repair, it is essential to relax the brain to minimize the retraction necessary during exposure of the surgical field. This involves reduction of the CSF volume in a controlled fashion by either placement of a lumbar drain or a ventriculostomy. Hyperventilation and mannitol administration prior to the skin incision also help reduce the intracranial pressure and help facilitate a safe dissection near the aneurysm. Procedural complications occur in 30% of the patients resulting from temporary vessel occlusion, brain retraction and intracerebral hemorrhage secondary to intraoperative aneurysm rupture [38]. Treatment at neurosurgical centers with high procedural volume is associated with a better outcome [8].

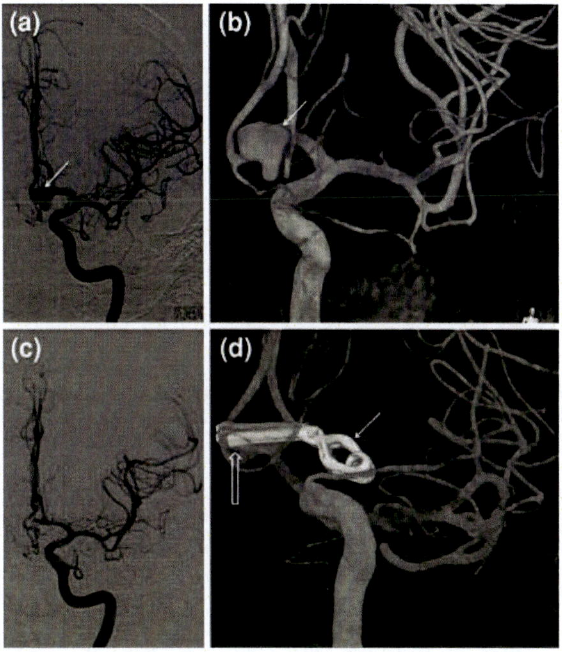

Fig. 5 A 60 year-old male had an incidental finding of a large anterior communicating artery aneurysm. Frontal DSA (**a**) and 3D rotational angiography (**b**) show the detailed structure of the aneurysm (*arrow*). After surgical clipping of the aneurysm (*arrow*, **d**), frontal DSA (**c**) demonstrates complete obliteration of the aneurysm with preservation of bilateral A2 segments. 3D angiography (**d**) however reveals a small remnant of the aneurysm neck (*hollow arrow*)

9 Aneurysm Embolization

The goal of endovascular embolization is to isolate and exclude the aneurysm from the cerebral circulation by decreasing the residual intra-aneurysmal blood flow, promoting thrombus organization in its lumen and endothelial cell proliferation across the neck with ultimate remodeling of the parent vessel [18, 104]. Numerous technological innovations in the past two decades have transformed this intervention as a mainstream treatment for IA. Endovascular treatment of IA involves treatment with embolic platinum coils with or without balloon remodeling and stent, flow divertors and liquid embolic agents.

9.1 Embolic Coil Technology

The embolic microcoil comprises of a biocompatible metal alloy that is made into a spring and further processed to have a secondary structure. The stock wire is the linear fabricated structure with a diameter D_1. The wire is wound around a mandrel with a diameter D_2, to conform to a secondary structure, also known as the primary wind of the coil. D_1 in conjunction with D_2 and the number of turns per unit length (n) of the primary wind, impact the coil stiffness.

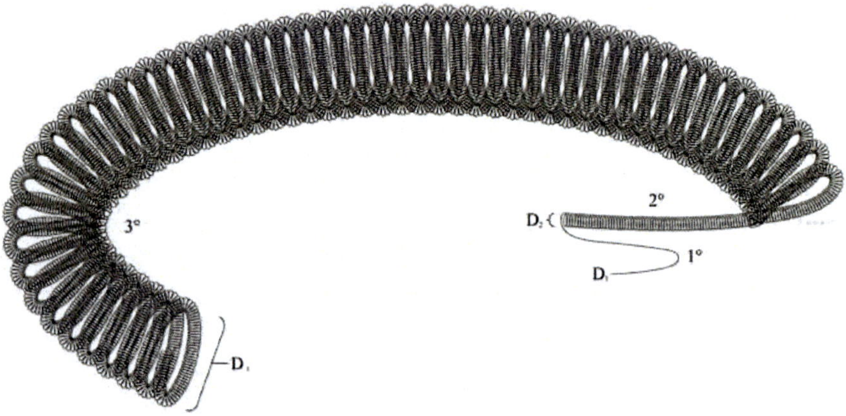

Fig. 6 Helical coil design. The stock wire with the primary (1°) diameter D_1 is wound around a mandrel with a diameter D_2 to a secondary (2°) configuration, and the tertiary (3°) configuration with D_3 [162]

Finally, the resulting spring structure can be conformed to a variety of designs (spherical, helical, complex) with a diameter D_3 (Fig. 6) [162].

The coil stiffness is directly proportional to the fourth power of D_1 and inversely proportional to n and the third power of D_2. G is the shear modulus or modulus of rigidity of the alloy.

$$stiffness \;\; \alpha \frac{D_1^4 G}{D_2^3 n}$$

Coil softness, an inverse measure of coil stiffness, is a desirable coil property and can be manipulated by fine tuning these parameters and has been the impetus for the emergence of wide array of choices available commercially.

Another fundamental property of the embolic coil, critical in the clinical context is packing density. Packing density is defined as the ratio of coil volume to the aneurysm volume. It offers increased biomechanical stability and is a well described predictor of treatment durability, and is inversely related to coil compaction and the rate and degree of aneurysm recurrence [62, 133]. Previous studies have demonstrated high packing density (≥24%) protects against recanalization in aneurysms with volumes smaller than 600 mm^3. The maximum packing density achieved, by the endovascular coil embolization of IA with the currently available coils is less than 50%, even in patients with documented complete angiographic occlusion. These low packing densities have been attributed to the quasi-random distribution and suboptimal geometric coil configurations achieved by the coil mass within the aneurysm [127].

9.2 Bare Platinum Coils

The GDC coil was the first endovascular device to get FDA approval for the treatment of an intracranial vascular pathology. It was a bare two dimensional platinum-alloy microcoil soldered to the ends of a stainless steel guidewire which electrolytically detached by the passage of DC current. The fundamental advantage of this coil type was the ability to introduce, reposition and evaluate the coil before detachment. However, these early prototypes suffered from multiple issues like premature detachment, detachment failure, coil stretching, fracture and prolapse into the vessel lumen. Also, occasionally the coil may become engaged with other indwelling coils and the resulting friction during detachment may result in proximal stretching and unraveling of the coil near the detachment site which may herniate into the vessel lumen [33]. The wire coil junction was subsequently reengineered to allow for a more rapid and consistent detachment. Subsequently, the coil technology has evolved by leaps and bounds with the emergence of anatomically more conformable 2D and 3D designs, stretch resistance and varying softness grades allowing for precise obliteration of the aneurysmal sac [3].

9.3 Coated Coil Technology

Platinum is a biologically inert material and illicits a limited inflammatory response when deployed in an aneurysm. Bioactive coil technology shifted the interest from pure mechanical aspects of endovascular devices to a biologically integrated approach in which the embolic agent is a tool to deliver active molecules or living cells to augment the biological healing of aneurysms. In this effort to accelerate the biological response within the aneurysmal sac, platinum coils were coated with collagen with the hypothesis that its addition would provide a substrate for the ingrowth of fibroblasts [23]. Also, direct endovascular delivery of the fibroblasts and recombinant human vascular endothelial growth factor into the aneurysmal lumen by the use of platinum coils as delivery vehicles was explored in preclinical aneurysm models [82].

The first commercially available bioactive coil, Matrix (Stryker Neurovascular, Fremont CA) was developed by coating a hydrolysable copolymer (90% polyglycolide, 10% polylactide: PGLA) over the platinum core. The hypothesis is that the copolymer may induce a mild inflammatory reaction accelerating the smooth muscle cell migration, leading to enhanced thrombus organization and scar retraction within the aneurysm lumen [68, 92, 93]. Another novel coil design of the Cerecyte (Codman & Shurtleff, Raynham, MA) incorporates PGA in the core of the platinum wind, which allows the bioactive polymer to be delivered while simultaneously achieving high packing densities.

However, the clinical experience with the Matrix coil has been disappointing with increased rates of recanalization in comparison to bare platinum coils [34, 98].

Also, of increased concern is its higher thromboembolic complication rate as compared to the GDC microcoils [144]. The poor outcomes with these bioactive coils have been attributed to the sub optimal packing achieved due to the increased friction induced by the PGLA exterior during coil placement, resulting in compartmentalization and aneurysm recanalization [30].

The Cerecyte Coil Trial (CCT) a multicenter, randomized, prospective clinical trial, 500 patients with ruptured and unruptured aneurysms designed to compare the Cerecyte to the bare platinum coils concluded recently. The trial recruited 500 patients with ruptured and unruptured aneurysms. The results announced recently at the 48th Annual American Society of Neuroradiology 2010 meeting in Boston showed that there was no statistical difference between the clinical outcomes and angiographic occlusion rates between the Cerecyte and the bare platinum coils. However, both coils combined revealed a 97% freedom from disability and dependence and an 86% investigator-reported angiographic occlusion rate at six months, demonstrating the remarkable safety profile of endovascular occlusion.

The HydroCoil Embolic System (Hydrogel, Microvention- Terumo, Tustin CA) features a platinum metallic core impregnated with an expandable polymeric gel composed of cross-linked acrylamide/sodium acrylate that absorbs water by diffusion in an optimal acid–base environment of whole blood, which results in a controlled volumetric expansion after an initial five minutes latency to allow for coil repositioning. In theory, this system should offer a tighter packing owing to the void-seeking properties of the hydrogel. The HydroCoil Endovascular Aneurysm Occlusion and Packing Study (HELPS) is a prospective randomized trial of HydroCoil versus the bare platinum microcoils in the endovascular treatment of IA. The periprocedural data reveal that the Hydrocoil has a safety profile equivalent to that of bare platinum [163]. The follow up data from the HELPS trial was presented at the Society of Neurointerventional Surgery (SNIS) 7th annual meeting in Carlsbad, California 2010. Primary poor outcome measured by the aneurysm remnant, recurrence and morbidity/mortality precluding angiographic follow up was lower for the Hydrocoil arm (28%) in comparison to the control (36%). Hyrocoil demonstrated a lower retreatment rate (2.9 versus 3.6%) but a higher risk of hydrocephalus (4.5 versus 0.9%) in comparison to the control; although there was no statistical difference between the arms.

9.4 Coil Assist Technologies

Durable occlusion of unfavorable aneurysmal morphology like a wide neck (>4 mm) and unfavorable dome to neck ratio (<2) remains challenging with conventional endovascular coiling alone. The migration and herniation of the coil mass from the aneurysmal sac into the arterial lumen can result in thromboembolic complications. To address this concern and to augment the packing density, a spectrum of ancillary techniques like balloon remodeling and stent-assisted coiling aimed to augment the packing density has emerged.

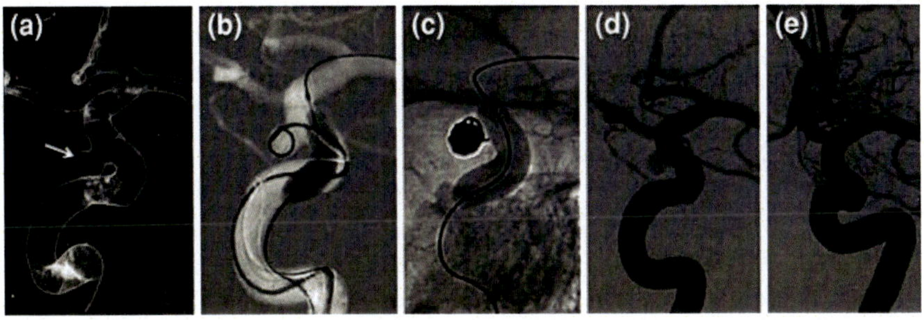

Fig. 7 A 49-year-old female with an incidental finding of a wide-neck left superior hypophyseal artery aneurysm (*arrow* in **a**, 3D angiogram in a frontal oblique projection). A compliant balloon is navigated into the left internal carotid artery in the working projection (**b**, frontal oblique) and inflated as the first coil is inserted into the aneurysm. Subsequently, the balloon is used to support the insertion of multiple coils into the aneurysm sac (**c**, frontal oblique) and immediate post-embolization angiography (**d**, frontal oblique) demonstrates excellent occlusion of the aneurysm with preservation of the parent artery. The 12-month follow up angiography (**e**, frontal oblique) shows complete exclusion of the aneurysm from the circulation with preservation of the left internal carotid artery

9.4.1 Balloon Remodeling

Balloon remodeling is a technique in which a temporary non-detachable balloon is inflated occluding the aneurysmal neck and parent artery during coil deployment [90] (Fig. 7). In one of the preliminary descriptions of this technique, 94% of 56 aneurysms were completely or partially occluded, with one thromboembolic complication and no deaths. This procedure gained wide acceptance in the interventional community. Another multicenter study achieved similar success using this technique and determined that the occlusion rate in this subset of aneurysms depended solely on the aneurysmal sac diameter and not on the neck width or dome to neck ratio [22]. The critiques of this adjunctive technique question its safety profile, and argue that it exposes the patient to an accentuated periprocedural risk of thromboembolic complications (9.8 versus 2.2%) and vessel rupture (4 versus 0.8%) in comparison to coil embolization alone [134]. They also reported a statistically significant increase in death and dependency with balloon assisted coiling in comparison to unassisted coiling [134, 149]. Using diffusion weighted MR imaging to assess the thromboembolic events, 79% of the balloon assisted coiling procedures demonstrated hyperintensities in comparison to 49% of the unassisted procedures, though predominantly the lesions were clinically silent [135]. However, a recent extensive literature review and meta analysis has failed to appreciate a statistically significant difference in thromboembolic events between balloon assisted coiling and coiling alone [131].

Using a more pragmatic approach, routine balloon embolization for uncomplicated aneurysms may help enhance the packing density. It may also be used as a rescue technique to push the coils herniating into the parent vessel lumen into the aneurysm sac or control aneurysm bleeding in the event that the aneurysm is perforated during coil embolization.

9.4.2 Stent Assisted Coil Embolization

Stent assisted coil embolization has revolutionized the treatment of wide necked and fusiform aneurysms. Conceptually, the stent mediated vascular reconstruction operates through many theoretical and technical mechanisms. Firstly, the protective mesh around the aneurysm neck allows for a more aggressive coil deployment while minimizing the risk of coil herniation and compromise to the parent vessel. Secondly, it can result in hemodynamic uncoupling of blood flow in the aneurysm-parent vessel complex, increasing particle residence times in the aneurysm that facilitate aneurysmal thrombosis [154]. It minimizes the impinging blood flow responsible for coil compaction and the shear stress on the untreated aneurysm wall. Lastly, it acts as a scaffold for endothelial growth for reconstruction of the parent vessel [32].

A stent is an extremely versatile device and allows for various ingenious strategies for remodeling intracranial vasculature. The advancements in the coronary stent delivery systems propelled experimentation with their deployment in the tortuous cerebrovasculature to treat intracranial vascular disease, achieving excellent results [78, 155]. These devices, however, lacked flexibility, possessed a larger profile and were deployed with balloon expansion which limited their use to the proximal aneurysms of the internal carotid and vertebrobasilar systems [69].

These limitations were overcome by the development of extremely versatile self expanding neurovascular stent technology, with better deliverability and ease of use and a lower tendency for vessel rupture and luminal damage during deployment [47]. They incorporate the fundamental mechanical properties of shape memory and superelasticity of the Nickel-Titanium alloy (Nitinol). They are manufactured to a size slightly larger than the target vessel size and constrained in a retractable sheath. At the treatment site, it is released from the delivery system and expands to conform to the vessel wall [140]. The two laser cut Nitinol stents available for stent assisted coil embolization are the open cell Neuroform (Stryker Neurovascular, Fremont, CA) and the closed cell Enterprise Vascular Reconstruction Device and Delivery System (Codman Neurovascular, Raynham MA) (Fig. 8a, b).

The Neuroform was approved for limited use by the United States FDA to treat wide necked ($\geq$4 mm or a dome to neck ratio $<$ 2) IA. Although the featured open cell design provides stent apposition to the vessel wall in curves, the stent delivery and deployment through the microcatheter remained challenging. Once deployed it is impossible to reposition the stent. One concern with this device is the propensity of the stent struts to open and prolapse into the side branches or the aneurysm, especially around the vascular bends [7]. Subsequent structural revisions of microcatheter delivery system and fusing some of the stent struts of the newer generation Neuroform 2 and 3 addressed these concerns. The open cell design does

Fig. 8 Intracranial nitinol self-expanding stents used for endovascular treatment of aneurysm in conjunction with coil embolization. **a** Closed cell design with flaring ends (Enterprise VRD; Codman Neurovascular, Raynham, MA). **b** Semiopen cell design (Neuroform; Stryker Neurovascular, Fremont, CA). **c** Braided Nitinol flow diverter with sinusoidal radiopaque wires for precise visualization (Silk stent; Balt Extrusion, Montmorency, France)

not allow for stent resheathing and repositioning in the event of a suboptimal deployment. The closed cell design of the Enterprise stent overcomes some of these shortcomings of the open cell design and allows for the partial deployment up to 70% of its length and to be reconstrained and recaptured if the initial deployment is not precise. It also reduces the tendency of coil protrusion through the stent interstices [47].

Another stent with a closed cell self expanding braided stent design approved for the treatment of wide necked post carotid siphon aneurysms in Europe is the Leo stent (Balt, Montmorency, France). It consists of a body with braided nitinol wires and two radio opaque markers which ensure the visibility of the entire diameter and length of the stent. Its novel braided design confers it a high radial force and homogenous wall coverage even at acute curves which reduces the possibility of stent displacement after deployment. Its innovative distal hook allows for repositioning and resheathing of the stent when it is as much as 90% deployed [108].

A recent single center experience in 107 wide necked aneurysms treated with the Neuroform stent assisted coiling revealed a procedure related morbidity and mortality of 5.6 and 0.9%, respectively. A favorable clinical outcome was achieved in 90.7% patients after a mean of 47 months [72]. The mean recanalization rate at 37 months was 13.7%. Another multicenter study of 142 wide necked aneurysms treated with the Enterprise stent reported 2.8% morbidity and 2% mortality [85]. Follow-up data from the study are awaited.

For a planned stent assisted coiling dual antiplatelet regimen needs to be ideally initiated three days prior to intervention and continued post procedure for at least 6 weeks and up to 6 months. One study observed up to a 50% prevelance of Clopidogrel resistance and low responders among patients receiving cerebrovascular stents [107]. It is suggested that antiplatelet-agent resistance be investigated before elective intracranial endovascular procedures to reduce peri-procedural in-stent thrombosis or delayed in-stent lumen reduction due to thrombus formation [83].

The inherent thrombogenic nature of the nitinol stent is a proverbial double edged sword. In the setting of an unruptured aneurysm the neutralization of the platelet function may be of little consequence. Conversely, in the context of acute SAH, this requirement may be catastrophic resulting in hemorrhagic complications requiring a ventriculoperitoneal shunt and an EVD placement. Reversal of the antiplatelet regimen may result in minimizing the hemorrhagic complications but has the potential to result in thomboembolic complications [26, 146]. A multicenter, nonrandomized study in Finland reported their experience in using stent assisted coiling in 61 patients with ruptured aneurysms. The technical success rate was 72% and complications related to stent assisted coiling occurred in 21%. The majority of complications were as a result of thromboembolism [142].

9.5 Follow-up Angiography

Even though the procedural complication rate is lower for endosaccular embolization, the initial clinical benefit may be offset by high rates of incomplete aneurysmal obliteration and late reoccurrences. The conformation of the coil mass and the flow within the aneurysm is dynamic. Post treatment stasis of blood flow may result in its thrombosis and exclusion from the circulation. Conversely, the impinging blood flow may result in coil compaction and recanalization, which has the potential to result in aneurysm regrowth and rupture. Significant factors determining the likelihood of aneurysm recanalization/coil compaction are aneurysm size, initial post-embolization angiographic outcome and packing density [62, 133]. This underscores the necessity for follow-up angiography post treatment. It also provides an opportunity to seek and assess for a change in other occult aneurysms, since these have an increased predilection to arise de novo and coexist in patients with a previously diagnosed SAH [169].

Long-term imaging follow-up of aneurysms treated primarily by coiling reveals that many first-time recanalizations requiring retreatment are detected at 6 months post angiography [129, 137]. In clinical practice this translates to an angiographic workup at 6 months [148]. Most centers perform an intra-arterial DSA for this initial assessment. Subsequent course of action depends on the aneurysm appearance. An incomplete occlusion may warrant retreatment and a more frequent follow-up. MR angiography is increasingly been advocated for the long term surveillance of these patients since it offers a non invasive and a safer alternative [130]. In our experience, non-dissecting aneurysms are surveiled by DSA at 6 months and 1 year. If the aneurysm is completely occluded after one year a third DSA is recommended at 3 years post-coiling. If there is a small aneurysm remnant or any changes to the coil mass, we choose to surveil the aneurysm annually with DSA. In the case of dissecting aneurysms, an earlier DSA is performed at 3-months post coiling due to the propensity of the dissection to expand.

9.5.1 Flow Divertors

The vast experience with the use of self-expanding intracranial stents in reconstructing the parent vessel has shown that they serve as an endoluminal buttress impeding the transfer of the flow momentum into the aneurysm by flow diversion into the parent artery. The porosity of the device and the pore density determine the resistance offered to the blood flow through the stent interstices, which prevents the recirculation of flow into the aneurysm resulting in flow stasis, thrombosis and its exclusion from the circulation [73, 74]. However sacrificing the porosity of the device increases its stiffness limiting its maneuverability and risk the patency of the side branches and perforators [128].

The existing intracranial self expanding stents possess limited metal surface area coverage of 6.5–9%. The Pipeline embolization device (Covedian, Mansfield, MA) is a flexible, self expanding microcatheter delivered flow diversion device engineered exclusively for the treatment of intracranial aneurysms. The cylindrical mesh device is composed of 48 individual cobalt-chromium and platinum strands having 30–35% metal surface area coverage when fully deployed [35]. Initial experience with the device in treating wide necked, large and giant aneurysms has shown promising results establishing its safety and durability with a 12 month angiographic occlusion rate of close to a 95% [79]. The device recently received pre-market approval from the U.S. FDA for the endovascular treatment of the large or giant wide-necked intracranial aneurysms in the internal carotid artery from the petrous to the superior hypophyseal segments. Another device in clinical use is the Silk stent (Balt Extrusion, Montmorency, France) (Fig. 8c), which is a 48 wire braided Nitinol stent. Once deployed, a sinusoidal radiopaque wire throughout the device assists in the precise visualization of this stent. Pre procedural treatment and continuation of an antiplatelet regimen similar to that required for intracranial stents is essential to prevent thromboembolic complications. Primary endovascular reconstruction of the parent vessel represents a paradigm shift from the conventional endosaccular occlusion approach. As a standalone therapy the deployment of flow diverters considerably simplifies the entire procedure. Since the aneurysm sac does not require to be embolized, the risk of intraprocedural perforation is also reduced. In a retrospective analysis of 12 patients with basilar artery aneurysm treated with the Silk flow diverter, one patient experienced an intraprocedural acute basilar thrombosis treated successfully with locally glycoprotein IIb/IIIa antagonist delivery. During a mean follow up of 16 weeks, 3 patients experienced a symptomatic neurological event. On short term follow up; 7 complete occlusions, 3 neck remnants and 2 patent aneurysms were observed [67]. The early results with these devices are encouraging, but long term data are needed to assess the thromboembolic occlusion, aneurysm recurrence rates, bleeding risk and the intimal reaction.

9.6 Liquid Embolic Agents

Another concept in aneurysmal obliteration involves the embolic injection of polymers into the aneurysm sac. The liquid embolics can precisely conform to the complex aneurysmal shapes and exclude it from the circulation, thus minimizing the risks of recanalization requiring retreatment (Fig. 9). There is significant potential of this technique in treating wide necked and giant aneurysms, since coil embolization achieves unsatisfactorily low packing densities, coil compaction and higher rates of aneurysmal recurrence in this subgroup. Although the experience with these agents in treating arteriovenous malformations is extensive, their use in cerebrovascular aneurysmal treatment is challenging due to concerns regarding reflux, distal embolization from polymer fragments and perforator and parent vessel occlusion.

The translation of this concept to the angiography suite for aneurysm treatment came with the development of Onyx (eV3 Covidien Vascular Therapies, Mansfield, MA), a liquid embolic agent composed of ethylene–vinyl-alcohol (EVOH) biocompatible polymer and dimethyl sulfoxide (DMSO) solvent with micronized tantalum for radiographic attenuation used in conjunction with a highly compliant DMSO-compatible occlusion balloon to maintain parent vessel patency.

Cerebral Aneurysm Multicenter European Onyx (CAMEO) trial, a prospective study conducted at 20 European centers reported their experience of 97 patients with 100 aneurysms [87]. At 12 months complete or subtotal aneurysmal occlusion was present in 92% of the patients. Delayed parent artery occlusion was observed in 9 patients, 5 of which were asymptomatic. Procedure or device related permanent neurological morbidity was observed in 8 of the 97 patients; 2 procedural and 1 disease related death. Another single center reported their experience of treating 22 large/giant wide necked ICA aneurysms treated with Onyx. At a mean of 13 months follow up, 17 (91%) aneurysms demonstrated total occlusion with 2 recanalizations. Parent vessel stenosis or an occlusion was observed in 16% of the patients [158]. Recent report of a large single center experience with 84 wide neck aneurysms (50 small, 30 large, 4 giant) revealed similar occlusion rates at 18 months follow up. 4.6% of the aneurysms recanalized (1 small, 2 large) with 3.3% requiring retreatment. Procedural mortality was 2.3% and overall morbidity 7.2% [105].

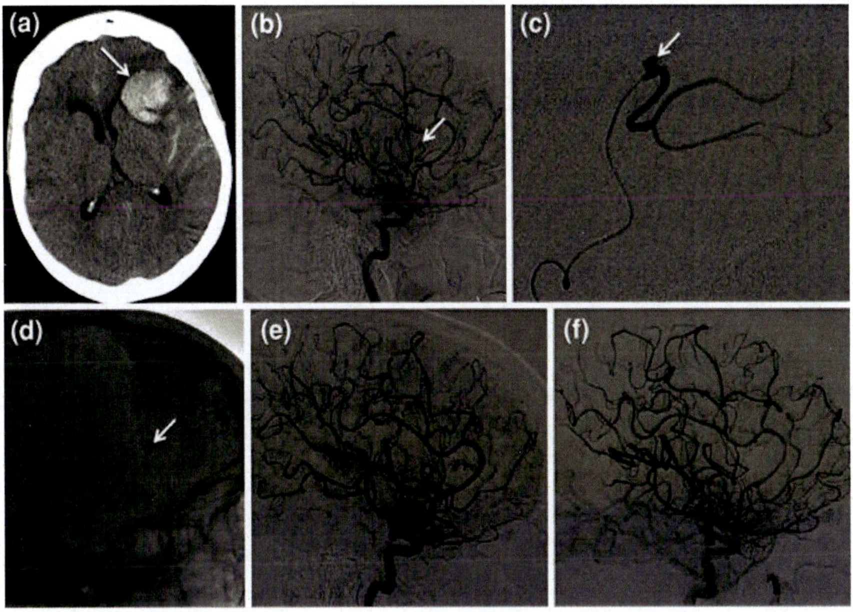

Fig. 9 A 62 year-old female presented with speech difficulty and right-sided weakness. A non-contrast head CT (**a**) reveals a large left fronto-temporal hematoma and acute subarachnoid hemorrhage (*arrow*). DSA (*arrow* in **b**, lateral) and selective microcatheterization and angiography **c** demonstrates a small mycotic aneurysm arising from the frontal branch of the anterior division of the left middle cerebral artery. A liquid embolic agent (n-butyl-2-cyanoacrylate) was injected through the microcatheter (glue cast shown in **d**, *arrow*) thereby occluding the aneurysm (DSA, **e** Follow-up angiography at 6-months **f** shows complete obliteration of the aneurysm and the recruitment of leptomeningeal anastomoses to the territory of the occluded branch

10 Complications

10.1 Aneurysm Rerupture

Intraprocedural aneurysm rupture of a previously ruptured aneurysm occurs in 5% of the endovascularly treated and 19% of the surgically treated aneurysms and is associated with greater periprocedural risk of death and disability [29]. In the cerebral aneurysm rerupture after treatment (CARAT) study, 1,000 patients treated with coil embolization or surgical clipping at high volume centers in the United States were followed for a mean of four years, it was concluded that both surgical clipping and endovascular coil embolization were extremely effective in reducing the risk of late rerupture, with annual hemorrhage rates of 0.11% per year for coil embolization, including those requiring retreatment and no late re-ruptures among those treated with surgical clipping [114]. Further interrogation of the data revealed that incomplete aneurysm occlusion was a strong predictor of rerupture risk [58]. This underscores the importance of the long term imaging followup of the endovascularly treated aneurysms.

10.2 Cerebral Vasospasm

The delayed narrowing of the large vessels at the base of the brain is thought to occur due to the spasmogenic substances in the lysed blood. Vasospasm affects 60–70% of the patients post SAH, half of which will experience ischemic symptoms [56]. Typically, it manifests between the third and the seventh day post hemorrhage as a decline in neurological status and/or a focal ischemic deficit [36]. The vasospasm may be monitored noninvasively at the bedside using transcranial Doppler ultrasonography (TCD) [132]. The sensitivity of TCD for the diagnosis of symptomatic vasospasm and subsequent cerebral infarction on CT scans varies from 70-80% [111, 170]. CT Angiography is extremely useful in situations where the clinical presentation is discordant with the TCD findings or if endovascular therapy is being contemplated. CT perfusion also has a role in determining severe vasospasm, with absolute cerebral blood flow values of less than 25 ml/100 g/min and relative mean transit times greater than 6.5 s associated with high risk of delayed ischemic deficits [141]. Hyperdynamic therapy which involves hemodilution, induced hypertension and hypervolemia is the mainstay for the prevention and treatment of the cerebral vasospasm after securing the ruptured aneurysm [91]. The underlying principle is to improve cerebral perfusion by raising the mean arterial pressure and overcome the increased vascular resistance imposed by the vasospasm. In situations where the aneurysm remains refractory to this hemodynamic augementation therapy, endovascular therapy like trasluminal balloon angioplasty and supraselective intra-arterial vasodilator administration is contemplated [9, 14].

10.3 Hydrocephalus

Hydrocephalus is a common complication post sub-arachanoid hemorrhage afflicting 20% of the patients. Obstruction of cerebrospinal fluid (CSF) flow by blood products or adhesions, or by a reduction of CSF absorption at the arachnoid granulations is thought to be responsible [43]. Since its presentation can be confounded by the neurological deterioration, its recognition is based primarily on radiographic findings. Ventriculostomy drain placement offers clinical benefit in patients with diminished level of consciousness [6]. Need for permanent CSF diversion needs assessment and may be required in 18–26% of the survivors [6].

10.4 Seizures

Routine prophylactic antieplileptic therapy has not proven to offer much therapeutic benefit. The indications for anticonvulsant therapy are a clinically or electrographically documented seizure [6, 111]. Phenytoin has been associated with an adverse clinical outcome post SAH and should be avoided [96].

10.5 Hyponatremia

Another complication to anticipate and aggressively correct is hyponatremia which is thought to occur due to excessive natriuresis and volume contraction mediated by an inappropriate secretion of antidiuretic hormone.

10.6 Thromboembolism

Intraprocedural thromboembolism may occur in response to intraluminal insertion and manipulation by thrombogenic devices like catheters, guidewires, implantation of stents and coils, endothelial denudation and intraneurysmal clot dislodgement. Subacute and delayed thromboembolism may occur due to discontinuation of antiplatelet therapy post stent implantation or coil prolapse into the parent vessel lumen [31]. Intraprocedural therapy with heparin should be initiated after baseline activated clotting time (ACT) measurement during intraluminal treatment of unruptured aneurysms [109]. When treating aneurysmal SAH, heparin therapy should be started after placement of the first coil. This diminishes the potential for a devastating bleed if the coil penetrates through the rupture site [115]. Intraprocedural ACT is measured subsequently and the heparin dose titrated to a target ACT of 250 s. Despite these measures thromboembolic complications occur with a frequency of 2.4–5.2% [31]. Recently, selective abciximab administration as rescue therapy during intraprocedural thrombus formation has demonstrated promising results without increasing the additional risk of intracranial hemorrhage or extracranial bleeding [121]. Intraoperative intravenous acetyl salicylic acid during the endovascular embolization of unruptured aneurysms and ruptured aneurysms is routinely administered in Europe and has demonstrated a significant reduction in the rate of thromboembolic events without increasing the rate and severity of intraoperative bleeding [120].

11 Future Perspectives

While the advancements in the treatment of this intracranial vascular anomaly continue to transform the natural history of the disease, there is an increased interest in being able to predict the risk of an aneurysm to rupture. Intramural aneurysm wall inflammation has been correlated with ruptured aneurysms [41]. Preliminary research has demonstrated the ability to image noninvasively molecular markers for intraneurysmal inflammation [24]. Future advancements of these techniques may lead to quantitative assays to identify patients at risk of rupture in unruptured aneurysms. Also, another intriguing idea is to halt progressive IA wall degeneration through systemic antiinflammatory drug treatments targeting the chemotactic and proinflammatory signaling to decrease the macrophage infiltration of the aneurysmal wall [145]. A better understanding of the aneurysmal biology is essential to conceive and design strategies to control the development of these lesions before their rupture.

References

1. Adams, H.P. Jr., Kassell, N.F., Wisoff, H.S., Drake, C.G.: Intracranial saccular aneurysm and moyamoya disease. Stroke **10**(2):174–179 (1979)
2. Agid, R., Lee, S.K., Willinsky, R.A., Farb, R.I., terBrugge, K.G.: Acute subarachnoid hemorrhage: using 64-slice multidetector CT angiography to "triage" patients' treatment. Neuroradiology **48**(11):787–794 (2006) doi:10.1007/s00234-006-0129-5
3. Agner, C., Dujovny, M.: Historical evolution of neuroendovascular surgery of intracranial aneurysms: from coils to polymers. Neurol. Res. **31**(6):632–637 (2009) doi: 10.1179/174313209X455790
4. Anson, J.A., Lawton, M.T., Spetzler, R.F.: Characteristics and surgical treatment of dolichoectatic and fusiform aneurysms. J. Neurosurg. **84**(2):185–193 (1996). doi: 10.3171/jns.1996.84.2.0185
5. Beck, J., Raabe, A., Szelenyi, A., Berkefeld, J., Gerlach, R., Setzer, M., Seifert, V.: Sentinel headache and the risk of rebleeding after aneurysmal subarachnoid hemorrhage. Stroke **37**(11):2733–2737 (2006).
6. Bederson, J.B., Connolly, E.S. Jr., Batjer, H.H., Dacey, R.G., Dion, J.E., Diringer, M.N., Duldner, J.E. Jr., Harbaugh, R.E., Patel, A.B., Rosenwasser, R.H.: Guidelines for the management of aneurysmal subarachnoid hemorrhage: a statement for healthcare professionals from a special writing group of the Stroke Council, American Heart Association. Stroke **40**(3):994–1025 (2009).
7. Benndorf, G., Claus, B., Strother, C.M., Chang, L., Klucznik, R.P.: Increased cell opening and prolapse of struts of a neuroform stent in curved vasculature: value of angiographic computed tomography technical: case report. Neurosurgery **58**(4 Suppl 2) (2006):ONS-E380; discussion ONS-E380.
8. Berman, M.F., Solomon, R.A., Mayer, S.A., Johnston, S.C., Yung, P.P.: Impact of hospital-related factors on outcome after treatment of cerebral aneurysms. Stroke **34**(9):2200–2207 (2003)
9. Biondi, A., Ricciardi, G.K., Puybasset, L., Abdennour, L., Longo, M., Chiras, J., Van Effenterre, R.: Intra-arterial nimodipine for the treatment of symptomatic cerebral

vasospasm after aneurysmal subarachnoid hemorrhage: preliminary results. AJNR Am. J. Neuroradiol. **25**(6):1067–1076 (2004).

10. Boesiger, B.M., Shiber, J.R.: Subarachnoid hemorrhage diagnosis by computed tomography and lumbar puncture: are fifth generation CT scanners better at identifying subarachnoid hemorrhage? J. Emerg. Med. **29**(1):23–27 (2005). doi: 10.1016/j.jemermed.2005.02.002
11. Bonita, R.: Cigarette smoking, hypertension and the risk of subarachnoid hemorrhage: a population-based case-control study. Stroke **17**(5):831–835 (1986)
12. Bor, A.S., Koffijberg, H., Wermer, M.J., Rinkel, G.J.: Optimal screening strategy for familial intracranial aneurysms: a cost-effectiveness analysis. Neurology **74**(21):1671–1679 (2010). doi: 10.1212/WNL.0b013e3181e04297
13. Brinker, T., Seifert, V., Stolke, D.: Acute changes in the dynamics of the cerebrospinal fluid system during experimental subarachnoid hemorrhage. Neurosurgery **27**(3):369–372 (1990)
14. Brisman, J.L., Eskridge, J.M., Newell, D.W.: Neurointerventional treatment of vasospasm. Neurol. Res. **28**(7):769–776 (2006). doi:10.1179/016164106X152043
15. Brisman, J.L., Song, J.K., Newell, D.W.: Cerebral aneurysms. N. Engl. J. Med. **355**(9):928–939 (2006) doi: 10.1056/NEJMra052760
16. Brown, R.D. Jr, Huston, J., Hornung, R., Foroud, T., Kallmes, D.F., Kleindorfer D, Meissner I, Woo D, Sauerbeck L, Broderick J (2008) Screening for brain aneurysm in the Familial Intracranial Aneurysm study: frequency and predictors of lesion detection. J. Neurosurg. **108**(6):1132–1138. doi:10.3171/JNS/2008/108/6/1132
17. Campi, A., Ramzi, N., Molyneux, A.J., Summers, P.E., Kerr, R.S., Sneade, M., Yarnold, J.A., Rischmiller, J., Byrne, J.V.: Retreatment of ruptured cerebral aneurysms in patients randomized by coiling or clipping in the International Subarachnoid Aneurysm Trial (ISAT). Stroke **38**(5):1538–1544 (2007). doi: 10.1161/STROKEAHA.106.466987
18. Cha, K.S., Balaras, E., Lieber, B.B., Sadasivan, C., Wakhloo, A.K.: Modeling the interaction of coils with the local blood flow after coil embolization of intracranial aneurysms. J. Biomech. Eng. **129**(6):873–879 (2007). doi:10.1115/1.2800773
19. Challa, V., Han, H.C.: Spatial variations in wall thickness, material stiffness and initial shape affect wall stress and shape of intracranial aneurysms. Neurol. Res. **29**(6):569–577 (2007).
20. Cloft, H.J., Joseph, G.J., Dion, J.E.: Risk of cerebral angiography in patients with subarachnoid hemorrhage, cerebral aneurysm, and arteriovenous malformation: a meta-analysis. Stroke **30**(2):317–320 (1999)
21. Cortnum, S., Sorensen, P., Jorgensen, J.: Determining the sensitivity of computed tomography scanning in early detection of subarachnoid hemorrhage. Neurosurgery **66**(5):900–902 (2010); discussion 903.
22. Cottier, J.P., Pasco, A., Gallas, S., Gabrillargues, J., Cognard, C., Drouineau, J., Brunereau, L., Herbreteau, D.: Utility of balloon-assisted Guglielmi detachable coiling in the treatment of 49 cerebral aneurysms: a retrospective, multicenter study. AJNR Am. J. Neuroradiol. **22**(2):345–351 (2001)
23. Dawson, R.C., Krisht, A.F., Barrow, D.L., Joseph, G.J., Shengelaia, G.G., Bonner, G.: Treatment of experimental aneurysms using collagen-coated microcoils. Neurosurgery **36**(1):133–139 (1995); discussion 139–140
24. DeLeo, M.J., 3rd, Gounis, M.J., Hong, B., Ford, J.C., Wakhloo, A.K., Bogdanov, A.A., Jr.: Carotid artery brain aneurysm model: in vivo molecular enzyme-specific MR imaging of active inflammation in a pilot study. Radiology **252**(3):696–703 (2009). doi: 10.1148/radiol.2523081426
25. Dorhout Mees, S.M., Rinkel, G.J., Feigin, V.L., Algra, A., van den Bergh, W.M., Vermeulen, M., van Gijn, J.: Calcium antagonists for aneurysmal subarachnoid haemorrhage. Cochrane Database Syst. Rev. **3**:CD000277 (2007). doi:10.1002/14651858.CD000277.pub3
26. Dumont, A.S., Evans, A.J., Jensen, M.E.: Hemorrhagic complications associated with stent-assisted coil embolization. J. Neurosurg. **108**(6):1119–1121 (2008). doi:10.3171/JNS/2008/108/6/1119

27. Edlow, J.A.: Diagnosis of subarachnoid hemorrhage. Neurocrit. Care **2**(2):99–109 (2005). doi: 10.1385/NCC:2:2:099
28. Edlow, J.A., Malek, A.M., Ogilvy, C.S.: Aneurysmal subarachnoid hemorrhage: update for emergency physicians. J. Emerg. Med. **34**(3):237–251 (2008). doi: 10.1016/j.jemermed.2007.10.003
29. Elijovich, L., Higashida, R.T., Lawton, M.T., Duckwiler, G., Giannotta, S., Johnston, S.C.: Predictors and outcomes of intraprocedural rupture in patients treated for ruptured intracranial aneurysms: the CARAT study. Stroke **39**(5):1501–1506 (2008). doi: 10.1161/STROKEAHA.107.504670
30. Feng, L., Vinuela, F., Murayama, Y.: Healing of intracranial aneurysms with bioactive coils. Neurosurg. Clin. N. Am. **16**(3):487–499 (2005), v-vi. doi: 10.1016/j.nec.2005.03.001
31. Fiehler, J., Ries, T.: Prevention and treatment of thromboembolism during endovascular aneurysm therapy. Klin. Neuroradiol. **19**(1):73–81 (2009). doi:10.1007/s00062-009-8029-9
32. Fiorella, D., Albuquerque, F.C., Han, P., McDougall, C.G.: Preliminary experience using the Neuroform stent for the treatment of cerebral aneurysms. Neurosurgery **54**(1):6–16 (2004); discussion 16–17
33. Fiorella, D., Albuquerque, F.C., Deshmukh, V.R., McDougall, C.G.: Monorail snare technique for the recovery of stretched platinum coils: technical case report. Neurosurgery **57**(1 Suppl):E210 (2005); discussion E210
34. Fiorella, D., Albuquerque, F.C., McDougall, C.G.: Durability of aneurysm embolization with matrix detachable coils. Neurosurgery **58**(1):51–59 (2006); discussion 51–59.
35. Fiorella, D., Woo, H.H., Albuquerque, F.C., Nelson, P.K.: Definitive reconstruction of circumferential, fusiform intracranial aneurysms with the pipeline embolization device. Neurosurgery **62**(5):1115–1120 (2008); discussion 1120–1111
36. Fisher, C.M., Kistler, J.P., Davis, J.M.: Relation of cerebral vasospasm to subarachnoid hemorrhage visualized by computerized tomographic scanning. Neurosurgery **6**(1):1–9 (1980)
37. Fontanarosa, P.B.: Recognition of subarachnoid hemorrhage. Ann. Emerg. Med. **18**(11):1199–1205 (1989).
38. Fridriksson, S., Saveland, H., Jakobsson, K.E., Edner, G., Zygmunt, S., Brandt, L., Hillman, J.: Intraoperative complications in aneurysm surgery: a prospective national study. J. Neurosurg. **96**(3):515–522 (2002). doi:10.3171/jns.2002.96.3.0515
39. Friedman, J.A., Piepgras, D.G., Pichelmann, M.A., Hansen, K.K., Brown, R.D. Jr., Wiebers, D.O.: Small cerebral aneurysms presenting with symptoms other than rupture. Neurology **57**(7):1212–1216 (2001)
40. Frizzell, R.T., Kuhn, F., Morris, R., Quinn, C., Fisher, W.S., 3rd.: Screening for ocular hemorrhages in patients with ruptured cerebral aneurysms: a prospective study of 99 patients. Neurosurgery **41**(3):529–533 (1997); discussion 533–524
41. Frosen, J., Piippo, A., Paetau, A., Kangasniemi, M., Niemela, M., Hernesniemi, J., Jaaskelainen, J.: Remodeling of saccular cerebral artery aneurysm wall is associated with rupture: histological analysis of 24 unruptured and 42 ruptured cases. Stroke **35**(10):2287–2293 (2004).
42. Gaetani, P., Tartara, F., Tancioni, F., Klersy, C., Forlino, A., Baena, R.R.: Activity of alpha 1-antitrypsin and cigarette smoking in subarachnoid haemorrhage from ruptured aneurysm. J. Neurol. Sci. **141**(1–2):33–38 (1996)
43. Germanwala, A.V., Huang, J., Tamargo, R.J.: Hydrocephalus after aneurysmal subarachnoid hemorrhage. Neurosurg. Clin. N. Am. **21**(2):263–270 (2010). doi: 10.1016/j.nec.2009.10.013
44. Guglielmi, G., Vinuela, F., Dion, J., Duckwiler, G.: Electrothrombosis of saccular aneurysms via endovascular approach. Part 2 Preliminary clinical experience. J. Neurosurg. **75**(1):8–14 (1991). doi:10.3171/jns.1991.75.1.0008
45. Hademenos, G.J., Massoud, T.F., Turjman, F., Sayre, J.W.: Anatomical and morphological factors correlating with rupture of intracranial aneurysms in patients referred for endovascular treatment. Neuroradiology **40**(11):755–760 (1998)

46. Hashimoto, T., Meng, H., Young, W.L.: Intracranial aneurysms: links among inflammation, hemodynamics and vascular remodeling. Neurol. Res. **28**(4):372–380 (2006). doi: 10.1179/016164106X14973
47. Higashida, R.T., Halbach, V.V, Dowd, C.F., Juravsky, L., Meagher, S.: Initial clinical experience with a new self-expanding nitinol stent for the treatment of intracranial cerebral aneurysms: the Cordis Enterprise stent. AJNR Am. J. Neuroradiol. **26**(7):1751–1756 (2005).
48. Higashida, R.T., Lahue, B.J., Torbey, M.T., Hopkins, L.N., Leip, E., Hanley, D.F.: Treatment of unruptured intracranial aneurysms: a nationwide assessment of effectiveness. AJNR Am. J. Neuroradiol. **28**(1):146–151 (2007).
49. Hillman, J., Fridriksson, S., Nilsson, O., Yu, Z., Saveland, H., Jakobsson, K.E.: Immediate administration of tranexamic acid and reduced incidence of early rebleeding after aneurysmal subarachnoid hemorrhage: a prospective randomized study. J. Neurosurg. **97**(4):771–778 (2002). doi:10.3171/jns.2002.97.4.0771
50. Hiratsuka, Y., Miki, H., Kiriyama, I., Kikuchi, K, Takahashi, S., Matsubara, I., Sadamoto, K., Mochizuki, T.: Diagnosis of unruptured intracranial aneurysms: 3T MR angiography versus 64-channel multi-detector row CT angiography. Magn. Reson. Med. Sci. **7**(4):169–178 (2008).
51. Housepian, E.M., Pool, J.L.: A systematic analysis of intracranial aneurysms from the autopsy file of the Presbyterian Hospital, 1914 to 1956. J. Neuropathol. Exp. Neurol. **17**(3):409–423 (1958)
52. Hunt, W.E., Hess, R.M.: Surgical risk as related to time of intervention in the repair of intracranial aneurysms. J. Neurosurg. **28**(1):14–20 (1968). doi:10.3171/jns.1968.28.1.0014
53. Inagawa, T.: What are the actual incidence and mortality rates of subarachnoid hemorrhage? Surg. Neurol. **47**(1):47–52 (1997); discussion 52–43.
54. Ingall, T., Asplund, K., Mahonen, M., Bonita, R.: A multinational comparison of subarachnoid hemorrhage epidemiology in the WHO MONICA stroke study. Stroke **31**(5):1054–1061 (2000)
55. Ishibashi, T., Murayama, Y., Urashima, M., Saguchi, T., Ebara, M., Arakawa, H., Irie, K., Takao, H., Abe, T.: Unruptured intracranial aneurysms: incidence of rupture and risk factors. Stroke **40**(1):313–316 (2009). doi: 10.1161/STROKEAHA.108.521674
56. Jabbour, P.M., Tjoumakaris, S.I., Rosenwasser, R.H.: Neuroendovascular management of vasospasm following aneurysmal subarachnoid hemorrhage. Neurosurg. Clin. N. Am. **20**(4):441–446. doi: 10.1016/j.nec.2009.07.006
57. Janoff, A., Dearing, R.: Alpha 1-proteinase inhibitor is more sensitive to inactivation by cigarette smoke than is leukocyte elastase. Am. Rev. Respir. Dis. **126**(4):691–694 (1982)
58. Johnston, S.C., Dowd, C.F., Higashida, R.T., Lawton, M.T., Duckwiler, G.R., Gress, D.R.: Predictors of rehemorrhage after treatment of ruptured intracranial aneurysms: the Cerebral Aneurysm Rerupture After Treatment (CARAT) study. Stroke **39**(1):120–125 (2008). doi: 10.1161/STROKEAHA.107.495747
59. Katayama, H., Yamaguchi, K., Kozuka, T., Takashima, T., Seez, P., Matsuura, K.: Adverse reactions to ionic and nonionic contrast media. A report from the Japanese Committee on the Safety of Contrast Media. Radiology **175**(3):621–628 (1990)
60. Kodama, N., Suzuki, J.: Moyamoya disease associated with aneurysm. J. Neurosurg. **48**(4):565–569 (1978). doi:10.3171/jns.1978.48.4.0565
61. Koivisto, T., Vanninen, R., Hurskainen, H., Saari, T., Hernesniemi, J., Vapalahti, M.: Outcomes of early endovascular versus surgical treatment of ruptured cerebral aneurysms. A prospective randomized study. Stroke **31**(10):2369–2377 (2000)
62. Kole, M.K., Pelz, D.M., Kalapos, P., Lee, D.H., Gulka, I.B., Lownie, S.P.: Endovascular coil embolization of intracranial aneurysms: important factors related to rates and outcomes of incomplete occlusion. J. Neurosurg. **102**(4):607–615 (2005). doi:10.3171/jns.2005.102.4.0607

63. Kowalski, R.G., Claassen, J., Kreiter, K.T., Bates, J.E., Ostapkovich, N.D., Connolly, E.S., Mayer, S.A.: Initial misdiagnosis and outcome after subarachnoid hemorrhage. JAMA **291**(7):866–869 (2004).
64. Krex, D., Schackert, H.K., Schackert, G.: Genesis of cerebral aneurysms—an update. Acta Neurochir (Wien) **143**(5):429–448 (2001); discussion 448–429
65. Kroon, M., Holzapfel, G.A.: A model for saccular cerebral aneurysm growth by collagen fibre remodelling. J. Theor. Biol. **247**(4):775–787 (2007). doi: 10.1016/j.jtbi.2007.03.009
66. Kroon, M., Holzapfel, G.A.: A theoretical model for fibroblast-controlled growth of saccular cerebral aneurysms. J. Theor. Biol. **257**(1):73–83 (2009). doi: 10.1016/j.jtbi.2008.10.021
67. Kulcsar, Z., Ernemann, U., Wetzel, S.G., Bock, A., Goericke, S., Panagiotopoulos, V., Forsting, M., Ruefenacht, D.A., Wanke, I.: High-profile flow diverter (silk) implantation in the basilar artery. Efficacy in the treatment of aneurysms and the role of the perforators. Stroke (2010). doi: 10.1161/STROKEAHA.110.580308
68. Lanzino, G., Kanaan, Y., Perrini, P., Dayoub, H., Fraser, K.: Emerging concepts in the treatment of intracranial aneurysms: stents, coated coils, and liquid embolic agents. Neurosurgery **57**(3):449–459 (2005); discussion 449–459
69. Lanzino, G., Wakhloo, A.K., Fessler, R.D., Hartney, M.L., Guterman, L.R., Hopkins, L.N.: Efficacy and current limitations of intravascular stents for intracranial internal carotid, vertebral, and basilar artery aneurysms. J. Neurosurg. **91**(4):538–546 (1999). doi: 10.3171/jns.1999.91.4.0538
70. Lasser, E.C., Berry, C.C., Mishkin, M.M., Williamson, B., Zheutlin, N., Silverman, J.M.: Pretreatment with corticosteroids to prevent adverse reactions to nonionic contrast media. AJR Am. J. Roentgenol. **162**(3):523–526 (1994)
71. Li, M.H., Cheng, Y.S., Li, Y.D., Fang, C., Chen, S.W., Wang, W., Hu, D.J., Xu, H.W.: Large-cohort comparison between three-dimensional time-of-flight magnetic resonance and rotational digital subtraction angiographies in intracranial aneurysm detection. Stroke **40**(9):3127–3129 (2009). doi: 10.1161/STROKEAHA.109.553800
72. Liang, G., Gao, X., Li, Z., Wei, X., Xue, H.: Neuroform stent-assisted coiling of intracranial aneurysms: a 5 year single-center experience and follow-up. Neurol. Res. **32**(7):721–727 (2010). doi: 10.1179/016164109X12445616596409
73. Lieber, B.B., Gounis, M.J.: The physics of endoluminal stenting in the treatment of cerebrovascular aneurysms. Neurol. Res. 24 (Suppl 1):S33–42 (2002)
74. Lieber, B.B., Stancampiano, A.P., Wakhloo, A.K.: Alteration of hemodynamics in aneurysm models by stenting: influence of stent porosity. Ann. Biomed. Eng. **25**(3):460–469
75. Linn, F.H., Rinkel, G.J., Algra, A., van Gijn, J.: Incidence of subarachnoid hemorrhage: role of region, year, and rate of computed tomography: a meta-analysis. Stroke **27**(4):625–629 (1996)
76. Longstreth, W.T. Jr., Nelson, L.M., Koepsell, T.D., van Belle, G.: Clinical course of spontaneous subarachnoid hemorrhage: a population-based study in King County, Washington. Neurology **43**(4):712–718 (1993)
77. Lozano, A.M., Leblanc, R.: Familial intracranial aneurysms. J. Neurosurg. **66**(4):522–528 (1987). doi:10.3171/jns.1987.66.4.0522
78. Lylyk, P., Cohen, J.E., Ceratto, R., Ferrario, A., Miranda, C.: Endovascular reconstruction of intracranial arteries by stent placement and combined techniques. J. Neurosurg. **97**(6):1306–1313 (2002). doi:10.3171/jns.2002.97.6.1306
79. Lylyk, P., Miranda, C., Ceratto, R., Ferrario, A., Scrivano, E., Luna, H.R., Berez, A.L., Tran, Q., Nelson, P.K., Fiorella, D.: Curative endovascular reconstruction of cerebral aneurysms with the pipeline embolization device: the Buenos Aires experience. Neurosurgery **64**(4):632–642 (2009); discussion 642–633; quiz N636
80. M, S., Lamont, A.C., Alias, N.A., Win, M.N.: Red flags in patients presenting with headache: clinical indications for neuroimaging. Br. J. Radiol. **76**(908):532–535 (2003)
81. MacDonald, D.J, Finlay, H.M., Canham, P.B.: Directional wall strength in saccular brain aneurysms from polarized light microscopy. Ann. Biomed. Eng. **28**(5):533–542 (2000)

82. Marx, W.E., Cloft, H.J., Helm, G.A., Short, J.G., Do, H.M., Jensen, M.E., Kallmes, D.E.: Endovascular treatment of experimental aneurysms by use of biologically modified embolic devices: coil-mediated intraaneurysmal delivery of fibroblast tissue allografts. AJNR Am. J. Neuroradiol. **22**(2):323–333 (2001)
83. Mazighi, M., Saint Maurice, J.P., Bresson, D., Szatmary, Z., Houdart, E.: Platelet aggregation in intracranial stents may mimic in-stent restenosis. AJNR Am. J. Neuroradiol. **31**(3):496–497 (2010). doi: 10.3174/ajnr.A1778
84. McKinney, A.M., Palmer, C.S., Truwit, C.L., Karagulle, A., Teksam, M.: Detection of aneurysms by 64-section multidetector CT angiography in patients acutely suspected of having an intracranial aneurysm and comparison with digital subtraction and 3D rotational angiography. AJNR Am. J. Neuroradiol. **29**(3):594–602 (2008). doi:ajnr.A0848[pii]; 10.3174/ajnr.A0848
85. Mocco, J., Snyder, K.V., Albuquerque, F.C., Bendok, B.R., Alan, S.B., Carpenter, J.S., Fiorella, D.J., Hoh, B.L., Howington, J.U., Jankowitz, B.T., Liebman, K.M., Rai, A.T., Rodriguez-Mercado, R., Siddiqui, A.H., Veznedaroglu, E., Hopkins, L.N., Levy, E.I.: Treatment of intracranial aneurysms with the Enterprise stent: a multicenter registry. J. Neurosurg. **110**(1):35–39 (2009). doi:10.3171/2008.7.JNS08322
86. Molyneux, A., Kerr, R., Stratton, I., Sandercock, P., Clarke, M., Shrimpton, J., Holman, R.: International Subarachnoid Aneurysm Trial (ISAT) of neurosurgical clipping versus endovascular coiling in 2143 patients with ruptured intracranial aneurysms: a randomised trial. Lancet **360**(9342):1267–1274 (2002)
87. Molyneux, A.J., Cekirge, S., Saatci, I., Gal, G.: Cerebral Aneurysm Multicenter European Onyx (CAMEO) trial: results of a prospective observational study in 20 European centers. AJNR Am. J. Neuroradiol. **25**(1):39–51 (2004)
88. Molyneux, A.J., Kerr, R.S., Yu, L.M., Clarke, M., Sneade, M., Yarnold, J.A., Sandercock, P.: International subarachnoid aneurysm trial (ISAT) of neurosurgical clipping versus endovascular coiling in 2143 patients with ruptured intracranial aneurysms: a randomised comparison of effects on survival, dependency, seizures, rebleeding, subgroups, and aneurysm occlusion. Lancet **366**(9488):809–817 (2005). doi:10.1016/S0140-6736(05)67214-5
89. Molyneux, A.J., Kerr, R.S., Birks, J., Ramzi, N., Yarnold, J., Sneade, M., Rischmiller, J.: Risk of recurrent subarachnoid haemorrhage, death, or dependence and standardised mortality ratios after clipping or coiling of an intracranial aneurysm in the International Subarachnoid Aneurysm Trial (ISAT): long-term follow-up. Lancet Neurol. **8**(5):427–433 (2009). doi: 10.1016/S1474-4422(09)70080-8
90. Moret, J., Cognard, C., Weill, A., Castaings, L., Rey, A.: Reconstruction technic in the treatment of wide-neck intracranial aneurysms. Long-term angiographic and clinical results. Apropos of 56 cases. J. Neuroradiol. **24**(1):30–44 (1997)
91. Muizelaar, J.P., Becker, D.P.: Induced hypertension for the treatment of cerebral ischemia after subarachnoid hemorrhage. Direct effect on cerebral blood flow. Surg. Neurol. **25**(4):317–325 (1986)
92. Murayama, Y., Vinuela, F., Tateshima, S., Gonzalez, N.R., Song, J.K., Mahdavieh, H., Iruela-Arispe, L.: Cellular responses of bioabsorbable polymeric material and Guglielmi detachable coil in experimental aneurysms. Stroke **33**(4):1120–1128 (2002)
93. Murayama, Y., Tateshima, S., Gonzalez, N.R., Vinuela, F.: Matrix and bioabsorbable polymeric coils accelerate healing of intracranial aneurysms: long-term experimental study. Stroke **34**(8):2031–2037 (2003)
94. Nael, K., Villablanca, J.P., Saleh, R., Pope, W., Nael, A., Laub, G., Finn, J.P.: Contrast-enhanced MR angiography at 3T in the evaluation of intracranial aneurysms: a comparison with time-of-flight MR angiography. AJNR Am. J. Neuroradiol. **27**(10):2118–2121 (2006)
95. Naidech, A.M., Janjua, N., Kreiter, K.T., Ostapkovich, N.D., Fitzsimmons, B.F., Parra, A., Commichau, C., Connolly, E.S., Mayer, S.A.: Predictors and impact of aneurysm rebleeding after subarachnoid hemorrhage. Arch. Neurol. **62**(3):410–416. doi: 10.1001/archneur.62.3.410

96. Naidech, A.M., Kreiter, K.T., Janjua, N., Ostapkovich, N., Parra, A., Commichau, C., Connolly, E.S., Mayer, S.A., Fitzsimmons, B.F.: Phenytoin exposure is associated with functional and cognitive disability after subarachnoid hemorrhage. Stroke **36**(3):583–587 (2005). doi: 10.1161/01.STR.0000141936.36596.1e
97. Nakatomi, H., Segawa, H., Kurata, A., Shiokawa, Y., Nagata, K., Kamiyama, H., Ueki, K., Kirino, T.: Clinicopathological study of intracranial fusiform and dolichoectatic aneurysms: insight on the mechanism of growth. Stroke **31**(4):896–900 (2000)
98. Niimi, Y., Song, J., Madrid, M., Berenstein, A.: Endosaccular treatment of intracranial aneurysms using matrix coils: early experience and midterm follow-up. Stroke **37**(4):1028–1032 (2006). doi: 10.1161/01.STR.0000206459.73897.a3
99. Ostergaard, J.R., Hog, E.: Incidence of multiple intracranial aneurysms. Influence of arterial hypertension and gender. J Neurosurg **63**(1):49–55 (1985). doi:10.3171/jns.1985.63.1.0049
100. Peters, D.G., Kassam, A.B., Feingold, E., Heidrich-O'Hare, E., Yonas H., Ferrell, R.E., Brufsky, A.: Molecular anatomy of an intracranial aneurysm: coordinated expression of genes involved in wound healing and tissue remodeling. Stroke **32**(4):1036–1042 (2001)
101. Petruk, K.C., West, M., Mohr, G., Weir, B.K., Benoit, B.G., Gentili, F., Disney, L.B., Khan, M.I., Grace, M., Holness, R.O., et al.: Nimodipine treatment in poor-grade aneurysm patients. Results of a multicenter double-blind placebo-controlled trial. J. Neurosurg. **68**(4):505–517 (1988). doi:10.3171/jns.1988.68.4.0505
102. Phillips, L.H. 2nd., Whisnant, J.P., O'Fallon, W.M., Sundt, T.M., Jr.: The unchanging pattern of subarachnoid hemorrhage in a community. Neurology **30**(10):1034–1040 (1980)
103. Pickard, J.D., Murray, G.D., Illingworth, R., Shaw, M.D., Teasdale, G.M., Foy, P.M., Humphrey, P.R., Lang, D.A., Nelson, R., Richards, P. et al.: Effect of oral nimodipine on cerebral infarction and outcome after subarachnoid haemorrhage: British aneurysm nimodipine trial. BMJ **298**(6674):636–642 (1989)
104. Piotin, M., Iijima, A., Wada, H., Moret, J.: Increasing the packing of small aneurysms with complex-shaped coils: an in vitro study. AJNR Am. J. Neuroradiol. **24**(7):1446–1448 (2003)
105. Piske, R.L., Kanashiro, L.H., Paschoal, E., Agner, C., Lima, S.S., Aguiar, P.H.: Evaluation of Onyx HD-500 embolic system in the treatment of 84 wide-neck intracranial aneurysms. Neurosurgery **64**(5):E865–E875 (2009); discussion E875
106. Polmear, A.: Sentinel headaches in aneurysmal subarachnoid haemorrhage: what is the true incidence? A systematic review. Cephalalgia **23**(10):935–941 (2003)
107. Prabhakaran, S., Wells, K.R., Lee, V.H., Flaherty, C.A., Lopes, D.K.: Prevalence and risk factors for aspirin and clopidogrel resistance in cerebrovascular stenting. AJNR Am. J. Neuroradiol. **29**(2):281–285 (2008). doi: 10.3174/ajnr.A0818
108. Pumar, J.M., Blanco, M., Vazquez, F., Castineira, J.A., Guimaraens, L., Garcia-Allut, A.: Preliminary experience with Leo self-expanding stent for the treatment of intracranial aneurysms. AJNR Am. J. Neuroradiol. **26**(10):2573–2577 (2005)
109. Qureshi, A.I., Luft, A.R., Sharma, M., Guterman, L.R., Hopkins, L.N.: Prevention and treatment of thromboembolic and ischemic complications associated with endovascular procedures: Part II—clinical aspects and recommendations. Neurosurgery **46**(6):1360–1375 (2000); discussion 1375–1366
110. Raaymakers, T.W.: Aneurysms in relatives of patients with subarachnoid hemorrhage: frequency and risk factors. MARS Study Group. Magnetic Resonance Angiography in Relatives of patients with Subarachnoid hemorrhage. Neurology **53**(5):982–988 (1999)
111. Rabinstein, A.A., Lanzino, G., Wijdicks, E.F.: Multidisciplinary management and emerging therapeutic strategies in aneurysmal subarachnoid haemorrhage. Lancet Neurol. **9**(5):504–519 (2010). doi: 10.1016/S1474-4422(10)70087-9
112. Racadio, J.M., Fricke, B.L., Jones, B.V., Donnelly, L.F.: Three-dimensional rotational angiography of neurovascular lesions in pediatric patients. AJR Am. J. Roentgenol. **186**(1):75–84 (2006). doi: 10.2214/AJR.04.1595
113. Raghavan, M.L., Ma, B., Harbaugh, R.E.: Quantified aneurysm shape and rupture risk. J. Neurosurg. **102**(2):355–362 (2005). doi:10.3171/jns.2005.102.2.0355

114. Rates of delayed rebleeding from intracranial aneurysms are low after surgical and endovascular treatment. Stroke **37**(6):1437–1442 (2006). doi: 10.1161/01.STR.0000221331.01830.ce
115. Raymond, J., Roy, D.: Safety and efficacy of endovascular treatment of acutely ruptured aneurysms. Neurosurgery **41**(6):1235–1245 (1997); discussion 1245-1236
116. Raymond, J., Meder, J.F., Molyneux, A.J., Fox, A.J., Johnston, S.C., Collet, J.P., Rouleau, I.: Unruptured intracranial aneurysms: the unreliability of clinical judgment, the necessity for evidence, and reasons to participate in a randomized trial. J. Neuroradiol. **33**(4):211–219 (2006)
117. Raymond, J., Molyneux, A.J., Fox, A.J., Johnston, S.C., Collet, J.P., Rouleau, I.: The TEAM trial: safety and efficacy of endovascular treatment of unruptured intracranial aneurysms in the prevention of aneurysmal hemorrhages: a randomized comparison with indefinite deferral of treatment in 2002 patients followed for 10 years. Trials **9**:43 (2008)
118. Regli, L., Uske, A., de Tribolet, N.: Endovascular coil placement compared with surgical clipping for the treatment of unruptured middle cerebral artery aneurysms: a consecutive series. J. Neurosurg. **90**(6):1025–1030 (1999). doi:10.3171/jns.1999.90.6.1025
119. Report of World Federation of Neurological Surgeons Committee on a Universal Subarachnoid Hemorrhage Grading Scale. J. Neurosurg. **68**(6):985–986 (1988)
120. Ries, T., Buhk, J.H., Kucinski, T., Goebell, E., Grzyska, U., Zeumer, H., Fiehler, J.: Intravenous administration of acetylsalicylic acid during endovascular treatment of cerebral aneurysms reduces the rate of thromboembolic events. Stroke **37**(7):1816–1821 (2006). doi: 10.1161/01.STR.0000226933.44962.a6
121. Ries, T., Siemonsen, S., Grzyska, U., Zeumer, H., Fiehler, J.: Abciximab is a safe rescue therapy in thromboembolic events complicating cerebral aneurysm coil embolization: single center experience in 42 cases and review of the literature. Stroke **40**(5):1750–1757 (2009). doi: 10.1161/STROKEAHA.108.539197
122. Rinkel, G.J., Djibuti, M., Algra, A., van Gijn, J.: Prevalence and risk of rupture of intracranial aneurysms: a systematic review. Stroke **29**(1):251–256 (1998)
123. Ronkainen, A., Hernesniemi, J., Puranen, M., Niemitukia, L., Vanninen, R., Ryynanen, M., Kuivaniemi, H., Tromp, G.: Familial intracranial aneurysms. Lancet **349**(9049):380–384 (1997). doi: 10.1016/S0140-6736(97)80009-8
124. Roos, Y.B., Rinkel, G.J., Vermeulen, M., Algra, A., van Gijn, J.: Antifibrinolytic therapy for aneurysmal subarachnoid haemorrhage. Cochrane Database Syst. Rev. **2**:CD001245 (2003). doi:10.1002/14651858.CD001245
125. Ruigrok, Y.M., Buskens, E., Rinkel, G.J.: Attributable risk of common and rare determinants of subarachnoid hemorrhage. Stroke **32**(5):1173–1175 (2001)
126. Sacco, R.L., Wolf, P.A., Bharucha, N.E., Meeks, S.L., Kannel, W.B., Charette, L.J., McNamara, P.M., Palmer, E.P., D'Agostino, R.: Subarachnoid and intracerebral hemorrhage: natural history, prognosis, and precursive factors in the Framingham Study. Neurology **34**(7):847–854 (1984)
127. Sadasivan, C., Lieber, B.B.: Numerical Investigation of Coil Configurations That Provide Ultra-High Packing Density of Saccular Aneurysms. J. Med. Device 3(4):41005 (2009). doi: 10.1115/1.4000453
128. Sadasivan, C., Lieber, B.B., Cesar, L., Miskolczi, L., Seong, J., Wakhloo, A.K.: Angiographic assessment of the performance of flow divertors to treat cerebral aneurysms. Conf. Proc. IEEE Eng. Med. Biol. Soc. **1**:3210–3213 (2006). doi: 10.1109/IEMBS.2006.260043
129. Schaafsma, J.D., Sprengers, M.E., van Rooij, W.J., Sluzewski, M., Majoie, C.B., Wermer, M.J., Rinkel, G.J.: Long-term recurrent subarachnoid hemorrhage after adequate coiling versus clipping of ruptured intracranial aneurysms. Stroke **40**(5):1758–1763 (2009). doi: 10.1161/STROKEAHA.108.524751
130. Shankar, J.J., Lum, C., Parikh, N., Dos Santos, M.: Long-term prospective follow-up of intracranial aneurysms treated with endovascular coiling using contrast-enhanced MR angiography. AJNR Am. J. Neuroradiol. (2010). doi: 10.3174/ajnr.A2064

131. Shapiro, M., Babb, J., Becske, T., Nelson, P.K.: Safety and efficacy of adjunctive balloon remodeling during endovascular treatment of intracranial aneurysms: a literature review. AJNR Am. J. Neuroradiol. **29**(9):1777–1781 (2008). doi:ajnr.A1216[pii]; 10.3174/ajnr.A1216
132. Sloan, M.A., Alexandrov, A.V., Tegeler, C.H., Spencer, M.P., Caplan, L.R., Feldmann, E., Wechsler, L.R., Newell, D.W., Gomez, C.R., Babikian, V.L., Lefkowitz, D., Goldman, R.S., Armon, C., Hsu, C.Y., Goodin, D.S.: Assessment: transcranial Doppler ultrasonography: report of the Therapeutics and Technology Assessment Subcommittee of the American Academy of Neurology. Neurology **62**(9):1468–1481 (2004)
133. Sluzewski, M., van Rooij, W.J., Slob, M.J., Bescos, J.O., Slump, C.H., Wijnalda, D.: Relation between aneurysm volume, packing, and compaction in 145 cerebral aneurysms treated with coils. Radiology **231**(3):653–658 (2004)
134. Sluzewski, M., van Rooij, W.J., Beute, G.N., Nijssen, P.C.: Balloon-assisted coil embolization of intracranial aneurysms: incidence, complications, and angiography results. J. Neurosurg. **105**(3):396–399 (2006). doi:10.3171/jns.2006.105.3.396
135. Soeda, A., Sakai, N., Sakai, H., Iihara, K., Yamada, N., Imakita, S., Nagata, I.: Thromboembolic events associated with Guglielmi detachable coil embolization of asymptomatic cerebral aneurysms: evaluation of 66 consecutive cases with use of diffusion-weighted MR imaging. AJNR Am. J. Neuroradiol. **24**(1):127–132 (2003)
136. Solomon, R.A., Fink, M.E., Pile-Spellman, J.: Surgical management of unruptured intracranial aneurysms. J. Neurosurg. **80**(3):440–446 (1994). doi:10.3171/jns.1994.80.3.0440
137. Sprengers, M.E., Schaafsma, J., van Rooij, W.J., Sluzewski, M., Rinkel, G.J., Velthuis, B.K., van Rijn, J.C, Majoie, C.B.: Stability of intracranial aneurysms adequately occluded 6 months after coiling: a 3T MR angiography multicenter long-term follow-up study. AJNR Am. J. Neuroradiol. **29**(9):1768–1774 (2008). doi: 10.3174/ajnr.A1181
138. Starke, R.M., Kim, G.H., Fernandez, A., Komotar, R.J., Hickman, Z.L., Otten, M.L., Ducruet, A.F., Kellner, C.P., Hahn, D.K.. Chwajol, M., Mayer, S.A., Connolly, E.S. Jr.: Impact of a protocol for acute antifibrinolytic therapy on aneurysm rebleeding after subarachnoid hemorrhage. Stroke 39(9):2617–2621 (2008). doi: 10.1161/STROKEAHA.107.506097
139. Stehbens WE (1972) Pathology of the Cerebral Blood Vessels. The C. V. Mosby Company, Saint Louis
140. Stoeckel, D., Pelton, A., Duerig, T.: Self-expanding nitinol stents: material and design considerations. Eur Radiol **14**(2):292–301 (2004). doi:10.1007/s00330-003-2022-5
141. Sviri, G.E., Britz, G.W., Lewis, .D.H., Newell, D.W., Zaaroor, M., Cohen, W.: Dynamic perfusion computed tomography in the diagnosis of cerebral vasospasm. Neurosurgery **59**(2):319–325 (2006); discussion 319–325
142. Tahtinen, O.I., Vanninen, R.L., Manninen, H.I., Rautio, R., Haapanen, A., Niskakangas, T., Rinne, J., Keski-Nisula, L.: Wide-necked intracranial aneurysms: treatment with stent-assisted coil embolization during acute (<72 hours) subarachnoid hemorrhage–experience in 61 consecutive patients. Radiology **253**(1):199–208 (2009). doi: 10.1148/radiol.2531081923
143. Tartaglino, L.M., Gorniak, R.J.: Advanced imaging applications for endovascular procedures. Neurosurg Clin N Am **20**(3):297–313 (2009). doi: 10.1016/j.nec.2009.01.001
144. Taschner, C.A., Leclerc, X., Rachdi, H., Barros, A.M., Pruvo, J.P.: Matrix detachable coils for the endovascular treatment of intracranial aneurysms: analysis of early angiographic and clinical outcomes. Stroke **36**(10):2176–2180 (2005). doi: 10.1161/01.STR.0000181770.14869.ce
145. Tulamo, R., Frosen, J., Hernesniemi, J., Niemela, M.: Inflammatory changes in the aneurysm wall: a review. J. NeuroInterv. Surg. **2**(2):120–130 (2010). doi: 10.1136/jnis.2009.002055
146. Tumialan, L.M., Zhang, Y.J., Cawley, C.M., Dion, J.E., Tong, F.C., Barrow, D.L.: Intracranial hemorrhage associated with stent-assisted coil embolization of cerebral

aneurysms: a cautionary report. J. Neurosurg. **108**(6):1122–1129 (2008). doi: 10.3171/JNS/2008/108/6/1122

147. van der Wee, N., Rinkel, G.J., Hasan, D., van Gijn, J.: Detection of subarachnoid haemorrhage on early CT: is lumbar puncture still needed after a negative scan? J. Neurol. Neurosurg. Psychiatry **58**(3):357–359 (1995)
148. van Rooij, W.J., Sluzewski, M.: Opinion: imaging follow-up after coiling of intracranial aneurysms. AJNR Am. J. Neuroradiol. **30**(9):1646–1648 (2009)
149. van Rooij, W.J., Sluzewski, M., Beute, G.N., Nijssen, P.C.: Procedural complications of coiling of ruptured intracranial aneurysms: incidence and risk factors in a consecutive series of 681 patients. AJNR Am. J. Neuroradiol. **27**(7):1498–1501 (2006)
150. Vanninen, R., Koivisto, T., Saari, T., Hernesniemi, J., Vapalahti, M.: Ruptured intracranial aneurysms: acute endovascular treatment with electrolytically detachable coils—a prospective randomized study. Radiology **211**(2):325–336 (1999)
151. Vermeulen, M., Hasan, D., Blijenberg, B.G., Hijdra, A., van Gijn, J.: Xanthochromia after subarachnoid haemorrhage needs no revisitation. J. Neurol. Neurosurg. Psychiatry **52**(7):826–828 (1989)
152. Vernooij, M.W., Ikram, M.A., Tanghe, H.L., Vincent, A.J., Hofman, A., Krestin, G.P., Niessen, W.J., Breteler, M.M., van der Lugt, A.: Incidental findings on brain MRI in the general population. N. Engl. J. Med. **357**(18):1821–1828 (2007). doi: 10.1056/NEJMoa070972
153. Vinuela, F., Duckwiler, G., Mawad, M.: Guglielmi detachable coil embolization of acute intracranial aneurysm: perioperative anatomical and clinical outcome in 403 patients. J. Neurosurg. **86**(3):475–482 (1997). doi:10.3171/jns.1997.86.3.0475
154. Wakhloo, A.K., Schellhammer, F., de Vries, J., Haberstroh, J., Schumacher, M.: Self-expanding and balloon-expandable stents in the treatment of carotid aneurysms: an experimental study in a canine model. AJNR Am. J. Neuroradiol. **15**(3):493–502 (1994)
155. Wakhloo, A.K., Mandell, J., Gounis, M.J., Brooks, C., Linfante, I., Winer, J., Weaver, J.P.: Stent-assisted reconstructive endovascular repair of cranial fusiform atherosclerotic and dissecting aneurysms: long-term clinical and angiographic follow-up. Stroke **39**(12):3288–3296 (2008). doi: 10.1161/STROKEAHA.107.512996
156. Wardlaw, J.M., White, P.M.: The detection and management of unruptured intracranial aneurysms. Brain **123**(Pt 2):205–221 (2000)
157. Watton, P.N., Ventikos, Y., Holzapfel, G.A.: Modelling the growth and stabilization of cerebral aneurysms. Math Med Biol **26**(2):133–164 (2009). doi: 10.1093/imammb/dqp001
158. Weber, W., Siekmann, R., Kis, B., Kuehne, D.: Treatment and follow-up of 22 unruptured wide-necked intracranial aneurysms of the internal carotid artery with Onyx HD 500. AJNR Am. J. Neuroradiol. **26**(8):1909–1915 (2005)
159. Wermer, M.J., Rinkel, G.J., van Gijn, J.: Repeated screening for intracranial aneurysms in familial subarachnoid hemorrhage. Stroke **34**(12):2788–2791 (2003)
160. Wermer, M.J., van der Schaaf, I.C., Van Nunen, P., Bossuyt, P.M., Anderson, C.S., Rinkel, G.J.: Psychosocial impact of screening for intracranial aneurysms in relatives with familial subarachnoid hemorrhage. Stroke **36**(4):836–840 (2005a). doi:10.1161/01.STR.0000158906.79898.3a
161. Wermer, M.J., van der Schaaf, I.C., Velthuis, B.K., Algra, A., Buskens, E., Rinkel, G.J.: Follow-up screening after subarachnoid haemorrhage: frequency and determinants of new aneurysms and enlargement of existing aneurysms. Brain **128**(Pt 10):2421–2429 (2005)
162. White, J.B., Ken, C.G., Cloft, H.J., Kallmes, D.F.: Coils in a nutshell: a review of coil physical properties. AJNR Am. J. Neuroradiol. **29**(7):1242–1246 (2008). doi: 10.3174/ajnr.A1067
163. White, P.M., Lewis, S.C., Nahser, H., Sellar, R.J., Goddard, T., Gholkar, A.: HydroCoil Endovascular Aneurysm Occlusion and Packing Study (HELPS trial): procedural safety and operator-assessed efficacy results. AJNR Am. J. Neuroradiol. **29**(2):217–223 (2008b). doi: 10.3174/ajnr.A0936

164. Wiebers, D.O.: Neuroepidemiology of unruptured intracranial aneurysms: implications for decision making regarding patient management. Neurosurg Clin N Am **16**(2):309–312, ix (2005). doi: 10.1016/j.nec.2004.08.018
165. Wiebers, D.O., Whisnant, J.P., Huston, J. 3rd., Meissner, I., Brown, R.D. Jr., Piepgras, D.G., Forbes, G.S., Thielen, K., Nichols, D., O'Fallon, W.M., Peacock, J., Jaeger, L., Kassell, N.F., Kongable-Beckman, G.L., Torner, J.C.: Unruptured intracranial aneurysms: natural history, clinical outcome, and risks of surgical and endovascular treatment. Lancet **362**(9378):103–110 (2003)
166. Wiebers, D.O., Piepgras, D.G., Meyer, F.B., Kallmes, D.F., Meissner, I., Atkinson, J.L., Link, M.J., Brown, R.D. Jr.: Pathogenesis, natural history, and treatment of unruptured intracranial aneurysms. Mayo Clin Proc **79**(12):1572–1583 (2004)
167. Wiesmann, M., Mayer, T.E., Yousry, I., Medele, R., Hamann, G.F., Bruckmann, H.: Detection of hyperacute subarachnoid hemorrhage of the brain by using magnetic resonance imaging. J. Neurosurg. **96**(4):684–689 (2002). doi:10.3171/jns.2002.96.4.0684
168. Wills, S., Ronkainen, A., van der Voet, M., Kuivaniemi, H., Helin, K., Leinonen, E., Frosen, J., Niemela, M., Jaaskelainen, J., Hernesniemi, J., Tromp, G.: Familial intracranial aneurysms: an analysis of 346 multiplex Finnish families. Stroke **34**(6):1370–1374 (2003). doi:10.1161/01.STR.0000072822.35605.8B
169. Winn, H.R., Almaani, W.S., Berga, S.L., Jane, J.A., Richardson, A.E.: The long-term outcome in patients with multiple aneurysms. Incidence of late hemorrhage and implications for treatment of incidental aneurysms. J. Neurosurg. **59**(4):642–651 (1983). doi:10.3171/jns.1983.59.4.0642
170. Wintermark, M., Ko, N.U., Smith, W.S., Liu, S., Higashida, R.T., Dillon, W.P.: Vasospasm after subarachnoid hemorrhage: utility of perfusion CT and CT angiography on diagnosis and management. AJNR Am. J. Neuroradiol. **27**(1):26–34 (2006)

Modelling Cerebral Aneurysm Evolution

Paul N. Watton, Yiannis Ventikos and Gerhard A. Holzapfel

Abstract We present a fluid–solid-growth model for the evolution of a saccular cerebral aneurysm on the internal carotid sinus artery. The model utilises a realistic constitutive model of the arterial wall that accounts for the structural arrangement of collagen fibres in the medial and adventitial layers, the natural reference configurations in which the collagen fibres are recruited to load bearing and the mass of the elastin and collagenous constituents. The structural model is integrated into a patient-specific geometry of the internal carotid artery. This enables growth and remodelling (G&R) of the aneurysmal section to be explicitly linked to physiologically realistic haemodynamic stimuli. In addition, a quasi-static approach is used to obtain the geometry of the aneurysmal section at systolic and diastolic pressures, enabling G&R to be explicitly linked to the magnitude of the cyclic stretches. To our knowledge, this is the first patient-specific model of cerebral aneurysm evolution that incorporates a realistic constitutive model of the arterial wall and explicitly links G&R to the pulsatile mechanical environment. It will provide the basis for further investigating and elucidating the aetiology of the disease.

P. N. Watton (✉) · Y. Ventikos
Department of Engineering Science, Institute of Biomedical Engineering,
University of Oxford, Oxford, UK
e-mail: Paul.Watton@eng.ox.ac.uk

Y. Ventikos
e-mail: Yiannis.Ventikos@eng.ox.ac.uk

G. A. Holzapfel
Center of Biomedical Engineering, Institute of Biomechanics,
Graz University of Technology, 8010 Graz, Austria
e-mail: holzapfel@tugraz.at

G. A. Holzapfel
Department of Solid Mechanics, School of Engineering Sciences,
Royal Institute of Technology, Osquars Backe 1,
100 44 Stockholm, Sweden

Stud Mechanobiol Tissue Eng Biomater (2011) 7: 373–399
DOI: 10.1007/8415_2011_77

Published Online: 31 May 2011

1 Introduction

Intracranial aneurysms (ICAs) are a focal disease of the brain vasculature. They appear as sac-like outpouchings of the arterial wall, and are inflated by the pressure of the blood that fills them. They are relatively common and affect 2–5% of the adult population. Their precise cause is still not known, but it is associated with elevated blood pressure, complex blood flow conditions [1–4], cigarette smoking, heavy alcohol consumption and genetic factors [5]. Fortunately, most remain asymptomatic. However, there is a small but inherent risk of rupture: 0.1–1% of detected aneurysms rupture every year [6]. If subarachnoid haemorrhage does occur there is a 30–50% chance of fatality. Of those that survive, an estimated 30% will face moderate to severe disability [7]. Consequently, if an aneurysm is detected, clinical intervention may be deemed appropriate. Therapy is currently aimed at pre-rupture detection and preventative treatment. However, interventional procedures, i.e., minimally invasive endovascular approaches or more traditional open cranium neurosurgery, are not without risk to the patient; interventional repair of an unruptured aneurysm has related mortality and morbidity rates of up to 2.5 and 6%, respectively [8].

Relatively recent developments in imaging technology have led to new healthcare policies in terms of application of medical imaging modalities. This trend has led to a dramatic increase in coincidentally detected asymptomatic cerebral aneurysms. Moreover, there has been an increase in the number of cases where intervention is deemed necessary, in spite of the lack of robustly founded rupture risk indicators. The improvement and optimisation of interventional techniques is an important concern for patient welfare. Moreover, it is necessary for rationalisation of healthcare priorities. Hence there is a need to develop methodologies to assist in identifying those ICAs most at risk of rupture. Biomechanics has an essential role to play in this respect. It offers the potential for computational tools to assist clinical diagnostic procedures. To date, ICA modelling has focused on: conceptual structural models of cerebral aneurysms which use idealised geometries (spheres, spherical caps) [9–15]; numerous computational fluid dynamic studies that utilise patient-specific geometries [16–35] but relatively few structural analyses for patient-specific geometries [36–38]; lastly, in recent years there has been a focus on modelling the evolution of an ICA [3, 39–57].

In this article, we focus on the mathematical modelling and computational simulation of ICA evolution. Models must take into consideration: (1) the bio-mechanics of the arterial wall; (2) the biology of the arterial wall and (3) the complex interplay between (1) and (2), i.e., the mechanobiology of the arterial wall. The ultimate ambition of such models is to aid clinical diagnosis on a patient-specific basis. However, due to the significant biological complexity coupled with limited histological information such models are still in their relative infancy. Current research focuses on simulating the evolution of an ICA with an aim to yield insight into the growth and remodelling (G&R) processes that give rise to inception, enlargement, stabilisation and rupture.

2 Mathematical/Computational Modelling of Cerebral Aneurysm Evolution

The physiological mechanisms that give rise to the development of an ICA involve the complex interplay between the local mechanical forces acting on the arterial wall and the biological processes occurring at the cellular level. To understand ICA formation we must solidify our understanding of the physiological mechanisms that regulate the maintenance and G&R of arterial tissue in healthy and diseased states.

The structure of the artery is continually maintained by cells within the arterial wall. The morphology and functionality of vascular cells is intimately linked to their mechanical environment. Haemodynamic forces give rise to: cyclic stretching of the extra-cellular matrix (ECM); frictional forces acting on the endothelial layer of the arterial wall; a normal hydrostatic pressure and interstitial fluid forces due to movement of the fluid through the ECM. Mechanosensors convert the mechanical stimuli into chemical signals which lead to activation of genes that regulate cell functionality.

Development of an ICA is associated with apoptosis of the medial smooth muscle cells [58–60], disruptured internal elastic laminae, the breakage and elimination of the elastin fibres within the aneurysmal wall [61–64], a thinned medial layer and G&R of the collagen fabric. The site of origin is strongly related to hemodynamic wall shear stress (WSS) [58]. In fact, it is postulated that high WSS is related to the initiation of ICA formation [65–68]. It is also suggested that high spatial gradients of the WSS may lead to destructive remodelling [69, 70]. However, whilst high WSS is associated with the inception of a cerebral aneurysm, low WSS is thought to give rise to its continued enlargement [71].

The first requirement to modelling the evolution of an aneurysm is to be able to model the healthy arterial wall. Given that the arterial wall is a highly complex integrated structure [72] open challenges still remain. However, in recent years, sophisticated theoretical models that depict the mechanical response of the arterial tissue have been developed [73, 74]. These account for the mechanical response of the individual layers of the arterial wall and model each as a fibre-reinforced composite. The (passive) mechanical response is governed by the elastinous and collagenous constituents (for a comprehensive review on the constitutive modelling of arteries [75]). Elastin is a relatively stable protein with a long half-life (approximately 50 years) whereas collagen has much faster turnover rates, i.e., two months [76, 77]. Consequently, understanding the G&R of the collagen is critical to predict arterial adaption during ICA formation.

In recent years, mathematical models of ICA evolution have been receiving much greater attention. Such models generally owe their foundations to a simple conceptual model proposed by Humphrey [78] which considered remodelling of a collagenous tissue at (altered) fixed length. In this model it was assumed that: collagen fibres are in a continual state of deposition and degradation; newly deposited collagen fibres configure with an *attachment stretch* λ_{AT}^{C} equal to unity.

In fact, fibroblasts work on the collagen they have secreted, acting to attach the collagen fibres to the extra-cellular matrix in a state of stretch [79]. Hence in subsequent studies it has been assumed that the attachment stretch $\lambda_{AT}^{C} > 1$ and the model has been applied to consider the adaption of arteries (maintaining cylindrical geometries) during altered flow, pressure, axial stretch [80–88] and the evolution of fusiform cerebral aneurysms [41, 42].

Kroon and Holzapfel were the first to model the evolution of axisymmetric [43] and saccular [44] ICA of the middle cerebral artery. The methodology has been applied to examine the influence of axial pre-stretch on aneurysm evolution [46] and the G&R framework extended to consider the effect of cyclic stretch on the adaption of the collagen fabric [47]. The theoretical framework for the remodelling of collagen is similar to that of [41]: both methodologies compute new natural reference configurations for newly deposited collagen fibres each computational time-step. However, a subtle difference is that Kroon & Holzapfel [43] assume a lifetime for the collagen fibres, consequently the maximum number of reference configurations to keep track of is bounded, whereas Baek et al. [41] assume a half-life for the collagen fibres and thus the number of reference configurations increases linearly with the number of time steps of a computational simulation.

One limitation of the above mentioned ICA evolution models is that whilst they simulated the structural adaption of the tissue, they did not relate this to the evolving haemodynamic environment. Feng et al. [39] appear to be the first to explicitly couple changes in mechanical parameters to haemodynamic stimuli to simulate evolution. However, they adopt simple linear elastic models of the arterial wall and do not explicitly model the G&R of the individual constituents [45, 49]. Consequently, the models offer little insight into the G&R of arterial tissue during the evolution of an aneurysm. Humphrey & Taylor [89] emphasised the need for a new class of *Fluid–Solid-Growth* models to study ICA evolution and propose the terminology *FSG* models. These combine computations of complex fluid–solid interactions during a cardiac cycle with detailed analyses of the solid mechanics of the vascular wall and descriptions of the kinetics of biological G&R. The approach has been applied to model the evolution of an axisymmetric ICA on a cylindrical section of an artery, linking G&R to haemodynamic stimuli [48].

Watton et al. [90] proposed the first mathematical model of aneurysm evolution and applied it to consider the evolution of abdominal aortic aneurysms (AAA). This model utilises a realistic structural model for the arterial wall [73] which is adapted to incorporate mesoscopic variables which relate to the concentration of the elastinous and collagenous constituents and the reference configuration in which the collagenous constituents are recruited to load bearing. The key hypotheses of the model are: (1) collagen fibres are in a continual state of deposition and degradation; (2) fibres attach to the ECM in a state of stretch, denoted the *attachment stretch*, λ_{AT}^{C}. The implication of these assumptions is that the collagen fabric is effectively continually remodelling to maintain its stretch to a homeostatic value, i.e., λ_{AT}^{C}, in the physiological (loaded) configuration. To simulate this behaviour, differential equations are employed to evolve the recruitment configurations (reference configurations that collagen fibres are recruited to load bearing) and concentration of collagen fibres to maintain the stretch of the collagen (in the systolic configuration) towards equilibrium levels as the aneurysm enlarges. This approach enables the G&R of the arterial microstructure during aneurysm development to be simulated. An advantage of this approach to that of, for example, [41, 43] is that it is not computationally expensive, i.e., the number of reference configurations to keep track of is fixed and does not increase with the number of computational time-steps of a simulation. It captures the gross effect of fibre deposition and degradation in altered configurations, i.e., evolving the configuration in which the collagen fabric begins to bear load, with lower theoretical and computational expense. It is relatively straightforward to implement into finite element formulations and could easily be applied to more complex geometries, for example, modelling evolution of patient-specific aneurysm geometries, or extended to a thick-walled G&R formulation [55, 91]. The model predicts evolution of AAA mechanical parameters consistent with clinical observations [92]. The G&R framework has subsequently been applied to consider conceptual 1D models of ICA evolution, i.e., enlarging (and stabilising) cylindrical and spherical membranes [52] and the evolution of saccular ICAs of the internal carotid artery [53]. The model has been integrated into a novel FSG framework to couple the G&R of aneurysmal tissue to local haemodynamic stimuli, for example, WSS and spatial WSS gradients [54]. Assuming that degradation of elastinous constistuents is driven by low WSS, the FSG model can simulate the evolution of sidewall saccular aneurysms of the internal carotid artery that stabilise in size [56]. However, these FSG models [54, 56] idealise the geometry of the parent artery to be a cylindrical tube. Consequently, the evolving haemodynamics within the aneurysmal region are, in general, not physiologically realistic. To overcome this limitation, we have recently extended our FSG modelling framework to integrate the structural model of the evolving aneurysm within a realistic physiological geometry (see Ref. [93] for methodology). In the following two sections, we summarise our computational framework for modelling ICA evolution (Sect. 3) and illustrate the evolution of an ICA on physiological vasculature in Sect. 4

3 Patient-Specific Fluid–Solid-Growth Model for Aneurysm Evolution

In this section, we describe our FSG computational framework for modelling ICA evolution. Figure 1 depicts the methodology. The computational modelling cycle begins with a structural analysis of the aneurysm to solve the equilibrium deformation fields for given pressure and boundary conditions. The structural analysis quantifies the stress and stretch, and the cyclic deformation, of the ECM components and the cells (each of which may have different natural reference configurations). The (systolic) geometry of the aneurysm is exported to be prepared for haemodynamic analysis: first, the aneurysm geometry is integrated into a

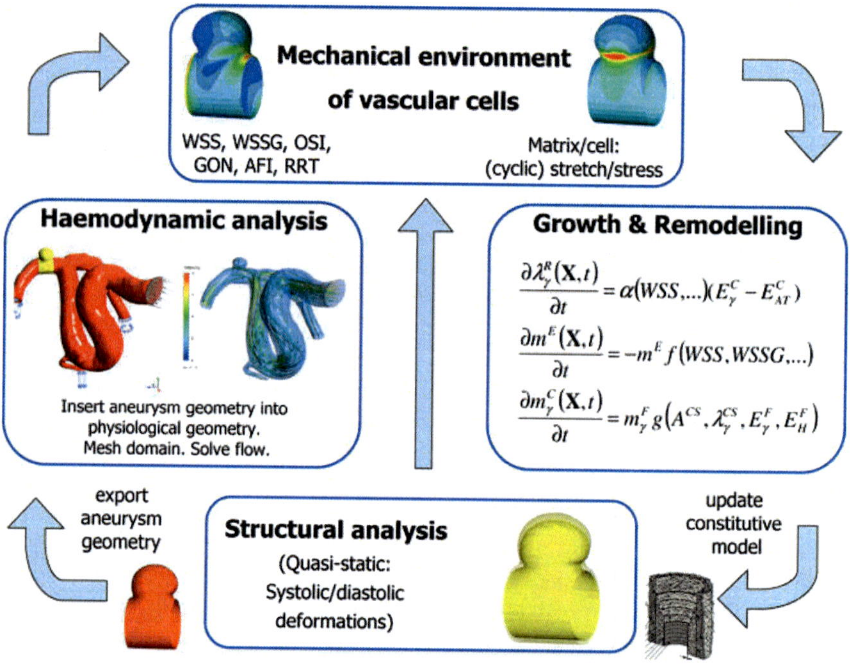

Fig. 1 Fluid–Solid-Growth computational framework for modelling aneurysm evolution

physiological geometrical domain, which is then automatically meshed; physiological flow rate and pressure boundary conditions are applied. The flow is solved assuming rigid boundaries for the haemodynamic domain. The haemodynamic quantities of interest, for example, WSS, WSSG are then exported and interpolated onto the nodes of the structural mesh: each node of the structural mesh contains information regarding the mechanical stimuli obtained from the haemodynamic and structural analyses. G&R algorithms simulate cells responding to the mechancial stimuli and adapting the tissue: the constitutive model of the aneurysmal tissue is updated. The structural analysis is executed to calculate the new equilbrium deformation fields. The updated geometry is exported for haemodynamic analysis. The cycle continues and as the tissue adapts an ICA evolves. The stages of the FSG framework, i.e., the structural modelling, haemodynamics modelling and the G&R methodology are detailed in the subsequent subsections, i.e., Sects. 3.1, 3.2 and 3.3, respectively.

3.1 Structural Model of Aneurysm Evolution

A geometric nonlinear membrane theory [94] is adopted to model the steady deformation of the arterial wall. The unloaded internal carotid artery is treated as a thin cylinder of undeformed radius R, length L_1, and thickness H. The thickness of the media H_M is assumed to be equal to 2/3 the thickness of the arterial wall, i.e., $H_M = H/3$, and thus the thickness of the adventitia $H_A = H/3$. The artery is subject to a physiological axial pre-stretch λ_z and a constant systolic pressure p equal to 16 kPa which causes a circumferential stretch of λ_0. A body fitted coordinate system is used to describe the cylindrical membrane with axial and azimuthal Lagrangian coordinates $\theta_1 \in [0, L_1]$ and $\theta_2 \in [0, 2\pi R]$, respectively. Formation and development of the aneurysm is assumed to be a consequence of G&R of the material constituents of the artery; the distal and proximal ends are considered fixed as the aneurysm evolves. The principle of stationary potential energy is the governing equation for the steady deformation of the arterial wall. It requires that the first variation of the total potential energy vanishes,

$$\delta\Pi_{\text{int}} - \delta\Pi_{\text{ext}} = 0, \tag{1}$$

where $\delta\Pi_{\text{int}}$ represents the variation of the internal potential energy Π_{int} stored in the arterial wall, whilst $\delta\Pi_{\text{ext}}$ is the variation of the external potential energy Π_{ext} caused by the normal pressure that acts on the artery. Appropriate functional forms for the spatially and temporally heterogeneous strain-energy functions (SEFs) for the media Ψ_M, and the adventitia Ψ_A must be specified so that $\delta\Pi_{\text{int}}$ can be computed. Details of the theoretical formulation to describe the deformation of the arterial wall and the numerical formulation to solve Eq. 1 can be found in [90]: the equilibrium displacement field is solved by a finite element method coded in FORTRAN 77.

3.1.1 Strain-Energy Functions for Heterogeneous Aneurysmal Tissue

The arterial wall is modelled as two layers. The inner layer models the mechanical response of the media (and intima), with contributions from the elastinous constituents (ground substance, elastin fibres and passive smooth muscle cells) and a double helical pitch of collagen fibres with orientations γ_{M_p} to the azimuthal axis: $p = +, p = -$ denote positively ($\gamma_{M_+} > 0$) and negatively ($\gamma_{M_-} < 0$) wound fibres, respectively.

The outer layer models the mechanical response of the adventitia, which is considered to have a small elastinous contribution and a double helical pitch of collagen fibres with orientations γ_{A_p} ($p = \pm$) to the azimuthal axis. The mechanical response of each layer is modelled as the sum of a neo-Hookean strain energy function (SEF) [95] and a highly nonlinear SEF which represents the mechanical response of the collagen [73]. Spatially and temporally dependent functions are introduced for the concentration of the elastinous and collagenous constituents and the configuration in which the collagen fibres begin to be recruited to load bearing.

Recruitment stretch variables define the factor the tissue must be stretched relative to the unloaded configuration in the direction of a collagen fibre for the fibre to begin to bear load. Remodelling the recruitment stretches enables the remodelling of the collagenous fabric during aneurysm development to be simulated. Constituent concentration variables define the ratio of the mass density of a constituent at time t to the mass density at time $t = 0$ and enable the growth/atrophy of the constituent to be simulated. For further details of the theoretical formulation to describe the G&R the interested reader is referred to [52].

The SEFs for the elastinous contributions in the medial and adventitia are multiplied by a normalised spatially and temporally dependent concentration function, denoted $m^E(\theta_1, \theta_2, t)$. This is employed to prescribe the degradation of the elastinous constituents, where $m^E(\theta_1, \theta_2, t = 0) = 1$. Fields of spatially and temporally dependent fibre recruitment stretch $\lambda^R_{J_p}(\theta_1, \theta_2, t)$ and concentration $m^C_{J_p}(\theta_1, \theta_2, t)$ variables are defined throughout the midplane of the arterial wall, where the subindex J denotes the media M or the adventitia A. The fibres within each layer are orientated at an angle of γ_{J_p} to the azimuthal axis, where p denotes the pitch $\pm\gamma_J$ relative to the azimuthal axis in the unloaded reference configuration. Hence, the SEFs are

$$\begin{aligned}\Psi_J &= m^E K^E_J (E_{11} + E_{22} + E_{33}) \\ &+ \sum_{p=\pm,\, E^C_{J_p} > 0} m^C_{J_p} K^C_J \left\{ \exp\left[A^C (E^C_{J_p})^2 \right] - 1 \right\}, \quad J = M, A,\end{aligned} \tag{2}$$

where the material parameters for the elastinous constituent are denoted by K^E_J, whilst K^C_J and A^C are parameters that relate to the collagen fabric. The Green-Lagrange (GL) strains of the elastin, i.e., E_{11}, E_{22} and E_{33}, are defined relative to the unloaded configuration; in the initial cylindrical configuration, these represent strains in the axial, azimuthal and radial directions, respectively. The GL strains in the collagen fibres are denoted by $E^C_{J_p}(\theta_1, \theta_2, t)$ and are defined relative to the configuration in which the collagen fibres are recruited to load bearing. More specifically, the GL strains $E^C_{J_p}$ of the collagen fibres are a function of the GL strains of the elastin resolved in the directions of the collagen fibres, denoted E_{J_p}. Thus,

$$E^C_{J_p} = \frac{E_{J_p} - E^R_{J_p}}{1 + 2E^R_{J_p}}, \tag{3}$$

where $E^R_{J_p} = [(\lambda^R_{J_p})^2 - 1]/2, \lambda^R_{J_p}(\theta_1, \theta_2, t)$ are the recruitment stretches and $E_{J_p} = E_{11} \sin^2 \gamma_{J_p} + E_{22} \cos^2 \gamma_{J_p} + 2E_{12} \sin \gamma_{J_p} \cos \gamma_{J_p}$.

3.1.2 Geometry, Physiological Data and Material Parameters

The material parameters for the media and the adventitia as well as all other values which serve the basis for our subsequent computation are summarized in Table 1.

Table 1 Geometry, physiological data used for modelling the human internal carotid artery

Radii		
At systole	r_S	1.26 mm
Reference configuration	R	r_S/λ_0
Wall thickness		
Total	H	$R/5$
Media	H_M	$2H/3$
Adventitia	H_A	$H/3$
Fibre orientation		
Media	$\gamma_{M_+}, \gamma_{M_-}$	$+30°, -30°$
Adventitia	$\gamma_{A_+}, \gamma_{A_-}$	$+60°, -60°$
Applied pressure, kinematics		
Systolic pressure	p	16 kPa
Axial pre-stretch	λ_z	1.3
Circumferential stretch at $t = 0$	λ_0	1.25
Ratio of systolic to diastolic diameters at $t = 0$	λ_D^S	1.1
Attachment stretch	λ_{AT}^C	1.07
Recruitment stretches ($t = 0$)		
Media	$\lambda_{M_+}^R, \lambda_{M_-}^R$	1.18
Adventitia	$\lambda_{A_+}^R, \lambda_{A_-}^R$	1.20

In the systolic configuration, the artery has a radius r_S, an axial pre-stretch of 1.3, a circumferential stretch of 1.25 and we assume that the elastinous constituents bear 80% of the load. We assume that the elastinous response of the adventitia is an order of magnitude weaker than that of the media, i.e., $K_A^E = K_M^E/10$ [73] and specify the ratio of the medial and adventitial collagen material parameters to be $K_A^C = K_M^C/4$. The remaining three independent material parameters, namely K_M^E, K_M^C, A^C are determined so that the SEFs adequately model the mechanical behaviour of the artery [92].

The magnitude of the attachment stretch λ_{AT}^C is defined to be the stretch in the collagen fabric at systole at $t = 0$. For the healthy artery, the collagen fabric does not bear significant load until the upper end of physiological pressures. To determine a suitable value for λ_{AT}^C we assume that the medial collagen fibres (with predominant azimuthal orientation) are recruited to load bearing at diastole, i.e., when the circumferential stretch is λ_0/λ_D^S, where λ_D^S denotes the ratio of the systolic to diastolic diameters. This implies initial values for the medial recruitment stretches $(\lambda_{M_+}^R, \lambda_{M_-}^R)$ of

$$\lambda_{M_+}^R|_{t=0} = \sqrt{(\lambda_0/\lambda_D^S)^2 \cos^2 \gamma_{M_+} + \lambda_z^2 \sin^2 \gamma_{M_+}}$$
$$= \sqrt{(\lambda_0/\lambda_D^S)^2 \cos^2 \gamma_{M_-} + \lambda_z^2 \sin^2 \gamma_{M_-}} = \lambda_{M_-}^R|_{t=0}. \quad (4)$$

Consequently, the attachment stretch λ_{AT}^{C}, which is defined to be equivalent to the stretch of the medial collagen fibres at systole at $t = 0$, i.e., $\lambda_{M_+}^{C}|_{t=0}(= \lambda_{M_-}^{C}|_{t=0})$, can be deduced as

$$\lambda_{AT}^{C} \equiv \lambda_{M_+}^{C}|_{t=0} = \sqrt{\lambda_0^2 \cos^2 \gamma_{M_+} + \lambda_z^2 \sin^2 \gamma_{M_+}} / \lambda_{M_+}^{R}|_{t=0} = \lambda_{M_-}^{C}|_{t=0}. \quad (5)$$

Finally, the initial values of the adventitial recruitment stretches are determined so that the stretch of the adventitial collagen fibres at systole at $t = 0$ equals the attachment stretch, i.e.,

$$\lambda_{A_+}^{R}|_{t=0} = \sqrt{\lambda_0^2 \cos^2 \gamma_{A_+} + \lambda_z^2 \sin^2 \gamma_{A_+}} / \lambda_{AT}^{C} = \lambda_{A_-}^{R}|_{t=0}. \quad (6)$$

3.2 Haemodynamic Modelling

To achieve physiologically realistic flow in the region where the aneurysm develops, extensions are attached to the structural model of the artery/aneurysm using vasculature geometry obtained from clinical imaging data. MRI data is automatically segmented using algorithms [96–98] integrated into @neufuse software; a software suite, developed as part of the @neurIST EU project, which integrates medical imaging, biomedical measurements and simulation. A small aneurysm developing on the right internal carotid artery (see Fig. 2a) is identified. The geometry is subsequently manipulated with ANSYS ICEM: the aneurysm (located on the right internal carotid sinus artery) is removed and replaced by a short cylindrical section which is reconnected to the upstream and downstream arterial sections so that the surface gradients are continuous [93]. It is within this inserted section, hereon referred to as *aneurysmal section*, that the formation of an ICA is simulated. This approach enables the structural model of Sect. 3.1 to be integrated within a realistic physiological geometry (Fig. 2b) with physiological boundary conditions so that the G&R of the aneurysmal section can be explicitly linked to physiologically realistic haemodynamic stimuli.

The methodological approach to solve the haemodynamics as the aneurysm evolves proceeds as follows. The geometry of the aneurysmal section is exported (see Fig. 1) from the structural solver to the meshing suite ANSYS ICEM (ANSYS Inc, Canonsburg, PA). ANSYS ICEM automatically integrates the aneurysmal section into the physiological domain and generates an unstructured tetrahedral mesh with prism layers lining the boundary in a scripted-automated manner for the fluid domain. After meshing, appropriate boundary conditions are applied (see Fig. 2b) and the flow is solved by ANSYS CFX (ANSYS Inc, Canonsburg, PA) which solves the incompressible Navier-Stokes equations using a finite volume formulation [99, 100]. The solver is based on a coupled approach (i.e., velocities and pressure are cast and solved as a single system) and a fully

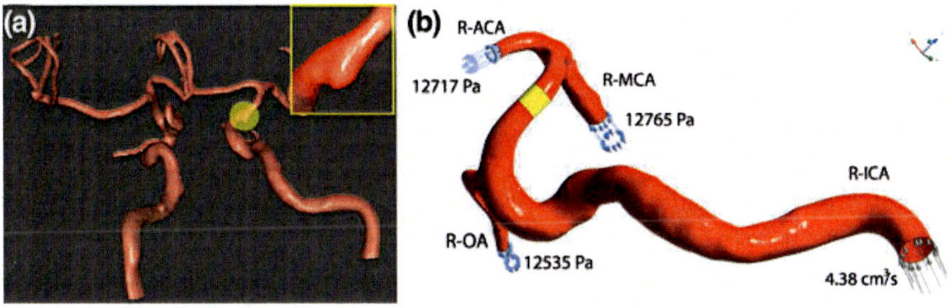

Fig. 2 **a** Clinical MRI data is automatically segmented using @neufuse software. A small sidewall saccular aneurysm developing on the right internal carotid artery (R-ICA) is identified. **b** The original aneurysm has been removed and replaced with a short cylindrical section (aneurysmal section) which is then reconnected to the vasculature. It is on this section that aneurysm evolution is simulated. The flow rate at the inlet (R-ICA) is prescribed and pressure boundary condtions are prescribed at the outlets, i.e., the Right opthalmic artery (R-OA), the Right middle cerebral artery (R-MCA), and Right anterior cerebral artery (R-ACA)

implicit time discretisation, where needed. An algebraic multigrid variant is used for convergence acceleration [101]. Blood is modelled as a Newtonian fluid [102, 103] with constant density $\rho = 1069\,\mathrm{kgm}^{-3}$ [3, 104] and constant viscosity $\eta = 0.0035$ Pa s. At the arterial wall a no slip condition is applied for a rigid wall. For the FSG model presented in this paper we adopted a steady flow analysis to reduce the cost of the computational simulations. The similarity of the spatial WSS/WSSG distributions for steady and pulsatile flow simulations [56, 105, 106] suggests that adopting steady flow may be a reasonable approach for the purposes of investigating phenomenological hypotheses that explore the link between G&R of aneurysmal tissue and deviations of the WSS/WSSG from normotensive values. However, note that a pulsatile flow analysis is necessary to explore more sophisticated G&R hypotheses given that pulsatile flow yields additional quantification of the mechanical stimuli that act on the endothelial cells.

3.3 Growth and Remodelling

Models of aneurysm evolution must predict how the main load bearing constituents, i.e., elastin and collagen, evolve. Growth relates to changes in the mass of the constituents whereas remodelling relates to changes in the structural organisation, for example, natural reference configuration or orientation of fibres, with no net change in mass. An ICA typically appears as a thin collagenous structure, whilst the parent artery still has a well-defined structure with an intact medial layer: there is a relatively abrupt transition from healthy to aneurysmal tissue in the neck of the aneurysm. Hence, models of ICA evolution must model the degradation of elastin and the G&R of the collagen fabric.

3.3.1 Elastin Degradation

The inception of the aneurysm is prescribed to create a small outpouching of the arterial domain. Details of the methodology to achieve this can be found in [54]. Briefly, a localised degradation of the elastinous constituents is prescribed whilst simultaneously the collagen fabric adapts to compensate for the loss of load bearing by the elastin. The artery achieves a new homeostasis, i.e., the stretch of the collagen fabric is restored to λ_{AT}^{C} throughout the domain. The loss of elastin and adaption of the collagen yields a small localised outpouching of the artery. This alters the spatial distribution of the WSS. Subsequent degradation of elastin is then linked to deviations of the WSS from homeostatic values.

We suppose that there is a maximum rate of degradation, say D_{Max}, that the elastinous constituents can degrade per year. The concentration of elastin m^E evolves according to

$$\frac{\partial m^E}{\partial t} = -\mathscr{F}_D D_{\text{Max}} m^E \tag{7}$$

where t is in years and $\mathscr{F}_D(\theta_1, \theta_2, t) : 0 \leq \mathscr{F}_D \leq 1$ is a spatially-dependent function of the haemodynamic quantities to be linked to elastin degradation. Clearly, if $\mathscr{F}_D = 1$, elastin degrades at a maximum rate whilst if $\mathscr{F}_D = 0$ no degradation occurs.

As an illustrative example, we suppose elastin degradation is linked to low levels of WSS. More specifically, we assume that if the maximum value of the WSS is greater than a critical value, say τ_{Crit}, no degradation of elastin occurs whilst lower values of WSS give rise to degradation of elastin. Furthermore we suppose that there exists a value of WSS, say $\tau_X : 0 \leq \tau_X < \tau_{\text{Crit}}$, at which maximum degradation occurs. We assume a simple quadratic functional form for the degradation function $\mathscr{F}_D$ that describes the relation between the local WSS and the degree of the degradation of elastin, i.e.,

$$\mathscr{F}_D(\tau(\theta_1, \theta_2, t)) = \begin{cases} 0, & \tau \geq \tau_{\text{Crit}}, \\ \left(\frac{\tau_{\text{Crit}} - \tau}{\tau_{\text{Crit}} - \tau_X}\right)^2, & \tau_X < \tau < \tau_{\text{Crit}}, \\ 1, & \tau \leq \tau_X. \end{cases} \tag{8}$$

In this study, $D_{\text{Max}} = 0.75$ and the elastin degradation is confined to the specific region of aneurysm inception (see Ref. [56] for methodology) otherwise the evolving aneurysms propagate along the arterial domain [54]. The magnitudes of the parameters τ_{Crit} and τ_X are determined by inspecting the initial WSS distribution following development of the small outpouching: τ_{Crit} must exceed the minimum level of WSS in the region of inception for elastin degradation to occur. For this study, $\tau_{\text{Crit}} = 15$ Pa and $\tau_X = 0.5$ Pa.

3.3.2 Collagen Adaption

The adaptive response of the collagen fabric consists of growth and remodelling. Remodelling the reference configurations of the fibres simulates the effect of fibre deposition and degradation in altered configurations. The growth/atrophy of the collagen fabric is simulated by evolving the fibre concentration. Increases (decreases) of the fibre concentration simulates fibroblasts up-regulating (down-regulating) collagen synthesis and down-regulating (up-regulating) enzymes that degrade the collagenous matrix.

3.3.3 Remodelling the Recruitment Configuration of Collagen

The turnover of collagen and the consequence of fibres attaching in a fixed state of stretch λ_{AT}^{C} is simulated by proposing evolution equations that act to maintain the GL strain in the collagen fibres to an equilibrium value E_{AT}^{C} (equivalently remodelling the stretch of the collagen fibres to λ_{AT}^{C}). This is achieved by remodelling the reference configuration of the collagen fibres throughout the arterial domain, i.e., remodelling the recruitment stretches (see Fig. 2 in [54] for physiological interpretation of remodelling of the recruitment stretch). Linear differential equations are proposed for the remodelling of the recruitment stretches, i.e.,

$$\frac{\mathrm{d}\lambda_{\mathrm{J_p}}^{\mathrm{R}}}{\mathrm{d}t} = \alpha \frac{E_{J_p}^{C} - E_{AT}^{C}}{E_{AT}^{C}}, \tag{9}$$

with $E_{AT}^{C} = [(\lambda_{AT}^{C})^2 - 1]/2$ where $\alpha = 0.6\,\mathrm{years}^{-1}$.

3.3.4 Growth/Atrophy of the Collagen Fabric

Fibroblasts adhere to the ECM via specialised cell surface receptors, in particular integrins [107]. The integrins physically link the ECM to the cytoskeleton of the fibroblast. They transduce mechanical signals to the fibroblast interior [108]. Evidence suggests that the integrins act as stretch sensors [109] and enable the fibroblasts to sense changes in their mechanical environment. Fibroblasts deposit collagen fibres and secrete proteases to degrade the collagenous material. We assume that the rate of change of the concentration of the collagenous constituents is dependent on the concentration of fibroblasts $m_{J_p}^{F}$ (ratio of the density of the fibroblast cells at time t to the density at time $t = 0$) in the arterial wall and rate of synthesis and degradation of collagen. Let $\mathscr{F}_S$ and $\mathscr{F}_D$ be functions depicting how the rate of collagen synthesis and the secretion of matrix metalloproteases are related to the deformation of the fibroblast cell during a cardiac cycle, respectively. Furthermore, to reduce theoretical complexity, restrict $\mathscr{F}_S$ and $\mathscr{F}_D$ to be functions of the maximum strains of the cells $E_{J_p}^{F}$ and suitable indices characterising the magnitude of cyclic deformation experience by the cells, for example, magnitude of cyclic stretch $\lambda_{J_p}^{F-CS}$, or magnitude of the cyclic areal stretch A^{F-CS} (areal stretch at systole divided by areal stretch at diastole). Under these assumptions, the rate of change of the collagen concentration is

$$\frac{\mathrm{d}m_{J_p}^C}{\mathrm{d}t} = m_{J_p}^F \mathscr{F}_S(E_{J_p}^F, \lambda_{J_p}^{F-CS}, A^{F-CS}) - m_{J_p}^F \mathscr{F}_D(E_{J_p}^F, \lambda_{J_p}^{F-CS}, A^{F-CS}). \tag{10}$$

In response to increased stretch, fibroblasts attempt to reduce their stretch and reach a new equilibrium by restructuring their cytoskeleton and ECM contacts, i.e., fibroblasts reconfigure their natural reference configuration. To simplify the mathematical analysis, we assume that the local (natural) reference configuration of a fibroblast cell is identical to that of the collagen fabric it is maintaining. Hence the GL strain of the fibroblast cell, say $E_{J_p}^F$, is assumed to be equal to the GL strain of the collagen, i.e., $E_{J_p}^F \equiv E_{J_p}^C$. Furthermore, we assume that the concentration of fibroblasts in the arterial tissue is proportional to the concentration of collagenous constituents, i.e., $m_{J_p}^F = \xi_0 m_{J_p}^C : \xi_0 > 0$. These assumptions imply that the rate of evolution of the collagen fibre concentration $m_{J_p}^C$ can be expressed in terms of the current fibre concentration, i.e.,

$$\begin{aligned} \frac{\mathrm{d}m_{J_p}^C}{\mathrm{d}t} &= \xi_0 m_{J_p}^C \left[\mathscr{F}_S(E_{J_p}^C, \lambda_{J_p}^{F-CS}, A^{F-CS}) - \mathscr{F}_D(E_{J_p}^C, \lambda_{J_p}^{F-CS}, A^{F-CS})\right] \\ &= m_{J_p}^C \mathscr{F}_G(E_{J_p}^C, \lambda_{J_p}^{F-CS}, A^{F-CS}) \end{aligned} \tag{11}$$

where $\mathscr{F}_G = \xi_0(\mathscr{F}_S - \mathscr{F}_D)$.

If $E_{J_p}^C = E_{AT}^C$ the collagen fabric is in homeostasis, i.e., the secretion of ECM is balanced by the degradation and there is no change in concentration. Hence, for $E_{J_p}^C = E_{AT}^C$ it is required that $\mathscr{F}_G(E_{J_p}^C) = 0$. The exact functional form of $\mathscr{F}_G$ is unknown. However, if the substrate is stretched, a net positive force acts on the cell and signalling to the nucleus results in an up-regulation of ECM protein expression and a down-regulation of collagenase expression. Conversely, relaxation of the substrate can trigger different signals resulting in a reversed pattern of protein expression [109], i.e., down regulation of ECM protein expression and up-regulation of collagenase expression. Although we do not explicitly model protein synthesis and enzymatic degradation, the net result is to stimulate increases/decreases in the ECM. The simplest functional form for $\mathscr{F}_G$ that satisfies these requirements is linear, and hence we propose the following evolution equation for the collagen fibre concentration

$$\frac{\mathrm{d}m_{J_p}^C}{\mathrm{d}} t = \beta m_{J_p}^C \frac{E_{J_p}^C - E_{AT}^C}{E_{AT}^C}, \tag{12}$$

where β is a phenomenological growth parameter that relates to the rate at which the fibroblasts increase or decrease the mass of the collagenous constituents in response to deviations of the magnitude of cyclic deformation from normotensive levels. Rates of activity of vascular cells increase with the magnitude of the cyclic deformation [110, 111]. For a healthy artery, vascular cells experience cyclic stretching with magnitudes of 10% whilst for older, more collagenous vessels, the magnitudes may decrease to 2%. We select a functional form for β which increases rates of production if the cyclic stretch enviroment increases from its initial values for the healthy artery at $t = 0$. We propose a function of the form suggested in [56], i.e.,

$$\beta = \beta_0 \exp\left[\beta^{CS} \max\left(\frac{A^{F-CS}}{A^{F-CS}\,|_{t=0}} - 1, 0\right)\right] \exp\left[\beta^{CS}_{J_p} \max\left(\frac{\lambda^{F-CS}_{J_p}}{\lambda^{F-CS}_{J_p}\,|_{t=0}} - 1, 0\right)\right]. \tag{13}$$

This choice is somewhat arbitrary, however it acts to prevent unrealistically large cyclic deformations in regions of the tissue as the aneurysm evolves. For the analysis in this paper, $\beta_0 = 0.1\,\text{years}^{-1}, \beta^{CS} = 10$ and $\beta^{CS}_{J_p} = 0$.

To simplify the presentation of the results, we define the average fibre concentration m^C of the medial and adventitial layers, where

$$m^C = \frac{1}{H}\left(H_M \frac{m^C_{M_+} + m^C_{M_-}}{2} + H_A \frac{m^C_{A_+} + m^C_{A_-}}{2}\right). \tag{14}$$

4 Physiological Growth Model: Sidewall Saccular Aneurysm of the Internal Carotid Sinus Artery

We briefly review the framework of the model prior to presenting the results. Equation 2 defines the (spatially and temporally heterogeneous) SEFs for the medial and adventitial layers of the aneurysmal tissue. An ICA develops as the material constituents of the artery evolve. The variation of the total potential energy (1) governs the equilibrium displacement field, and is solved by the finite element method [90]; volume meshes of the computational domain are generated and the haemodynamics are solved automatically. The evolving geometry alters the haemodynamics and the spatial distribution of the WSS and the WSSGs that act on the endothelial layer of the artery. Initially, the degradation of elastin is prescribed to create a small outpouching of the computational domain to perturb the haemodynamic environment [54, 56]. Subsequent elastin degradation is then linked to low WSS (see (7) and (8)) in this localised region of the computational domain, whilst the collagen fabric adapts (throughout the arterial domain) to restore its strain to the homeostatic value (see (9) and (12)).

Figure 3 illustrates the WSS, WSSG and pressure distributions on the arterial domain following the development of a sidewall saccular aneurysm ($t = 9$ years).

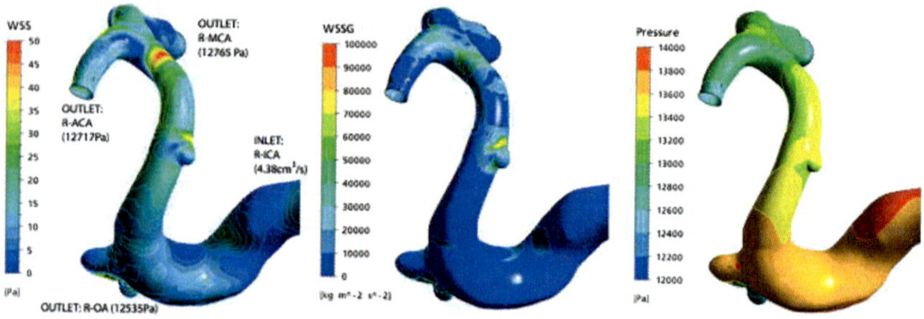

Fig. 3 WSS, WSS spatial gradients and pressure distributions following the evolvement of a model of an ICA on patient-specific vasculature. The flow rate at the inlet, the right internal carotid artery (R-ICA), is prescribed and pressure boundary condtions are prescribed at the outlets, i.e., the Right opthalmic artery (R-OA), the Right middle cerebral artery (R-MCA), and Right anterior cerebral artery (R-ACA). Patient-specific inlets and outlet geometries yield physiologically realistic haemodynamics within the aneurysmal region. Consequently, the haemodynamic stimuli that are incorporated into the G&R algorithms have realistic magnitudes and spatial distributions

Elevated regions of WSS/WSSG are observed distal to the aneurysm and close to the bifurcation. The variation in pressure is negligible over the aneurysm region. Figures 4–6 zoom in on the location where the aneurysm develops and depict the aneurysm evolving at $t = 0, 3, 6 \text{ and } 9$ years following the initial inception that creates a localised perturbation to the geometry. The prescribed inception of the aneurysm, i.e., localised degradation of elastin, creates a perturbation to the geometry and a local reduction in WSS to approximately 8 Pa.

As the aneurysm enlarges it can be seen that a region of elevated WSS occurs distal to the aneurysm and decreased WSS occurs within the proximal region of the dome and at the side regions of the aneurysm neck, see Fig. 4a. Figure 4b illustrates the evolution of the degradation factor $\mathscr{F}_{\mathscr{D}}$. This has a maximum initial value of 0.3. As the aneurysm enlarges, the WSS distribution changes and thus so does the distribution of $\mathscr{F}_{\mathscr{D}}$: the region it encompasses (with values >0) enlarges initially but then becomes restricted to the proximal region of the aneurysm dome. This is due to the pronounced asymmetry in the evolution of the WSS. Due to the asymmetry in the distribution $\mathscr{F}_{\mathscr{D}}$, the region of elastin degradation evolves asymmetrically (see Fig. 4c), i.e., the elastin degrades at a greater rate in the proximal region of the aneurysm. Noticeably, as the aneurysm enlarges the region of degradation is constrained to the proximal region. The asymmetry in the evolution of the elastin degradation creates a substantial asymmetry in the evolution of the aneurysm geometry: the proximal side of the dome develops a well-defined aneurysm neck whereas the distal region of the aneurysm flattens to connect with the downstream section of the artery smoothly.

Figure 5a, b illustrate the evolution of the Green-Lagrange (GL) strains of the elastin, i.e., E_{11} and E_{22}, respectively. The strains increase to the greatest extent at the upper proximal region of the dome. As the aneurysm enlarges the asymmetry

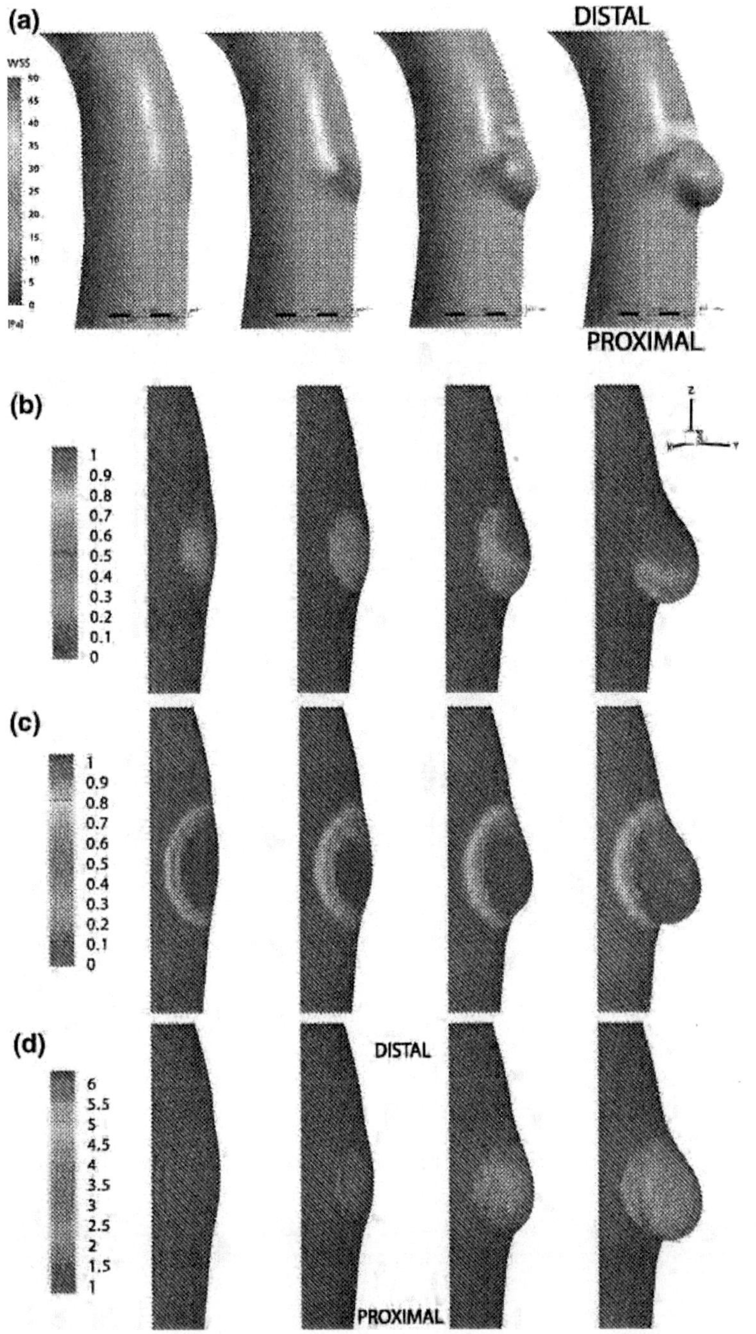

Fig. 4 The inception of the aneurysm is prescribed creating a perturbation to the WSS distribution. Subsequent degradation of elastin is explicitly linked to low levels of WSS. Evolvement of **a** the WSS, **b** the degradation factor $\mathcal{F}_{\mathcal{D}}$ **c** the elastin concentration m^E and **d** the average collagen fibre concentration m^C at $t = 0$, 3, 6 and 9 years

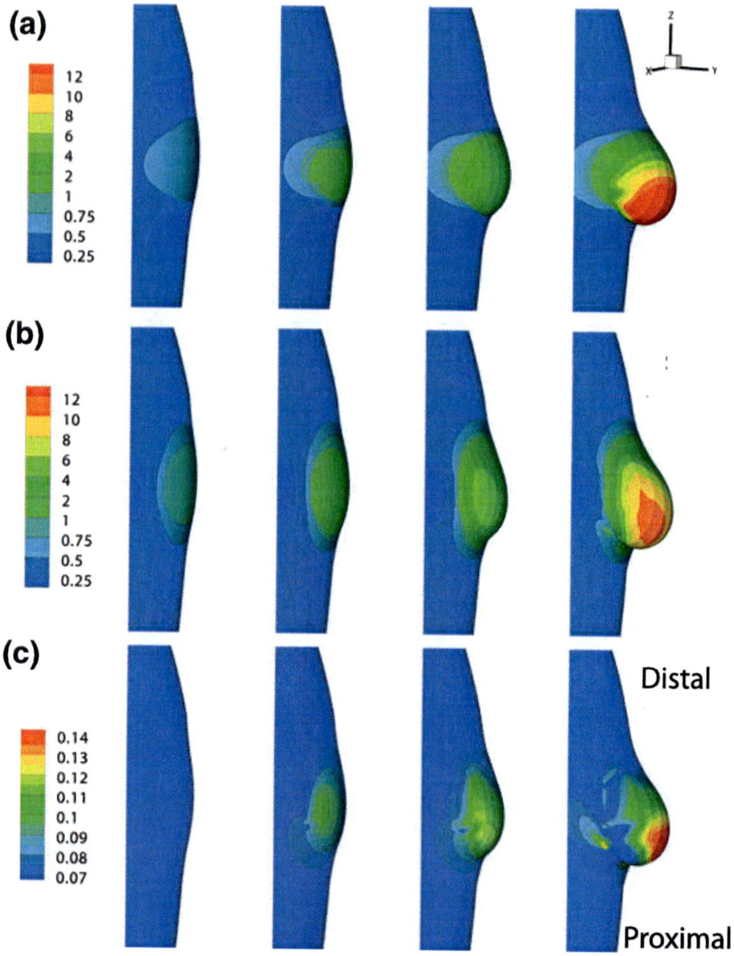

Fig. 5 Evolution of the Green-Lagrange strains of the elastin, **a** E_{11} and **b** E_{22}. **c** Evolution of the medial collagen fibre strain $E^C_{M_+}$. Note that the magnitudes are low even though there is large deformation. This is due to the evolvement of the reference configurations that the fibres are recruited to load bearing

in the strain distribution increases. In contrast to the evolution of the elastin strain, the collagen fibre GL strains increase negligibly even though there is large deformation, see Fig. 5c, i.e., maximum values of 0.14 for collagen as opposed to 12 for elastin. This is due to the evolution of the natural reference configurations that the fibres are recruited to load bearing.

Finally, Fig. 6a illustrates the evolution of the systolic and diastolic geometries. It can be seen that the evolution of the diastolic geometry follows closely the evolution of the systolic geometry. Figure 6b illustrates the evolution of the magnitude of the cyclic areal stretch (ratio of the magnitude of a differential area

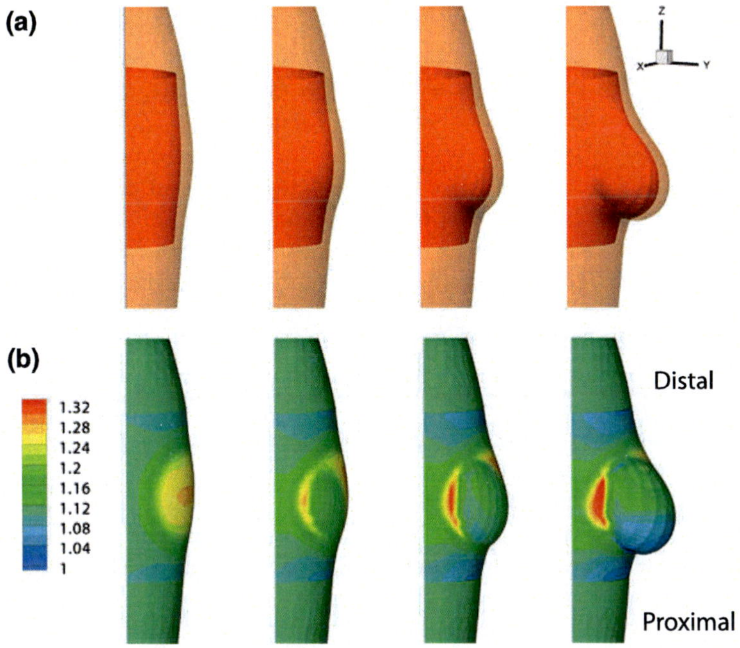

Fig. 6 **a** Evolution of the systolic and diastolic geometries. **b** Evolution of the cyclic areal stretch (ratio of the magnitude of a differential area element at systole to its magnitude at diastole)

element at systole to its magnitude at diastole). Initially, the metric increases at the apex of the outpouching where there has been loss of elastin. However, as the aneurysm enlarges, elevated changes in the cyclic area metric occur at the side neck regions of the aneurysm. The proximal region of the aneurysm has lost the most elastin and contains increased collagen—in this region the pulsation of the tissue deceases.

5 Discussion

In this chapter, we have presented a fluid–solid-growth (FSG) model for the evolution of an ICA on the internal carotid sinus artery. The model utilises a realistic constitutive model of the arterial wall that accounts for the structural arrangement of collagen fibres in the medial and adventitial layers, the configuration the collagen fibres are recruited to load bearing and the concentration of the elastinous and collagenous constituents. A quasi-static structural analysis is employed to obtain the geometry of the aneurysm at systolic and diastolic pressures, enabling G&R to be explicitly linked to the cyclic deformation of the arterial tissue. The structural model is integrated into a patient-specific geometry of the internal carotid artery. This enables G&R of the aneurysmal section to be explicitly linked to physiologically realistic haemodynamic stimuli. Degradation of elastin is linked to low magnitudes of WSS and the collagen fabric adapts to compensate for the loss of load bearing by the elastin and restores its stretch to λ_{AT}^{C}. This is the first patient-specific model of cerebral aneurysm evolution that incorporates a realistic constitutive model of the arterial wall. It will provide the basis for further investigating and elucidating the aetiology of the disease.

ICA evolution models, such as the one presented in this chapter, are proving useful exploring the G&R of the microstructure of the arterial wall during ICA development. However, they are in need of substantial further development and sophistication to represent the underlying biology more accurately and to connect with clinical findings; the latter requirement being directly related with practical predictive use in a clinical framework. It is the cells, for example, endothelial, vascular smooth muscle and fibroblasts, that maintain, remodel and repair the ECM to maintain the artery so that it functions in an optimum manner. Whilst in recent years significant advances have been made developing arterial constitutive models [73] further sophistications are needed. In particular, given that apoptosis of vascular SMCs is associated with the formation of ICAs [60] and that SMCs synthesise and secrete connective tissue, matrix degrading enzymes [112] and proteoglycans [113] the SMCs must play a critical role in the development of an ICA. Consequently, there is a need to: (1) sophisticate arterial constitutive models to explicitly represent the active and passive response of the SMCs, relating their mechanical response to the chemical environment, for example, calcium concentration [114]; (2) develop G&R frameworks to relate the SMC orientation, growth/atrophy and phenotype to their evolving mechanical environment [115]. Hypotheses linking the protein expressions of the cells to their local mechanical stimuli, for example, stress [116], cyclic stretch [117], WSS [118], WSS spatial gradients [69] need to be explored. However, whilst it is well recognised that the mechanical environment of the cells influences their functionality [109, 118, 119] the complex interplay remains poorly understood.

Obtaining data to follow the temporal progression of an individual's ICA, for example, histological analysis of the evolving structure of the tissue is not possible in vivo in humans. Consequently, there is a need for experimental models that can provide such quantification. Although the aetiology of experimental ICA models differs from human ICA, there are similarities that can be exploited. For instance, the underlying biological mechanisms that control the G&R of the collagen fabric, bare important similarities across species and therefore merit further investigation. Indeed, given that a developed ICA consists predominately of collagen [63], understanding how this structure evolves is essential for understanding the pathobiology of ICAs. Computational models provide an ideal basis to explore hypotheses for how ICAs form and evolve and have the potential to yield insight into the aetiology of the disease. The long term objective is that such models may ultimately have diagnostic application, for example, patient-specific stress analysis and prediction of future growth/rupture. In the shorter term however, they present us with excellent in silico testbeds for theory and hypothesis evaluation.

Acknowledgments Financial support for this research was partly provided by @neurIST, an Integrated EU Project (Call Identifier FP6-2004-IST-4) and The Centre of Excellence in Personalized Healthcare (funded by the Wellcome Trust and EPSRC, grant number WT 088877/Z/09/Z). This support is gratefully acknowledged. We acknowledge the Harwell Software Library (http://www.hsl.ac.uk) for granting UK academics the free use of its FORTRAN subroutines in non-commercial applications. MA38 was employed to solve the linear system that arises in the Newton iteration, which is required to update the deformation at successive time steps.

References

1. Resnick, N., Yahav, H., Shay-Salit, A., Shushy, M., Schubert, S., Zilberman, L.C.M., Wofovitz, E.: Fluid shear stress and the vascular endothelium: for better and for worse. Prog. Biophys. **81**, 177–199 (2003)
2. Chatziprodromou, I., Poulikakos, D., Ventikos, Y.: On the influence of variation in haemodynamic conditions on the generation and growth of cerebral aneurysm and atherogenesis: a computational model. J. Biomech. **40**, 3626–3640 (2007)
3. Chatziprodromou, I., Tricoli, A., Poulikakos, D., Ventikos, Y.: Haemodynamics and wall remodelling of a growing cerebral aneurysm: A computational model. J. Biomech. **40**, 412–426 (2007)
4. Tateshima, S., Tanishita, K., Omura, H., Villablanca, J.P., Vinuela, F.: Intra-aneurysmal hemodynamics during the growth of an unruptured aneurysm: In vitro study using longitudinal CT angiogram database. Am. J. Neuroradiol. **28**, 622–627 (2007)
5. Krex, D., Schackert, H.K., Schackert, G.: Genesis of cerebral aneurysms—an update. Acta Neurochirurgica (Wien) **143**, 429–448 (2001)
6. Juvela, S.: Treatment options of unruptured intracranial aneurysms. Stroke **35**, 372–374 (2004)
7. Brisman, J.L., Song, J.K., Newell, D.W.: Cerebral aneurysms. New Eng. J. Med. **355**, 928–939 (2006)
8. Peters, D.G., Kassam, A.B., Feingold, E., Heidrich-O'Hare, E., Yonas H., Ferrell, R.E., Brufsky, A.: Molecular anatomy of an intracranial aneurysm coordinated expression of genes involved in wound healing and tissue remodeling. Stroke **32**, 1036–1042 (2001)
9. Humphrey, J.D., Kyriacou, S.K.: The use of laplace's equation in aneurysm mechanics. Neurol. Res. **18**, 204–208 (1996)
10. Kyriacou, S.K., Humphrey, J.D.: Influence of size, shape and properties on the mechanics of axisymmetric saccular aneurysms. J. Biomech. **29**, 1015–1022 (1996)
11. Ryan, J.M., Humphrey, J.D.: Finite element based predictions of preferred material symmetries in saccular aneurysms. Ann. Biomed. Eng. **27**, 641–647 (1999)
12. Shah, A.D., Humphrey, J.D.: Finite strain elastodynamics of intracranial saccular aneurysms. J. Biomech. **32**, 593–599 (1999)
13. Seshaiyer, P., Humphrey, J.D.: On the potentially protective role of contact constraints on saccular aneurysms. J. Biomech. **34**, 607–612 (2001)
14. David, G., Humphrey, J.D.: Further evidence for the dynamic stability of intracranial saccular aneurysms. J. Biomech. **36**, 1143–1150 (2003)
15. Daniel, J.C., Tongen, A., Warne, D.A., Warne, P.G.: A 3D nonlinear anisotropic spherical inflation model for intracranial saccular aneurysm elastodynamics. Math. Mech. Solid. **15**, 279–307 (2010)
16. Figueroa, C.A., Vignon-Clementel, I.E., Jansen, K.E., Hughes, T.J.R., Taylor, C.A.: A coupled momentum method for modeling blood flow in three-dimensional deformable arteries. Comput. Meth. Appl. Mech. Eng. **195**, 5685–5706 (2006)

17. Hoi, Y., Woodward, S.H., Kim, M., Taulbee, D.B., Meng, H.: Validation of CFD simulations of cerebral aneurysms with implication of geometric variations. ASME J. Biomech. Eng. **128**, 844–851 (2006)
18. Valencia, A.A., Guzmn, A.M., Finol, E.A., Amon, C.H.: Blood flow dynamics in saccular aneurysm models of the basilar artery. ASME J. Biomech. Eng. **128**, 516–526 (2006)
19. Liou, T.M., Li, Y.C., Juan, W.C.: Numerical and experimental studies on pulsatile flow in aneurysms arising laterally from a curved parent vessel at various angles. J. Biomech. **40**, 1268–1275 (2007)
20. Utter, B., Rossmann, J.S.: Numerical simulation of saccular aneurysm hemodynamics: Influence of morphology on rupture risk. J. Biomech. **40**, 2716–2722 (2007)
21. Ford, M.D., Lee, S.W., Lownie, S.P., Holdsworth, D.W., Steinman, D.A.: On the effect of parent aneurysm angle on flow patterns in basilar tip aneurysms: Towards a surrogate geometric marker of intra-aneurismal hemodynamics. J. Biomech. **41**, 241–248 (2008)
22. Ford, M.D., Nikolov, H.N., Milner, J.S., Lownie, S.P., DeMont, E.M., Kalata, W., Loth, F., Holdsworth, D.W., Steinman, D.A.: PIV-measured versus CFD-predicted flow dynamics in anatomically realistic cerebral aneurysm models. J. Biomech. Eng. **130**, 021015 (2008)
23. Rayz, V.L., Boussel, L., Acevedo-Bolton, G., Martin, A.J., Young, W.L., Lawton, M.T., Higashida, R., Saloner, D.: Numerical simulations of flow in cerebral aneurysms: Comparison of CFD results and in vivo MRI measurements. ASME J. Biomech. Eng. **130**, 051011 (2008)
24. Rayz, V.L., Lawton, M.T., Martin, A.J., Young, W.L., Saloner, D.: Numerical simulation of pre- and postsurgical flow in a giant basilar aneurysm. ASME J. Biomech. Eng. **130**, 021004 (2008)
25. Imai, Y., Sato, K., Ishikawa, T., Yamaguchi, T.: Inflow into saccular cerebral aneurysms at arterial bends. Ann. Biomed. Eng. **36**, 1489–1495 (2008)
26. Rayz, V.L., Boussel, L., Lawton, M.T., Acevedo-Bolton, G., Ge, L., Young, W.L., Higashida, R.T., Saloner, D.: Numerical modeling of the flow in intracranial aneurysms: Prediction of regions prone to thrombus formation. Ann. Biomed. Eng. **36**, 1793–1804 (2008)
27. Funamoto, K., Suzuki, Y., Hayase, T., Kosugi, T., Isoda, H.: Numerical validation of mr-measurement-integrated simulation of blood flow in a cerebral aneurysm. Ann. Biomed. Eng. **37**, 1105–1116 (2009)
28. Blanco, P.J., Pivello, M.R., Urquiza, S.A., Feijo, R.A.: On the potentialities of 3D-1D coupled models in hemodynamics simulations. J. Biomech. **42**, 919–930 (2009)
29. Shimogonya, Y., Ishikawa, T., Imai, Y., Matsuki, N., Yamaguchi, T.: Can temporal fluctuation in spatial wall shear stress gradient initiate a cerebral aneurysm? A proposed novel hemodynamic index, the gradient oscillatory number (GON). J. Biomech. **42**, 550–554 (2009)
30. Jiang, J., Strother, C.: Computational fluid dynamics simulations of intracranial aneurysms at varying heart rates: A 'patient-specific' study. ASME J. Biomech. Eng. **131**, 091001 (2009)
31. Mantha, A.R., Benndorf, G., Hernandez, A., Metcalfe, R.W.: Stability of pulsatile blood flow at the ostium of cerebral aneurysms. J. Biomech. **42**, 1081–1087 (2009)
32. Valencia, A.A., Munoz, F., Araya, S.: Comparison between computational fluid dynamics, fluid-structure interaction and computational structural dynamics predictions of flow-induced wall mechanics in an anatomically realistic cerebral aneurysm model. Int. J. Comput. Fluid Dyn. **23**, 649–666 (2009)
33. Robertson, A.M., Zeng, Z., Kallmes, D.F., Durika, M., Ding, Y., Lewis, D.A., Kadirvel, R.: Sensitivity of CFD based hemodynamic results in rabbit aneurysm models to idealizations in surrounding vasculature. ASME J. Biomech. Eng. **132**, 091009 (2010)
34. Imai, Y., Sato, K., Ishikawa, T., Yamaguchi, T.: ATP transport in saccular cerebral aneurysms at arterial bends. Ann. Biomed. Eng. **38**, 927–934 (2010)

35. Bazilevs, Y., Hsu, M.-C., Zhang, Y., Wang, W., Kvamsdal, T., Hentschel, S., Isaksen, J.G.: Computational vascular fluid-structure interaction: methodology and application to cerebral aneurysms. Biomech. Model. Mechanobiol. **9**, 481–498 (2010)
36. Raghavan, M.L., Ma, B., Fillinger, M.F.: Non-invasive determination of zero-pressure geometry of arterial aneurysms. Ann. Biomed. Eng. **34**, 1414–1419 (2006)
37. Ma, B., Lu, J., Harbaugh, R.E., Raghavan, M.L.: Nonlinear anisotropic stress analysis of anatomically realistic cerebral aneurysms. ASME J. Biomech. Eng. **129**, 88–96 (2007)
38. Zhou, X., Raghavan, M.L., Harbaugh, R.E., Lu, J.: Patient-specific wall stress analysis in cerebral aneurysms using inverse shell model. Ann. Biomed. Eng. **38**, 478–489 (2010)
39. Feng, Y., Wada, S., Tsubota, K., Yamaguchi, T.: Growth of intracranial aneurysms arising from curved vessels under the influence of elevated wall shear stress—a computer simulation study. Jpn. Soc. Mech. Eng. (JSME) Int. J. Ser. C **47**, 1035–1042 (2004)
40. Wulandana, R., Robertson, A.M.: An inelastic multi-mechanism constitutive equation for cerebral arterial tissue. Biomech. Model. Mechanobiol. **4**, 235–248 (2005)
41. Baek, S., Rajagopal, K.R., Humphrey, J.D.: Competition between radial expansion and thickening in the enlargement of an intracranial saccular aneurysm. J. Elast. **80**, 13–31 (2006)
42. Baek, S., Rajagopal, K.R., Humphrey, J.D.: A theoretical model of enlarging intracranial fusiform aneurysms. J. Biomech. Eng. **128**, 142–149 (2006)
43. Kroon, M., Holzapfel, G.A.: A model for saccular cerebral aneurysm growth by collagen fibre remodelling. J. Theor. Biol. **247**, 775–787 (2007)
44. Kroon, M., Holzapfel, G.A.: Modeling of saccular aneurysm growth in a human middle cerebral artery. ASME J. Biomech. Eng. **130**, 051012 (2008)
45. Feng, Y., Wada, S., Ishikawa, T., Tsubota, K., Yamaguchi, T.: A rule-based computational study on the early progression of intracranial aneurysms using fluid-structure interaction: Comparison between straight model and curved model. J. Biomech. Sci. Eng. **3**, 124–137 (2008)
46. Eriksson, T., Kroon, M., Holzapfel, G.A.: Influence of medial collagen organization and axial in situ stretch on saccular cerebral aneurysm growth. ASME J. Biomech. Eng. **131**, 101010 (2009)
47. Kroon, M., Holzapfel, G.A.: A theoretical model for fibroblast-controlled growth of saccular cerebral aneurysms. J. Theor. Biol. **26**, 133–164 (2009)
48. Figueroa, C.A., Baek, S., Taylor, C.A., Humphrey, J.D.: A computational framework for coupled solid-fluid-growth mechanics in cardiovascular simulations. Comput. Methods Appl. Mech. Eng. **198**, 3583–3602 (2009)
49. Shimogonya, Y., Ishikawa, T., Imai, Y., Matsuki, N., Yamaguchi, T.: A realistic simulation of saccular cerebral aneurysm formation: focussing on a novel haemodynamic index, the gradient oscillatory number. Int. J. Comput. Fluid Dyn. **23**, 583–589 (2009)
50. Li, D.L., Robertson, A.M.: A structural multi-mechanism constitutive equation for cerebral arterial tissue. Int. J. Solids Struct. **46**, 2920–2928 (2009)
51. Li, D.L., Robertson, A.M.: A structural multi-mechanism damage model for cerebral arterial tissue. ASME J. Biomech. Eng. **131**, 101013 (2009)
52. Watton, P.N., Ventikos, Y., Holzapfel, G.A.: Modelling the growth and stabilisation of cerebral aneurysms. Math. Med. Biol. **26**, 133–164 (2009)
53. Watton, P.N., Ventikos, Y.: Modelling evolution of saccular cerebral aneurysms. J. Strain Anal. **44**, 375–389 (2009)
54. Watton, P.N., Raberger, N.B., Holzapfel, G.A., Ventikos, Y.: Coupling the hemodynamic environment to the evolution of cerebral aneurysms: Computational framework and numerical examples. ASME J. Biomech. Eng. **131**, 101003 (2009)
55. Schmid, H., Watton, P.N., Maurer, M.M., Wimmer, J., Winkler, P., Wang, Y.K., Roehrle, O., Itskov, M.: Impact of transmural heterogeneities on arterial adaptation: application to aneurysm formation. Biomech. Model. Mechanobiol. **9**, 295–315 (2010)
56. Watton, P.N., Selimovic, A., Raberger, N.B., Huang, P., Holzapfel, G.A., Ventikos, Y.: Modelling Evolution and the Evolving Mechanical Environment of Saccular Cerebral Aneurysms. Biomech. Model. Mechanobiol. **10**, 109–132 (2011)

57. Machyshyn, I.M., Bovendeerd, P.H.M., van de Ven, A.A.F., Rongen, P.M.J., van de Vosse, F.N.: A model for arterial adaptation combining microstructural collagen remodeling and 3D tissue growth. Biomech. Model. Mechanobiol. **9**, 671–687 (2010)
58. Kondo, S., Hashoimotot, N., Kikuchi, H., Hazama, F., Nagata, I., Kataoka, H.: Cerebral aneurysms arising at nonbranching sites. an experimental study. Stroke **28**, 398–404 (1997)
59. Sakaki, T., Kohmura, E., Kishiguchi, T., Yuguchi, T., Yamashita, T., Hayakawa, T.: Loss and apoptosis of smooth muscle cells in intracranial aneurysms-studies with in situ DNA end labeling and antibody against single-stranded DNA. Acta Neurochirurgica **139**, 469–474 (1997)
60. Kondo, S., Hashoimotot, N., Kikuchi, H., Hazama, F., Nagata, I., Kataoka, H.: Apoptosis of medial smooth muscle cells in the development of saccular cerebral aneurysms in rats. Stroke **29**, 181–188 (1998)
61. Mimata, C., Kitaoka, M., Nagahiro, S., Iyama, K., Hori, H., Yoshioka, H., Ushio, Y.: Differential distribution and expressions of collagens in the cerebral aneurysmal wall. Acta Neuropathologica **94**, 197–206 (1997)
62. Gaetani, P., Tartara, F., Grazioli, V., Tancioni, F., Infuso, L., Rodriguez, y., Baena, R.: Collagen cross-linkage, elastolytic and collagenolytic activities in cerebral aneurysms: a preliminary investigation. Life Sci. **63**, 285–292 (1998)
63. Humphrey, J.D., Canham, P.B.: Structure, mechanical properties and mechanics of intracranial saccular aneurysms. J. Elast. **61**, 49–81 (2000)
64. Frosen, J., Piippo, A., Paetau, A., Kangasniemi, M., Niemela, M., Hernesniemi, J., Jaaskelelainen, J.: Remodelling of saccular cerebral artery aneurysm wall is associated with rupture. histological analysis of 24 unruptured and 42 ruptured cases. Stroke **35**, 2287–2293 (2004)
65. Ahn, S., Shin, D., Tateshima, S., Tanishita, K., Vinuela, F., Sinha, S.: Fluid-induced WSS in anthropomorphic brain aneurysm models: MR phase-contrast study at 3T. J. Magn. Reson. Imaging **25**, 1120–1130 (2007)
66. Hoi, Y., Meng, H., Woodward, S.H., Bendok, B.R., Hamel, R.A., Guterman, L.R., Hopkins, L.N.: Effects of arterial geometry on aneurysm growth: three dimensional computational fluid dynamics study. J. Neurosurg. **101**, 676–681 (2004)
67. Nakatani, H., Hashimoto, N., Kang, Y., Yamazoe, N., Kikuchi, H., Yamaguchi, S., Niimi, H.: Cerebral blood flow apatterns at major vessel bifurcations and aneurysms in rats. J. Neurosurg. **74**, 258–262 (1991)
68. Tateshima, S., Murayama, Y., Villablanca, J.P., Morino, T., Nomura, K., Tanishita, K., Viuela, F.: In vitro measurement of fluid-induced WSS in unruptured cerebral aneurysms harboring blebs. Stroke **34**, 187–192 (2003)
69. Meng, H., Wang, Z., Hoi, Y., Gao, L., Metaxa, E., Swart, D.D., Kolega, J.: Complex hemodynamics at the apex of an arterial bifurcation induces vascular remodelling resembling cerebral aneurysm initiation. Stroke **38**, 1924–1931 (2007)
70. DePaola, N., Gimbrone, M.A. Jr, Davies, P.F., Dewey, C.F. Jr: Vascular endothelium responds to fluid shear stress gradients. Arterioscler. Thromb. **12**, 1254–1257 (1992)
71. Shojima, M., Oshima, M., Takagi, K., Torii, R., Hayakawa, M., Katada, K., Morita, A., Kirino, T.: Magnitude and role of WSS on cerebral aneurysm: computational fluid dynamic study of 20 middle cerebral artery aneurysms. Stroke **35**, 2500–2505 (2004)
72. O'Connell, M.K., Murthy, S., Phan, S., Xu, C., Buchanan, J., Spilker, R., Dalman, R.L., Zarins, C.K., Denk, W., Taylor, C.A.: The three-dimensional micro—and nanostructure of the aortic medial lamellar unit measured using 3D confocal and electron microscopy imaging. Matrix Biol. **27**, 171–181 (2008)
73. Holzapfel, G.A., Gasser, T.C., Ogden, R.W.: A new constitutive framework for arterial wall mechanics and a comparative study of material models. J. Elast. **61**, 1–48 (2000)
74. Gasser, T.C., Ogden, R.W., Holzapfel, G.A.: Hyperelastic modelling of arterial layers with distributed collagen fibre orientations. J. R. Soc. Interface **3**, 15–35 (2006)
75. Holzapfel, G.A., Ogden, R.W.: Constitutive modelling of arteries. Proc. R. Soc. Lond. A **466**, 1551–1597 (2010)

76. Nissen, R., Cardinale, G.J., Udenfriend, S.: Increased turnover of arterial collagen in hypertensive rats. Proc. Natl. Acad. Sci. U.S.A Med. Sci.**75**, 451–453 (1978)
77. Alford, P.W., Humphrey, J.D., Taber, L.A.: Growth and remodeling in a thick-walled artery model: Effects of spatial variations in wall constituents. Biomech. Model. Mechanobiol. **7**, 245–262 (2008)
78. Humphrey, J.D.: Remodelling of a collagenous tissue at fixed lengths. J. Biomech. Eng. **121**, 591–597 (1999)
79. Alberts, B., Bray, D., Lewis, J., Raff, M., Roberts, K., Watson, J.D.: Molecular Biology of the Cell. Garland Publishing. 4th ed (1994)
80. Gleason, R.L., Taber, L.A., Humphrey, J.D.: A 2-D model of flow-induced alterations in the geometry, structure and properties of carotid arteries. J. Biomech. Eng. **126**, 371–381 (2004)
81. Gleason, R.L., Humphrey, J.D.: A mixture model of arterial growth and remodeling in hypertension: Altered muscle tone and tissue turnover. J. Vasc. Res. **41**, 352–363 (2004)
82. Gleason, R.L., Humphrey, J.D.: Effects of a sustained extension on arterial growth and remodeling: a theoretical study. J. Biomech. Eng. **38**, 1255–1261 (2007)
83. Gleason, R.L., Humphrey, J.D.: A 2d constrained mixture model for arterial adaptations to large changes in flow, pressureand axial stretch. Math. Med. Biol. **22**, 347–369 (2005)
84. Gleason, R.L., Wilson, E., Humphrey, J.D.: Biaxial biomechanical adaptations of mouse carotid arteries cultured at altered axial extension. J. Biomech. Eng. **40**, 766–776 (2007)
85. Valentín, A., Cardamone, L., Baek, S., Humphrey, J.D.: Complementary vasoactivity and matrix remodelling in arterial adaptations to altered flow and pressure. J. R. Soc. Interface **6**, 293–306 (2009)
86. Valentín, A., Humphrey, J.D.: Modeling effects of axial extension on arterial growth and remodeling. Med. Biol. Eng. Comput. **47**, 979–987 (2009)
87. Valentín, A., Humphrey, J.D.: Evaluation of fundamental hypotheses underlying constrained mixture models of arterial growth and remodelling. Philos. Transact. R. Soc. Lond. A **367**, 3585–3606 (2009)
88. Valentín, A., Humphrey, J.D.: Parameter sensitivity study of a constrained mixture model of arterial growth and remodeling. J. Biomech. Eng. **10**, 101006 (2009)
89. Humphrey, J.D., Taylor, C.A.: Intracranial and abdominal aortic aneurysms: Similarities, differences, and need for a new class of computational models. Annu. Rev. Biomed. Eng. **10**, 221–246 (2008)
90. Watton, P.N., Hill, N.A., Heil, M.: A mathematical model for the growth of the abdominal aortic aneurysm. Biomech. Model. Mechanobiol. **3**, 98–113 (2004)
91. Schmid, H., Watton, P.N., McCormick, M., Lanir, Y., Ho, H., Lloyd, C., Hunter, P., Ehret, A., Itskov, M.: Computational modeling of cerebral aneurysm formation-framework formodeling the interaction between fluid dynamics, signal transduction pathways and arterial wall mechanics. In: Sloten, J.V., Verdonck, P., Nyssen, M., Haueisen, J., (eds.) vol. 22, pp. 1894–1898. ECIFMBE 2008, IFMBE Proceedings (2008)
92. Watton, P.N., Hill, N.A.: Evolving mechanical properties of a model of abdominal aortic aneurysm. Biomech. Model. Mechanobiol. **8**, 25–42 (2009)
93. Selimovic, A., Villa-Uriol, M.-C., Holzapfel, G.A., Ventikos, Y., Watton, P.N. (2010) A computational framework to explore the role of the pulsatile haemodynamic environment on the development of cerebral aneurysms for patient-specific arterial geometries. In: Lim, C.T., Goh, J.C.H. (eds.) 6th World Congress of Biomechanics (WCB 2010), IFMBE Proceedings, vol. 31, pp. 759–762. Springer (2010)
94. Heil, M.: The stability of cylindrical shells conveying viscous flow. J. Fluids Struct. **10**, 173–196 (1996)
95. Watton, P.N., Ventikos, Y., Holzapfel, G.A.: Modelling the mechanical response of elastin for arterial tissue. J. Biomech. **42**, 1320–1325 (2009)
96. Hernandez, M., Frangi, A.F.: Non-parametric geodesic active regions: Method and evaluation for cerebral aneurysms segmentation in 3DRA and CTA. Med. Image Anal. **11**, 224–241 (2007)

97. Mellado, X., Larrabide, I., Hernandez, M., Frangi, A.F.: Flux driven medial curve extraction. Insight J. http://hdl.handle.net/1926/560 (2007)
98. Bogunović, H., Radaelli, A., de Craene, M., Delgado, D., Frangi, A.F.: Image intensity standardization in 3D rotational angiography and its application to vascular segmentation vol. 6914. Society of Photo-Optical Instrumentation Engineers Bellingham, WA San Diego CA USA (2008)
99. Patakar, H.S.: Numerical Heat Transfer and Fluid Flow (Hemisphere Series on Computational Methods in Mechanics and Thermal Science). Taylor & Francis, London (1980)
100. Ferziger, J.H., Peric, M.: Computational Methods for Fluid Dynamics 3rd ed. Springer-Verlag, Heidelberg (2002)
101. Hutchinson, B.R., Raithby, G.D.: A multigrid method based on the additive correction strategy. Numer. Transf. **9**, 511–537 (1986)
102. Cebral, J.R., Castro, M.A., Appanaboyina, S., Putman, C.M., Millan, D., Frangi, A.F.: Efficient pipeline for image-based patient-specific analysis of cerebral aneurysm hemodynamics: technique and sensitivity. IEEE Transact. Med. Imaging **24**, 457–467 (2005)
103. Fisher, C., Rossmann, J.S.: Effect of non-newtonian behavior on hemodynamics of cerebral aneurysms. ASME J. Biomech. Eng. **131**, 091004 (2009)
104. Oshima, M., Torii, R., Kobayashia, T., Taniguchic, N., Takagid, K.: Finite element simulation of blood flow in the cerebral artery. Comput. Methods Appl. Mech. Eng. **191**, 661–671 (2001)
105. Myers, J.G., Moore, J.A., Ojha, M., Johnston, K.W., Ethier, C.R.: Factors influencing blood flow patterns in the human right coronary artery. Ann. Biomed. Eng. **29**, 109–120 (2001)
106. Mantha, A., Karmonik, C., Benndorf, G., Strother, C., Metcalfe, R.: Hemodynamics in a cerebral artery before and after the formation of an aneurysm. Am. J. Neuroradiol. **27**, 1113–1118 (2006)
107. Wang, J.H.C., Thampaty, B.P.: An introductory review of cell mechanobiology. Biomech. Model. Mechanobiol. **5**, 1–16 (2006)
108. Gupta, V., Grande-Allen, K.J.: Effects of static and cyclic loading in regulating extracellular matrix synthesis by cardiovascular cells. Cardiovasc. Res. **72**, 375–383 (2006)
109. Chiquet, M., Renedo, A.S., Huber, F., Flück, M.: How do fibroblasts translate mechanical signals into changes in extracellular matrix production?. Matrix Biol. **22**, 73–80 (2003)
110. Sotoudeh, M., Jalali, S., Usami, S., Shyy, J.-Y., Chien, S.: A strain device imposing dynamic and uniform equi-biaxial strain to cultured cells. Ann. Biomed. Eng. **26**, 181–189 (1998)
111. Shin, H.Y., Gerritsen, M.E., Bizios, R.: Regulation of endothelial cell proliferation and apoptosis by cyclic pressure. Ann. Biomed. Eng. **30**, 297–304 (2002)
112. Asanuma, K., Magid, R., Johnson, C., Nerem, R.M., Galis, Z.S.: Uniaxial strain regulates matrix-degrading enzymes produced by human vascular smooth muscle cells. Am. J. Physiol. Heart Circ. Physiol. **284**, H1778–H1784 (2003)
113. Lee, R.T., Yamamoto, C., Feng, Y., Potter-Perigo, S., Briggs, W.H., Landschulz, K.T., Turi, T.G., Thompson, J.F., Libby, P., Wight, T.N.: Mechanical strain induces specific changes in the synthesis and organization of proteoglycans by vascular smooth muscle cells. J. Biochem. **276**, 13847–13851 (2001)
114. Hill, M.A., Zou, H., Potocnik, S.J., Meininger, G.A., Davis, M.J.: Signal transduction in smooth muscle invited review: Arteriolar smooth muscle mechanotransduction: Ca2+ signalling pathways underlying myogenic reactivity. J. Appl. Physiol. **91**, 973–983 (2001)
115. Williams, B.: Mechanical influences on vascular smooth muscle function. J. Hypertens. **16**, 1921–1929 (1998)
116. Kim, Y.S., Galis, Z.S., Rachev, A., Han, H.C., Vito, R.P.: Matrix metalloproteinase-2 and −9 are associated with high stresses predicted using a nonlinear heterogeneous model of arteries. ASME J. Biomech. Eng. **131**, 011009 (2009)

117. Cummins, P.M., von Offenberg Sweeney, N., Killeen, M.T., Birney, Y.A., Redmond, E.M., Cahill, P.A.: Cyclic strain-mediated matrix metalloproteinase regulation within the vascular endothelium: a force to be reckoned with. Am. J. Physiol. Heart Circ. Physiol. **292**, H28–H42 (2007)
118. Chien, S.: Mechanotransduction and endothelial cell homeostasis: the wisdom of the cell. Am. J. Physiol. Heart Circ. Physiol. **292**, H1209–H1224 (2007)
119. Davies, P.F.: Flow mediated endothelial mechanotransduction. Physiol. Rev. **75**, 519–560 (1995)

Biomechanical Considerations of Animal Models of Aortic Aneurysm

Darren Haskett, Mohamad Azhar and Jonathan P. Vande Geest

Abstract Aortic aneurysm is a focal enlargement of the aorta developing over years and carrying the risk of rupture and death. As aneurysms are a chronic disease, animal models have come to be used in both determining the underlying mechanisms that cause aneurysm formation and in designing new treatments for the disease. These models include mechanically and chemically induced methods used for both refining surgical techniques and stent graft device characterization. The latter are typically implemented in large animals (dog, pig, and sheep). Other models employ methods that rely on genetic manipulation, often with the addition of chemical induction, to induce aneurysm formation in small animals, predominantly mice. Recent efforts have also aimed at determining both the biomechanical alterations that occur with aneurysm formation and the potential for rupture. However, many animal models for aortic aneurysm do not exhibit some of the

D. Haskett · J. P. Vande Geest
Graduate Interdisciplinary Program in Biomedical Engineering,
The University of Arizona, Tucson, AZ, USA

M. Azhar · J. P. Vande Geest
BIO5 Institute, The University of Arizona, Tucson, AZ, USA

M. Azhar
Department of Cell Biology and Anatomy, The University of Arizona,
Tucson, AZ, USA

J. P. Vande Geest
Department of Aerospace and Mechanical Engineering, The University of Arizona,
Tucson, AZ, USA

J. P. Vande Geest (✉)
Department of Biomedical Engineering, The University of Arizona, 1130 N. Mountain,
PO Box 210119, Tucson, AZ 85721-0119, USA
e-mail: jpv1@email.arizona.edu

Stud Mechanobiol Tissue Eng Biomater (2011) 7: 401–421
DOI: 10.1007/8415_2011_76

Published Online: 28 April 2011

native characteristics of the disease and as such are not suitable for investigating disease initiation and progression. The current review summarizes the various approaches of animal models for aortic aneurysm in the context of their appropriateness for biomechanical investigation.

1 Introduction

It is generally accepted that the formation of an aneurysm in the aorta is a complex and multi-factorial disease. Some factors which have been identified as playing a role in the disease process include genetic predisposition [1, 2], autoimmunity [3], and environmental influences such as smoking [4]. Over the past 15 years there has been an explosion of information regarding how both fluid and solid mechanics might also play a role in aneurysm development, assessment, and rupture prediction [5, 6]. Current research has aimed at improving treatment options for aneurysm (e.g., endovascular repair) and predicting rupture, and although it is known that aneurismal tissue is remodeled in the disease process in humans [5] and that such reorganization leads to alterations in mechanical properties [7], there is a need to understand the underlying mechanisms and how they develop. Animal models of aneurismal disease can be useful for studying alterations during disease development (e.g., in the tissues mechanical response) and also for studying novel and new endovascular treatments.

Development of animal models of aortic aneurysms can be broken down into two basic categories based on their purpose. The purpose of some animal models is to assist in the development of novel endovascular prostheses and their deployment techniques for the treatment of aneurysms in humans. These models are focused on the clinical application of the device and therefore require large animals and induce aneurysm through either mechanical or chemical means. Beginning in 1986, Balko et al. [8] inserted a prosthesis into the abdominal aorta of adult sheep creating an abdominal aortic aneurysm (AAA) which was then endoluminally excluded using a stent graft. Since then several large animal models including dogs [9–12], pigs [13, 14], and sheep [8, 15, 16] have been used for the study and development of endovascular grafts with dogs being the most common animal model. The use of small animals (e.g., mice) as a means to study endovascular repair has not largely been used likely due to size challenges and significant differences between human and murine aortic hemodynamics. The choice of animal for such models is very important and is often a tradeoff between availability, aortic geometry, physiological differences with humans, and suitable characteristics of the animal.

The second purpose of an aortic aneurysm animal model is to study the mechanisms behind aneurysm formation enabling us to better understand aneurysm pathophysiology and propose novel drug therapies. These models for the most part rely on small animals for their low cost and ease of handling relative to large animals. Small animals with shorter life spans are also desirable for such studies, since these models may depend on spontaneous aneurysm formation and thus require an accelerated time course for disease. In small animal models aneurysm development is induced chemically, genetically predisposed, or imposed using a combination of both. The number of aneurysm models has increased significantly in the past few years. This is due to sequencing and annotation of the mouse genome and efforts such as the Knockout Mouse Project (KOMP) which are focused on creating genetically modified mouse knockout strains for every gene in order to study disease [17].

The purpose of this chapter is to review both the methods used in creating experimental aneurysms in animals and the several different animal models currently available, highlighting their typical means of use and also detailing how such models may be used to better understand the role of mechanics in aneurysm development.

2 Mechanisms for Inducing Aortic Aneurysms in Animals

Much like the choice of animal, the choice of mechanism used for inducing aneurysm is based on the model's purpose and is typically induced using either mechanical or chemical means or using a genetically predisposed animal.

2.1 Mechanically Induced Models

Open repair of AAAs has in recent years become less preferable to the alternative and less invasive procedure of endovascular repair. Since the first studies to use stents grafts to treat AAAs in the mid-1980s, large animal models of aortic aneurysm have been used to test the delivery, biocompatibility, and efficacy of endovascular grafts. Most of these models have used mechanical, otherwise known as surgical, techniques to induce aneurysm. Some techniques developed for inducing aneurysm such as weakening the aortic wall through injury [18] or the Blanton technique for creating dissecting aneurysms [19] may not be in common use. However, more common methods for mechanically induced aneurysm have been established and include the arterial wall patch model, the graft model, and transluminal models.

2.2 Arterial Wall Patch Model

The arterial wall patch model for inducing AAA has been one of the most widely used methods for studying endovascular grafts. The function of the patch model is to introduce an aneurismal pouch by repairing an incision in the aortic wall using a structurally weaker graft, also referred to as a patch. This patch is able to dilate postoperatively and remain patent until an endovascular graft can be deployed to bypass the experimental aneurysm. During the procedure the animal is heperinized, the aorta is isolated and cross-clamped between the renal arteries and above the iliac bifurcation, and the collateral arteries are secured using temporary hemoclips or vessel loops. The incision, known as an aortotomy, is created along the longitudinal axis of the aorta and an elliptical patch is then sutured into the opening. An intervening period of between 3 and 12 weeks is then required for animal recovery and tissue healing before the endovascular repair can then be attempted [20, 21]. This procedure has been used to create saccular aneurysms in dogs, pigs, and sheep.

The patch model is sometimes referred to as the vein-patch method, as venous tissue from either portions of the inferior vena cava, jugular, or iliac vein [11, 22–24] is often used. Various other materials have been used including jejunum [9], peritoneal [14], and fascia [25, 26] tissues, as well as synthetic materials such as Dacron [8, 10, 27] and ePTFE [15, 28]. Use of analogous tissue has been reported to be favorable as they are able to progressively expand and the tendency to rupture can be controlled based on the tissue type. Such studies have found that vein patches will enlarge but not rupture due to scar tissue formation [29] and that manipulating jejunum or peritoneum tissue can result in varying tendencies to rupture [9, 26]. Some work using synthetic patches has also shown the ability for the material to enlarge [28], however, this is not a critical feature for as long as the aneurysm remains patent the function of the patch is preserved. In the patch model, limited thrombus formation has been found with a few materials such as jejunum [9, 30], and the collateral arteries associated with the patch have remained patent. However, once the prosthesis is inserted into the experimental aneurysm it is often flush against the lumen and the collateral arteries bridged by the endovascular graft have been found to completely thrombose. Therefore, a modified technique is required for any study focused on the perfusion of the aneurysm sac by retrograde flow [31, 32].

Although widely used and often found to exhibit similarities to the human disease [33], the value of the arterial wall patch model for studying the mechanisms involved in aneurysm and determination of aneurysm mechanical properties is not seen as appropriate, since the processes behind their formation is artificial. However, this model is able to create large aneurysms with morphological similarities to human aneurysms, and thus it offers a valuable model for developing novel endovascular devices and techniques. The patch model has also shown its versatility by being modified to allow for studies into kinking of endovascular devices and into the different types of endoleak. Furthermore, the patch model provides valuable opportunities to gain surgical experience and will likely continue to be used in this capacity [34].

2.3 Graft Model

Almost as common as the patch model, the graft model for inducing AAA has also been used in dogs, pigs, and sheep. Similar materials have also been used, such as Dacron [10, 16, 35] and ePTFE [36] along with treated jugular vein [37]. Again the animal is heparinized, and the aorta is isolated and cross-clamped below the renal arteries and above the iliac bifurcation. However, the collateral arteries are ligated and divided as they are removed along with the section of the isolated aorta. Once the aorta has been excised, the graft representing the artificial aneurysm is anastomosed with each end of the aorta replacing the removed portion and forming a conduit [38]. Once the procedure is complete, the aneurysm is fully formed and there have been no reports of the graft losing patency prior to the placement of an endovascular graft.

The graft model has an advantage over the arterial patch model as it can more easily manipulate its size and configuration to more closely resemble that of aneurysms found in humans. The graft model has also been used to form fusiform AAAs through the replacement of a portion of infrarenal aorta with large crimped artificial grafts [39]. This model has been used for the evaluation of endovascular graft deployment techniques and delivery systems, as well as to determine the effectiveness of novel devices. However, this model is not appropriate for studying the biological effects of the aortic wall on the healing process, studying any degradation that occurs to the endovascular graft, or studying certain failure processes like reentrant endoleak, because the large animal model uses either synthetic material or treated tissue and the collateral arteries are ligated.

The graft model has also been used in rats by one research group to generate aneurysms based on the immune rejection of the implanted tissue which originally came from guinea pigs [40, 41]. Aneurysms were created through rejection of the foreign tissue within 4 weeks of transplant and included components of the human disease, such as loss of medial cells and extracellular matrix and an influx of macrophages, T cells, and B cells. Although the model may not seem suitable as a model to study aneurysm causation, seeding the transplanted tissue with syngenic rat vascular smooth muscle cells has shown significant reductions in elastin degradation, inflammatory infiltration, and aneurysm formation and may lead to new therapeutic treatments for AAA [41]. It is in this capacity that the graft model for aortic aneurysm may be used to study the mechanical response of experimental aneurysm tissue, something that has yet to be reported in the literature.

2.4 Transluminal Angioplasty Model

A more recent model for creating aneurysms in large animals is the transluminal angioplasty model. In this catheter based technique, a standard angioplasty balloon and a balloon-expandable intravascular metallic stent were used to dilate the infrarenal abdominal aorta to twice its diameter and were able to form a fusiform shaped AAA. Hallisey was able to demonstrate this model in dogs without the need for retroperitoneal dissection or major surgery and found that after 30 days the AAAs remained unchanged and the collateral arteries remained patent [42]. This method has also been adapted to work in sheep as well [43]. Although the model is easily created, it lacks physiological elastin breakdown and inflammatory cell infiltration, which are found in human AAAs. However, due to its ease of implementation and lack of using artificial material, this method has been able to be incorporated into use with an elastase/collagenase model (see Sect. 3.3) to better mimic the human disease and allow for biomechanical analysis of the aneurismal tissue.

3 Chemically Induced Models

Chemically induced aneurysm models are easily the most transferable between animals as they have been used in both small and large animals. The two common methods include the elastase and calcium chloride induced models. These models came about from attempts to mimic the early stages of human AAA disease which found that disruption of the elastic media and medial inflammation occurred [44, 45].

3.1 Elastase Model

The intraluminal perfusion of the abdominal aorta with elastase was the first anreusym model developed in small animals and was performed on rats in 1990 [46] and later transplanted into mice [47]. The procedure consisted of isolating a 1 cm segment of abdominal aorta by means of inserting a catheter through the femoral artery up to the level of the infrarenal aorta which was then clamped above and ligated below. Collateral arteries were also ligated and the segment was then perfused with the elastase solution for up to 2 h. The elastase degraded elastin within the aortic wall, where inflammation and aneurismal development was then accompanied by increased macrophage perfusion and matrix metalloproteinase (MMP) production. The timecourse for full aneurismal development had a delayed reaction time of 7–14 days after elastase infusion and revealed a number of similarities to aneurysm development in humans [48]. The elastase-induced AAA model has therefore been useful in examining the role of chronic inflammation and specific MMPs in aortic deterioration and to determine the effects of pharmacological agents or genetic alterations on aneurysm formation [49].

As for large animals, the elastase induced model for aneurysm has been around since the 1960s when elastase perfusion was first attempted on dogs [50]. However, Martin et al. demonstrated difficulty obtaining reproducible results in this large animal model which had a timeline of two months [51], and the method went largely unused except in small animal models where the model had much better reproducibility. In 1993, Boudghène et al. reported the use of elastase to create an AAA model in dogs for evaluation of endovascular grafting and found that the dose of elastase was critical to the formation of aneurysm [52]. Since then the use of elastase perfusion in large animals to induce experimental AAAs has grown as the model is able to mimic the pathology of the human disease better than other large animal models such as the arterial wall patch and graft models. The elastase model also offers the opportunity to study how MMP and inflammation accompany biomechanical changes in aneurismal tissue as the disease progresses.

3.2 Calcium Chloride Model

Another method for creating a chemically induced aortic aneurysm in animals is through the use of periarterial application of calcium chloride which was first performed in rabbits [53]. This model has also been successfully transferred to mice [54]. Unlike the elastase model, exposure to the calcium chloride was done by anesthetizing the animal, exposing the abdominal aorta between the renal arteries and the iliac bifurcation, and placing calcium chloride soaked gauze directly on the aortic section for 10 min. Marked dilation was reported after the second week with full aneurismal development occurring by 3 weeks. In accordance with aneurysm in humans, varying degrees of intimal fibromuscular hyperplasia and medial disorganization, along with disruption of the elastic network of the intima and the media were observed. The inflammatory cell infiltration includes neutrophils, lymphocytes, monocytes, and multinucleated giant cells. Like the elastase perfusion model, the time course of biomechanical changes during disease development can be studied using this model.

3.3 Elastase/Collagenase Model With Angioplasty

In an effort to better model native AAAs in large animals an elastase and collagenase perfusion method was developed to permit the evaluation of endovascular devices. The elastase/collagenase model with angioplasty combines both mechanical and chemical approaches to induce experimental AAAs [13]. The experimental aneurysms were created by surgically exposing the abdominal aorta of pigs, dilating the aorta with an angioplasty balloon, and perfusing the aorta with a solution of elastase and collagenase (Fig. 1). Serial MRI was used to determine the postoperative diameter, and the animals were sacrificed at different time points up to 6 weeks for histological evaluation. The MRI showed that there was a gradual increase postoperatively in aortic diameter, and histology showed partial endothelial loss, mural neutrophil infiltrate, and elastin disruption within the first 7 days. There was reendothelialization and smooth muscle repopulation between 3 and 6 weeks, however, the elastin degradation remained. The study was able to show that the creation of an AAA is possible using a combined mechanical and chemical approach. While further evaluation is necessary to characterize the nature of the aneurysm, this model shows potential for studying the biochemical and biomechanical alterations that occur during aneurysm development.

3.4 Genetic Models

Rather than relying on mechanical and/or chemical induction, various animal models have recently been introduced that stem from aneurysm development in genetically compromised animals.

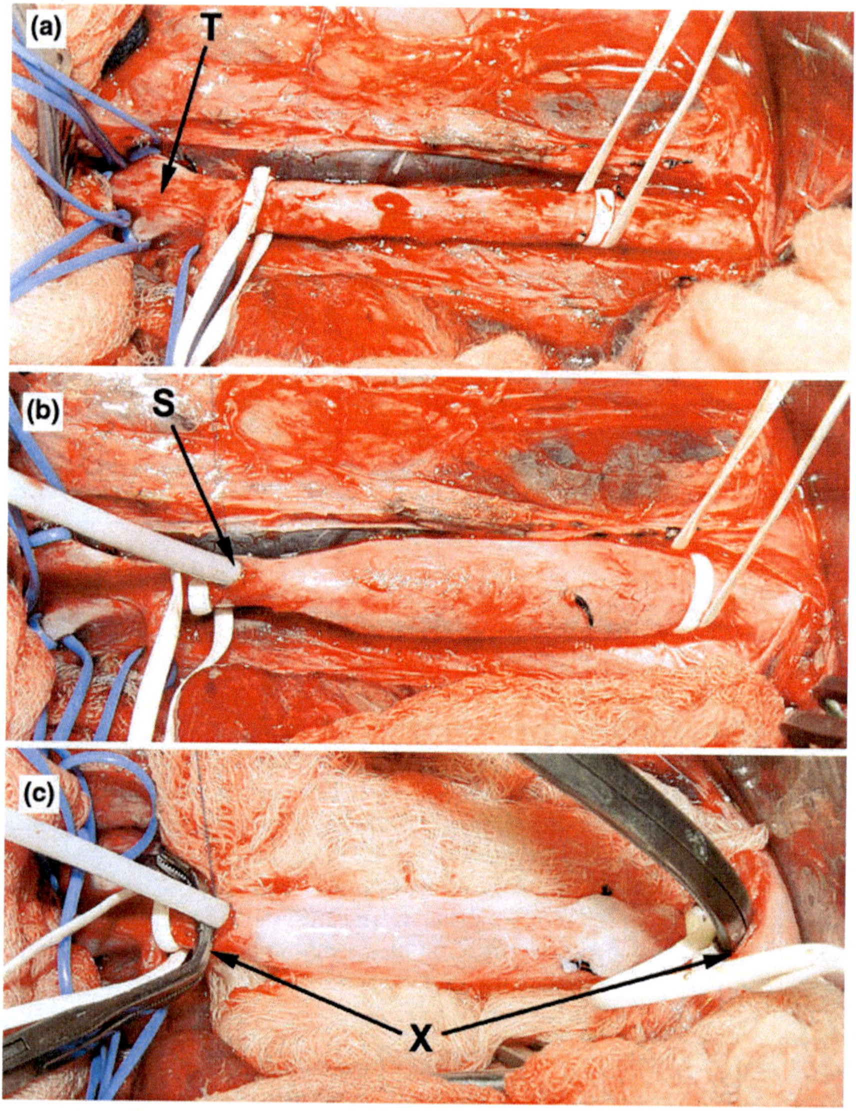

Fig. 1 Development of infrarenal AAA in swine using the elastase/collagenase model with angioplasty. In tile **a**, the abdominal aorta is exposed between the caudal aortic trifurcation (*T*) and the renal arteries (not shown). In tile **b**, an introducer sheath (S) is inserted into the abdominal aorta and angioplasty balloon is slowly inflated to dilate the aorta. In tile **c**, the aorta is cross-clamped (*X*) while the abdominal aorta is perfused with a solution of elastase and collagenase. Figure taken from Ref. [13]

3.4.1 Genetic Mouse Models of Thoracic Aneurysm

There are several genetic animal models of aneurysm that preferentially form in the thoracic region of the aorta. One of these models includes the 'blotchy' mouse that has a chromosomal mutation affecting copper metabolism causing abnormal cross-linking of collagen and elastin and weakening of connective tissue [55]. Aneurysms that formed did so generally in the thoracic aorta. However, this genetic defect results in large metabolic alterations that most closely resemble Menkes' kinky hair syndrome, otherwise known as copper transport disease, which also presents with degeneration of the nervous system and impaired growth [56] and does not truly mimic the general process of thoracic aneurismal formation in humans.

In addition to the 'blotchy' mouse, Lysyl oxidase (LOX)-deficient mice also form aneurysms in the thoracic aorta. This mouse model relies on a mutation that does not allow lysyl oxidase to initiate the crosslinking of collagen and elastin [57]. These crosslinks are essential to the mechanical properties of vascular tissue. Upon removing the *Lox* in mice, LOX-deficient neonate mice revealed large aortic aneurysms. Microscopy showed hazy and unruffled elastic lamellae and highly fragmented elastic fibers with discontinuity of the smooth muscle layers. The aortic wall of these animals was also found to be significantly thicker and their diameter significantly smaller. However, in the $Lox^{-/-}$ model, mice typically die at the end of gestation so such a model would not easily allow for the study of progression of the disease or allow for biomechanical investigations. Hence, mice with conditional null alleles (target gene is eliminated from specific tissue of interest at desired time in the body rather than the whole body as gene knockout technology would entail) of *Lox* will be useful for circumventing the embryonic lethality.

Another transgenic mouse model that is predisposed to aneurysm development in the thoracic region is the Tsukuba hypertensive mouse (THM), which over produces angiotensin II (AngII) [58]. The THM is defined by increased blood pressure but no other clear vascular pathology. However, when placed on a 1% sodium diet the incidence of aortic rupture increased dramatically [59], primarily occurring in the thoracic region. However, much like the blotchy mouse, this model does not accurately mimic the human disease or allow for the study of aneurysm progression as the aorta is prone to rupture following the second day on a high sodium diet [60].

Another transgenic model able to produce aortic aneurysm is the mouse that is heterozygous for *Fibrillin-1* ($Fbn1^{+/C1039G}$) mutation [61]. In these mice excess Transforming Growth Factor *beta* (TGF-β) signaling drives the formation and progression of aortic aneurysm in the aortic root. $Fbn1^{+/C1039}$ mice recapitulate many aspects of aortic pathology of human Marfan syndrome (MFS). Unlike the other animal models for thoracic aneurysm, the *Fbn1* model does allow for biomechanical investigations of thoracic aorta during disease progression (Fig. 2).

3.4.2 Hyperlipidemic Mice

Some of the mouse models now used for aneurysm models were first created for atherosclerosis research. These models include mice that are deficient in either apolipoprotein E (ApoE) or low density lipoprotein (LDL) receptors [62, 63]. Both $Ldlr^{-/-}$ and $ApoE^{-/-}$ mice have delayed clearance of vascular LDL and

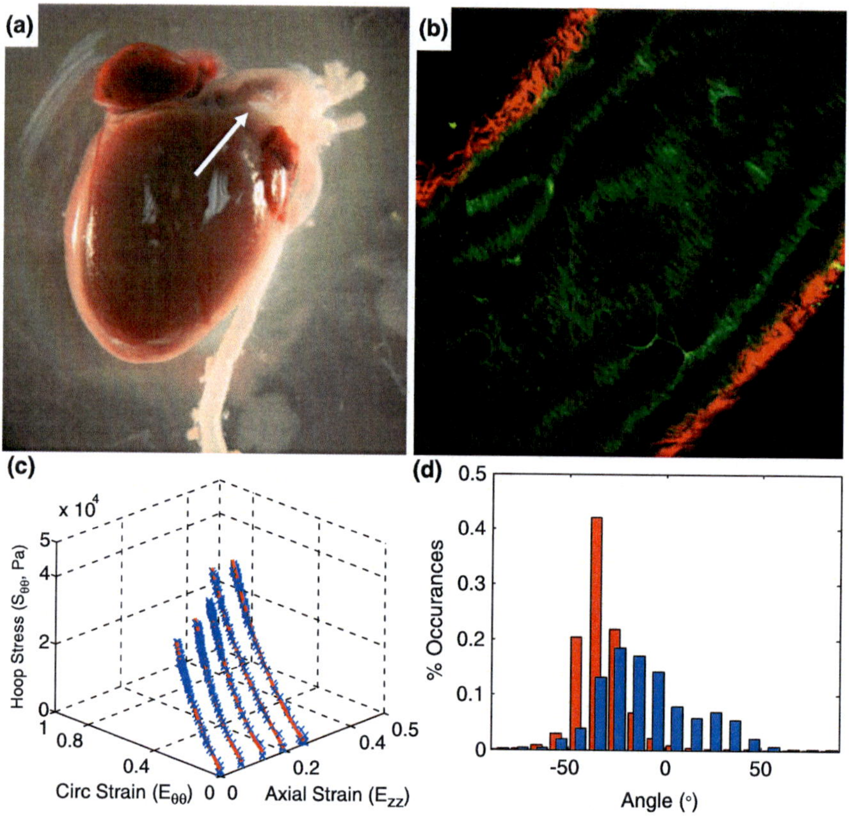

Fig. 2 Biomechanical and microstructural analysis of thoracic aorta of adult wild-type and *Fbn1*$^{+/C1039G}$ mice. Tile **a** demonstrates the presence of significant aortic root dilation (*arrow*) in an aged adult *Fbn1*$^{+/C1039G}$ mouse. Tile **b** is a multiphoton image showing both second harmonic generation (*red*, collagen) and autofluorescence (*green*, elastin) in the inflated thoracic aorta of *Fbn1*$^{+/C1039}$ mouse. Note that the cystic medial degeneration of the elastic lamina in *Fbn1*$^{+/C1039G}$ aortic tissue has a striking resemblance to diseased aortic tissue of patients of MFS. Tile **c** shows a representative 3-dimensional stress verses strain response for aged *Fbn1*$^{+/C1039G}$ from a biaxial mechanical test (*blue*, data; *red*, polynomial fit). Tile **d** shows the gross fiber angle response upon inflation (0 mmHg, *red*; 100 mmHg, *blue*) for an aged *Fbn1*$^{+/C1039G}$ mouse based off of analysis of the multiphoton images; −45° corresponds to fibers aligned with the longitudinal axis while 45° corresponds to circumferentially aligned fibers—Unpublished work

hyperlipidemia and atherosclerosis on normal diets [64]. Along with hypercholesterolemia and atherosclerosis, small dilations in the aorta have also been reported [65], though the vessel dilations observed in *ApoE*$^{-/-}$ mice are quite small. Upon prolonged feeding with high fat diets, large AAAs formed under the atherosclerotic lesions (Fig. 3). Some studies have been able to show that such large aneurysms also occur without the need for dietary stimulus [63]. However, both methods require aged mice that do not always exhibit aneurysm, and as such these models without further biochemical modification have not been widely embraced as AAA models.

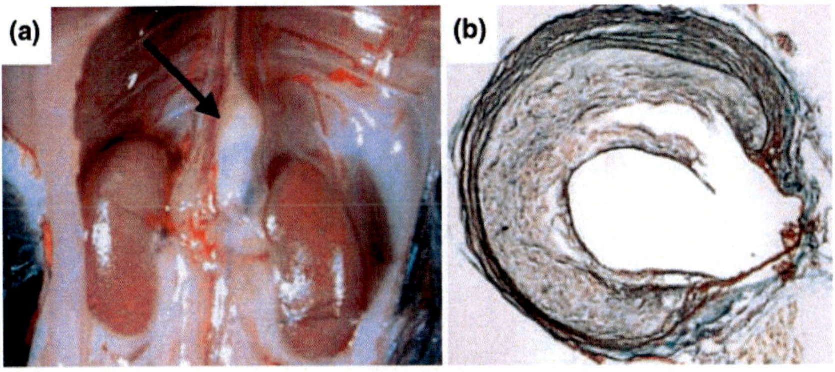

Fig. 3 Development of AAA in $ApoE^{-/-}$ mice. Tile **a** demonstrates the presence of an AAA in the suprarenal region of the aorta of an aged $ApoE^{-/-}$ mouse. Tile **b** shows a cross-section stained with gomori trichrome strain for muscle tissue demonstrating the dilation of the aorta and the thrombus formation in $ApoE^{-/-}$ mice that mimics human disease. Figure taken from Ref. [62]

3.4.3 Aneurysm Models with Genetic Combinations

In an effort to take advantage of the AAA formed in the hyperlipidemic mouse model which is more representative of human AAAs than other mouse models, several combined knockout strains of mice have been created. Van Herck et al. studied one such model using $ApoE^{-/-}$ combined with $Fbn1^{+/C1039G}$ mice fed with a high fat, Western-type diet for 20 weeks [66]. This study looked into both the pathophysiologic alterations (atherosclerotic development) along with the biomechanical response (arterial stiffness) and found them both to change simultaneously (Fig. 4). The $ApoE^{-/-}$ $Fbn1^{+/C1039G}$ mice showed that arterial stiffening allowed for the more rapid development of atherosclerotic plaques, while the presence of atherosclerotic plaques encouraged increased aortic stiffness. These mice demonstrated many of the pathologies of the human disease such as apoptosis of smooth muscle cells and an increase in macrophages. Although these changes occurred throughout the length of the aorta and did not present as aortic aneurysms, the model does allow for more straightforward biomechanical investigations of disease processes that co-exist with human aneurysm development.

Deguchi et al. also looked into another combined genetic model that paired $ApoE^{-/-}$ mice with one expressing a form of collagen resistant to collagenase in order to look into the effect of excess collagen deposition towards aneurysm formation [67]. The resulting $Col^{R/R}ApoE^{-/-}$ mouse was subsequently infused with AngII (see Sect. 3.4.4) for two weeks to induce AAA, and histological analysis and mechanical testing of the aortas were conducted. Picrosirius red and Masson trichrome staining showed $Col^{R/R}ApoE^{-/-}$ mouse aortas contained more interstitial collagen than control mice. Uniaxial tensile tests demonstrated that the $Col^{R/R}ApoE^{-/-}$ mouse aortas had greater stiffness and were more susceptible to mechanical failure than control mouse aortas (Fig. 5). This study detailed how

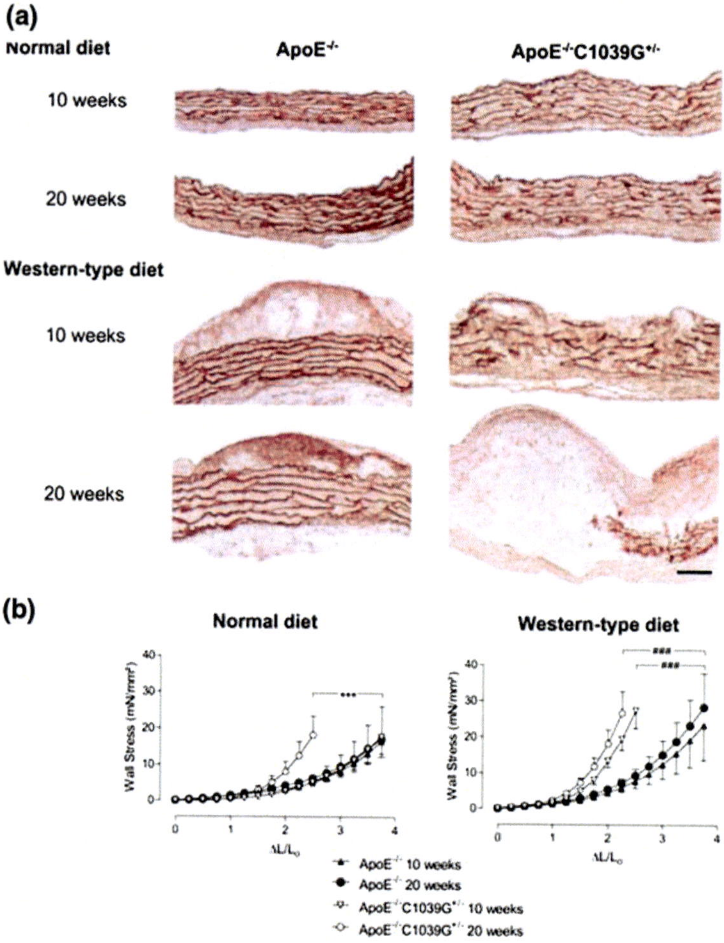

Fig. 4 Histological and biomechanical analysis of $ApoE^{-/-}Fbn1^{+/C1039G}$ mice. Tile **a** shows the disruption that occurs in $ApoE^{-/-}Fbn1^{+/C1039G}$ (labeled as $ApoE^{-/-}C1039G^{+/-}$) compared to standard $ApoE^{-/-}$ mice while on normal and Western-style diets. Histological analysis was done using orcein staining for elastic fibers of the ascending aorta. The scale bar is 100 μm. Tile **b** shows the resulting stress verses strain response at 10 and 20 weeks for both normal and Western-style diets. The $ApoE^{-/-}Fbn1^{+/C1039G}$ mouse aorta demonstrated significantly greater stiffness after 20 weeks on a normal diet (***) and demonstrated significantly greater stiffness after both 10 and 20 weeks on western-style diet (###). Figure taken from Ref. [66]

the biomechanical properties of the aorta are regulated by the amount of interstitial collagen and concluded that aneurysm formation can be influenced by the collagen content. The $Col^{R/R}ApoE^{-/-}$ model has thus exhibited the ability to investigate the biomechanical alterations that occur with aneurysm formation in an animal model.

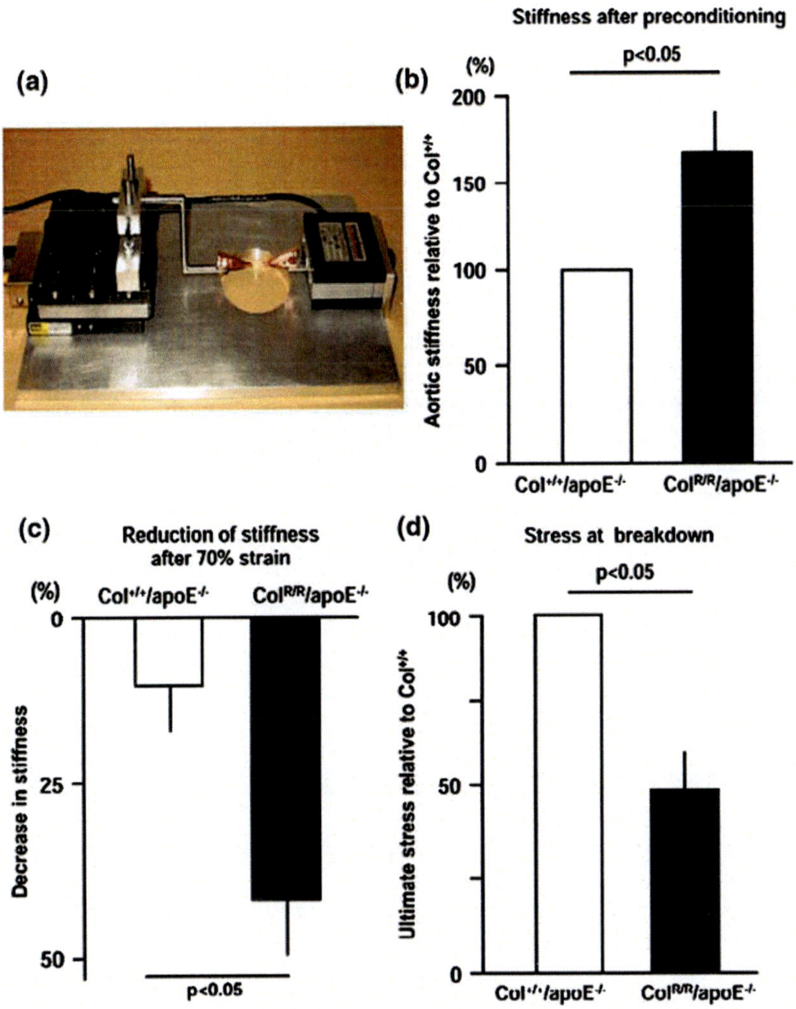

Fig. 5 Biomechanical testing of $Col^{R/R}ApoE^{-/-}$ aortas. Tile **a** is an image of the strain-device composed of a load cell, positioning system, and clamps used to biomechanically assess the mouse aortas through cyclical strain and strain to failure tests. Tiles **b**, **c**, and **d** show the results from mechanical tests conducted on $Col^{R/R}ApoE^{-/-}$ ($n = 4$) and control $ApoE^{-/-}$ mice ($n = 3$). In tile **b** the aortic stiffness of $Col^{R/R}ApoE^{-/-}$ mice was found to be greater than the aortic stiffness in $ApoE^{-/-}$ control mice. In tile **c** the aortic stiffness of $Col^{R/R}ApoE^{-/-}$ mice was found to decrease relative to $ApoE^{-/-}$ control mice after 10 cycles. While in tile **d**, the failure of $Col^{R/R}ApoE^{-/-}$ mouse aortas was found to occur at less than half that of $ApoE^{-/-}$ control mouse aortas. Figure taken from Ref. [67]

3.4.4 Chemical Infusion of Genetic Models

Much like combined genetic aneurysm models, chemically infused genetic models have also been created in order to more quickly induce aneurysm formation and better mimic the human disease. Most of these models again use hyperlipidemic mice and often use vasopressors such as AngII.

In one such model, infusing $ApoE^{-/-}$ mice with AngII using subcutaneously implanted Alzet pumps was found to not only affect atherosclerotic development, but there was striking aneurismal development in the region of the abdominal aorta. The same was true for $Ldl^{-/-}$ mice also subjected to AngII. Described by Daugherty et al., the model employs AngII infusion of $ApoE^{-/-}$ mice on a C57BL/6 J genetic background [68]. $ApoE^{-/-}$ mice were allowed to feed ad libitum on normal laboratory diet until 6 months of age, at which point Alzet ozmotic minipumps were implanted into the subcutaneous space through a small incision in the back of the neck which was then closed with surgical staples. The pumps contained either saline or 500 or 1000 ng/min/kg AngII solutions and delivered each solution for 28 days. Analysis of the abdominal region of the aorta revealed luminal dilation, medial degeneration, and thrombotic material and are features that have been proposed as requirements for an appropriate animal model that mimics human disease [59]. For this reason, the AngII infused $ApoE^{-/-}$ model has been used to study the effects of different pharmacological treatments of AAA [69–71].

In the AngII infused $ApoE^{-/-}$ model changes in aortic physiology were observed nearly immediately. Within 1–4 days after initiation of AngII infusion, medial accumulation of macrophages occurred in areas where aneurysms are known to develop [72]. Disruption of elastin fibers was also frequently observed at sites of macrophage accumulation, however, it is unknown whether elastin degradation is the event that leads to macrophage accumulation or whether macrophage accumulation leads to degradation of elastin [73].

The 4–10 days following AngII infusion found vascular hematomas in the majority of ApoE$^{-/-}$ mice and aortic dissections contained from rupture by the adventitia; during this time period 10% of mice died [72, 73]. Deaths during this period were found to be caused by AAA rupture with subsequent exsanguinations into the abdominal cavity. For those mice that survived the disruption of elastic lamina and medial dissection, formation of thrombus occurred and was thought to be a strong stimulus for inflammation [73]. After thrombus development, macrophages were found accumulated in areas near the thrombus in both intact and disrupted media and were even found within the thrombus. During this time period, remodeling of the aortic tissue began in the aneurismal arterial segments. This is one of the few animal models that describe macrophage infiltration into the thrombotic clot which is characteristic of human AAA [6].

Beyond 14 days of AngII infusion, aneurismal tissue was found to mature with increased deposition of extracellular matrix and inclusion of T and B lymphocytes, however, disruption of medial elastin fibers was still present even after 14 days of AngII infusion. By 28 days of AngII infusion, the aorta had completely re-endothelialized over the areas of medial disruption, and neovascularization occurred throughout the aneurismal tissue. Remodeling in this manner could then lead to a permanently enlarged aorta with all three arterial layers [73]. Only after 28 days of AngII infusion were atherosclerotic lesions observed in regions of aneurismal development [72]. Finally, Daugherty et al. [74] have also reported thoracic aortic aneurysms in AngII-infused $ApoE^{-/-}$ mice. This has broadened the utility of chemically-induced genetic mouse models to investigate the pathogenic mechanisms of both AAA and thoracic aortic aneurysms.

3.5 Drug Treatment and Prevention of Aneurysm in Genetic Animal Models

With the discovery of new disease pathways that lead to aneurysm several studies have looked into possible drug therapies for aneurysm treatment. From these investigations two possible drug treatment therapies have come to the forefront: losartan and doxycycline.

3.5.1 Losartan

The $Fbn1^{+/C1039G}$ genetic mouse model for Marfan syndrome which manifests thoracic aneurysm was found to have noticeable dysregulation of TGF-β activation and signaling [75]. For this reason Habashi et al. studied the effects of TGF-β neutralizing antibody (which blocks all three TGF-β ligands) or the AngII receptor blocker, losartan [61]. They determined that thoracic aneurysm in the mouse model was coupled with increased TGF-β signaling and that using TGF-β antagonists like losartan can prevent aneurysm formation. They also noted that losartan is already used clinically to treat hypertension and could very well have the potential to prevent aortic aneurysms from forming in Marfan patients. So far investigations into the effect that losartan has on the biomechanical response of the aorta has yet to be conducted.

3.5.2 Doxycycline

The discovery that aneurysm development was associated with increased local matrix metalloproteinase (MMP) production in aneurismal tissue [76–80], and that MMPs were potentially the cause of extracellular matrix degradation has initiated research into compounds able to suppress MMP activity. Specifically MMP-2, -9, and -12 have been found associated with human AAA disease [81, 82]. The family of antibiotics known as tetracyclines became known as inhibitors of MMPs when used at levels higher than antimicrobial doses [83–86]. One tetracycline in particular, doxycycline, has been found to be a non-selective inhibitor of MMPs through either a transcriptional or direct method [59], and is readily available.

The first animal models of AAA to incorporate doxycycline in order to suppress aneurysm formation used elastase-induced aortic injury in the rat [87, 88]. Petrinec et al. found that doxycycline did have an aneurysm-suppressing effect in vivo with a dosage of 25 mg/day administered via drinking water [87]. Curci et al. also found doxycycline inhibited aortic dilation was most likely dose dependent with effects starting at 6 mg/kg/day and maximal effects at 30 mg/kg/day again administered via drinking water [88]. Mouse models of AAA that also utilize doxcycline in aneurismal suppression include Manning et al., Pyo et al., and Vihn et al. [47, 71, 89]. While Manning et al. again used elastase infusion, both Pyo et al. and Vihn et al. used AngII infused $ApoE^{-/-}$ mice. All used a dosage of 30 mg/kg/day administer via drinking water, and all found doxycycline to greatly reduce the incidence of AAA formation as well as severity of the aneurysm.

Success in mice and rat studies has spurred human trials of doxycycline for the suppression of AAA [90–92]. Manning and Baxter conducted one of the first clinical trials that treated patients undergoing elective AAA repair for 7 days prior with 100 mg by mouth twice daily docycycline [92]. Examination of the tissue obtained from surgery found a great reduction in MMP-2, and -9 activities in treated patients compared to those who were not. A phase two study conducted by Baxter et al. observed 36 patients with AAA to evaluate treatment with 100 mg twice daily for 6 months [90]. Results from this study found no significant change in AAA diameter and substantially decreased MMP activity after 6 months, and that such doses were well tolerated. A placebo-controlled pilot study conducted by Mosorin et al. found that over 18 months, treatment with 150 mg/day doxycycline significantly reduced AAA expansion compared to the placebo group [91].

In order to address the questions of whether the doxycycline serum levels needed for AAA inhibition achieved in animal models can be safely achieved in humans with standard dosages, Prall et al. conducted a controlled study [93]. Wild type mice with elastase perfusion were dosed with 0, 10, 50, and 100 mg/kg doxycycline, while human patients were given 100 mg twice daily. AAA reduction was observed in mice at 10 mg/kg and reduction increased with increased dosage, and the study was able to show that circulating doxycycline values in mice were found to be similar to plasma levels of doxycycline in human patients.

Another study conducted by Bartoli et al. compared wild type elastase perfused mice dosed with 100 mg/kg/day oral doxycycline with localized continuous injection of 0.75–1.0 mg/kg/day via osmotic minipumps [94]. This study found that localized infusion suppressed AAA formation equivalent to or even greater than that of oral administration, while plasma levels of doxycycline in the mice receiving infusion were undetectable.

4 Summary and Conclusion

Over the past sixty years there has been significant effort devoted to creating an experimental animal model of aneurysm disease. The means of induction as well as the choice of approach is governed primarily by the questions being investigated by the researcher. Simple mechanical/surgical models have become the mainstay for studies in large animals and have been the most popular for studying endovascular stent-graft design and performance. Chemical and/or genetic models of aneurismal disease have arisen more recently. As our understanding of the mechanisms and pathways involved in aneurysm formation has improved, these new genetic models have become available to investigate potential causes for aneurysm and possible therapeutic targets. The chemically and/or genetically induced animal models are attractive from a biomechanical point of view, since these can be used to investigate how the mechanical properties of aneurysms alter during disease development. In addition, longitudinal studies based on the biomechanical analysis of mouse models that are susceptible to aneurysm formation will inform about the process of aneurysm formation. Such future research will provide a meaningful link between the matrix degradation and inflammatory response known to be present in human AAA and how such changes eventually lead to the mechanical failure of ruptured AAA. In summary, there are potential applications of prognostic utility of assessment of aortic wall biomechanics in prevention and medical treatment of aortic aneurysms and sudden aortic rupture.

5 Acknowledgments

The authors would like to acknowledge the support of the American Heart Association-Grant in Aid (10GRNT4580045 to JPVG), the National Science Foundation (0644570 to JPVG), the help of Tom Doetschman and Connie Gard, the National Institutes of Health (HL92508), The Stephen Michael Schneider & The William J. "Billy" Gieszl Investigator Award (University of Arizona), and Arizona Biomedical Research Commission (#0901).

References

1. Sandford, R.M., et al.: The genetic basis of abdominal aortic aneurysms: a review. Eur. J. Vasc. Endovasc. Surg. **33**(4), 381–390 (2007)
2. Thompson, A.R., et al.: Candidate gene association studies in abdominal aortic aneurysm disease: a review and meta-analysis. Eur. J. Vasc. Endovasc. Surg. **35**(1), 19–30 (2008)
3. Jagadesham, V.P., Scott, D.J., Carding, S.R.: Abdominal aortic aneurysms: an autoimmune disease? Trends Mol. Med. **14**(12), 522–529 (2008)
4. Kakafika, A.I., Mikhailidis, D.P.: Smoking and aortic diseases. Circ. J. **71**(8), 1173–1180 (2007)
5. Vorp, D.A.: Biomechanics of abdominal aortic aneurysm. J. Biomech. **40**(9), 1887–1902 (2007)
6. Vorp, D.A., Vande Geest, J.P.: Biomechanical determinants of abdominal aortic aneurysm rupture. Arterioscler. Thromb. Vasc. Biol. **25**(8), 1558–1566 (2005)
7. Vande Geest, J.P., Sacks, M.S., Vorp, D.A.: The effects of aneurysm on the biaxial mechanical behavior of human abdominal aorta. J. Biomech. **39**(7), 1324–1334 (2006)
8. Balko, A., et al.: Transfemoral placement of intraluminal polyurethane prosthesis for abdominal aortic aneurysm. J. Surg. Res. **40**(4), 305–309 (1986)

9. Criado, E., et al.: An aortic aneurysm model for the evaluation of endovascular exclusion prostheses. J. Vasc. Surg. **22**(3), 306–314 (1995). discussion 314–315
10. Laborde, J.C., et al.: Intraluminal bypass of abdominal aortic aneurysm: feasibility study. Radiology **184**(1), 185–190 (1992)
11. Ruiz, C.E., et al.: A novel method for treatment of abdominal aortic aneurysms using percutaneous implantation of a newly designed endovascular device. Circulation **91**(9), 2470–2477 (1995)
12. Strindberg, G., et al.: Experimental modifications to a canine infrarenal aortic aneurysm model for the validation of endovascular stent-grafts: an exploratory study. J. Invest. Surg. **11**(3), 185–197 (1998)
13. Hynecek, R.L., et al.: The creation of an infrarenal aneurysm within the native abdominal aorta of swine. Surgery **142**(2), 143–149 (2007)
14. Maynar, M., et al.: An animal model of abdominal aortic aneurysm created with peritoneal patch: technique and initial results. Cardiovasc. Intervent. Radiol. **26**(2), 168–176 (2003)
15. Murphy, E.H., Johnson, E.D., Arko, F.R.: Device-specific resistance to in vivo displacement of stent-grafts implanted with maximum iliac fixation. J. Endovasc. Ther. **14**(4), 585–592 (2007)
16. Gorin, D.R., et al.: A new generation endovascular graft for repair of abdominal aortic aneurysms. Am. J. Surg. **173**(3), 159–164 (1997)
17. Guan, C., et al.: A review of current large-scale mouse knockout efforts. Genesis **48**(2), 73–85
18. Trollope, A., et al.: Animal models of abdominal aortic aneurysm and their role in furthering management of human disease. Cardiovasc. Pathol. **20**(2), 114–123 (2011)
19. Blanton, F.S., Muller Jr., W.H., Warren, W.D.: Experimental production of dissecting aneurysms of the aorta. Surgery **45**(1), 81–90 (1959)
20. Argenta, R., Pereira, A.H.: Modelos animais de aneurisma de aorta. J. Vasc. Bras. **8**, 148–153 (2009)
21. Conn, P.M., et al.: Animal Models for Atherosclerosis, Restenosis, and Endovascular Aneurysm Repair. In: Sourcebook of Models for Biomedical Research, pp. 369–384. Humana Press, Totowa (2008)
22. Kajimoto, M., et al.: Basic fibroblast growth factor slow release stent graft for endovascular aortic aneurysm repair: a canine model experiment. J. Vasc. Surg. **48**(5), 1306–1314 (2008)
23. Lerouge, S., et al.: Endovascular aortic aneurysm repair with stent-grafts: experimental models can reproduce endoleaks. J. Vasc. Interv. Radiol. **15**(9), 971–979 (2004)
24. Eton, D., et al.: Results of endoluminal grafting in an experimental aortic aneurysm model. J. Vasc. Surg. **23**(5), 819–829 (1996). discussion 829–831
25. Benson, A.E., et al.: Polytetrafluoroethylene-encapsulated stent-grafts: use in experimental abdominal aortic aneurysm. J. Vasc. Interv. Radiol. **10**(5), 605–612 (1999)
26. Palmaz, J.C., et al.: Use of stents covered with polytetrafluoroethylene in experimental abdominal aortic aneurysm. J. Vasc. Interv. Radiol. **6**(6), 879–885 (1995)
27. Verbin, C., et al.: Anterior patch aortic aneurysm model for the study of endoluminal grafts. J. Invest. Surg. **8**(5), 381–388 (1995)
28. Arends, J., et al.: A new technique for the surgical creation of aneurysms in an in vivo tortuous vessel model to test neurovascular devices. J. Invest. Surg. **21**(1), 39–45 (2008)
29. Turk, A.S., et al.: Natural history of the canine vein pouch aneurysm model. AJNR Am. J. Neuroradiol. **28**(3), 531–532 (2007)
30. Marston, W.A., et al.: Reduction of aneurysm pressure and wall stress after endovascular repair of abdominal aortic aneurysm in a canine model. Ann. Vasc. Surg. **10**(2), 166–173 (1996)
31. Hiraki, T., et al.: Prophylactic residual aneurysmal sac embolization with expandable hydrogel embolic devices for endoleak prevention: preliminary study in dogs. Cardiovasc. Intervent. Radiol. **28**(4), 459–466 (2005)
32. Pavcnik, D., et al.: A canine model for studying endoleak after endovascular aneurysm repair. J. Vasc. Interv. Radiol. **14**(10), 1303–1310 (2003)

33. Molacek, J., et al.: Optimization of the model of abdominal aortic aneurysm–experiment in an animal model. J. Vasc. Res. **46**(1), 1–5 (2009)
34. Huynh, H., et al.: Totally laparoscopic aortic surgery: comparison of the apron and retrocolic techniques in a porcine model. Vasc. Endovasc. Surg. **41**(3), 230–238 (2007)
35. Piquet, P., et al.: Tantalum-dacron coknit stent for endovascular treatment of aortic aneurysms: a preliminary experimental study. J. Vasc. Surg. **19**(4), 698–706 (1994)
36. Hagen, B., et al.: Self-expandable macroporous nitinol stents for transfemoral exclusion of aortic aneurysms in dogs: preliminary results. Cardiovasc. Intervent. Radiol. **16**(6), 339–342 (1993)
37. Whitbread, T., et al.: A new animal model for abdominal aortic aneurysms:initial results using a multiple-wire stent. Eur. J. Vasc. Endovasc. Surg. **11**(1), 90–7 (1996)
38. Conn, P.M.: Sourcebook of Models for Biomedical Research. xvi, 778 p., [8] p. of plates. Humana Press, Totowa (2008)
39. Parodi, J.C., Palmaz, J.C., Barone, H.D.: Transfemoral intraluminal graft implantation for abdominal aortic aneurysms. Ann. Vasc. Surg. **5**(6), 491–499 (1991)
40. Allaire, E., et al.: Cell and extracellular matrix rejection in arterial concordant and discordant xenografts in the rat. Transplantation **62**(6), 794–803 (1996)
41. Allaire, E., et al.: Paracrine effect of vascular smooth muscle cells in the prevention of aortic aneurysm formation. J. Vasc. Surg. **36**(5), 1018–1026 (2002)
42. Hallisey, M.J., SCVIR Gary J Becker Young Investigator Award paper: A transluminally created abdominal aortic aneurysm model. J. Vasc. Interv. Radiol. **8**(3), 305–312 (1997)
43. Schoder, M., et al.: Small intestinal submucosa aneurysm sac embolization for endoleak prevention after abdominal aortic aneurysm endografting: a pilot study in sheep. J. Vasc. Interv. Radiol. **15**(1 Pt 1), 69–83 (2004)
44. Powell, J., Greenhalgh, R.M.: Cellularenzymatic and genetic factors in the pathogenesis of abdominal aortic aneurysms. J. Vasc. Surg. **9**(2), 297–304 (1989)
45. Thompson, R.W., Geraghty, P.J., Lee, J.K.: Abdominal aortic aneurysms: basic mechanisms and clinical implications. Curr. Probl. Surg. **39**(2), 110–230 (2002)
46. Anidjar, S., et al.: Elastase-induced experimental aneurysms in rats. Circulation **82**(3), 973–981 (1990)
47. Pyo, R., et al.: Targeted gene disruption of matrix metalloproteinase-9 (gelatinase B) suppresses development of experimental abdominal aortic aneurysms. J. Clin. Invest. **105**(11), 1641–1649 (2000)
48. Halpern, V.J., et al.: The elastase infusion model of experimental aortic aneurysms: synchrony of induction of endogenous proteinases with matrix destruction and inflammatory cell response. J. Vasc. Surg. **20**(1), 51–60 (1994)
49. Thompson, R.W., et al.: Pathophysiology of abdominal aortic aneurysms: insights from the elastase-induced model in mice with different genetic backgrounds. Ann. N. Y. Acad. Sci. **1085**, 59–73 (2006)
50. Economou, S.G., et al.: Persistent experimental aortic aneurysms in dogs. Surgery **47**, 21–28 (1960)
51. Martin, D.I., Nasbeth, D., Rowe, M.: Production of experimental aneurysms with pancreatic elastase. Surg. Forum **8**, 237–239 (1962)
52. Boudghene, F., et al.: Endovascular grafting in elastase-induced experimental aortic aneurysms in dogs: feasibility and preliminary results. J. Vasc. Interv. Radiol. **4**(4), 497–504 (1993)
53. Freestone, T., et al.: Influence of hypercholesterolemia and adventitial inflammation on the development of aortic aneurysm in rabbits. Arterioscler. Thromb. Vasc. Biol. **17**(1), 10–17 (1997)
54. Chiou, A.C., Chiu, B., Pearce, W.H.: Murine aortic aneurysm produced by periarterial application of calcium chloride. J. Surg. Res. **99**(2), 371–376 (2001)
55. Andrews, E.J., White, W.J., Bullock, L.P.: Spontaneous aortic aneurysms in blotchy mice. Am. J. Pathol. **78**(2), 199–210 (1975)

56. Danks, D.M., et al.: Menkes's kinky hair syndrome. An inherited defect in copper absorption with widespread effects. Pediatrics **50**(2), 188–201 (1972)
57. Maki, J.M., et al.: Inactivation of the lysyl oxidase gene Lox leads to aortic aneurysms, cardiovascular dysfunction, and perinatal death in mice. Circulation **106**(19), 2503–2509 (2002)
58. Sugiyama, F., et al.: Acceleration of atherosclerotic lesions in transgenic mice with hypertension by the activated renin-angiotensin system. Lab. Invest. **76**(6), 835–842 (1997)
59. Manning, M.W., et al.: Abdominal aortic aneurysms: fresh insights from a novel animal model of the disease. Vasc. Med. **7**(1), 45–54 (2002)
60. Nishijo, N., et al.: Salt-sensitive aortic aneurysm and rupture in hypertensive transgenic mice that overproduce angiotensin II. Lab. Invest. **78**(9), 1059–1066 (1998)
61. Habashi, J.P., et al.: Losartan, an AT1 antagonist, prevents aortic aneurysm in a mouse model of Marfan syndrome. Science **312**(5770), 117–121 (2006)
62. Daugherty, A., Cassis, L.A.: Mouse models of abdominal aortic aneurysms. Arterioscler. Thromb. Vasc. Biol. **24**(3), 429–434 (2004)
63. Ishibashi, S., et al.: Massive xanthomatosis and atherosclerosis in cholesterol-fed low density lipoprotein receptor-negative mice. J. Clin. Invest. **93**(5), 1885–1893 (1994)
64. Plump, A.S., et al.: Severe hypercholesterolemia and atherosclerosis in apolipoprotein E-deficient mice created by homologous recombination in ES cells. Cell **71**(2), 343–353 (1992)
65. Tangirala, R.K., Rubin, E.M., Palinski, W.: Quantitation of atherosclerosis in murine models: correlation between lesions in the aortic origin and in the entire aorta and differences in the extent of lesions between sexes in LDL receptor-deficient and apolipoprotein E-deficient mice. J. Lipid Res. **36**(11), 2320–2328 (1995)
66. Van Herck, J.L., et al.: Impaired fibrillin-1 function promotes features of plaque instability in apolipoprotein E-deficient mice. Circulation **120**(24), 2478–2487 (2009)
67. Deguchi, J.O., et al.: Genetically engineered resistance for MMP collagenases promotes abdominal aortic aneurysm formation in mice infused with angiotensin II. Lab. Invest. **89**(3), 315–326 (2009)
68. Daugherty, A., Manning, M.W., Cassis, L.A.: Angiotensin II promotes atherosclerotic lesions and aneurysms in apolipoprotein E-deficient mice. J. Clin. Invest. **105**(11), 1605–1612 (2000)
69. Cassis, L.A., et al.: Aldosterone does not mediate angiotensin II-induced atherosclerosis and abdominal aortic aneurysms. Br. J. Pharmacol. **144**(3), 443–448 (2005)
70. Henriques, T.A., et al.: Orchidectomy, but not ovariectomy, regulates angiotensin II-induced vascular diseases in apolipoprotein E-deficient mice. Endocrinology **145**(8), 3866–3872 (2004)
71. Manning, M.W., Cassis, L.A., Daugherty, A.: Differential effects of doxycycline a broad-spectrum matrix metalloproteinase inhibitor, on angiotensin II-induced atherosclerosis and abdominal aortic aneurysms. Arterioscler. Thromb. Vasc. Biol. **23**(3), 483–488 (2003)
72. Saraff, K., et al.: Aortic dissection precedes formation of aneurysms and atherosclerosis in angiotensin II-infused, apolipoprotein E-deficient mice. Arterioscler. Thromb. Vasc. Biol. **23**(9), 1621–1626 (2003)
73. Daugherty, A., Cassis, L.: Angiotensin II-mediated development of vascular diseases. Trends Cardiovasc. Med. **14**(3), 117–120 (2004)
74. Daugherty, A., et al.: Angiotensin II infusion promotes ascending aortic aneurysms: attenuation by CCR2 deficiency in apoE−/− mice. Clin. Sci. (Lond) **118**(11), 681–689 (2010)
75. Neptune, E.R., et al.: Dysregulation of TGF-beta activation contributes to pathogenesis in Marfan syndrome. Nat. Genet. **33**(3), 407–411 (2003)
76. Freestone, T., et al.: Inflammation and matrix metalloproteinases in the enlarging abdominal aortic aneurysm. Arterioscler. Thromb. Vasc. Biol. **15**(8), 1145–1151 (1995)
77. Curci, J.A., et al.: Expression and localization of macrophage elastase (matrix metalloproteinase-12) in abdominal aortic aneurysms. J. Clin. Invest. **102**(11), 1900–1910 (1998)

78. Davis, V., et al.: Matrix metalloproteinase-2 production and its binding to the matrix are increased in abdominal aortic aneurysms. Arterioscler. Thromb. Vasc. Biol. **18**(10), 1625–1633 (1998)
79. Thompson, R.W., et al.: Production and localization of 92-kilodalton gelatinase in abdominal aortic aneurysms. An elastolytic metalloproteinase expressed by aneurysm-infiltrating macrophages. J. Clin. Invest. **96**(1), 318–326 (1995)
80. Longo, G.M., et al.: Matrix metalloproteinases 2 and 9 work in concert to produce aortic aneurysms. J. Clin. Invest. **110**(5), 625–632 (2002)
81. Crowther, M., et al.: Increased matrix metalloproteinase 2 expression in vascular smooth muscle cells cultured from abdominal aortic aneurysms. J. Vasc. Surg. **32**(3), 575–583 (2000)
82. Hovsepian, D.M., et al.: Elevated plasma levels of matrix metalloproteinase-9 in patients with abdominal aortic aneurysms: a circulating marker of degenerative aneurysm disease. J. Vasc. Interv. Radiol. **11**(10), 1345–1352 (2000)
83. Golub, L.M., et al.: A non-antimicrobial tetracycline inhibits gingival matrix metalloproteinases and bone loss in porphyromonas gingivalis-induced periodontitis in rats. Ann. N. Y. Acad. Sci. **732**, 96–111 (1994)
84. Golub, L.M., et al.: Minocycline reduces gingival collagenolytic activity during diabetes. Preliminary observations and a proposed new mechanism of action. J. Periodontal Res. **18**(5), 516–526 (1983)
85. Greenwald, R.A.: Treatment of destructive arthritic disorders with MMP inhibitors. Potential role of tetracyclines. Ann. N. Y. Acad. Sci. **732**, 181–198 (1994)
86. Uitto, V.J., et al.: Doxycycline and chemically modified tetracyclines inhibit gelatinase A (MMP-2) gene expression in human skin keratinocytes. Ann. N. Y. Acad. Sci. **732**, 140–151 (1994)
87. Petrinec, D., et al.: Doxycycline inhibition of aneurysmal degeneration in an elastase-induced rat model of abdominal aortic aneurysm: preservation of aortic elastin associated with suppressed production of 92 kD gelatinase. J. Vasc. Surg. **23**(2), 336–346 (1996)
88. Curci, J.A., et al.: Pharmacologic suppression of experimental abdominal aortic aneurysms: acomparison of doxycycline and four chemically modified tetracyclines. J. Vasc. Surg. **28**(6), 1082–1093 (1998)
89. Vinh, A., et al.: A novel histone deacetylase inhibitor reduces abdominal aortic aneurysm formation in angiotensin II-infused apolipoprotein E-deficient mice. J. Vasc. Res. **45**(2), 143–152 (2008)
90. Baxter, B.T., et al.: Prolonged administration of doxycycline in patients with small asymptomatic abdominal aortic aneurysms: report of a prospective (phase II) multicenter study. J. Vasc. Surg. **36**(1), 1–12 (2002)
91. Mosorin, M., et al.: Use of doxycycline to decrease the growth rate of abdominal aortic aneurysms: a randomized, double-blind, placebo-controlled pilot study. J. Vasc. Surg. **34**(4), 606–610 (2001)
92. Thompson, R.W., Baxter, B.T.: MMP inhibition in abdominal aortic aneurysms. Rationale for a prospective randomized clinical trial. Ann. N. Y. Acad. Sci. **878**, 159–178 (1999)
93. Prall, A.K., et al.: Doxycycline in patients with abdominal aortic aneurysms and in mice: comparison of serum levels and effect on aneurysm growth in mice. J. Vasc. Surg. **35**(5), 923–929 (2002)
94. Bartoli, M.A., et al.: Localized administration of doxycycline suppresses aortic dilatation in an experimental mouse model of abdominal aortic aneurysm. Ann. Vasc. Surg. **20**(2), 228–236 (2006)

Author Index

A
Alberto Figueroa C., 221
Azhar Mohamad, 401

B
Bluestein D., 181
Broderick Stephen, 163

C
Callanan Anthony, 163
Conway Brian D., 285
Corbett Timothy, 247

D
David Vorp A., 67
Defraigne Jean-Olivier, 1
Doyle Barry J., 119
Durieux Rodolphe, 1

G
Gounis Matthew J., 331
Grace A. Pierce, 247
Greenberg Roy K., 285

H
Haskett Darren, 401
Holzapfel Gerhard A., 373
Hoskins Peter, 35

J
Julie Phillippi A., 67

K
Kavanagh G. Eamon, 247
Kuivaniemi Helena, 1

M
McGloughlin Timothy M., 119, 163, 247
Mehra Manik, 331
Molony David S., 163, 247

N
Nusgens Betty, 1

P
Pasta Salvatore, 67
Phillippi Julie A., 67

Q
Qureshi Moqueet A., 285

R
Raghavan Madhavan L., 139
Richards Jennifer, 35

Stud Mechanobiol Tissue Eng Biomater (2011) 7: 423–424
DOI: 10.1007/978-3-642-18095-8

S

Sakalihasan Natzi, 1
Salvatore Pasta, 67
Semple Scott, 35
Simão da Silva Erasmo, 139
Spilberg Gabriela, 331

V

Vande Geest Jonathan P., 401
Ventikos Yiannis, 373
Vorp David A., 67

W

Wakhloo Ajay K., 331
Walsh Michael T., 163, 247
Watton Paul N., 373
White Phil, 35

X

Xenos M., 181

Z

Zarins Christopher K., 221